Problems in Quantum Field Theory

This collection of problems in quantum field theory, accompanied by their complete solutions, aims to bridge the gap between learning the foundational principles and applying them practically. The carefully chosen problems cover a wide range of topics, starting from the foundations of quantum field theory and the traditional methods in perturbation theory, such as LSZ reduction formulas, Feynman diagrams and renormalization. Separate chapters are devoted to functional methods (bosonic and fermionic path integrals; worldline formalism), to non-Abelian gauge theories (Yang–Mills theory, quantum chromodynamics), to the novel techniques for calculating scattering amplitudes and to quantum field theory at finite temperature (including its formulation on the lattice, and extensions to systems out of equilibrium). The problems range from those dealing with QFT formalism itself to problems addressing specific questions of phenomenological relevance, and they span a broad range in difficulty, for graduate students taking their first or second course in QFT.

François Gelis is a researcher at the Institut de Physique Théorique of CEA-Saclay, France. He received his M.S. from École Normale Supérieure de Lyon and a Ph.D. from the Université de Savoie. His research area is the theoretical study of the extreme phases of matter produced in heavy ion collisions. This led him to develop new quantum field theory techniques to handle the strong fields encountered in these situations, for which he was awarded the Paul Langevin Prize of the French Physical Society in 2015. He has taught QFT at École Polytechnique since 2016.

Problems in Quantum Field Theory

With Fully-Worked Solutions

FRANÇOIS GELIS Commissariat à l'Energie Atomique (CEA), Saclay

CAMBRIDGE
UNIVERSITY PRESS

CAMBRIDGE
UNIVERSITY PRESS

University Printing House, Cambridge CB2 8BS, United Kingdom

One Liberty Plaza, 20th Floor, New York, NY 10006, USA

477 Williamstown Road, Port Melbourne, VIC 3207, Australia

314–321, 3rd Floor, Plot 3, Splendor Forum, Jasola District Centre, New Delhi – 110025, India

103 Penang Road, #05–06/07, Visioncrest Commercial, Singapore 238467

Cambridge University Press is part of the University of Cambridge.

It furthers the University's mission by disseminating knowledge in the pursuit of education, learning, and research at the highest international levels of excellence.

www.cambridge.org
Information on this title: www.cambridge.org/9781108838801
DOI: 10.1017/9781108976688

First published 2021

A catalogue record for this publication is available from the British Library.

ISBN 978-1-108-83880-1 Hardback
ISBN 978-1-108-97235-2 Paperback

Additional resources for this publication at cambridge.org/problemsinqft.

To Kanako, Nathan and Simon.

Reality is that which, when you stop believing in it, doesn't go away.

PHILIP K. DICK

How to build a universe that doesn't fall apart two days later (1978)

Contents

Figures and Tables

Figures

Tables

Preface

This project grew out as an extension to my previous *Quantum Field Theory* book (hereafter referred to as book I), in the form of an additional set of solved problems. The starting observation was that most textbooks (and mine is no exception) have a general inclination towards the exposition of the concepts, rather than the more practical aspects. This is of course quite understandable for such a vast subject, where one needs to absorb a quite large volume of concepts before becoming operational. The intended goal of the present book is to help fill the gap between theory and applications by providing a text almost exclusively geared towards actual practice (but occasionally new concepts are also introduced).

The set of problems included in this book cover most of the subjects treated in book I, and expose the reader to a broad variety of techniques. Occasionally, the same question is addressed by various methods, in order to shed light on it from different perspectives. The problems proposed in this set cover a broad range of difficulties and fall into several categories:

- More sophisticated and lengthier applications of the techniques and results of book I, that are too long to be treated as examples in a textbook
- Problems that extend book I towards new topics and concepts that are too specialized to fit reasonably in a textbook
- Classic results of quantum field theory (QFT) presented in a more modern fashion and with uniform notation, with the goal of making them more accessible than from a reading of the original literature
- More recent results, presented in a way that is accessible to the readers of book I; these are meant to be a bridge between the material of a typical textbook and contemporary research articles.

When deciding which problems to include in this collection, I made the deliberate choice to keep almost exclusively problems that can be worked out analytically "by hand" in a reasonable amount of time. Occasionally, straightforward but tedious calculations have been avoided by the use of a computer algebra system (for these problems, PYTHON notebooks will be provided as a separate online resource). This prejudice is of course an important limitation since these computer tools are a common aid in contemporary theoretical research, but my impression is that the didactical virtue of working out simpler problems by hand is higher.

Although this book has an obvious lineage with book I (e.g., the two books share the same notation), a significant effort has been made to ensure that it is self-contained and can be used on its own. Each chapter starts with a reminder of the important concepts and tools relevant for the problems of that chapter, but an important word of caution is in order here: these introductions are meant to be a concise refresher for a reader who has already studied the corresponding subjects, but they are not an appropriate source for learning a subject for the first time. This book is also self-contained by the inclusion of detailed solutions to all the problems. When necessary, a quick exposition of some relevant mathematical tools is included in the solutions, to avoid a detour via a mathematical textbook (this is intended to give plausibility to a given mathematical statement, not to provide a thorough and rigorous description of the underlying mathematics).

The intended readership of this book is of course primarily students who are in the process of learning quantum field theory, as well as their instructors. Roughly speaking, Chapters 1 and 2 cover the topics one would usually learn in a first QFT course. Chapter 3, on non-Abelian gauge theories, tends to be a bit more advanced and is often treated in a second course. Chapters 4 and 5 deal with more specialized subjects, respectively the newly developed tools for calculating scattering amplitudes, and aspects of quantum field theory that are on the border of many-body physics. Hopefully, more experienced readers will also find the book useful, both for the discussion of these more advanced topics and for a more modern treatment of some classic QFT results.

Acknowledgements

This book would not exist without the many questions from students to whom I had the pleasure of teaching quantum field theory during the past five years at École Polytechnique. Some of these questions were indicative of a demand to go beyond the formal developments that constitute the main body of the course. Occasionally, these questions would touch on subtle and not often discussed points of QFT. Many problems in this book are the result of my attempts to clarify these points, for myself first, before I could comfortably provide an answer to my students.

This book also owes a lot to Vince Higgs at Cambridge University Press, for his supportive reception of a preliminary draft and for his precious advice regarding how to improve it. I would like to extend these thanks to all the support staff at Cambridge University Press, who helped me a lot in the more technical aspects of making this book, with a special mention to Elle Ferns and Henry Cockburn, and to the anonymous referees whose feedback was very helpful. Last but not least, John King, who went through the painstaking task of copy-editing the manuscript, provided me with immensely valuable feedback that considerably improved its readability.

Finally, I would like to address my warmest thanks to my wife, Kanako, for her constant encouragement, support and patience. In normal times, writing a book is an activity which is already quite prone to interfering with family life. The covid pandemic, by forcing me to work from home most of the time during the preparation of this book, exacerbated this by blurring even more the boundary between work and leisure.

Notation and Conventions

We list here some notation and conventions that are used throughout this book:

- $c = 1$: length = time, energy = momentum = mass
- $\hbar = 1$: momentum = wavenumber, energy = frequency
- $p^\mu \equiv (p^0, \mathbf{p})$: 4-momentum; $\mathbf{p}$: three-dimensional spatial components
- $E_\mathbf{p} \equiv \sqrt{\mathbf{p}^2 + m^2}$: positive on-shell energy of a particle of momentum $\mathbf{p}$ and mass m
- $\mathcal{L}$: Lagrangian density (more rarely, the Lagrangian itself)
- $\mathcal{H}$: Hamiltonian; $\mathcal{S}$: action
- $\mu, \nu, \rho, \sigma, \ldots$ (more rarely $\alpha, \beta, \gamma, \ldots$): Lorentz indices in Minkowski space
- $i, j, k, l, \ldots$: Lorentz indices in Euclidean space
- $a, b, c, d, \ldots$: group indices in the adjoint representation
- $i, j, k, l, \ldots$: group indices in the fundamental representation
- $\epsilon_{\mu\nu\rho\cdots}$: Levi–Civita symbol. In situations where it makes sense to raise or lower the indices (e.g., for Lorentz indices in Minkowski space), the normalization convention is with lowered indices, $\epsilon_{012\cdots} = +1$
- The normalization of creation and annihilation operators is defined so that the canonical commutation relation reads $[a_\mathbf{p}, a_\mathbf{q}^\dagger] = (2\pi)^3 2E_\mathbf{p}\delta(\mathbf{p} - \mathbf{q})$ (except in Problems 80 and 81, where it is convenient to omit the factor $2E_\mathbf{p}$). Their dimension is $(\text{mass})^{-1}$.

CHAPTER 1

Quantum Field Theory Basics

Introduction

This chapter is devoted to basic aspects of quantum field theory, ranging from the foundations to perturbation theory and renormalization, and is limited to the canonical formalism (functional methods are treated in Chapter 2) and to the traditional workflow (Lagrangian → Feynman rules → time-ordered products of fields → scattering amplitudes) for the calculation of scattering amplitudes (the spinor-helicity formalism and on-shell recursion are considered in Chapter 4). The problems of this chapter deal with questions in scalar field theory and quantum electrodynamics, while non-Abelian gauge theories are discussed in Chapter 3.

Non-interacting Field Theory

A non-interacting field theory may be defined by a quadratic Lagrangian. In the simplest case of a scalar field theory, it reads

$$\mathcal{L} \equiv \int d^3x \left\{ \tfrac{1}{2}(\partial_\mu \phi)(\partial^\mu \phi) - \tfrac{1}{2} m^2 \phi^2 \right\}. \tag{1.1}$$

Such a Lagrangian defines a dynamical system with infinitely many degrees of freedom, corresponding to the values taken by $\phi(\boldsymbol{x})$ at every point $\boldsymbol{x}$ of space. The momentum canonically conjugate to $\phi(\boldsymbol{x})$ is given by

$$\Pi(\boldsymbol{x}) \equiv \frac{\partial \mathcal{L}}{\partial(\partial^0 \phi(\boldsymbol{x}))} = \partial^0 \phi(\boldsymbol{x}), \tag{1.2}$$

which leads to the Hamiltonian

$$\mathcal{H} \equiv \int d^3x\, \Pi(\boldsymbol{x}) \partial^0 \phi(\boldsymbol{x}) - \mathcal{L} = \int d^3x \left\{ \tfrac{1}{2}\Pi^2 + \tfrac{1}{2}(\boldsymbol{\nabla}\phi)^2 + \tfrac{1}{2} m^2 \phi^2 \right\}. \tag{1.3}$$

From the Hamiltonian or Lagrangian, one obtains the equation of motion of the field, which in the present example reads

$$(\square_x + m^2)\,\phi(x) = 0, \tag{1.4}$$

known as the *Klein–Gordon equation.* Generic real solutions of this linear equation are superpositions of plane waves:

$$\phi(x) = \int \frac{d^3\mathbf{k}}{(2\pi)^3 2E_{\mathbf{k}}} \left\{\alpha^*_{\mathbf{k}}\, e^{-ik\cdot x} + \alpha_{\mathbf{k}}\, e^{ik\cdot x}\right\}, \tag{1.5}$$

where $E_{\mathbf{k}} \equiv \sqrt{\mathbf{p}^2 + m^2}$ is the dispersion relation associated with the wave equation (1.4), and $\alpha_{\mathbf{k}}$ is a function of momentum that depends on the boundary conditions imposed on the solution.

Canonical quantization consists in promoting the coefficients $\alpha_{\mathbf{k}}$, $\alpha^*_{\mathbf{k}}$ into annihilation and creation operators $a_{\mathbf{k}}$, $a^\dagger_{\mathbf{k}}$ that obey the following commutation relations:

$$\left[a_{\mathbf{p}}, a^\dagger_{\mathbf{q}}\right] \equiv (2\pi)^3 2E_{\mathbf{p}}\, \delta(\mathbf{p} - \mathbf{q}). \tag{1.6}$$

The normalization in Eqs. (1.5) and (1.6) is chosen so that $\left[\mathcal{H}, a^\dagger_{\mathbf{p}}\right] = E_{\mathbf{p}} a^\dagger_{\mathbf{p}}$ and $\left[\mathcal{H}, a_{\mathbf{p}}\right] = -E_{\mathbf{p}} a_{\mathbf{p}}$, which means that $a^\dagger_{\mathbf{p}}$ increases the energy of the system by $E_{\mathbf{p}}$ while $a_{\mathbf{p}}$ decreases it by the same amount. As a consequence, this setup describes a collection of non-interacting particles. The commutation relation (1.6) implies the following equal-time commutation relation between the field operator and its conjugate momentum:

$$\left[\phi(x), \Pi(y)\right] \underset{x^0=y^0}{=} i\,\delta(\mathbf{x} - \mathbf{y}), \tag{1.7}$$

which one may view as the quantum version of the classical Poisson bracket between a coordinate and its conjugate momentum.

Interacting Field Theory and Interaction Representation

Interactions are introduced via terms of degree higher than two in the Lagrangian:

$$\mathcal{L} \equiv \int d^3x \left\{ \underbrace{\tfrac{1}{2}(\partial_\mu\phi)(\partial^\mu\phi) - \tfrac{1}{2}m^2\phi^2}_{\mathcal{L}_0,\ \text{non-interacting theory}} \underbrace{-V(\phi)}_{\text{interactions}} \right\}. \tag{1.8}$$

(In order to have a causal theory, the potential $V(\phi)$ must be a local function of the field $\phi(x)$; see Problem 4.) In the presence of interactions, the Klein–Gordon equation of motion becomes

$$(\square_x + m^2)\,\phi(x) + V'(\phi(x)) = 0. \tag{1.9}$$

Since the degree of $V(\phi)$ is higher than two, this equation is non-linear, which induces a mixing between the Fourier modes of the field and prevents writing its solutions as superpositions of plane waves.

By assuming that the interactions are turned off at large times, $x^0 \to \pm\infty$, we may define free fields ϕ_{in} and ϕ_{out} that coincide with the interacting field ϕ of the Heisenberg representation, respectively when $x^0 \to -\infty$ and $x^0 \to +\infty$. For instance, ϕ and ϕ_{in} are related by

$$\phi(x) = U(-\infty, x^0)\, \phi_{in}(x)\, U(x^0, -\infty),$$
$$U(t_2, t_1) \equiv T \exp\left(-i\int_{t_1}^{t_2} dx^0 d^3x\; V(\phi_{in}(x))\right). \tag{1.10}$$

In this representation, the time dependence of the field $\phi(x)$ is split into a trivial one that comes from the free field ϕ_{in} and the time evolution operator U that depends on the interactions. Since they are free fields obeying Eq. (1.4), ϕ_{in} and ϕ_{out} can be written as superpositions of plane waves, with coefficients $a_{p,in}$, $a^\dagger_{p,in}$ and $a_{p,out}$, $a^\dagger_{p,out}$, respectively. These two sets of creation and annihilation operators define two towers of *Fock states*, i.e., states with a definite particle content at $x^0 = -\infty$ and $x^0 = +\infty$, respectively.

Lehmann–Symanzik–Zimmermann Reduction Formulas

Experimentally measurable quantities, such as cross-sections, may be related to correlation functions of the field operator as follows. An intermediate step involves the transition amplitudes between *in* and *out* states,

$$\left\langle q_1 \cdots q_{n\,out} \middle| p_1 \cdots p_{m\,in} \right\rangle \equiv (2\pi)^4 \delta\Big(\sum_i p_i - \sum_j q_j\Big)\, \mathcal{T}(q_{1\cdots n}|p_{1\cdots m}), \tag{1.11}$$

in terms of which a cross-section in the center of momentum frame is given by

$$\sigma_{12\to 1\cdots n}\Big|_{\substack{\text{center of}\\ \text{momentum}}} = \frac{1}{4\sqrt{s}\,|p_1|} \int d\Gamma_n(p_{1,2})\, \left|\mathcal{T}(q_{1,\cdots,n}|p_{1,2})\right|^2, \tag{1.12}$$
$$\text{with } d\Gamma_n(p_{1,2}) \equiv \prod_j \frac{d^3q_j}{(2\pi)^3 2E_{q_j}}\, (2\pi)^4\delta\Big(p_1 + p_2 - \sum_j q_j\Big), \quad s \equiv (p_1+p_2)^2.$$

In turn, the transition amplitudes from *in* to *out* states are expressed in terms of expectation values of time-ordered products of field operators by the Lehmann–Symanzik–Zimmermann (LSZ) reduction formulas:

$$\left\langle q_1 \cdots q_{n\ out} \middle| p_1 \cdots p_{m\ in} \right\rangle = \frac{i^{m+n}}{Z^{\frac{m+n}{2}}} \int \prod_{i=1}^m d^4x_i\; e^{-ip_i\cdot x_i}\, (\square_{x_i} + m^2)$$
$$\times \int \prod_{j=1}^n d^4y_j\; e^{iq_j\cdot y_j}\, (\square_{y_j} + m^2)\, \left\langle 0_{out} \middle| T\,\phi(x_1)\cdots\phi(x_m)\phi(y_1)\cdots\phi(y_n) \middle| 0_{in} \right\rangle, \tag{1.13}$$

where Z is the wavefunction renormalization factor.

Generating Functional, Feynman Propagator

The calculation of expectation values of time-ordered products of field operators is usually organized by encapsulating them in a generating functional

$$\left\langle 0_{out} \middle| T\,\phi(x_1)\cdots\phi(x_n) \middle| 0_{in} \right\rangle = \left.\frac{\delta^n Z[j]}{i\delta j(x_1)\cdots i\delta j(x_n)}\right|_{j\equiv 0}, \tag{1.14}$$

$$\text{with } Z[j] \equiv \left\langle 0_{\rm out} \middle| T \exp i \int d^4x\, j(x)\phi(x) \middle| 0_{\rm in} \right\rangle \tag{1.15}$$

$$= \exp\left(-i \int d^4x\, V\left(\frac{\delta}{i\delta j(x)}\right)\right) \underbrace{\left\langle 0_{\rm in} \middle| T \exp i \int d^4x\, j(x)\phi_{\rm in}(x) \middle| 0_{\rm in} \right\rangle}_{Z_0[j],\ \text{non-interacting theory}} . \tag{1.16}$$

The last factor, the generating functional of the non-interacting theory, is a Gaussian in the auxiliary source j:

$$Z_0[j] = \exp\left(-\frac{1}{2} \int d^4x d^4y\, j(x) j(y) G^0_{\rm F}(x,y) \right), \tag{1.17}$$

where $G^0_{\rm F}(x,y)$ is the free *Feynman propagator*, which can be expressed in various equivalent ways:

$$G^0_{\rm F}(x,y) = \left\langle 0_{\rm in} \middle| T\, \phi_{\rm in}(x)\phi_{\rm in}(y) \middle| 0_{\rm in} \right\rangle \tag{1.18}$$

$$= \int \frac{d^3p}{(2\pi)^3 2E_p} \left(\theta(x^0 - y^0)\, e^{-ip\cdot(x-y)} + \theta(y^0 - x^0)\, e^{+ip\cdot(x-y)} \right), \tag{1.19}$$

$$G^0_{\rm F}(p) = \frac{i}{p^2 - m^2 + i0^+} . \tag{1.20}$$

Feynman Rules of Scalar Field Theory

The effect of interactions can be calculated order-by-order by expanding the first exponential in Eq. (1.16). The successive terms of this expansion are obtained from a diagrammatic expansion, where each diagram is converted into a formula by means of *Feynman rules*. Below we list these rules in momentum space, for a scalar field theory:

1. Draw all the graphs with as many external lines as field operators in the correlation function, and a number of vertices equal to the desired order. The vertices allowed in these graphs must have valences equal to the degrees of the terms in $V(\phi)$. Graphs with multiple connected components need not be considered in the calculation of scattering amplitudes.

2. A 4-momentum k is assigned to each internal line of the graph, and the associated Feynman rule is a free propagator $G^0_{\rm F}(k)$:

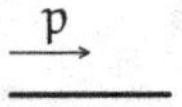

$$= \frac{i}{p^2 - m^2 + i0^+} .$$

No propagator should be assigned to the external lines of a graph when calculating a scattering amplitude (because of the factors $\square + m^2$ in the reduction formulas).

3. For an interaction $\frac{\lambda}{n!}\phi^n$, each vertex of valence n brings a factor $-i\lambda(2\pi)^4\delta\left(\sum_i k_i\right)$, where the k_i are the momenta incoming into this vertex:

$$= -i\lambda .$$

3. All the internal momenta that are not constrained by the delta functions at the vertices should be integrated over with a measure $d^4k/(2\pi)^4$. In a connected graph with n_I internal lines and n_V vertices, there are $n_L = n_I - n_V + 1$ of them, which is also the number of *loops* in the graph.

4. Each graph must be weighted by a *symmetry factor*, defined as the inverse of the order of the discrete symmetry group of the graph (assuming interaction terms properly symmetrized, as in $V(\phi) = \phi^n/n!$).

Dimensional Regularization

The momentum integrals that correspond to loops in Feynman diagrams may be divergent at large momentum. Convergence may be assessed from the *superficial degree of divergence* of a graph, $\omega(\mathcal{G}) \equiv 4n_L - 2n_I$ for a graph with n_L loops and n_I internal lines in a scalar field theory with quartic coupling in four spacetime dimensions: the graph $\mathcal{G}$ is convergent if $\omega(\mathcal{G}) < 0$ and the superficial degree of divergence of all its subgraphs is negative as well. In order to safely manipulate possibly divergent loop integrals, the first step is to introduce a *regularization*, i.e., a modification of the Feynman rules such that all loop integrals become well defined. Many regularization methods are possible: Pauli–Villars subtraction, lattice discretization, momentum cutoff, dimensional regularization.

Dimensional regularization, based on the observation that loop integrals calculated in an arbitrary number D of dimensions have an analytical continuation which is well defined at all D's except a discrete set of values, is particularly adapted to analytical calculations. With this regularization scheme, some common (Euclidean) loop integrals are given by

$$\int \frac{d^Dk_E}{(2\pi)^D}\,\frac{1}{(k_E^2+\Delta)^n} = \frac{\Delta^{\frac{D}{2}-n}}{(4\pi)^{\frac{D}{2}}}\,\frac{\Gamma(n-\frac{D}{2})}{\Gamma(n)},$$

$$\int \frac{d^Dk_E}{(2\pi)^D}\,\frac{k_E^\mu k_E^\nu}{(k_E^2+\Delta)^n} = \frac{g^{\mu\nu}}{2}\,\frac{\Delta^{\frac{D}{2}+1-n}}{(4\pi)^{\frac{D}{2}}}\,\frac{\Gamma(n-1-\frac{D}{2})}{\Gamma(n)},$$

$$\int \frac{d^Dk_E}{(2\pi)^D}\,\frac{k_E^\mu k_E^\nu k_E^\rho k_E^\sigma}{(k_E^2+\Delta)^n} = \frac{g^{\mu\nu}g^{\rho\sigma}+g^{\mu\rho}g^{\nu\sigma}+g^{\mu\sigma}g^{\nu\rho}}{4}\,\frac{\Delta^{\frac{D}{2}+2-n}}{(4\pi)^{\frac{D}{2}}}\,\frac{\Gamma(n-2-\frac{D}{2})}{\Gamma(n)},$$

$$\int \frac{d^Dk_E}{(2\pi)^D}\,\frac{k_E^{\mu_1}\cdots k_E^{\mu_{2n+1}}}{(k_E^2+\Delta)^n} = 0. \tag{1.21}$$

The first of these equations is obtained by integration in D-dimensional spherical coordinates, and the subsequent equations follow from Lorentz invariance.

Renormalization

The list of correlation functions that exhibit ultraviolet divergences can be obtained from the superficial degree of divergence $\omega(\mathcal{G})$ (except in situations where a symmetry produces cancellations that cannot be anticipated by power counting). For a scalar field theory with a quartic coupling, one has $\omega(\mathcal{G}) = 4 - n_E + (D-4)n_L$ in D spacetime dimensions, where

n_E is the number of external points and n_L the number of loops. The *Weinberg convergence theorem* states that a graph is ultraviolet convergent if and only if the superficial degree of divergence of the graph, and of any of its subgraphs, is negative.

In $D = 4$ spacetime dimensions, $\omega(\mathcal{G})$ is negative for all correlation functions with $n_E > 4$ points, implying that only a finite number of correlation functions have intrinsic divergences. Moreover, these divergent correlation functions are the expectation values of the operators already present in the Lagrangian, $(\partial_\mu\phi)^2, \phi^2, \phi^4$. The divergences that appear in these functions can be subtracted order-by-order via a redefinition of their coefficients in the Lagrangian, i.e., Z (this one is usually not explicit in the bare Lagrangian because it is set to 1), m^2 and λ, respectively. Such a quantum field theory is called *renormalizable.*

In $D > 4$ dimensions, $\omega(\mathcal{G})$ increases with the number of loops at fixed n_E. This implies that any correlation function exhibits intrinsic ultraviolet divergences beyond a certain loop order. Removing these divergences would require that one adds arbitrarily many new terms in the Lagrangian, reducing considerably the predictive power of such a theory (but it may nevertheless be of some use in an effective sense, at low loop orders). It is called *non-renormalizable.*

When $D < 4$, the superficial degree of divergence of any correlation function eventually becomes negative after a certain loop order. These theories have a finite number of ultraviolet divergent Feynman graphs, whose calculation is sufficient to determined the renormalized Lagrangian once and for all. These theories are called *super-renormalizable.*

For general interactions in arbitrary dimensions, the above criteria can be expressed in terms of the mass dimension of the prefactor that accompanies the operator in the Lagrangian. The corresponding operator is renormalizable if the mass dimension of its coupling constant is zero, non-renormalizable if this dimension is negative, super-renormalizable if it is positive.

Renormalization Group

In a renormalized quantum field theory, one may still freely choose the *renormalization scale* μ at which the conditions that define the parameters of the renormalized Lagrangian (masses, couplings, etc.) are imposed. Physical results should not depend on this scale. The dependence of various renormalized quantities with respect to μ is controlled by the *Callan–Symanzik equations,* also known as *renormalization group* equations. For the renormalized n-point correlation function G_n, this equation reads

$$\big(\underbrace{\mu\partial_\mu + \beta\partial_\lambda + \gamma_m\, m\partial_m}_{\equiv \mathcal{D}_\mu} + n\gamma\big)\, G_n = 0, \tag{1.22}$$

$$\text{with } \gamma \equiv \frac{1}{2}\frac{\partial \ln(Z)}{\partial \ln(\mu)}, \quad \beta \equiv \frac{\partial \lambda}{\partial \ln(\mu)}, \quad \gamma_m \equiv \frac{\partial \ln(m)}{\partial \ln(\mu)} \tag{1.23}$$

(γ is called an *anomalous dimension,* and β is the β *function*). Physical quantities are invariant under the action of $\mathcal{D}_\mu$, i.e., under the simultaneous change of the scale μ and of the parameters Z, λ, m as prescribed by the above differential equations (the solutions $\lambda(\mu)$ and $m(\mu)$ are called the running coupling and running mass, respectively). The curves $(Z(\mu), \lambda(\mu), m(\mu))$ in the parameter space of the renormalized theory, along which physical quantities are invariant, define a vector field called the *renormalization flow.*

From the Callan–Symanzik equation satisfied by the propagator, $(\mathcal{D}_\mu + 2\gamma)G_2 = 0$, one obtains the corresponding flow equations for the pole mass m_p (defined from the value of p^2

at the pole of the propagator) and for the residue Z at the pole:

$$\mathcal{D}_\mu m_p = 0, \quad (\mathcal{D}_\mu + 2\gamma)\, Z = 0. \tag{1.24}$$

Thus, a n-point scattering amplitude $\mathcal{A}_n \sim Z^{-n/2} G_n$ also satisfies $\mathcal{D}_\mu \mathcal{A}_n = 0$. Amputated correlation functions $\Gamma_n \equiv (G_2)^{-n} G_n$ obey $(\mathcal{D}_\mu - n\gamma)\Gamma_n = 0$.

Spin-1/2 Fields

The representation of the Lorentz algebra of lowest even dimension is defined by the generators $M^{\mu\nu}_{1/2} \equiv \frac{i}{4}[\gamma^\mu, \gamma^\nu]$, where the γ^μ are the *Dirac* 4×4 *matrices*, which obey $\{\gamma^\mu, \gamma^\nu\} = 2\, g^{\mu\nu}$. Under a Lorentz transformation $\Lambda \equiv \exp\left(-\frac{i}{2}\omega_{\mu\nu} M^{\mu\nu}\right)$, a *Dirac spinor* is a four-component field that transforms as

$$\psi(x) \quad \to \quad \exp\left(-\tfrac{i}{2}\omega_{\mu\nu} M^{\mu\nu}_{1/2}\right) \psi(\Lambda^{-1} x). \tag{1.25}$$

In the absence of interactions, such a field obeys the – Lorentz invariant – *Dirac equation*,

$$\left(i\gamma^\mu \partial_\mu - m\right)\psi = 0, \tag{1.26}$$

which can be obtained as the equation of motion that results from the following Lagrangian:

$$\mathcal{L} = \overline{\psi}\left(i\gamma^\mu \partial_\mu - m\right)\psi, \quad \text{with } \overline{\psi} \equiv \psi^\dagger \gamma^0. \tag{1.27}$$

The canonical quantization of a free spinor (i.e., a solution of the Dirac equation (1.26)) consists in replacing its Fourier coefficients by creation and annihilation operators:

$$\psi(x) \equiv \sum_{s=\pm} \int \frac{d^3\mathbf{p}}{(2\pi)^3 2E_\mathbf{p}} \left\{ d^\dagger_{s\mathbf{p}} v_s(\mathbf{p}) e^{+ip\cdot x} + b_{s\mathbf{p}} u_s(\mathbf{p}) e^{-ip\cdot x} \right\}. \tag{1.28}$$

Since ψ is not Hermitian, the two operators in this decomposition need not be mutual conjugates (except in the special case of Majorana fermions). The spinors u_s, v_s are a basis of free spinors in momentum space defined by

$$(\gamma^\mu p_\mu - m) u_s(\mathbf{p}) = 0, \quad (\gamma^\mu p_\mu + m) v_s(\mathbf{p}) = 0, \tag{1.29}$$

$$u^\dagger_r(\mathbf{p}) u_s(\mathbf{p}) = 2E_\mathbf{p} \delta_{rs}, \quad v^\dagger_r(\mathbf{p}) v_s(\mathbf{p}) = 2E_\mathbf{p} \delta_{rs}. \tag{1.30}$$

For the Hamiltonian of this system to have a well-defined ground state, these creation and annihilation operators must obey anti-commutation relations. The non-zero ones read

$$\{d_{s\mathbf{p}}, d^\dagger_{s'\mathbf{p}'}\} = \{b_{s\mathbf{p}}, b^\dagger_{s'\mathbf{p}'}\} = (2\pi)^3 2E_\mathbf{p} \delta_{ss'} \delta(\mathbf{p} - \mathbf{p}'), \tag{1.31}$$

or, equivalently,

$$\{\psi_\alpha(x), \psi^\dagger_\beta(y)\} \underset{x^0 = y^0}{=} \delta_{\alpha\beta} \delta(\mathbf{x} - \mathbf{y}). \tag{1.32}$$

The Dirac Lagrangian has a $U(1)$ symmetry, $\psi \to e^{-i\alpha}\psi$, which by Noether's theorem leads to a conserved current $J^\mu \equiv \overline{\psi}\gamma^\mu\psi$ and conserved charge

$$Q \equiv \int d^3x\, J^0 = \sum_{s=\pm} \int \frac{d^3p}{(2\pi)^3 2E_p} \left(b^\dagger_{sp} b_{sp} - d^\dagger_{sp} d_{sp}\right). \tag{1.33}$$

Spin-1 Fields

A *vector field* $A^\mu(x)$ is a four-component field that transforms as $A^\mu(x) \to \Lambda^\mu{}_\nu A^\nu(\Lambda^{-1}x)$ under a Lorentz transformation Λ. The simplest such (massless) field is the electromagnetic field, whose Lagrangian reads

$$\mathcal{L} = -\tfrac{1}{4} F_{\mu\nu}F^{\mu\nu}, \quad \text{with } F^{\mu\nu} \equiv \partial^\mu A^\nu - \partial^\nu A^\mu. \tag{1.34}$$

The corresponding equation of motion, $\partial_\mu F^{\mu\nu} = 0$, has several remarkable properties:

- *Gauge invariance*: for any function θ, $A^\mu - \partial^\mu\theta$ is a solution if A^μ is a solution.
- The field A^0 is not dynamical, but given by a constraint from the spatial components A^i.
- Only the transverse (i.e., transverse to the momentum k^i in Fourier space) components of A^i are constrained by the equation of motion.

The unphysical redundancy due to gauge invariance is removed by imposing a *gauge condition* – e.g., $\partial_\mu A^\mu = 0$ (Lorenz gauge), $\partial_i A^i = 0$ (Coulomb gauge), $A^0 = 0$ (temporal gauge) – leaving only two independent dynamical solutions per Fourier mode. The quantization of the vector field A^μ amounts to replacing the coefficients in its Fourier decomposition by creation and annihilation operators:

$$A^\mu(x) \equiv \sum_{\lambda=1,2} \int \frac{d^3p}{(2\pi)^3 2|\mathbf{p}|} \left\{ a^\dagger_{\lambda p}\, \epsilon^{\mu *}_\lambda(\mathbf{p})\, e^{ip\cdot x} + a_{\lambda p}\, \epsilon^\mu_\lambda(\mathbf{p})\, e^{-ip\cdot x} \right\}, \tag{1.35}$$

with the canonical commutation relation $\left[a_{\lambda p}, a^\dagger_{\lambda' p'}\right] = (2\pi)^3 2|\mathbf{p}|\delta_{\lambda\lambda'}\delta(\mathbf{p} - \mathbf{p}')$ and where the objects $\epsilon^\mu_\lambda(\mathbf{p})$ are *polarization vectors* that encode the Lorentz indices of a vector field of polarization λ and momentum $\mathbf{p}$. The polarization vectors may depend on the choice of gauge condition, but always satisfy $p_\mu \epsilon^\mu_\lambda(\mathbf{p}) = 0$.

Quantum Electrodynamics

The conserved charge of the Dirac fermions can be interpreted as an electrical charge. Interactions between these fermions and photons are introduced by *minimal coupling*, i.e., by requesting that the modified Dirac Lagrangian is invariant under spacetime-dependent $U(1)$ transformations, $\psi(x) \to e^{-ie\theta(x)}\psi(x)$. This is achieved by replacing the ordinary derivative by a *covariant derivative*, $D_\mu \equiv \partial_\mu - ieA_\mu$. Perturbation theory in QED has the following Feynman rules:

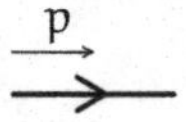

$= \dfrac{i(\not{p} + m)}{p^2 - m^2 + i0^+}$

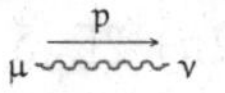

$= \dfrac{i\,C^{\mu\nu}(p)}{p^2 + i0^+}$

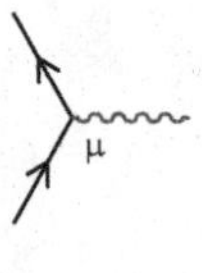

$= -ie\gamma^\mu$

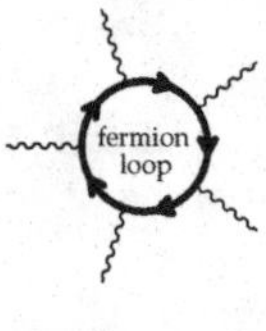

$=$ (minus sign)

The numerator $C^{\mu\nu}$ in the photon propagator depends on the gauge fixing (for instance, $C^{\mu\nu} = -g^{\mu\nu}$ in Feynman gauge).

Ward–Takahashi Identity

A crucial property of QED amplitudes with external photons is the *Ward–Takahashi identity*, namely

$$p_\mu \Gamma^{\mu\cdots}(p,\dots) = 0, \tag{1.36}$$

where $\Gamma^{\mu\cdots}(p,\dots)$ is an amplitude amputated of its external propagators, containing a photon of momentum p with Lorentz index μ. The dots represent the other external lines, either photons or charged particles. The conditions of validity of this identity, which follows from the conservation of the electrical current, are the following:

- All the external lines corresponding to charged particles must be on-shell, and contracted in the appropriate spinors if they are fermions.
- The gauge fixing condition must be linear in the gauge potential, in order not to have three- and four-photon vertices.

The Ward–Takahashi identity plays a crucial role in ensuring that QED scattering amplitudes are gauge invariant, and that they fulfill the requirements of unitarity despite the presence of non-physical photon polarization in certain gauges.

Unitarity, the Optical Theorem and Cutkosky's Cutting Rules

The time evolution operator from $x^0 = -\infty$ to $x^0 = +\infty$ (also called the S-matrix) is unitary, $SS^\dagger = 1$. Writing it as $S \equiv 1 + iT$ to separate the interactions, this implies the *optical theorem*:

$$\mathrm{Im}\,\langle \alpha_{\mathrm{in}}|T|\alpha_{\mathrm{in}}\rangle = \frac{1}{2}\sum_{\text{states }\beta} \left|\langle \alpha_{\mathrm{in}}|T|\beta_{\mathrm{in}}\rangle\right|^2.$$

This relation implies that the total probability of scattering from the state α to any state β (with at least one interaction) equals twice the imaginary part of the forward scattering amplitude $\alpha \to \alpha$. In perturbation theory, the imaginary part of a transition amplitude Γ can be obtained by means of *Cutkosky's cutting rules*:

$$\mathrm{Im}\,\Gamma = \frac{1}{2}\sum_{\text{cuts }\gamma} [\Gamma]_\gamma,$$

where a cut is a fictitious line that divides the graph into two subgraphs, with at least one external leg on each side of the cut. A cut graph $[\Gamma]_\gamma$ is calculated with the following rules:

- Left of the cut: use the propagator $G^0_{++}(p) = \frac{i}{p^2-m^2+i0^+}$ and the vertex $-i\lambda$,
- Right of the cut: use the propagator $G^0_{--}(p) = \frac{-i}{p^2-m^2-i0^+}$ and the vertex $+i\lambda$,
- The propagators traversing the cut should be $G^0_{+-}(p) = 2\pi\,\theta(-p^0)\delta(p^2)$.

About the Problems of this Chapter

- **Problem 1** establishes a crucial relationship between the field operators ϕ (Heisenberg representation) and $\phi_{\rm in}$ (interaction representation), namely that the former obeys the interacting equation of motion if the latter obeys the free Klein–Gordon equation.
- In **Problem 2**, we derive an explicit form of the elements of the *little group* for massless particles. This is then used in **Problem 9** in order to show that, in a theory with massless spin-1 bosons, the Lorentz invariance of scattering amplitudes implies a property that may be viewed as a weak form of the Ward–Takahashi identity. This observation, due to Weinberg, is extended to gravity in **Problem 10**.
- **Problem 3** establishes some formal relationships between various expressions for the time evolution operator and the S-matrix. Then, **Problem 4** shows that the expression for the S-matrix as the time-ordered exponential of a local interaction term is to a large extent a consequence of causality.
- In **Problem 5**, we derive a set of conditions, known as the *Landau equations*, for a given loop integral to have infrared or collinear singularities. An explicit multi-loop integration is studied in **Problem 6**, which provides another point of view on these conditions.
- **Problem 7** establishes *Weinberg's convergence theorem* in the simple case of scalar field theory, a crucial result in the discussion of renormalization since it clarifies the role of the superficial degree of divergence in assessing whether a particular diagram is ultraviolet divergent.
- The electron anomalous magnetic moment is calculated at one loop in **Problem 8**. This is a classic QED calculation of great historical importance, which has now been pushed to five loops and provides one of the most precise agreements between theory and experiment in all of physics.
- **Problem 11** derives the *Lee–Nauenberg theorem*, an important result about soft and collinear singularities which states that such divergences are removed by summing transition probabilities over degenerate states, thereby providing a link between the finiteness of a quantity and its practical measurability.
- In **Problem 12**, we discuss the external classical field approximation, thanks to which a heavy charged object may be replaced by its classical Coulomb field.
- **Problems 13** and **14** are devoted to a derivation of the *Low–Burnett–Kroll theorem*, a result that states that the emission probability of a soft photon is proportional to the probability of the underlying hard process, at the first two orders in the energy of the emitted photon.
- *Coherent states* are introduced in **Problem 15** and their main properties established. They will be discussed further in **Problems 20, 21** and **22**.
- **Problems 16** and **17** study the running coupling in a scalar field theory with two fields, and in a QCD-like theory at two-loop order.

1. Relationship between the Equations of Motion of ϕ and ϕ_{in} Recall that the field operators in the Heisenberg representation (ϕ) and in the interaction representation (ϕ_{in}) are related by

$$\phi(x) = U(-\infty, x^0)\,\phi_{in}(x)\,U(x^0, -\infty). \tag{1.37}$$

The goal of the problem is to show that, if $V(\phi)$ is the interaction potential, this implies the following relationship between the left-hand sides of their respective equations of motion:

$$(\square + m^2)\phi(x) + V'(\phi(x)) = U(-\infty, x^0)\Big[(\square + m^2)\phi_{in}(x)\Big]U(x^0, -\infty),$$

provided that $U(t_2, t_1) \equiv T\,\exp -i\int_{t_1}^{t_2} d^4x\, V(\phi_{in}(x))$. This is an important consistency check, since it implies that ϕ_{in} is a free field while ϕ evolves as prescribed by the self-interaction term in the Lagrangian.

1.a Apply a derivative ∂_μ to (1.37). Note that spatial derivatives do not act on the U's. In particular, show that $\partial_0\Big[U(-\infty, x^0)\phi_{in}(x)U(x^0, -\infty)\Big] = U(-\infty, x^0)\Pi_{in}(x)U(x^0, -\infty)$. How did the terms coming from the time derivative of the U's cancel?

1.b Apply a second time derivative to this result, to obtain

$$\partial_0^2\Big[U(-\infty, x^0)\phi_{in}(x)U(x^0, -\infty)\Big]$$
$$= U(-\infty, x^0)\Big[\partial_0^2\phi_{in}(x) - i\int d^3y\ \big[\Pi_{in}(x), V(\phi_{in}(x^0, y))\big]\Big]U(x^0, -\infty).$$

1.c Calculate the commutator on the right-hand side (one may prove that if $[A, B]$ is an object that commutes with all other operators, then $[A, f(B)] = f'(B)\,[A, B]$).

1.a Let us start from

$$\phi(x) = U(-\infty, x^0)\phi_{in}(x)U(x^0, -\infty).$$

Since the evolution operators depend only on time, we have trivially

$$(-\nabla^2 + m^2)\phi(x) = U(-\infty, x^0)\Big[(-\nabla^2 + m^2)\phi_{in}(x)\Big]U(x^0, -\infty),$$

and the main difficulty is to deal with the time derivatives. The first time derivative reads

$$\begin{aligned}\partial_0\phi(x) &= \Big[\partial_0 U(-\infty, x^0)\Big]\phi_{in}(x)U(x^0, -\infty) + U(-\infty, x^0)\phi_{in}(x)\Big[\partial_0 U(x^0, -\infty)\Big] \\ &\quad + U(-\infty, x^0)\Big[\partial_0\phi_{in}(x)\Big]U(x^0, -\infty) \\ &= i\,U(-\infty, x^0)\big[\phi_{in}(x), \mathcal{I}(x^0)\big]U(x^0, -\infty) \\ &\quad + U(-\infty, x^0)\Big[\partial_0\phi_{in}(x)\Big]U(x^0, -\infty),\end{aligned}$$

where we denote $\mathcal{I}(x^0) \equiv -\int d^3x\ V(\phi_{in}(x^0, x))$. The first line contains an equal-time commutator of ϕ_{in} with some functional of ϕ_{in}, which is zero, leaving only the non-vanishing term of the second line.

1.b A second differentiation with respect to time gives

$$\partial_0^2\phi(x) = i\,U(-\infty,x^0)\big[\partial_0\phi_{in}(x),\mathcal{I}(x^0)\big]U(x^0,-\infty)$$
$$+\,U(-\infty,x^0)\Big[\partial_0^2\phi_{in}(x)\Big]U(x^0,-\infty).$$

1.c The commutator in the first line is an equal-time commutator between the canonical momentum $\partial_0\phi_{in}$ and a functional of the field ϕ_{in}. In order to evaluate it, we need the following result:

$$[A,f(B)] = [A,B]\,f'(B)\,, \quad \text{valid when } [[A,B],B]=0.$$

This can be shown by using the Taylor series of $f(B)$, by first showing by recursion that $[A,B^n] = n[A,B]B^{n-1}$. Then, we can write

$$\begin{aligned} i\left[\partial_0\phi_{in}(x),\mathcal{I}(x^0)\right] &= -i\int d^3y\ \left[\partial_0\phi_{in}(x),V(\phi_{in}(x^0,y))\right] \\ &= -i\int d^3y\ \underbrace{\left[\partial_0\phi_{in}(x),\phi_{in}(x^0,y)\right]}_{-i\delta(x-y)}\ V'(\phi_{in}(x^0,y)) \\ &= -V'(\phi_{in}(x)). \end{aligned}$$

Now, using $U(-\infty,x^0)V'(\phi_{in}(x))U(x^0,-\infty) = V'(\phi(x))$, we get

$$(\Box+m^2)\phi(x) - \mathcal{L}'(\phi(x)) = U(-\infty,x^0)\Big[(\Box+m^2)\phi_{in}(x)\Big]U(x^0,-\infty).$$

Therefore, the left-hand sides of the equations of motion for the Heisenberg representation (interacting) field ϕ and for the interaction representation (free) field ϕ_{in} are related by a unitary transformation identical to the formula that relates the field operators themselves.

2. Little-Group Elements for Massless Particles The *little group* is the subgroup of the Lorentz group that leaves a fixed reference vector q^μ invariant. In this problem, we derive a particularly convenient explicit form of the elements of the little group in the case where q^μ is the light-like vector $q^\mu \equiv (\omega,0,0,\omega)$.

2.a First, show that an infinitesimal little-group transformation of this kind can be written as follows:

$$R \approx 1 - i\theta\,J^3 + i\alpha_1\,B^1 + i\alpha_2\,B^2,$$

with three generators J^3, B^1, B^2 (the first one being the generator of rotations about the third direction of space) that one should determine explicitly. Check also that they satisfy the following commutation relations (after an appropriate normalization):

$$\big[J^3,B^i\big] = i\epsilon_{ij}\,B^j, \quad \big[B^1,B^2\big] = 0.$$

2.b Show that any finite element R of the massless little group can be written as

$$R^{\mu}{}_{\nu} = \begin{pmatrix} 1 & 0 & 0 & 0 \\ 0 & \cos\theta & -\sin\theta & 0 \\ 0 & \sin\theta & \cos\theta & 0 \\ 0 & 0 & 0 & 1 \end{pmatrix} \times \begin{pmatrix} 1+\frac{\beta_1^2+\beta_2^2}{2} & \beta_1 & \beta_2 & -\frac{\beta_1^2+\beta_2^2}{2} \\ \beta_1 & 1 & 0 & -\beta_1 \\ \beta_2 & 0 & 1 & -\beta_2 \\ \frac{\beta_1^2+\beta_2^2}{2} & \beta_1 & \beta_2 & 1-\frac{\beta_1^2+\beta_2^2}{2} \end{pmatrix}.$$

Hint: use the exact Baker–Campbell–Hausdorff formula:

$$\ln\left(e^{iX}e^{iY}\right) = iX + i\int_0^1 dt\, F\left(e^{t\,\mathrm{ad}_Y}\, e^{\mathrm{ad}_X}\right)\, Y,$$

where $\mathrm{ad}_X(Y) \equiv -i[X,Y]$ *and* $F(z) \equiv \dfrac{\ln(z)}{z-1}$,

in order to exponentiate the infinitesimal form.

2.a This question is not difficult to solve by "brute force," i.e., by looking for the most general little-group transformation

$$R^{\mu}{}_{\nu} \equiv \delta^{\mu}{}_{\nu} + \omega^{\mu}{}_{\nu}$$

such that $\omega^{\mu}{}_{\nu}q^{\nu} = 0$ for all values of the index μ, with the additional constraint that $\omega_{\mu\nu} = 0$ is antisymmetric (so we have a legitimate infinitesimal Lorentz transformation). If we parameterize

$$\omega_{\mu\nu} \equiv \begin{pmatrix} 0 & a & b & c \\ -a & 0 & d & e \\ -b & -d & 0 & f \\ -c & -e & -f & 0 \end{pmatrix},$$

the condition that $q^{\nu} \equiv (\omega, 0, 0, \omega)$ is invariant is equivalent to

$$c = 0, \quad a = e, \quad b = f,$$

implying that there is a three-parameter family of $\omega^{\mu}{}_{\nu}$'s that fulfill all the requirements:

$$\omega^{\mu}{}_{\nu} = \theta\begin{pmatrix} 0 & 0 & 0 & 0 \\ 0 & 0 & -1 & 0 \\ 0 & 1 & 0 & 0 \\ 0 & 0 & 0 & 0 \end{pmatrix} + \alpha_1\begin{pmatrix} 0 & 1 & 0 & 0 \\ 1 & 0 & 0 & -1 \\ 0 & 0 & 0 & 0 \\ 0 & 1 & 0 & 0 \end{pmatrix} + \alpha_2\begin{pmatrix} 0 & 0 & 1 & 0 \\ 0 & 0 & 0 & 0 \\ 1 & 0 & 0 & -1 \\ 0 & 0 & 1 & 0 \end{pmatrix}$$
$$= -i\theta\,(\underbrace{M^{12}}_{J^3})^{\mu}{}_{\nu} + i\alpha_1(\underbrace{M^{10} - M^{13}}_{\equiv B^1})^{\mu}{}_{\nu} + i\alpha_2(\underbrace{M^{20} - M^{23}}_{\equiv B^2})^{\mu}{}_{\nu}.$$

(The identification in the second line follows from $\omega_{\mu\nu} = -\frac{i}{2}\omega_{\alpha\beta}\left(M^{\alpha\beta}\right)_{\mu\nu}$, valid for the spin-1 representation of the Lorentz algebra, where $(M^{\alpha\beta})_{\mu\nu} = i(\delta^{\alpha}{}_{\mu}\delta^{\beta}{}_{\nu} - \delta^{\alpha}{}_{\nu}\delta^{\beta}{}_{\mu})$.) Note that the first term corresponds to rotations about the x^3 axis, which trivially leaves invariant any vector whose only non-zero spatial component is along the third direction.

The announced commutation relations can be checked by an explicit evaluation of the corresponding matrix products. Alternatively, they can also be obtained from

$$J^3 = M^{12}, \quad B^1 = M^{10} - M^{13}, \quad B^2 = M^{20} - M^{23},$$

and by using the defining commutation relation of the Lorentz algebra,

$$\left[M^{\mu\nu}, M^{\rho\sigma}\right] = i(g^{\nu\rho}M^{\mu\sigma} - g^{\mu\rho}M^{\nu\sigma}) - i(g^{\nu\sigma}M^{\mu\rho} - g^{\mu\sigma}M^{\nu\rho}).$$

2.b Given the infinitesimal form of little-group transformations for massless particles derived above, any finite little-group transformation R can be obtained by exponentiating the inifinitesimal ones:

$$R \equiv e^{i(-\theta J^3 + \alpha_1 B^1 + \alpha_2 B^2)}.$$

Note first that the factor on the left of the proposed formula is nothing but the rotation $e^{-i\theta J^3}$, i.e., a rotation by an angle θ about the x^3 axis, which affects only the coordinates 1, 2:

$$\left(e^{-i\theta J^3}\right)^\mu{}_\nu = \begin{pmatrix} 1 & 0 & 0 & 0 \\ 0 & \cos\theta & -\sin\theta & 0 \\ 0 & \sin\theta & \cos\theta & 0 \\ 0 & 0 & 0 & 1 \end{pmatrix}.$$

Thus, the form proposed in the statement of the problem suggests that R may also be written as

$$R = e^{-i\theta J^3}\, e^{i(\beta_1 B^1 + \beta_2 B^2)}.$$

Our ansatz for the form of the second factor is based on the case $\theta = 0$ (in this case, we should of course have $\beta_i = \alpha_i$, but this is not necessarily true for $\theta \neq 0$). Verifying that the two expressions for R are equivalent could in principle be performed by calculating both expressions and equating them in order to find the relationship between the coefficients $\alpha_{1,2}$ and $\beta_{1,2}$, but this is rather challenging.

A much more efficient method is to use the exact form of the Baker–Campbell–Hausdorff formula, recalled in the statement of the problem. In the present case, the commutation relations among J^3, B^1, B^2 lead to

$$\mathrm{ad}_{-\theta J^3}(\beta_i B^i) = -\theta\, \beta_i \epsilon_{ij} B^j, \quad e^{\mathrm{ad}_{-\theta J^3}}\, \beta_i B^i = \beta_i \left(e^{-\theta\epsilon}\right)_{ij} B^j, \quad \mathrm{ad}_{\beta_i B^i}(B^j) = 0,$$

and the Baker–Campbell–Hausdorff formula gives

$$\ln\left(e^{-i\theta J^3} e^{i\beta_i B^i}\right) = -i\theta J^3 - i\beta_i \left[\frac{\theta\epsilon}{e^{-\theta\epsilon} - 1}\right]_{ij} B^j.$$

Using the fact that $\epsilon_{ij}\epsilon_{jk} = -\delta_{ik}$, we have

$$e^{-\theta\epsilon} - 1 = \cos\theta - 1 - \epsilon\sin\theta,$$

$$\ln\left(e^{-i\theta J^3} e^{i\beta_i B^i}\right) = -i\theta J^3 - i\,\underbrace{\theta\beta_i\left[\epsilon(\cos\theta - 1 - \epsilon\sin\theta)^{-1}\right]_{ij}}_{-\alpha_j} B^j.$$

This formula shows that $e^{i(-\theta J^3 + \alpha_i B^i)} = e^{-i\theta J^3} e^{i\beta_i B^i}$ with the following relationship between the coefficients $\alpha_{1,2}$ and $\beta_{1,2}$:

$$\beta_i = \theta^{-1}\left(\sin\theta + \epsilon(1 - \cos\theta)\right)_{ij} \alpha_j.$$

(Note that $\beta_i = \alpha_i$ when $\theta \to 0$, as expected trivially in this limit.)

In order to calculate the factor $e^{i\beta_i B^i}$, one should first note that $e^{i\beta_i B^i} = e^{i\beta_1 B^1} e^{i\beta_2 B^2}$ since B^1 and B^2 commute. Using the explicit representations of $B^{1,2}$, simple algebra shows that

$$((iB^1)^2)^\mu{}_\nu = ((iB^2)^2)^\mu{}_\nu = \begin{pmatrix} 1 & 0 & 0 & -1 \\ 0 & 0 & 0 & 0 \\ 0 & 0 & 0 & 0 \\ 1 & 0 & 0 & -1 \end{pmatrix},$$

and

$$((iB^1)^n)^\mu{}_\nu = ((iB^2)^n)^\mu{}_\nu = 0 \quad \text{for } n \geq 3.$$

Therefore, we have

$$(e^{i\beta_1 B^1})^\mu{}_\nu = \left(1 + i\beta_1 B^1 + \tfrac{1}{2}\beta_1^2 (iB^1)^2\right)^\mu{}_\nu = \begin{pmatrix} 1+\frac{\beta_1^2}{2} & \beta_1 & 0 & -\frac{\beta_1^2}{2} \\ \beta_1 & 1 & 0 & -\beta_1 \\ 0 & 0 & 1 & 0 \\ \frac{\beta_1^2}{2} & \beta_1 & 0 & 1-\frac{\beta_1^2}{2} \end{pmatrix},$$

$$(e^{i\beta_2 B^2})^\mu{}_\nu = \left(1 + i\beta_2 B^2 + \tfrac{1}{2}\beta_2^2 (iB^2)^2\right)^\mu{}_\nu = \begin{pmatrix} 1+\frac{\beta_2^2}{2} & 0 & \beta_2 & -\frac{\beta_2^2}{2} \\ 0 & 1 & 0 & 0 \\ \beta_2 & 0 & 1 & -\beta_2 \\ \frac{\beta_2^2}{2} & 0 & \beta_2 & 1-\frac{\beta_2^2}{2} \end{pmatrix},$$

and finally

$$(e^{i\beta_i B^i})^\mu{}_\nu = \begin{pmatrix} 1+\frac{\beta_1^2+\beta_2^2}{2} & \beta_1 & \beta_2 & -\frac{\beta_1^2+\beta_2^2}{2} \\ \beta_1 & 1 & 0 & -\beta_1 \\ \beta_2 & 0 & 1 & -\beta_2 \\ \frac{\beta_1^2+\beta_2^2}{2} & \beta_1 & \beta_2 & 1-\frac{\beta_1^2+\beta_2^2}{2} \end{pmatrix},$$

which establishes the announced result.

---

3. S-matrix in Terms of $\phi_{\rm in}$ and $\phi_{\rm out}$ Given an interaction Lagrangian $\mathcal{L}_I$, the field operators in the Heisenberg representation (ϕ) and in the interaction representation ($\phi_{\rm in}$) are related by means of a time evolution operator

$$U_{\rm in}(t_2, t_1) \equiv T \exp\left(i \int_{t_1}^{t_2} d^4x\, \mathcal{L}_I(\phi_{\rm in}(x))\right)$$

expressed in terms of the free field $\phi_{\rm in}$ (we have added a subscript *in* to this evolution operator, in order to recall that it is defined in terms of $\phi_{\rm in}$, since we are about to introduce its counterpart defined in terms of $\phi_{\rm out}$). Likewise, we define a similar evolution operator in terms of $\phi_{\rm out}$, the field operator in the interaction picture that coincides with the Heisenberg picture at $x^0 = +\infty$:

$$U_{\rm out}(t_2, t_1) \equiv T \exp\left(i \int_{t_1}^{t_2} d^4x\, \mathcal{L}_I(\phi_{\rm out}(x))\right).$$

3.a Show that $U_{in}(+\infty,-\infty) = U_{out}(+\infty,-\infty)$. In other words, the S-matrix (i.e., the time evolution operator over the entire time range) does not depend on whether it is defined in terms ϕ_{in} or ϕ_{out}.

3.b Are U_{in} and U_{out} identical in general? Find the relationship between the two.

3.c Show that the S-matrix is also given by $S = U_{in}(x^0,-\infty)U_{out}(+\infty,x^0)$, for any intermediate time x^0. Note that, on the surface, this expression does not seem to be properly time-ordered. Why is it nevertheless a correct formula?

3.a Recall the relationship between the interacting field ϕ and the free field ϕ_{in} of the interaction representation:

$$\phi(x) = U_{in}(-\infty,x^0)\,\phi_{in}(x)\,U_{in}(x^0,-\infty),$$
$$U_{in}(t_2,t_1) \equiv T\,\exp\left(i\int_{t_1}^{t_2} d^4x\,\mathcal{L}_I(\phi_{in}(x))\right). \tag{1.38}$$

(Since we shall shortly write the analogous relationship with ϕ_{out}, it is important to have a subscript *in* on the evolution operator to avoid confusion, since it depends on ϕ_{in}.) By taking the limit $x^0 \to +\infty$ in this equation, and using the fact that in this limit the interacting field becomes identical to the free field ϕ_{out}, we obtain a first relationship between ϕ_{in} and ϕ_{out}:

$$\phi_{out}(x) \underset{x^0\to+\infty}{=} U_{in}(-\infty,+\infty)\,\phi_{in}(x)\,U_{in}(+\infty,-\infty).$$

Strictly speaking, this limiting procedure gives a relationship between the two fields only for large x^0. Then, we use the fact that two fields obeying the same equation of motion (here, ϕ_{in} and ϕ_{out} both obey the Klein–Gordon equation) and identical in some region of time are equal at all times (this argument relies on the uniqueness of the solutions of the Klein–Gordon equation, if their value and that of their first time derivative are prescribed at some time). Therefore, the above equation is in fact valid at all times.

Note that the right-hand side of this equation depends only on ϕ_{in}, but in a completely non-linear and non-local fashion because of the evolution operators. Another noteworthy aspect of this equation is that, despite the fact that both ϕ_{in} and ϕ_{out} are free fields, the relationship between the two involves the interactions.

The easiest way to invert the relationship between ϕ_{in} and ϕ_{out} is to write the analogue of (1.38) for the free field ϕ_{out}:

$$\phi(x) = U_{out}(+\infty,x^0)\,\phi_{out}(x)\,U_{out}(x^0,+\infty),$$
$$U_{out}(t_2,t_1) \equiv T\,\exp\left(i\int_{t_1}^{t_2} d^4x\,\mathcal{L}_I(\phi_{out}(x))\right). \tag{1.39}$$

Taking the limit $x^0 \to -\infty$ in this equation leads to a second form of the formula that relates

ϕ_{in} and ϕ_{out}:

$$\phi_{in}(x) = U_{out}(+\infty, -\infty)\, \phi_{out}(x)\, U_{out}(-\infty, +\infty),$$

or, equivalently,

$$\phi_{out}(x) = U_{out}(-\infty, +\infty)\, \phi_{in}(x)\, U_{out}(+\infty, -\infty).$$

In order for the two relations we have obtained to be consistent, we must have

$$U_{in}(+\infty, -\infty) = U_{out}(+\infty, -\infty).$$

Therefore, the evolution operators *over the entire time range* are identical, regardless of whether they are constructed with the fields ϕ_{in} or ϕ_{out}. For this combination of time arguments, we may drop the subscripts *in/out* on the evolution operators.

3.b But it is important to realize that this property is not true for arbitrary time intervals. By requesting that (1.38) and (1.39) give the same interacting field ϕ, we must have

$$\begin{aligned} & U_{in}(-\infty, x^0)\, \phi_{in}(x)\, U_{in}(x^0, -\infty) \\ &\quad = U_{out}(+\infty, x^0)\, \phi_{out}(x)\, U_{out}(x^0, +\infty) \\ &\quad = U_{out}(+\infty, x^0) U(-\infty, +\infty)\, \phi_{in}(x)\, U(+\infty, -\infty) U_{out}(x^0, +\infty), \end{aligned}$$

implying that in general we have

$$U_{in}(x^0, -\infty) = U(+\infty, -\infty) U_{out}(x^0, +\infty). \tag{1.40}$$

Writing $U_{in}(x^0, -\infty) = U_{in}(x^0, y^0) U_{in}(y^0, -\infty)$ and using the same identity with x^0 replaced by y^0, we obtain

$$U_{out}(x^0, y^0) = U(-\infty, +\infty) U_{in}(x^0, y^0) U(+\infty, -\infty).$$

(This relation could also have been obtained from the definition of U_{out}, by performing its Taylor expansion in powers of ϕ_{out}, replacing every occurrence of ϕ_{out} by its expression in terms of ϕ_{in}, and at the end repackaging the series to obtain a U_{in}.)

3.c By multiplying (1.40) on the right by the inverse of $U_{out}(x^0, +\infty)$, we obtain another formula for the full evolution operator,

$$U(+\infty, -\infty) = U_{in}(x^0, -\infty) U_{out}(+\infty, x^0),$$

which is rather counterintuitive since the order of the operators on the right-hand side may (wrongly) suggest that it is inconsistent with the time ordering. The resolution of this paradox is that $U(+\infty, -\infty)$ is time-ordered when expressed entirely in terms of ϕ_{in} or entirely in terms of ϕ_{out}; but the right-hand side of the above formula mixes ϕ_{in} and ϕ_{out}, and the relationship between ϕ_{in} and ϕ_{out} is non-local in time, which obscures the actual time ordering of the operators.

4. Constraints on the S-matrix from Causality The goal of this problem is to derive general constraints on the S-matrix from causality. To that end, let us assume that the coupling constant λ that controls the interactions is a function of spacetime, $\lambda(x)$. With this modification, the S-matrix becomes a functional $S[\lambda]$.

4.a Consider two regions of spacetime, Ω_1 and Ω_2, such that Ω_2 lies *in the future light-cone* of Ω_1, and denote by $\lambda_{1,2}(x)$ the coupling function restricted to these domains (we assume it is zero outside of $\Omega_1 \cup \Omega_2$). Show that $S[\lambda_1 + \lambda_2] = S[\lambda_2]S[\lambda_1]$.

4.b Generalize this result to the more general situation where Ω_1 and Ω_2 are simply separated by a locally space-like surface.

4.c Consider now the case where the separation between any pair of points of Ω_1 and Ω_2 is space-like. Show that $\big[S[\lambda_1], S[\lambda_2]\big] = 0$.

4.d If two coupling functions λ and λ' coincide for $x^0 \le y^0$, show that $S[\lambda']S^\dagger[\lambda]$ does not depend on the behavior of the coupling at times $\le y^0$. By considering an infinitesimal variation of the coupling function, show that

$$\frac{\delta}{\delta\lambda(y)}\left(\frac{\delta S[\lambda]}{\delta\lambda(x)}\, S^\dagger[\lambda]\right) = 0 \quad \text{if } y \text{ is not in the future light-cone of } x.$$

4.e Solve the constraints of causality and unitarity to obtain the form of $S[\lambda]$ up to $\mathcal{O}(\lambda^2)$.

4.a Let us consider two regions Ω_1 and Ω_2 of spacetime, as shown in Figure 1.1. Thus, for any pair of points $x_1 \in \Omega_1, x_2 \in \Omega_2$, we have $x_2^0 > x_1^0$ and $(x_1 - x_2)^2 > 0$. We assume that interactions exist only in $\Omega_1 \cup \Omega_2$ and are zero elsewhere, and we denote by $\lambda_{1,2}$ the coupling functions in these two domains. The coupling function over the entire spacetime is thus

$$\lambda(x) = \lambda_1(x) + \lambda_2(x),$$

and the full S-matrix is $S[\lambda_1 + \lambda_2]$. Recall that the S-matrix is the operator that connects the *in* states at $x^0 = -\infty$ and the *out* states at $x^0 = +\infty$:

$$\big\langle \alpha_{\text{out}}\big| = \big\langle \alpha_{\text{in}}\big|\, S_{\text{in}}[\lambda_1 + \lambda_2].$$

For the time being, it is safer to add a subscript *in* on the S-matrix in order to indicate that it is expressed in terms of the field operator ϕ_{in}. Given the relative configuration of the domains Ω_1 and Ω_2, we could also construct another version of the interaction representation, where the fields coincide with the Heisenberg representation ones at some intermediate time located between Ω_1 and Ω_2. Let us call *intermediate* this representation, and ϕ_{inter} the corresponding free field operator. We have

$$\big\langle \alpha_{\text{inter}}\big| = \big\langle \alpha_{\text{in}}\big|\, S_{\text{in}}[\lambda_1], \quad \big\langle \alpha_{\text{out}}\big| = \big\langle \alpha_{\text{inter}}\big|\, S_{\text{inter}}[\lambda_2].$$

Thus, we obtain

$$S_{\text{in}}[\lambda_1 + \lambda_2] = S_{\text{in}}[\lambda_1]\, S_{\text{inter}}[\lambda_2].$$

Note that this equation is subject to the same paradox regarding the ordering of the operators as in the last equation in Problem 3: the seemingly unnatural ordering between $S_{\text{in}}[\lambda_1]$ and

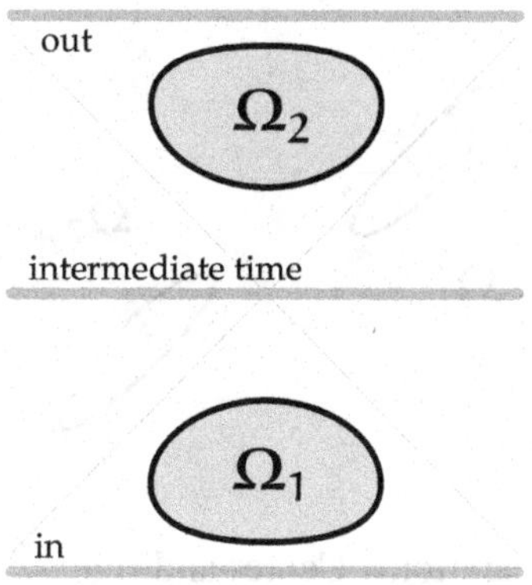

Figure 1.1 Domains Ω_1 and Ω_2, with Ω_2 under the causal influence of Ω_1.

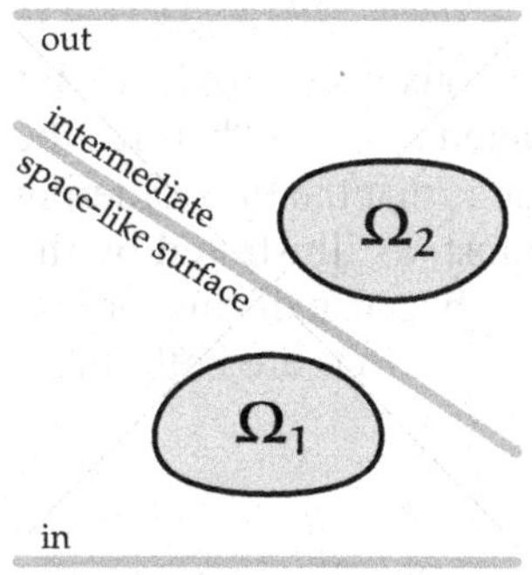

Figure 1.2 Domains $\Omega_{1,2}$, such that Ω_2 does not causally influence Ω_1.

$S_{inter}[\lambda_2]$ is due to the fact that the latter is implicitly expressed in terms of the field ϕ_{inter} while the former depends on the field ϕ_{in}. Using exactly the same manipulations as in Problem 3, we can rewrite this expression solely in terms of ϕ_{in}, which gives

$$\begin{aligned} S_{inter}[\lambda_2] &= S_{out}[\lambda_2], \\ S_{in}[\lambda_1 + \lambda_2] &= S_{in}[\lambda_1]S_{out}[\lambda_2], \\ S_{in}[\lambda_1 + \lambda_2] &= S_{in}[\lambda_2]\, S_{in}[\lambda_1]. \end{aligned}$$

From now on, we implicitly assume that all S-matrix operators are expressed in terms of ϕ_{in}, and we suppress the subscript *in*.

4.b In words, the previous setup could be described by saying that Ω_2 *is under the influence of* Ω_1. A much more general situation would be to simply request that Ω_2 *does not influence* Ω_1. This is achieved by dividing spacetime with a hyper-surface Σ located between the domains Ω_1 and Ω_2, provided that this surface is locally space-like (i.e., no signal can travel from the surface towards the domain below it). This setup is shown in Figure 1.2. This ordering of the domains Ω_1 and Ω_2 is sufficient to reproduce the preceding arguments, leading again to $S[\lambda_1 + \lambda_2] = S[\lambda_2]\, S[\lambda_1]$.

4.c Consider now the situation where the interval between any point in Ω_1 and any point in Ω_2 is space-like, as illustrated in Figure 1.3 (left panel). Since it is possible to find an

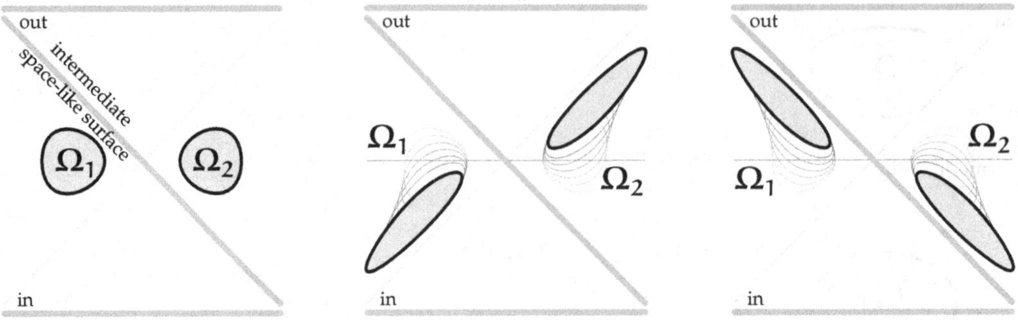

Figure 1.3 Left: domains Ω_1 and Ω_2 with mutual space-like separations. Center and right: displacements of the domains under a Lorentz boost.

appropriate space-like surface separating the two domains, the previous result applies. But now there is an ambiguity regarding which domain should be considered as being "before" the other one. In fact, what is special about this kinematical configuration is that the time ordering between the two domains can be altered by applying a Lorentz boost, as illustrated by the second ($\beta < 0$) and third panels ($\beta > 0$) of Figure 1.3. In this case, because of this lack of absolute time ordering between the two domains, the following two equations are both true:

$$S[\lambda_1 + \lambda_2] = S[\lambda_1]\, S[\lambda_2] \quad \text{and} \quad S[\lambda_1 + \lambda_2] = S[\lambda_2]\, S[\lambda_1],$$

which implies that $\big[S[\lambda_1], S[\lambda_2]\big] = 0$.

4.d Consider now two distinct coupling functions in the domain Ω_2, namely λ_2 and λ_2'. Thus, we have two realizations of the coupling function:

$$\lambda \equiv \lambda_1 + \lambda_2, \quad \lambda' \equiv \lambda_1 + \lambda_2'.$$

For these two coupling functions, the S-matrix is given by

$$S[\lambda] = S[\lambda_2]\, S[\lambda_1], \quad S[\lambda'] = S[\lambda_2']\, S[\lambda_1],$$

and we therefore have

$$S[\lambda']\, S^\dagger[\lambda] = S[\lambda_2']\, S[\lambda_1]\, S^\dagger[\lambda_1]\, S^\dagger[\lambda_2] = S[\lambda_2']\, S^\dagger[\lambda_2].$$

This combination is independent of the function λ_1, i.e., independent of the behavior of the coupling function in the portion of spacetime that cannot receive any causal influence from Ω_2. Let us now assume that the difference $\delta\lambda \equiv \lambda_2' - \lambda_2$ is infinitesimal. To first order in $\delta\lambda$, we have

$$S[\lambda'] = S[\lambda] + \int d^4x\, \delta\lambda(x)\, \frac{\delta S[\lambda]}{\delta\lambda(x)} + \mathcal{O}(\delta\lambda^2),$$

and

$$S[\lambda']\, S^\dagger[\lambda] = 1 + \int d^4x\, \delta\lambda(x)\, \frac{\delta S[\lambda]}{\delta\lambda(x)}\, S^\dagger[\lambda] + \mathcal{O}(\delta\lambda^2).$$

Since this should be independent of λ_1 for any variation $\delta\lambda$, we must have

$$\frac{\delta S[\lambda]}{\delta\lambda(x)}\, S^\dagger[\lambda] \quad \text{is independent of } \lambda_1 \text{ if } \Omega_1 \text{ is not under the causal influence of } x.$$

This condition can also be phrased as

$$\frac{\delta}{\delta\lambda(y)} \left(\frac{\delta S[\lambda]}{\delta\lambda(x)}\, S^\dagger[\lambda] \right) = 0 \quad \text{if } y \text{ is not in the future light-cone of } x. \tag{1.41}$$

Note that the same identity would be true in a theory where the fields are coupled to some external source J_{ext}, if we replace λ by J_{ext}, again thanks to causality.

4.e In order to see the consequences of this constraint on the S-matrix, let us write a formal Taylor expansion of the functional $S[\lambda]$:

$$S[\lambda] = 1 + \int d^4x\, S_1(x)\, \lambda(x) + \frac{1}{2} \int d^4x d^4y\, S_2(x,y)\, \lambda(x)\lambda(y) + \cdots.$$

In this expansion, the objects $S(x_1, \ldots, x_n)$ are operator-valued symmetric functions of their arguments. In addition to the constraint (1.41), the S-matrix must be unitary, and also satisfy $\big[S[\lambda_1], S[\lambda_2]\big] = 0$ when the supports of λ_1 and λ_2 have purely space-like separations. The last constraint implies that the coefficients S_n are *multi-local* operators (i.e., $S_n(x_1, \ldots, x_n)$ depends only on the field operator and its derivatives at the points $x_1, \ldots, x_n$) constructed with the field operator and its derivatives (non-locality would lead to violations of this commutation relation). In the first two orders, the unitarity of S implies that

$$S_1(x) + S_1^\dagger(x) = 0, \quad S_2(x,y) + S_2^\dagger(x,y) + S_1^\dagger(x) S_1(y) + S_1^\dagger(y) S_1(x) = 0.$$

(In deriving the second equation, we must be careful to symmetrize the coefficient that multiplies $\lambda(x)\lambda(y)$ in the integrand of the second-order term.) These equations can be rewritten as

$$S_1^\dagger(x) = -S_1(x), \quad S_2(x,y) + S_2^\dagger(x,y) = S_1(x) S_1(y) + S_1(y) S_1(x).$$

Note that unitarity can only constrain the Hermitian part of the coefficients S_n, and does not say anything about their anti-Hermitian part. To put constrains on the latter, we need to make use of (1.41). It is straightforward to check that

$$\frac{\delta}{\delta\lambda(y)}\left(\frac{\delta S[\lambda]}{\delta\lambda(x)}S^{\dagger}[\lambda]\right)=S_2(x,y)+S_1(x)S_1^{\dagger}(y)+\mathcal{O}(\lambda).$$

Therefore, at lowest order, (1.41) tells us that

$$S_2(x,y)=-S_1(x)S_1^{\dagger}(y)=S_1(x)S_1(y)\quad \text{if } y \text{ is not in the future light-cone of } x.$$

Using the fact that $S_2(x,y)$ should be symmetric, we also have

$$S_2(x,y)=S_2(y,x)=S_1(y)S_1(x)\quad \text{if } x \text{ is not in the future light-cone of } y.$$

(Note that these two conditions on x, y are both satisfied if their separation is space-like. For such a space-like separation, we could thus use either of the two formulas. This does not lead to any contradiction, provided that $S_1(x)$ is a local function of the field operator and its derivatives.) Therefore, the answer valid for any x, y can be written as

$$S_2(x,y)=\mathrm{T}\left(S_1(x),S_1(y)\right).$$

(One may check a posteriori that the unitarity constraint is satisfied.) Using the same constraints, one could show by induction that the coefficient of order n, $S_n(x_1,\ldots,x_n)$, is the time-ordered product of n factors S_1. Therefore, we see that unitarity and causality provide an almost closed form for the S-matrix,

$$S[\lambda]=\mathrm{T}\exp\left(\int d^4x\; S_1(x)\lambda(x)\right),$$

in which the only remaining unknown is the first coefficient $S_1(x)$. The latter can be related to the interaction Lagrangian by considering a $\lambda(x)$ which is non-zero in an infinitesimal region of spacetime. Therefore, regardless of the microscopic details of a given theory – which control the first coefficient S_1 –, the general structure of the S-matrix is governed to a large extent by the constraints provided by unitarity and causality.

5. Landau Equations for Soft and Collinear Singularities The goal of this problem is to study the singularities that may occur in a Feynman integral due to vanishing denominators (not to be confused with ultraviolet divergences, due to an integrand that decreases too slowly at large momentum). Consider a Feynman integral with L loops and m denominators:

$$\mathcal{I}(\{p_i\})\equiv\int\prod_{j=1}^{L}\frac{d^D\ell_j}{(2\pi)^D}\,\frac{\mathcal{N}(\{p_i\},\{\ell_j\})}{(q_1^2-m_1^2+i0^+)\cdots(q_m^2-m_m^2+i0^+)}.$$

In this integral, the p_i are the external momenta, the ℓ_j the loop momenta, and the q_r the momenta of the propagators in the loops, i.e., linear combinations of loop momenta and external momenta of the form $q_r\equiv\sum_{j=1}^{L}\epsilon_{rj}\ell_j+\Delta_r$ (where the coefficients ϵ_{rj} take values in $\{-1,0,+1\}$ and the Δ_r depend only on the external momenta).

5.a Use Feynman parameters x_r to combine the m denominators into a single one, $\mathcal{D}^m$.

5.b By considering the following elementary examples,

$$\int_{-1}^{+1} \frac{dx}{x+i0^+}, \quad \int_{0}^{+1} \frac{dx}{x+i0^+}, \quad \int_{-1}^{+1} \frac{dx}{(x+i0^+)(x-i0^+)},$$

explain why singularities occur only when a pole is located on the boundary of the integration manifold, or when multiple poles "pinch" the integration manifold.

5.c Study the variations of $\mathcal{D}$ as a function of the loop momenta ℓ_j, and explain why the zeroes of $\mathcal{D}$ must also be extrema of $\mathcal{D}$ in order to produce singularities.

5.d Now study the dependence with respect to the Feynman parameters and show that, for each r, one must have either $x_r = 0$ or $q_r^2 = m_r^2$ in order to have a singularity.

5.e Conclude that the conditions for a singularity are given by the *Landau equations*:

$$\text{For each propagator } r\text{: } x_r(q_r^2 - m_r^2) = 0; \quad \text{For each loop } j\text{: } \sum_{r=1}^{m} \epsilon_{rj} x_r q_r^\mu = 0.$$

5.f Determine the singularities of the one-loop Feynman diagram shown on the right (assume $k^2 = k'^2 = 0$, $(k+k')^2 > 0$ and all internal particles are massless)

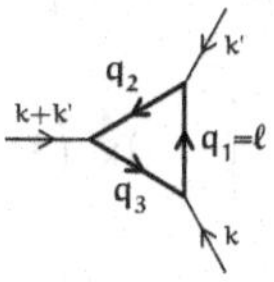

5.a Consider a Feynman integral with L loops and m denominators. Generically, such an integral may be written as

$$\mathcal{I}(\{p_i\}) \equiv \int \prod_{j=1}^{L} \frac{d^D \ell_j}{(2\pi)^D} \frac{\mathcal{N}(\{p_i\},\{\ell_j\})}{(q_1^2 - m_1^2 + i0^+)\cdots(q_m^2 - m_m^2 + i0^+)},$$

where the p_i are the momenta external to the loops and the ℓ_j are L independent loop momenta. The momenta q_r are the momenta carried by the various propagators along the loops. They are all of the form

$$q_r \equiv \sum_{j=1}^{L} \epsilon_{rj} \ell_j + \Delta_r,$$

where the coefficients ϵ_{rj} take values in $\{-1, 0, +1\}$ (a propagator may belong to a loop or not, and may be oriented in the same way as the loop momentum or in the opposite direction) and where the Δ_r depend only on the external momenta (they are thus constants from the point of view of evaluating the loop integrals). $\mathcal{N}(\{p_i\},\{\ell_j\})$ is a numerator that comprises all the momentum dependence that may arise, e.g., from three-gluon vertices in QCD or from the Dirac traces if there are fermion loops. This factor plays no role in analyzing the singularities of the integral, except in those rare situations where a singularity due to a vanishing denominator may be canceled by an accidental concomitant vanishing of the numerator.

The first step is to combine the m denominators into a single one thanks to Feynman parameterization:

$$\mathcal{I}(\{p_i\}) = \Gamma(m)\int\prod_{j=1}^{L}\frac{d^D\ell_j}{(2\pi)^D}\int_{x_r\geq 0}\prod_{r=1}^{m}dx_r\,\delta(1-\sum_r x_r)\,\frac{\mathcal{N}(\{p_i\},\{\ell_j\})}{\big(\mathcal{D}(\{q_r\},\{x_r\})+i0^+\big)^m}, \quad (1.42)$$

$$\mathcal{D}(\{q_r\},\{x_r\}) \equiv \sum_{r=1}^{m} x_r(q_r^2 - m_r^2).$$

5.b In this problem, we assume that the ultraviolet divergences have been properly disposed of by some regularization, and we are chiefly interested in the possibility of additional singularities that may arise from a vanishing denominator. Obviously, that the equation

$$\mathcal{D}(\{q_r\},\{x_r\}) = 0$$

has solutions in the integration domain is a necessary condition for having such singularities. But a zero of the denominator of the integrand does not always lead to a singularity in the integral. In order to see this, let us consider the following toy examples:

$$\int_{-1}^{+1}\frac{dx}{x+i0^+} = \int_{-1}^{+1}dx\left(P\left(\frac{1}{x}\right) - i\pi\,\delta(x)\right) = -i\pi,$$

$$\int_{0}^{+1}\frac{dx}{x+i0^+} = \infty,$$

$$\int_{-1}^{+1}\frac{dx}{(x+i0^+)(x-i0^+)} = \lim_{\epsilon\to 0^+}\int_{-1}^{+1}\frac{dx}{x^2+\epsilon^2} = \infty.$$

In the first example, the integral is finite despite the fact that denominator vanishes at $x = 0$, because the integration contour is slightly shifted from the singularity by the presence of the $i0^+$. An infinite result would be obtained when it is impossible to shift the contour to avoid the pole. In the second example, we cannot avoid the singularity because the pole occurs on the boundary (here, at an *endpoint*) of the integration domain. In the third example, there are two poles in the interior of the integration range, but these poles *pinch* the integration contour, which prevents moving the contour to avoid the poles and also leads to an infinite result.

5.c Although the conditions of occurrence of a genuine singularity are the same, the situation we have to analyze is arguably more complex than these toy examples because of the multivariate nature of the denominator. For a multi-dimension integral, the above condition is that the poles of the integrand cannot be avoided by deforming the integration manifold. We can make the following observations:

- Note first that the fact that $\mathcal{D}$ is raised to the m-th power in (1.42) is irrelevant for this discussion: every zero of $\mathcal{D}$ leads to a pole of order m of the integrand, but this is equivalent to having m poles all on the same side of the integration domain, so this cannot produce a pinch.

- For the L loop momenta ℓ_j, the integration domain is $\mathbb{R}^{DL}$. We may add to this domain a point at infinity (assuming an ultraviolet regularization, the integrand goes to zero in all directions at infinity), which leads to an integration domain topologically equivalent to a DL-dimensional sphere S_{DL}. This domain is boundaryless and therefore the only possibility of singularities when integrating over the loop momenta is to have a pinch.

- For the Feynman parameters x_r, the integration domain is a $(m-1)$-dimensional simplex,

$$\Big\{(x_1,\ldots,x_m)\Big|x_r \geq 0, \sum_r x_r = 1\Big\},$$

  i.e., a line segment for $m = 2$, a triangle for $m = 3$, a tetrahedron for $m = 4$, etc. Clearly, this domain has a boundary and therefore endpoint singularities may occur.

Let us now study the behavior of the denominator $\mathcal{D}$. This function is quadratic in the loop momenta, and linear in the Feynman parameters. Note also that, after a Wick rotation of the loop momenta, $\mathcal{D}$ is a negative definite quadratic form in the Euclidean loop momenta, at fixed $\{x_r\}$ (this is obvious from the fact that the second-degree part of $\mathcal{D}$ is a sum of squares weighted by the negative coefficients $-x_r$). Therefore, when varying the loop momenta at fixed $\{x_r\}$, the denominator $\mathcal{D}$ has a maximum. We can distinguish the following cases:

- If $\max_{\{\ell_j\}}(\mathcal{D}) < 0$, the denominator is always non-zero and there is no singularity.

- If $\max_{\{\ell_j\}}(\mathcal{D}) > 0$, the denominator can vanish, but the zeroes are simple zeroes that cannot pinch the integration manifold for the variables $\{\ell_j\}$. The integral is also finite in this case.

- The dangerous situation is when $\max_{\{\ell_j\}}(\mathcal{D}) = 0$, because the location of the maximum is then a double zero (as in the third of the toy examples considered earlier) in all the ℓ_j^μ variables.

Therefore, we are seeking zeroes of $\mathcal{D}$ that are also extrema of its dependence with respect to all the loop momenta:

$$\mathcal{D} = 0, \qquad \frac{\partial \mathcal{D}}{\partial \ell_j^\mu} = 0.$$

Note that these conditions are still not sufficient for a genuine singularity since we have only discussed what happens at fixed $\{x_r\}$, and we have not yet analyzed whether it may be avoided by a deformation of the integration domain for the x_r's. This discussion can be divided into two cases:

- First, note that if $q_r^2 \neq m_r^2$, a small variation of x_r will change the value of $\mathcal{D}$ and move the denominator away from zero. For x_r in the interior of its allowed range, this means that there is no actual singularity. The only exception is at $x_r = 0$, since this is on the boundary of the integration range.

- In contrast, when $q_r^2 = m_r^2$, the denominator $\mathcal{D}$ is independent of x_r, and a zero of $\mathcal{D}$ persists at all values of x_r.

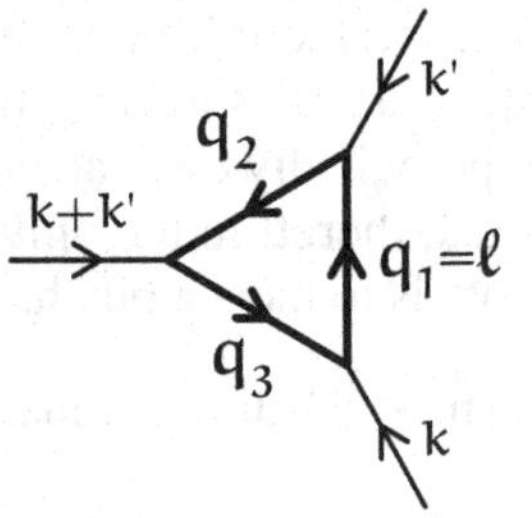

Figure 1.4 One-loop example of application of the Landau equations.

5.d We now have all the information to give the conditions of occurrence of a singularity in the Feynman integral $\mathcal{I}(\{p_i\})$:

$$\mathcal{D} = 0, \quad \frac{\partial \mathcal{D}}{\partial \ell_j^\mu} = 0; \quad \text{for each } r, \text{ either } x_r = 0 \text{ or } q_r^2 = m_r^2.$$

(Note that the last condition can be phrased as $x_r(q_r^2 - m_r^2) = 0$ for each r, which makes the first one, $\mathcal{D} = 0$, redundant.) These conditions are known as the *Landau equations*. By explicitly evaluating the derivative of the denominator with respect to the loop momenta, these conditions can be written as

$$\text{for each propagator } r\text{: } x_r(q_r^2 - m_r^2) = 0; \qquad \text{for each loop } j\text{: } \sum_{r=1}^{m} \epsilon_{rj} x_r q_r^\mu = 0.$$

5.e Consider now the example of Figure 1.4. The corresponding Landau equations read

$$x_1 \ell^2 = x_2(\ell + k')^2 = x_3(\ell - k)^2 = 0, \quad x_1 \ell^\mu + x_2(\ell^\mu + k'^\mu) + x_3(\ell^\mu - k^\mu) = 0.$$

A first solution is obtained for

$$\ell^2 = (\ell - k)^2 = 0, \quad x_2 = 0, \quad x_1 \ell^\mu + x_3(\ell^\mu - k^\mu) = 0,$$

which (since we have assumed that $k^2 = 0$) is equivalent to

$$\ell^2 = k^2 = \ell \cdot k = 0, \quad \ell^\mu = \frac{x_3}{x_1 + x_3} k^\mu.$$

This type of singularity is called a *collinear singularity*, since it occurs when the loop momentum is aligned with one of the external momenta. A similar type of singularity is obtained at $x_3 = 0$, i.e.,

$$\ell^2 = k'^2 = \ell \cdot k' = 0, \quad \ell^\mu = -\frac{x_2}{x_1 + x_2} k'^\mu.$$

Another way to fulfill the Landau equations is to have $\ell^\mu \ll k^\mu, k'^\mu$. In this limit, the first three conditions are automatically satisfied, and the last one becomes $x_2 k'^\mu = x_3 k^\mu$. Since k^μ and k'^μ are a priori not collinear, this implies that $x_2 = x_3 = 0$. This last type of singularity is called a *soft* or *infrared singularity*, since it occurs when all the components of the loop momentum go to zero.

Coleman–Norton Interpretation: A more intuitive physical interpretation of the Landau equations was found by Coleman and Norton (see Coleman, S. and Norton, R. E. (1965), *Nuovo Cimento* 38: 438). They propose to interpret a Feynman graph literally, as if the graph represents a process in spacetime, with the vertices being the locations where some instantaneous interactions happen. They further propose that the spacetime separation dx_r^μ between two vertices is parallel to the momentum q_r^μ, and proportional to the Feynman parameter x_r:

$$dx_r^\mu = x_r q_r^\mu .$$

(The constant of proportionality could be different from 1, but should be the same for all propagators. Its value is not important, as one may freely rescale the entire diagram without affecting the Landau equations.) Based on this identification, one may make the following observations:

- If a singularity happens at $x_r = 0$, the spacetime separation is zero, and we may shrink the corresponding propagator to a point. The resulting graph is called a *reduced graph.* A similar reduction is possible if $q_r^\mu = 0$, in the case of a soft divergence.
- If $x_r \neq 0$, the Landau equations tell us that we should have $q_r^2 = m_r^2$ instead. Thus, in this interpretation, the propagators of a reduced graph represent the on-shell propagation of a particle between two interactions. In this case, we may also write

  $$x_r = \frac{dx_r^0}{q_r^0}, \quad dx_r^\mu = dx_r^0 \frac{q_r^\mu}{q_r^0}, \quad \text{i.e.,} \quad \frac{dx_r^\mu}{dx_r^0} = \frac{q_r^\mu}{q_r^0} = v_r^\mu ,$$

  where v_r^μ is the 4-velocity of an on-shell particle of momentum q_r^μ.
- The last of the Landau equations becomes

  $$\sum_r \epsilon_{rj}\, dx_r^\mu = 0, \qquad \text{for every loop } j.$$

  This equation is consistent with dx_r^μ being a separation in spacetime, since adding these separations along a closed loop should obviously give zero. This also implies that the spacetime separation between two vertices does not depend on which path we follow on the graph to connect them, as it should for this interpretation to make sense.

As an illustration, let us show the reduced diagrams for the three singularities we have found for the triangle one-loop graph studied above:

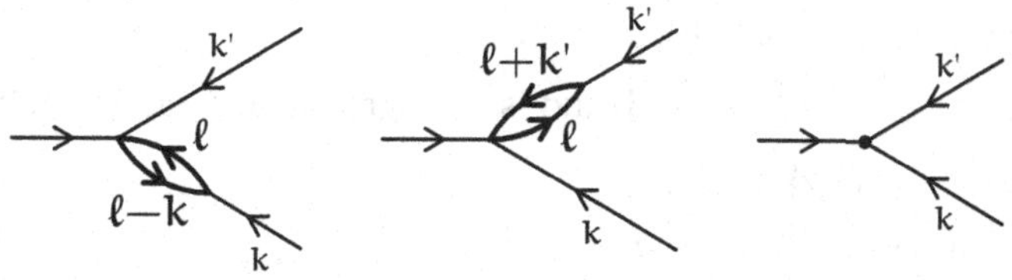

In more complicated cases, where solving the Landau equations may be difficult, the Coleman–Norton interpretation and the associated reduced diagrams can be used as guidance for identifying the possible solutions. This is based on the fact that there is a one-to-one correspondence between the solutions of the Landau equations and the spacetime diagrams proposed by Coleman and Norton. For instance, in the case of the above example, we could use it as a way of checking that there are no other solutions. Indeed, there could in principle be two additional reduced diagrams:

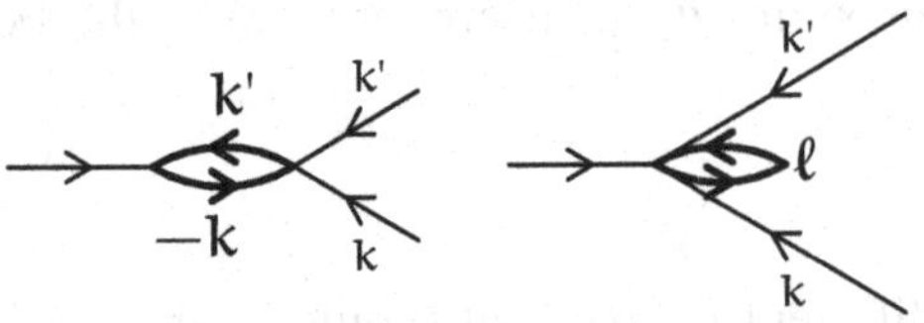

However, neither of them is kinematically allowed; the left graph would have two non-collinear particles propagating along the same spacetime interval, and the right graph would have a physical free particle making a closed loop.

6. Multi-loop Integration in D Dimensions Start from equation (1.42) in Problem 5, in the special case where the numerator is trivial, i.e., $\mathcal{N} \equiv 1$. Note that the denominator $\mathcal{D}$ is a quadratic form in the components of the loop momenta, which we may arrange as follows:

$$\mathcal{D} \equiv L_\mu^t A L^\mu + L_\mu^t B^\mu + B_\mu^t L^\mu + C, \quad L^\mu \equiv (\ell_1^\mu, \ell_2^\mu, \ldots, \ell_L^\mu).$$

6.a Perform explicitly the integration over the L loop momenta in D dimensions in order to obtain

$$\mathcal{I}(\{p_i\}) = \frac{i^L \Gamma(m - \frac{DL}{2})}{(-4\pi)^{\frac{DL}{2}}} \int\limits_{x_r \geq 0} \prod_{r=1}^{m} dx_r \, \frac{\delta(1 - \sum_r x_r)\,(\det A)^{m - \frac{(L+1)D}{2}}}{\underbrace{\left(C \det A - B_\mu^t C_A^t B^\mu + i0^+\right)}_{\equiv \Delta}{}^{m - \frac{DL}{2}}},$$

where C_A is the matrix of co-factors of A.

6.b Show that $\Delta = \det \begin{pmatrix} A_{11} & \ldots & A_{1L} & B_1^\mu \\ A_{21} & \ldots & A_{2L} & B_2^\mu \\ \vdots & & \vdots & \vdots \\ A_{L1} & \ldots & A_{LL} & B_L^\mu \\ B_{1\mu} & \ldots & B_{L\mu} & C \end{pmatrix}$ and is of degree two in each variable x_r.

6.a Using the notation of Problem 5, $q_r \equiv \sum_j \epsilon_{rj}\ell_j + \Delta_r$, the denominator $\mathcal{D}$ can be rewritten as

$$\begin{aligned}
\mathcal{D} &= \sum_r x_r(q_r^2 - m_r^2) \\
&= \sum_{j,k} \underbrace{\Big(\sum_r x_r \epsilon_{rj}\epsilon_{rk}\Big)}_{\equiv A_{jk}} \ell_j^\mu \ell_{k\mu} + 2\sum_j \underbrace{\Big(\sum_r x_r \epsilon_{rj}\Delta_r^\mu\Big)}_{\equiv B_j^\mu} \ell_{j\mu} + \underbrace{\sum_r x_r(\Delta_r^2 - m_r^2)}_{\equiv C} \\
&= L_\mu^t\, A\, L^\mu + L_\mu^t B^\mu + B_\mu^t L^\mu + C.
\end{aligned}$$

In this expression, note that:

- A is a real symmetric $L \times L$ matrix, whose coefficients are linear in the Feynman parameters x_r and are independent of the external momenta. This matrix is positive since $z_j A_{jk} z_k = \sum_r x_r (\sum_j \epsilon_{rj} z_j)^2$, and positive definite if all $x_r > 0$ (depending on the details of the Feynman graph under consideration, it can still be positive definite if some – but not all – of the x_r are zero).
- B^μ is a column vector with L components, linear in the Feynman parameters and linear in the external momenta.
- C is linear in the Feynman parameters, and also contains the squared masses and Lorentz invariant scalar products of external momenta.

Let us first rearrange the quadratic form $\mathcal{D}$ as follows:

$$\begin{aligned}
\mathcal{D} &= L_\mu^t\, A\, L^\mu + L_\mu^t B^\mu + B_\mu^t L^\mu + C \\
&= (L_\mu + A^{-1}B_\mu)^t \underbrace{A}_{\Omega^t D \Omega} \underbrace{(L^\mu + A^{-1}B^\mu)}_{K^\mu} + C - B_\mu^t A^{-1} B^\mu \\
&= (\Omega K_\mu)^t\, D\, (\Omega K^\mu) + C - B_\mu^t A^{-1} B^\mu \\
&= R_\mu R^\mu + C - B_\mu^t A^{-1} B^\mu, \quad \text{with } R_j^\mu \equiv \sqrt{D_{jj}}\, (\Omega K)_j^\mu.
\end{aligned}$$

The second line is the standard manipulation that eliminates the linear terms from the quadratic form (the Jacobian is 1, since this is just a translation). In the third line, we used the fact that the symmetric matrix A_{jk} (of size $L \times L$) is diagonalizable by an orthogonal transformation Ω (the Jacobian is also 1). In the last line, we have rescaled the various loop momenta in order to absorb the diagonal elements of D. In the final form, we may view $R_\mu R^\mu$ as the norm of a unique vector with DL components (D being the dimension of spacetime). The overall

Jacobian of this sequence of transformations is given by

$$d^{DL}L = \left(\det A\right)^{-D/2} \times d^{DL}R.$$

Therefore, the integration over the loop momenta reads

$$\begin{aligned}
\int \frac{d^{DL}L}{(2\pi)^{DL}}\,\frac{1}{\mathcal{D}^m} &= \frac{1}{(\det A)^{D/2}} \int \frac{d^{DL}R}{(2\pi)^{DL}}\,\frac{1}{\left(R_\mu R^\mu + C - B^t_\mu A^{-1} B^\mu\right)^m} \\
&= \frac{i^L}{(\det A)^{D/2}} \int \frac{d^{DL}R_E}{(2\pi)^{DL}}\,\frac{(-1)^m}{\left(R_E^2 + B^t_\mu A^{-1} B^\mu - C\right)^m} \\
&= \frac{(-1)^m\, i^L}{(4\pi)^{\frac{DL}{2}}}\,\frac{\Gamma(m-\frac{DL}{2})}{\Gamma(m)}\,\frac{\left(B^t_\mu A^{-1} B^\mu - C\right)^{\frac{DL}{2}-m}}{(\det A)^{D/2}}.
\end{aligned}$$

In the second line, we have applied a Wick rotation to the temporal components of the L loop momenta, and the result of the last line is a standard integration in DL dimensions (see the first of Eqs. (1.21)). Therefore, the expression for the L-loop integral is

$$\mathcal{I}(\{p_i\}) = \frac{(-1)^m i^L \Gamma(m-\frac{DL}{2})}{(4\pi)^{\frac{DL}{2}}} \int\limits_{x_r \geq 0} \prod_{r=1}^{m} dx_r\, \delta(1 - \sum_r x_r) \times \frac{\left(B^t_\mu A^{-1} B^\mu - C - i0^+\right)^{\frac{DL}{2}-m}}{(\det A)^{D/2}}.$$

(We have reinstated the $i0^+$ prescription of Feynman propagators.) Using $A^{-1} = \left(\det A\right)^{-1} C^t_A$, where C_A is the matrix of co-factors of A, we may rewrite this as

$$\mathcal{I}(\{p_i\}) = \frac{i^L \Gamma(m-\frac{DL}{2})}{(-4\pi)^{\frac{DL}{2}}} \int\limits_{x_r \geq 0} \prod_{r=1}^{m} dx_r\, \delta(1-\sum_r x_r)\, \frac{(\det A)^{m-\frac{(L+1)D}{2}}}{\underbrace{\left(C \det A - B^t_\mu C^t_A B^\mu + i0^+\right)}_{\equiv \Delta}{}^{m-\frac{DL}{2}}}.$$

Written in this form, the fraction in the integrand is a rational function of the Feynman parameters, with $\Delta \equiv C \det A - B^t_\mu C^t_A B^\mu$ a homogeneous polynomial of degree $L+1$ and $\det A$ a polynomial of degree L.

6.b This expression for the loop integral can shed extra light on the discussion of singularities in Problem 5, which led to the Landau equations. First, note that the determinant of A is zero only in accidental situations, since in general $\mathcal{D}$ is a negative definite quadratic form of the Euclidean loop momenta. Thus, the singularities in the above expression come from zeroes in Δ. This quantity is precisely the maximum of the denominator $\mathcal{D}$, viewed as a function of the Euclidean loop momenta ℓ^μ_j. Therefore, we recover the fact that the singularities of $\mathcal{I}(\{p_i\})$ may only occur when $\mathcal{D}$ has a vanishing maximum.

Recall now that $C \det A - B^t_\mu C^t_A B^\mu = C \det A - B^t_\mu (\det A)\, A^{-1} B^\mu$. Consider now the linear equation $A_{jk} X^\mu_k = B^\mu_j$ (where μ is treated as a fixed parameter). Its solution may be written as

$$(\det A)\,(A^{-1})_{jk}B^{\mu}_{k} = (\det A)\,X^{\mu}_{j} = \det\left(A_{[1,j-1]}\,B^{\mu}\,A_{[j+1,L]}\right),$$

where the notation $A_{[1,j-1]}\,B^{\mu}\,A_{[j+1,L]}$ stands for the $L \times L$ matrix obtained with columns 1 to $j-1$ of A, the column vector B^{μ}, and columns $j+1$ to L of A. Therefore, we have

$$\begin{aligned}
\Delta &= C\det A - B^{t}_{j\mu}\det\left(A_{[1,j-1]}\,B^{\mu}\,A_{[j+1,L]}\right)\\
&= C\det A - B^{t}_{j\mu}\,(-1)^{L-j}\det\left(A_{[1,j-1]}\,A_{[j+1,L]}\,B^{\mu}\right)\\
&= \det\begin{pmatrix} A_{11} & \dots & A_{1L} & B^{\mu}_{1}\\ A_{21} & \dots & A_{2L} & B^{\mu}_{2}\\ \vdots & & \vdots & \vdots\\ A_{L1} & \dots & A_{LL} & B^{\mu}_{L}\\ B_{1\mu} & \dots & B_{L\mu} & C \end{pmatrix}. \qquad (1.43)
\end{aligned}$$

In the second line, we just need to count the number of column permutations necessary to bring the column B^{μ} to the rightmost position. The last equality can be checked by expanding the determinant on the right-hand side according to the minors of the last line. Observe now that the function of $x_1, x_2, \dots, x_m$ obtained after integrating over the loop momenta does not depend on how we labeled the loop momenta (possibly up to a permutation of the $\{x_r\}$). Therefore, let us assume that, for the internal propagator $r = 1$, we have made the choice

$$q^{\mu}_{1} = \ell^{\mu}_{1}, \quad \text{i.e., } \epsilon_{1j} = \delta_{1j},\ \Delta^{\mu}_{1} = 0.$$

With this choice, we have

$$A_{jk} = x_1\delta_{1j}\delta_{1k} + \sum_{r\geq 2} x_r\epsilon_{rj}\epsilon_{rk}, \quad B^{\mu}_{j} = \sum_{r\geq 2} x_r\epsilon_{rj}\Delta^{\mu}_{r},$$
$$C = -m_1^2x_1 + \sum_{r\geq 2} x_r(\Delta_r^2 - m_r^2).$$

In particular, the only coefficients that depend (linearly) on x_1 are A_{11} and C. By expanding the determinant in (1.43), we see that Δ is a quadratic function of x_1 (with the other x_r fixed). Since our choice of setting $q_1 = \ell_1$ was arbitrary, we conclude that Δ is of degree two in every variable x_r (and overall homogeneous of degree $L+1$).

7. Weinberg Convergence Theorem The goal of this problem is to establish a criterion for ultraviolet convergence for loop integrals in scalar field theories (this restriction simplifies the problem a bit, since the integrands have a trivial numerator equal to one). This result, known as *Weinberg's convergence theorem*, states that a loop integral is ultraviolet convergent if the superficial degree of divergence of the loop integral, and of any of its restrictions to hyperplanes obtained by setting linear combinations of the loop momenta to constants, is negative. To establish this, we consider general Euclidean integrals of the form

$$\mathcal{I}(C, q, m) \equiv \int \frac{d^4\ell_1 \cdots d^4\ell_m}{\prod_{j=1}^{n}\left((C_{ji}\ell_i + q_j)^2 + m_j^2\right)},$$

where C_{ji} is an $n \times m$ constant matrix.

7.a Why can one consider instead the simpler integrals

$$\mathcal{J}(C) \equiv \int_D \frac{d^4\ell_1 \cdots d^4\ell_m}{\prod_{j=1}^n (C_{ji}\ell_i)^2}, \quad D \equiv \{(\ell_1^\mu, \ldots, \ell_m^\mu) | 1 \le (C\ell)^2\},$$

that do not depend on the masses or on the external momenta?

7.b Show that $\mathcal{I}(C, q, m)$ is absolutely convergent if $4m - 2n < 0$ and if the restriction of this integral to any hyperplane of co-dimension 4 is also absolutely convergent.

Hint: define $k_j \equiv C_{ji}\ell_i$ *and write the integral* $\mathcal{J}(C)$ *as a sum of terms* $\mathcal{J}_j(C)$ *in which the squared norm* k_j^2 *is the smallest among all the* k_ℓ^2*'s, then rescale all the* $k_{\ell \neq j}$ *by* $|k_j|$*. Show that the integral* $\mathcal{J}_j(C)$ *factorizes into a one-dimensional integral whose convergence is determined by power counting, and an integral over a subspace of co-dimension 4 of the same type as the original integral.*

7.c Show that an equivalent convergence criterion is that the superficial degree of divergence of $\mathcal{I}(C, q, m)$, and of restrictions of $\mathcal{I}$ to any hyperplane defined by setting some linear combinations of the momenta to constants, is negative.

7.a Since we are interested only in the ultraviolet convergence of this integral, the masses m_j and the shifts q_j do not play any role at large momenta (even though they matter for the convergence in the infrared and for the precise value of the integral). As far as ultraviolet convergence is concerned, we may as well set these parameters to zero, but we need to cut out a small region around $\ell_j = 0$ in order to avoid infrared problems. Thus, we may consider instead

$$\mathcal{J}(C) \equiv \int_D \frac{d^4\ell_1 \cdots d^4\ell_m}{\prod_{j=1}^n (C_{ji}\ell_i)^2}, \quad D \equiv \{(\ell_1^\mu, \ldots, \ell_m^\mu) | 1 \le (C\ell)^2\}, \tag{1.44}$$

which has exactly the same ultraviolet behavior as the original integral. (For technical reasons that will become clear later, it turns out to be a bit simpler to remove a ball of radius unity in the space of the variables $k_j \equiv C_{ji}\ell_j$ rather than ℓ_i.)

7.b Note that the first of the two conditions listed in the statement of the problem for the ultraviolet convergence of this integral, namely $4m - 2n < 0$, is nothing but the demand that the superficial degree of divergence of this integral be negative. This is a necessary condition because the integrand cannot decrease faster than ξ^{4m-2n} when all the ℓ_i's are rescaled according to $\ell_i \to \xi\ell_i$. The reason why this condition alone is insufficient to ensure the convergence is that, depending on the matrix C_{ji}, there could be directions in $\mathbb{R}^{4m}$ along which the decrease of the integrand is slower, for instance if one or more of the ℓ_i's do not appear in one of the k_j's. As we shall see, the second condition ensures that this problematic situation does not occur.

Then, we may write the integral $\mathcal{I}(C)$ as a sum of integrals $\mathcal{I}_j(C)$ in which the squared norm k_j^2 is the smallest among all the k_l^2's:

$$\mathcal{I}(C) = \sum_{j=1}^{n} \mathcal{I}_j(C), \quad \mathcal{I}_j(C) \equiv \int_{D_j} \frac{d^4\ell_1 \cdots d^4\ell_m}{\prod_{l=1}^{n} k_l^2},$$
$$D_j \equiv \big\{(\ell_1^\mu, \dots, \ell_m^\mu) \big| 1 \le k_j^2 \le \cdots\big\}.$$

The next step is to perform a linear change of the integration variables, $t \equiv A\ell$, such that

$$\det A = 1, \quad t_1 = k_j.$$

In terms of the t_i^μ, the integral $\mathcal{I}_j(C)$ may be rewritten as

$$\mathcal{I}_j(C) = \int_{t_1^2 \ge 1} d^4t_1 \int_U \frac{d^4t_2 \cdots d^4t_m}{\prod_{l=1}^{n} (C_{li} A_{ik}^{-1} t_k)^2}, \quad U \equiv \big\{(t_2^\mu, \dots, t_n^\mu) \big| t_1^2 \le (CA^{-1}t)^2\big\}.$$

Then, we rescale $t_2, \dots, t_n$ by the Euclidean norm of t_1,

$$t_i^\mu \equiv |t_1|\, \ell_i'^\mu \quad \big(|t_1| \equiv \sqrt{t_1^2}\big),$$

which leads to the following form of the integral $\mathcal{I}_j(C)$:

$$\mathcal{I}_j(C) = \int_{t_1^2 \ge 1} d^4t_1\, |t_1|^{4m-2n-4} \times \int_{U'} \frac{d^4\ell_2' \cdots d^4\ell_m'}{\prod_{l \ne j} (C_{lk}' \ell_k')^2}, \tag{1.45}$$
$$\text{with } C_{lk}' \equiv C_{li} A_{ik}^{-1}, \quad U' \equiv \big\{(\ell_2'^\mu, \dots, \ell_n'^\mu) \big| 1 \le (C'\ell')^2\big\}.$$

In this form, it is clear that:

- The first integral on the right-hand side is absolutely convergent if $4m - 2n < 0$.
- The second integral is of the form (1.44) with one less integration variable. Since the two integrals on the right-hand side are independent, it must also be convergent for $\mathcal{I}_j(C)$ to be finite.

This proves the announced convergence criterion.

7.c If we apply this criterion recursively, we get integrals of lower and lower dimensionality, until the last step where the analogue of (1.45) contains only the first factor (this corresponds to a one-loop integral). When we reach this point, we can conclude about the convergence from power counting only. Thus, the convergence of the original integral is ensured if its degree of divergence $4m - 2n$, and the degree of divergence of all the sub-integrals obtained by restricting the integration domain to hyperplanes of lower dimension, are negative. In this form, the convergence criterion is known as *Weinberg's convergence theorem* (note that here we have studied a less general situation than in the original theorem, which also considers the possibility of a polynomial of the integration variables in the numerator of the integrand). Let us add a final remark: the criterion derived in this problem is a criterion for *absolute convergence*. When it is not satisfied, it could still happen that the integral is nevertheless

(weakly) convergent because of cancellations among various parts of the integration domain. And of course, there could also be cancellations among the contributions of various Feynman graphs. This is typically what happens in gauge theories, where the gauge symmetry can induce cancellations among graphs that form a gauge invariant set (a single graph is usually not gauge invariant).
Source: Hahn, Y. and Zimmermann, W. (1968), *Commun Math Phys* 10: 330.

8. Electron Anomalous Magnetic Moment Consider the amputated renormalized photon–electron–positron vertex function $\Gamma_r^\mu(k,p,q)$ (with the convention that all momenta are incoming).

8.a Using the Dirac equation, prove the following relationship:

$$(p^\mu - q^\mu)\,\overline{u}(-q)u(p) = 2m\,\overline{u}(-q)\left[\gamma^\mu + \frac{i}{m}M^{\mu\nu}(p+q)_\nu\right]u(p),$$

known as *Gordon's identity* (with $M^{\mu\nu} \equiv \frac{i}{4}[\gamma^\mu,\gamma^\nu]$).

8.b Show that its contribution to electron scattering off an external field $\mathcal{A}^\mu$ can be parameterized as follows:

$$\mathcal{A}_\mu(k)\overline{u}(-q)\Gamma_r^\mu(k,p,q)u(p) = e_r\mathcal{A}_\mu(k)\overline{u}(-q)\left[F_1(k^2)\gamma^\mu + i\,F_2(k^2)\frac{M^{\mu\nu}k_\nu}{m_r}\right]u(p).$$

Hint: the most general form of Γ_r^μ *is*

$$\Gamma_r^\mu = C_1^\mu 1 + C_2^{\mu\alpha}\gamma_\alpha + C_3^\mu\gamma_5 + C_4^{\mu\alpha}\gamma_\alpha\gamma_5 + C_5^{\mu\alpha\beta}M_{\alpha\beta}.$$

Then, use Lorentz invariance and the Ward–Takahashi identity to show that $C_1^\mu \propto (p^\mu - q^\mu)$, *and then the Gordon identity to bring this term to the announced form. The same reasoning can be used to bring the remaining four terms in* Γ_r^μ *to the announced form.*

8.c Approximate this formula for a constant magnetic field. In particular, for an electron at rest in a homogeneous magnetic field B in the x^3 direction, show that

$$\mathcal{A}_\mu(k)\overline{u}(-q)\Gamma_r^\mu(k,p,q)u(p) \quad \text{becomes} \quad \frac{e_r B}{m_r}\left(1+F_2(0)\right)\overline{u}(0)M^{12}u(0).$$

8.d Calculate the relevant parts of Γ_r^μ at one loop in order to show that $F_2(0) = \frac{\alpha}{2\pi}$. (This result led to one of the first experimental verifications of quantum electrodynamics.)

8.a Let us rewrite the Poincaré algebra generator $M^{\mu\nu}$ as follows:

$$M^{\mu\nu} = \frac{i}{4}(\gamma^\mu\gamma^\nu - \gamma^\nu\gamma^\mu) = \frac{i}{2}(\gamma^\mu\gamma^\nu - g^{\mu\nu}) = \frac{i}{2}(g^{\mu\nu} - \gamma^\nu\gamma^\mu).$$

Then, consider

$$\begin{aligned}\overline{u}(-q)M^{\mu\nu}(p_\nu + q_\nu)u(p) &= \frac{i}{2}\overline{u}(-q)(\gamma^\mu\gamma^\nu - g^{\mu\nu})p_\nu u(p)\\ &\quad + \frac{i}{2}\overline{u}(-q)(g^{\mu\nu} - \gamma^\nu\gamma^\mu)q_\nu u(p)\\ &= im\,\overline{u}(-q)\gamma^\mu u(p) - \frac{i(p^\mu - q^\mu)}{2}\overline{u}(-q)u(p).\end{aligned}$$

This relation, known as *Gordon's identity*, can be rewritten as

$$(p^\mu - q^\mu)\,\overline{u}(-q)u(p) = 2m\,\overline{u}(-q)\left[\gamma^\mu + \frac{i}{m}M^{\mu\nu}(p+q)_\nu\right]u(p). \tag{1.46}$$

8.b The object $\overline{u}(-q)\Gamma_r^\mu u(p)$ is a radiative correction to an electromagnetic current. Since the incoming and outgoing electrons are on-shell, it must satisfy the following Ward–Takahashi identity:

$$k_\mu\,\overline{u}(-q)\Gamma_r^\mu(k,p,q)u(p) = 0.$$

Moreover, this identity must be satisfied even if the photon is off-shell, $k^2 \neq 0$. Obviously, the right-hand side of the announced formula obeys this identity, after one uses the Dirac equation and the antisymmetry of $M^{\mu\nu}$.

Let us now give a glimpse of how one would prove the converse, namely that this form is the only possible one. The starting point is to note that Γ_r^μ is a 4×4 matrix carrying a pair of Dirac indices, which we may decompose on the basis $\{1, \gamma_\alpha, \gamma_5, \gamma_\alpha\gamma_5, M_{\alpha\beta}\}$:

$$\Gamma_r^\mu = C_1^\mu 1 + C_2^{\mu\alpha}\gamma_\alpha + C_3^\mu\gamma_5 + C_4^{\mu\alpha}\gamma_\alpha\gamma_5 + C_5^{\mu\alpha\beta}M_{\alpha\beta}.$$

Since these terms are linearly independent, they must fulfill the Ward–Takahashi identity independently. From charge conjugation and parity symmetry, the term in γ_5 should be zero, $C_3^\mu \equiv 0$. The Lorentz indices of the remaining coefficients must be carried by the vectors p, q (or $k = -p - q$), the metric tensor, and possibly the Levi–Civita symbol in the case of $\gamma_\alpha\gamma_5$, with prefactors that depend only on Lorentz invariant quantities. In fact, all Lorentz invariant quantities can be expressed in terms of the electron mass and the photon virtuality k^2, since

$$p \cdot q = k^2 - m^2, \quad p \cdot k = q \cdot k = -\tfrac{k^2}{2}.$$

Consider for instance the coefficient C_1^μ of the identity. It may be written as

$$C_1^\mu = C_{1a}(k^2)\,p^\mu + C_{1b}(k^2)\,k^\mu.$$

The Ward–Takahashi identity implies

$$0 = C_{1a}\,k \cdot p + C_{1b}\,k^2 = k^2\left(C_{1b} - \tfrac{1}{2}C_{1a}\right).$$

Therefore, this coefficient must have the following form:

$$C_1^\mu = C_{1b}(k^2)(2p^\mu + k^\mu) = C_{1b}(k^2)\,(p^\mu - q^\mu).$$

Thanks to the Gordon identity (1.46), this term of Γ_r^μ indeed has the general form quoted in the statement of the problem. This turns out to be true for all the terms in this decomposition, as one may check by first writing the most general Lorentz structure allowed by the Ward–Takahashi identity and then by using the Dirac equation to simplify the result after insertion between $\overline{u}(-q)\cdots u(p)$.

8.c The coefficients $F_1(k^2)$ and $F_2(k^2)$, called the *form factors* of the electron, describe the properties of the cloud of photons and virtual pairs that surround the electron, as one varies the virtuality k^2 of the photon that probes the electron (in a sense, k^2 plays the role of a the resolution scale at which the photon probes this cloud). In the limit $k^2 \to 0$, the photon probes this cloud on very large distance scales. Given its similarity with the bare vertex, the term proportional to γ^μ encodes the electrical charge seen by the photon. On very large distance scales, this must be the usual charge of the electron as we know it from atomic physics, which means that $F_1(0) = 1$. Let us now discuss the meaning of $F_2(0)$. Consider a constant external magnetic field $\mathcal{B}$ in the x^3 direction, corresponding to

$$\mathcal{A}_2(x) = \mathcal{B}x^1, \quad \mathcal{A}_2(k) = i\mathcal{B}\partial_{k^1}, \quad \mathcal{F}_{12}(x) = -\mathcal{F}_{12}(x) = \mathcal{B},$$

where $\mathcal{F}_{\mu\nu}$ is the field strength. When an electron at rest is embedded in this field, the dressed coupling of the photon to the current reads (in the limit $k \to 0$)

$$e_r\mathcal{B}\,\overline{u}(-p)\left[i\gamma^2\partial_{p_1} + F_2(0)\frac{M^{12}}{m_r}\right]u(p)\Big|_{p=0}.$$

The derivative ∂_{p_1} acting on a spinor of small momentum acts like the boost K^1 in the direction x^1. More precisely, we have

$$\begin{aligned}\overline{u}(-p)i\gamma^2\partial_{p_1}u(p)\Big|_{p=0} &= \overline{u}(-p)i\gamma^2\frac{iK^1}{m_r}u(p)\Big|_{p=0} = -\overline{u}(-p)i\gamma^2\frac{\gamma^1\gamma^0}{2m_r}u(p)\Big|_{p=0}\\ &= \frac{1}{m_r}\overline{u}(-p)M^{12}u(p)\Big|_{p=0}.\end{aligned}$$

(The Dirac equation for zero momentum spinors reduces to $\gamma^0 u(p) = u(p)$.) Therefore, the above coupling becomes

$$\frac{e_r\mathcal{B}}{m_r}\left(1 + F_2(0)\right)\overline{u}(-p)M^{12}u(p)\Big|_{p=0}.$$

The 1 in $1 + F_2(0)$ encodes the bare coupling of the magnetic field to the electron, and $F_2(0)$ is therefore a correction to this coupling.

8.d At one loop, the QED Feynman rules lead to the following expression for the vertex function contracted between the spinors of the incoming and outgoing fermions:

$$\begin{aligned}\overline{u}(-q)\Gamma^\mu(k,p,q)u(p) &= -i\,e^3\int\frac{d^D\ell}{(2\pi)^D}\frac{\overline{u}(-q)\gamma^\sigma(\not{\ell}-\not{q}+m)\gamma^\mu(\not{\ell}+\not{p}+m)\gamma_\sigma u(p)}{\ell^2((\ell-q)^2-m^2)((\ell+p)^2-m^2)}\\ &\equiv -i\,e^3\int\frac{d^D\ell}{(2\pi)^D}\frac{\mathcal{N}^\mu}{\mathcal{D}}.\end{aligned}$$

For the time being, we use dimensional regularization to make all intermediate expressions finite, but we shall see shortly that the form factor F_2 is ultraviolet finite. The denominators of the three propagators can be combined into a single one by introducing Feynman parameters:

$$\frac{1}{ABC} = 2\int_0^1 dx_1 dx_2 dx_3 \, \frac{\delta(x_1 + x_2 + x_3 - 1)}{(x_1 A + x_2 B + x_3 C)^3}.$$

In the situation of interest here, the resulting denominator reads

$$\begin{aligned} & x_1((\ell+p)^2 - m^2) + x_2((\ell-q)^2 - m^2) + x_3\,\ell^2 \\ &= L^2 + x_1(1-x_1)p^2 + x_2(1-x_2)q^2 - (x_1+x_2)m^2 + x_1x_2\,p\cdot q \\ &= L^2 \underbrace{-(x_1+x_2)^2m^2 + x_1x_2k^2}_{\equiv -\Delta}, \end{aligned}$$

where we have defined $L \equiv \ell + x_1 p - x_2 q$ and where in the final line we have assumed the electrons to be on-shell ($p^2 = q^2 = m^2$, $k = -(p+q)$).

The next step is to express the numerator $\mathcal{N}^\mu$ in terms of the new integration variable L:

$$\begin{aligned} \mathcal{N}^\mu = & \,\overline{u}(-q)\gamma^\sigma \not{L}\gamma^\mu \not{L}\gamma_\sigma u(p) \\ & + \text{terms linear in } L \\ & + \overline{u}(-q)\gamma^\sigma(m - x_1\not{p} - (1-x_2)\not{q})\gamma^\mu(m + (1-x_1)\not{p} + x_2\not{q})\gamma_\sigma u(p). \end{aligned} \quad (1.47)$$

The terms linear in L can be dropped since the denominator is even in L. The combination of Dirac matrices in the first line can be rewritten as

$$\gamma^\sigma \not{L}\gamma^\mu \not{L}\gamma_\sigma = \frac{(2-D)^2}{D} L^2 \gamma^\mu.$$

Since it is quadratic in L, this terms leads to a logarithmic ultraviolet divergence, but the proportionality to γ^μ indicates that it contributes only to the form factor F_1 (the ultraviolet divergence in F_1 leads to a renormalization of the electron electrical charge). Since our goal is to evaluate F_2, we can disregard this term from now on. The term on the last line of (1.47) does not depend on L and therefore gives an ultraviolet finite integral over L, which implies that we can perform the Dirac algebra for this term in $D = 4$. Using

$$\gamma^\mu\gamma^\nu\gamma_\mu = -2\gamma^\nu, \quad \gamma^\mu\gamma^\nu\gamma^\rho\gamma_\mu = 4\,g^{\nu\rho}, \quad \gamma^\mu\gamma^\nu\gamma^\rho\gamma^\sigma\gamma_\mu = -2\gamma^\sigma\gamma^\rho\gamma^\nu$$

and the Dirac equations obeyed by the two spinors, a straightforward but somewhat tedious calculation leads to

$$\begin{aligned} & \overline{u}(-q)\gamma^\sigma(m - x_1\not{p} - (1-x_2)\not{q})\gamma^\mu(m + (1-x_1)\not{p} + x_2\not{q})\gamma_\sigma u(p) \\ &= 4m\left(\overline{u}(-q)u(p)\right)\left(\tfrac{p^\mu - q^\mu}{2}(x_3 - x_3^2) + \tfrac{k^\mu}{2}(x_1 + x_1^2 - x_2 - x_2^2)\right) \\ &\quad + 2\left(\overline{u}(-q)\gamma^\mu u(p)\right)\left(m^2(x_3^2 + 2x_3 - 1) - k^2(x_3 + x_1x_2)\right). \end{aligned}$$

The term of the last line, in $\overline{u}(-q)\gamma^\mu u(p)$, contributes only to F_1 and can be ignored. In the first line on the right-hand side, the term in k^μ is odd under the exchange of x_1 and x_2, while the denominator is even. Its integral over the Feynman parameters is therefore zero. Thus, we need only consider further the term in $p^\mu - q^\mu$, which can be rearranged thanks to the Gordon identity (1.46):

$$\begin{aligned}
&\overline{u}(-q)\gamma^\sigma(m - x_1\not{p} - (1-x_2)\not{q})\gamma^\mu(m + (1-x_1)\not{p} + x_2\not{q})\gamma_\sigma u(p) \\
&= \frac{i}{m}(\overline{u}(-q)M^{\mu\nu}u(p))k_\nu(-4m^2)(x_3 - x_3^2) \\
&\quad + \text{terms in } \overline{u}(-q)\gamma^\mu u(p) \\
&\quad + \text{terms that integrate to zero.}
\end{aligned}$$

By performing the integral over L (using a Wick rotation to have a Euclidean integration momentum),

$$\int \frac{d^D L}{(2\pi)^D}\frac{1}{(L^2-\Delta)^3} = -i\frac{\Delta^{D/2-3}}{(4\pi)^{D/2}}\frac{\Gamma(3-\frac{D}{2})}{\Gamma(3)} \underset{D=4}{=} -\frac{i}{32\pi^2\,\Delta},$$

and then by comparing the term in $M^{\mu\nu}$ with the expression for the vertex function in terms of the electron form factors, we obtain

$$F_2(k^2) = \frac{e^2}{4\pi^2}\int_0^1 dx_1\,dx_2\,dx_3\;\delta(x_1+x_2+x_3-1)\,\frac{x_3 - x_3^2}{(1-x_3)^2 - x_1x_2(k^2/m^2)}.$$

Noting that

$$\int_0^1 dx_1\,dx_2\;\delta(x_1+x_2+x_3-1) = 1 - x_3,$$

we finally get

$$F_2(0) = \frac{e^2}{8\pi^2} = \frac{\alpha}{2\pi}.$$

This is the celebrated result obtained by Schwinger in 1948. As of 2019, the QED calculation of this quantity has been pushed to order α^5, allowing a comparison with experimental measurements of the electron anomalous magnetic moment with an unprecedented accuracy.

---

9. Ward–Takahashi Identities and Lorentz Invariance The goal of this problem is to discuss the relationship between the Lorentz invariance of scattering cross-sections involving photons and the Ward–Takahashi identities obeyed by amplitudes with external photons. We will see that the Ward–Takahashi identities imply the Lorentz invariance, but also that the converse is true to some extent: namely a theory of massless spin-1 particles must obey at least some weak form of Ward–Takahashi identities in order to be Lorentz invariant. To that end, we consider an S-matrix element with an external photon, written as $\epsilon^\mu_\lambda(p)\mathcal{M}_\mu(p,\ldots)$.

9.a What is the transformation law of $\mathcal{M}_\mu(p,\ldots)$ under a Lorentz transformation Λ?

9.b Show that, under the same Lorentz transformation, the polarization vectors satisfy

$$\Lambda_\nu{}^\mu \epsilon_\pm^\nu(\Lambda p) = e^{\mp i\theta} \epsilon_\pm^\mu(p) + \text{terms in } p^\mu,$$

where we use the basis of positive and negative helicities as the two physical polarizations. *Hint: show that the polarization vectors* $\epsilon_\pm^\mu(p)$ *can be obtained from those of the reference momentum* $q^\mu \equiv (\omega, 0, 0, \omega)$ *by applying the Lorentz transformation* $L(p)$ *that maps* q^μ *into* p^μ. *Then, use the fact that* $R \equiv L^{-1}(p)\Lambda^{-1}L(\Lambda p)$ *belongs to the little group, and the result of Problem 2.*

9.c Show that

$$\left|\epsilon_\pm^\mu(\Lambda p)\,\mathcal{M}_\mu(\Lambda p, \dots)\right|^2 = \left|\epsilon_\pm^\mu(p)\,\mathcal{M}_\mu(p, \dots)\right|^2 + \text{terms in } p^\mu \mathcal{M}_\mu(p, \dots).$$

Thus, the S-matrix element is Lorentz invariant up to a phase, provided that $\mathcal{M}_\mu(p, \dots)$ satisfies the Ward–Takahashi identity. Discuss whether the converse is true.

9.a The matrix element $\mathcal{M}_\mu(p, \dots)$ transforms covariantly under a Lorentz transformation, i.e.,

$$\mathcal{M}_\mu(p, \dots) = \left[\Lambda^{-1}\right]_\mu{}^\nu \mathcal{M}_\nu(\Lambda p, \dots) = \Lambda^\nu{}_\mu \mathcal{M}_\nu(\Lambda p, \dots).$$

(Here, we are writing only the factor Λ that corresponds to the external line of momentum p^μ; there should be one extra such factor for each other external particle with non-zero spin.)

9.b In order to construct a basis of physical polarization vectors associated with the momentum p^μ, let us start from the two helicity polarization vectors of the reference momentum $q^\mu \equiv (\omega, 0, 0, \omega)$,

$$\epsilon_\pm^\mu(q) \equiv \tfrac{1}{\sqrt{2}}(0, 1, \pm i, 0).$$

The polarization vectors of an arbitrary momentum p^μ can then be obtained by applying the spatial rotation $\mathcal{R}(\widehat{\mathbf{p}})$ that brings the axis 3 into the direction $\widehat{\mathbf{p}}$:

$$\epsilon_\pm^\mu(p) = \left[\mathcal{R}(\widehat{\mathbf{p}})\right]^\mu{}_\nu\, \epsilon_\pm^\nu(q). \tag{1.48}$$

Let us introduce the Lorentz transformation $L(p)$ that transforms the reference vector q^μ into the vector p^μ. This transformation can be decomposed into a boost $\mathcal{B}_3(p^0/\omega)$ in the direction 3 that rescales the vector q^μ so that it has the same temporal component as p^μ, followed by the rotation $\mathcal{R}(\widehat{\mathbf{p}})$ that rotates its spatial components into the correct orientation:

$$L(p) = \mathcal{R}(\widehat{\mathbf{p}})\,\mathcal{B}_3(p^0/\omega).$$

Note that the boost $\mathcal{B}_3$ does not affect the reference polarization vectors, since their only non-zero components are in the directions 1, 2. Therefore, we also have

$$\epsilon_\pm^\mu(p) = \left[L(p)\right]^\mu{}_\nu\, \epsilon_\pm^\nu(q).$$

Recall now that, for any momentum p^μ and Lorentz transformation Λ, $R \equiv L^{-1}(p)\Lambda^{-1}L(\Lambda p)$

is an element of the little group since it leaves the reference vector q^μ invariant:

$$q^\mu \underset{L(\Lambda p)}{\longrightarrow} (\Lambda p)^\mu \underset{\Lambda^{-1}}{\longrightarrow} p^\mu \underset{L^{-1}(p)}{\longrightarrow} q^\mu.$$

Let us apply this transformation to the reference polarization vectors, which leads to

$$[\Lambda^{-1}]^\mu{}_\nu \underbrace{[L(\Lambda p)]^\nu{}_\rho\, \epsilon^\rho_\pm(q)}_{\epsilon^\nu_\pm(\Lambda p)} = [L(p)]^\mu{}_\nu\, R^\nu{}_\rho\, \epsilon^\rho_\pm(q).$$

In order to evaluate the right-hand side, we use the explicit representation of little-group elements obtained in Problem 2, which leads to

$$R^\nu{}_\rho\, \epsilon^\rho_\pm(q) = e^{\mp i\theta} \epsilon^\nu_\pm(q) + \beta_\pm q^\nu,$$

where we denote $\beta_\pm \equiv (\beta_1 \pm i\beta_2)/(\omega\sqrt{2})$. Multiplying this equation on the left by $[L(p)]^\mu{}_\nu$, we obtain

$$[\Lambda^{-1}]^\mu{}_\nu\, \epsilon^\nu_\pm(\Lambda p) = e^{\mp i\theta}\, \epsilon^\mu_\pm(p) + \beta_\pm p^\mu. \tag{1.49}$$

This formula shows that the physical polarization vectors do not transform as Lorentz vectors: they transform as 4-vectors only up to a phase and up to an additive term proportional to the momentum.

9.c By multiplying Eqs. (1.48) and (1.49), we obtain

$$\begin{aligned}\left(e^{\mp i\theta}\, \epsilon^\mu_\pm(p) + \beta_\pm p^\mu\right) \mathcal{M}_\mu(p,\dots) &= \underbrace{\Lambda^\nu{}_\mu\, [\Lambda^{-1}]^\mu{}_\rho}_{\delta^\nu{}_\rho}\, \epsilon^\rho_\pm(\Lambda p) \mathcal{M}_\nu(\Lambda p,\dots)\\ &= \epsilon^\mu_\pm(\Lambda p)\, \mathcal{M}_\mu(\Lambda p,\dots).\end{aligned}$$

The right-hand side is simply the S-matrix element evaluated in a boosted frame. The first term on the left-hand side is the same matrix element in the original frame, multiplied by a phase. Thus, we have

$$\left|\epsilon^\mu_\pm(\Lambda p)\, \mathcal{M}_\mu(\Lambda p,\dots)\right|^2 = \left|\epsilon^\mu_\pm(p)\, \mathcal{M}_\mu(p,\dots)\right|^2 + \text{terms in } p^\mu \mathcal{M}_\mu(p,\dots).$$

From this equation, we see immediately that the Ward–Takahashi identities (i.e., the fact that the last term on the right-hand side is zero) imply the Lorentz invariance of squared matrix elements with photons.

Let us now consider the converse. The Lorentz invariance of cross-sections requires that

$$\left|\epsilon^\mu_\pm(\Lambda p)\, \mathcal{M}_\mu(\Lambda p,\dots)\right|^2 = \left|\epsilon^\mu_\pm(p)\, \mathcal{M}_\mu(p,\dots)\right|^2,$$

which by the above identity implies that $p^\mu \mathcal{M}_\mu(p,\dots) = 0$. This looks like the usual Ward–Takahashi identity, except for an important restriction: all the other lines are on-shell and contracted with physical polarizations if they are photons (while in the Ward–Takahashi identities derived from current conservation, the photons do not need to be on-shell, nor do they need to have physical polarizations). In other words, any theory of massless spin-1 particles must satisfy a weak form of the Ward–Takahashi identities (in fact, identical to the non-Abelian version of the Ward–Takahashi identity), in order for its physical predictions to be consistent with Lorentz symmetry.

Additional Note: In the limit of soft massless spin-1 particles, the eikonal approximation leads to the following universal (i.e., independent of the spin of the emitters) form for the amplitude for producing a photon of momentum k in addition to hard particles of momenta p_i:

$$\mathcal{M}^\mu(k;p_1,\dots,p_n) = \mathcal{M}(p_1,\dots,p_n)\sum_{i=1}^{n}\frac{e_i\,p_i^\mu}{p_i\cdot k},$$

where the e_i are the couplings of the spin-1 particles to the emitters (conventionally defined to be all outgoing). In this limit, the condition $k_\mu\mathcal{M}^\mu = 0$ is equivalent to $\sum_i e_i = 0$. In other words, the Lorentz invariance of scattering amplitudes implies that any soft massless spin-1 particle couples to other fields via a conserved charge. The limitation of this argument to *soft* momenta (i.e., long-distance interactions) is easy to understand: a hard spin-1 would probe the substructure of composite objects, instead of its total charge (this does not mean that the charge is not conserved on these shorter-distance scales, but that we need to change our description of the emitters to make it manifest).

10. Equivalence Principle and Lorentz Invariance Recall that the emission of a soft photon of momentum k^μ off a hard scattering amplitude $\mathcal{M}(p_1,\dots,p_n)$ is given by

$$\mathcal{M}^\mu(k;p_1,\dots,p_n) \underset{k\to 0}{=} \mathcal{M}(p_1,\dots,p_n)\sum_{i=1}^{n}\frac{e_i\,p_i^\mu}{p_i\cdot k}, \tag{1.50}$$

where the p_i^μ are the momenta of the hard external particles (with the convention that they are all outgoing) and e_i the electrical charge they carry.

10.a Generalize the formula (1.50) to the emission of a soft graviton in a hard scattering process.

10.b Extend the arguments of Problem 9 to show that an S-matrix element with an external soft graviton is Lorentz invariant up to a phase, provided that soft gravitons couple with the same strength to all the other fields (i.e., provided that long-distance gravitational interactions satisfy the equivalence principle).

10.a Since they have spin 2, gravitons carry two Lorentz indices. Moreover, since they are massless, they have only two physical polarizations. It is possible to choose a gauge in which their "polarization tensors" are obtained from two copies of the photon polarization vectors:

$$\epsilon_\pm^{\mu\nu}(k) = \epsilon_\pm^\mu(k)\,\epsilon_\pm^\nu(k). \tag{1.51}$$

The only information needed to generalize (1.50) is to recall that gravitons couple to the energy–momentum tensor. Therefore, instead of a single factor p^μ, the coupling of a soft graviton should contain $p^\mu p^\nu$. The denominators in (1.50) are unchanged since they come from the propagators of the other external lines (fermions, scalars, etc.). Therefore, (1.50) becomes

$$\mathcal{M}^{\mu\nu}(k; p_1, \ldots, p_n) \underset{k\to 0}{=} \mathcal{M}(p_1, \ldots, p_n) \sum_{i=1}^{n} \frac{g_i\, p_i^\mu p_i^\nu}{p_i \cdot k}, \tag{1.52}$$

where g_i is the strength of the coupling of a soft graviton to the particle on the i-th external line.

10.b Thanks to Eq. (1.51) and the results of Problem 9, the Lorentz transform of the polarization tensor of a graviton is given by

$$[\Lambda^{-1}]^\mu{}_\rho [\Lambda^{-1}]^\nu{}_\sigma\, \epsilon_\pm^\rho(\Lambda k)\epsilon_\pm^\sigma(\Lambda k) = e^{\mp 2i\theta}\, \epsilon_\pm^\mu(k)\epsilon_\pm^\nu(k)$$
$$+ \text{ terms in } k^\mu \epsilon_\pm^\nu,\ \epsilon_\pm^\mu k^\nu \text{ or } k^\mu k^\nu.$$

Then, the reasoning of Problem 9 indicates that S-matrix elements with external gravitons are Lorentz invariant (up to a phase), provided that the corresponding amplitude satisfies the following identities:

$$k_\mu\, \mathcal{M}^{\mu\nu}(k, \ldots) = k_\nu\, \mathcal{M}^{\mu\nu}(k, \ldots) = 0.$$

Specializing to the case where the graviton is soft and all other external lines are hard, we can use (1.52), and this becomes

$$\sum_{i=1}^{n} g_i p_i^\nu = 0. \tag{1.53}$$

This identity must be true for any allowed configuration of the momenta p_i of the hard external lines. Obviously, this holds if we have $g_i = \text{const}$ (the identity then follows from energy–momentum conservation). This corresponds to the situation where soft gravitons couple to all other fields with the same strength, i.e., to the *equivalence principle.* Thus, we have shown that the equivalence principle implies the Lorentz invariance of cross-sections involving gravitons.

As in Problem 9, the converse is also true. Indeed, $g = \text{const}$ is the only way to satisfy (1.53) for all p_i. Therefore, *the Lorentz invariance of cross-sections with soft external gravitons implies the equivalence principle for long-distance gravitational interactions.* (Here, "long-distance" should be understood as long compared to the other distance scales of the process, but certainly short compared to the size of the Universe, since translation invariance does not hold on such large scales.) This observation was first made by Weinberg (see Weinberg, S. (1964), *Phys Rev* 135: 1049).

Higher Spins: This argument can be extended in order to exclude the possibility of interacting higher-spin massless particles. For instance, for a spin-3 massless particle, Lorentz invariance would imply

$$\sum_{i=1}^{n} g_i\, p_i^\mu p_i^\nu = 0.$$

One may check that the only way this can be true for all p_i's and all μ, ν is to have $g_i = 0$. (But interacting *massive* spin-3 particles fall out of the scope of this argument, and are in fact allowed.)

11. Lee-Nauenberg Theorem In quantum mechanics and quantum field theory, there exist states that are practically undistinguishable (e.g., states that differ by features that are too dim to be experimentally detected). Transition amplitudes with such initial or final states are in general plagued by singularities. The goal of this problem is to discuss these pathologies in the simple setting of quantum mechanics, and to show that they can be avoided by averaging transition probabilities over all degenerate states. To that end, consider a Hamiltonian $\mathcal{H} = \mathcal{H}_0 + V$ decomposed into a free Hamiltonian $\mathcal{H}_0$ and an interaction potential V. We recall that the scattering matrix S can be written as

$$S_{\beta\alpha} = \delta_{\beta\alpha} - 2\pi i\, \delta(E_\alpha - E_\beta)\, T_{\beta\alpha},$$
$$\text{with } T_{\beta\alpha} \equiv V_{\beta\alpha} + \int d\gamma\, \frac{V_{\beta\gamma} T_{\gamma\alpha}}{E_\alpha - E_\gamma + i0^+}, \qquad V_{\beta\alpha} \equiv \big\langle \Psi_\beta^{(0)} \big| V \big| \Psi_\alpha^{(0)} \big\rangle,$$

where the E_α's and $\big|\Psi_\alpha^{(0)}\big\rangle$'s are the eigenvalues and eigenstates of $\mathcal{H}_0$.

11.a Show that $\left(\Omega_{\text{in}}\right)_{\beta\alpha} \equiv \delta_{\beta\alpha} + \frac{T_{\beta\alpha}}{E_\alpha - E_\beta + i0^+}$ is a unitary matrix. *Hint: define* $\big|\Psi_{\alpha,\text{in}}\big\rangle \equiv \int d\gamma\, \left(\Omega_{\text{in}}^\dagger\right)_{\alpha\gamma} \big|\Psi_\gamma^{(0)}\big\rangle$ *and consider the matrix element* $\big\langle\Psi_{\beta,\text{in}}\big|V\big|\Psi_{\alpha,\text{in}}\big\rangle$.

11.b Solve formally the equation for $T_{\beta\alpha}$, to obtain

$$T_{\beta\alpha} = V_{\beta\alpha} + \sum_{n=1}^{+\infty} \int d\gamma_1 \cdots d\gamma_n\, \frac{V_{\beta\gamma_n} V_{\gamma_n\gamma_{n-1}} \cdots V_{\gamma_1\alpha}}{(E_\alpha - E_{\gamma_1} + i0^+) \cdots (E_\alpha - E_{\gamma_n} + i0^+)}.$$

11.c Denote by $\mathcal{D}(\alpha)$ the set of states obtained from α by adding/removing ultrasoft particles, by splitting collinearly hard massless particles or by recombining collinear particles, i.e., the set of states that are undistinguishable from the state α. Explain why $T_{\beta\alpha}$ has divergences if $\gamma_1, \gamma_2, \ldots, \gamma_i$ all belong to $\mathcal{D}(\alpha)$ or if $\gamma_i, \gamma_{i+1}, \ldots, \gamma_n$ all belong to $\mathcal{D}(\beta)$.

11.d Introduce P_α, P_β, P_3, the projectors on $\mathcal{D}(\alpha), \mathcal{D}(\beta)$ and on the rest of the Hilbert space, respectively. Show that squared matrix elements of T become finite if they are summed over all the states in $\mathcal{D}(\alpha)$ and $\mathcal{D}(\beta)$, i.e., $\sum_{a\in\mathcal{D}(\alpha), b\in\mathcal{D}(\beta)} |T_{ba}|^2 < \infty$.

Reminders on Scattering Theory in Quantum Mechanics: The setting of this study is the formulation of scattering in quantum mechanics. Let us consider a Hamiltonian $\mathcal{H}$, which we decompose into a free Hamiltonian and an interaction potential:

$$\mathcal{H} = \mathcal{H}_0 + V.$$

Denote by $\big|\Psi_\alpha^{(0)}\big\rangle$ the eigenstates of the free Hamiltonian,

$$\mathcal{H}_0 \left|\Psi_\alpha^{(0)}\right\rangle = E_\alpha \left|\Psi_\alpha^{(0)}\right\rangle,$$

and by $\left|\Psi_\alpha\right\rangle$ those of the full Hamiltonian,

$$\mathcal{H} \left|\Psi_\alpha\right\rangle = E_\alpha \left|\Psi_\alpha\right\rangle.$$

The eigenvalues of the free and interacting Hamiltonians are the same, provided we disregard the formation of bound states that may happen with certain interactions, and provided the free Hamiltonian is expressed in terms of the physical masses of the particles. The last equation may be rewritten as

$$(E_\alpha - \mathcal{H}_0) \left|\Psi_\alpha\right\rangle = V \left|\Psi_\alpha\right\rangle. \tag{1.54}$$

Note that the operator $E_\alpha - \mathcal{H}_0$ is not invertible, since $\left|\Psi_\alpha^{(0)}\right\rangle$ lies in its kernel (as well as any other state degenerate with it). Consider the following states:

$$\begin{aligned}\left|\Psi_{\alpha,\text{in}}\right\rangle &\equiv \left|\Psi_\alpha^{(0)}\right\rangle + \left(E_\alpha - \mathcal{H}_0 + i0^+\right)^{-1} V \left|\Psi_{\alpha,\text{in}}\right\rangle,\\ \left|\Psi_{\alpha,\text{out}}\right\rangle &\equiv \left|\Psi_\alpha^{(0)}\right\rangle + \left(E_\alpha - \mathcal{H}_0 - i0^+\right)^{-1} V \left|\Psi_{\alpha,\text{out}}\right\rangle.\end{aligned}$$

These states are obviously solutions of (1.54). These states are time-independent (i.e., they are states in the Heisenberg representation), but may look different to observers in different frames. In particular, in a frame that differs by a translation by a time T, these states are multiplied by $e^{-i\mathcal{H}T}$. Since they are energy eigenstates, the action of this operator is just a multiplication by an irrelevant phase $\exp(-iE_\alpha T)$. But these phases have a non-trivial effect on superpositions of states. Consider, for instance,

$$\left|\Psi_{h,\text{in}}\right\rangle \equiv \int d\alpha\; h(\alpha)\; \left|\Psi_{\alpha,\text{in}}\right\rangle$$

(the notation $d\alpha$ is a shorthand for the integration/summation measure over all the quantum numbers of the particles in the state α, such as 3-momenta, spins, etc.), where $h(\alpha)$ is a smooth function that spans a small range of energies E_α. Under such a translation in time, this state becomes

$$\begin{aligned}\left|\Psi_{h,\text{in}}\right\rangle \to \left|\Psi_{h,\text{in}}\right\rangle_T &= \int d\alpha\; h(\alpha)\; e^{-iE_\alpha T} \left|\Psi_{\alpha,\text{in}}\right\rangle\\ &= e^{-i\mathcal{H}_0 T} \left|\Psi_h^{(0)}\right\rangle + \int d\alpha\, h(\alpha)\; e^{-iE_\alpha T} \left(E_\alpha - \mathcal{H}_0 + i0^+\right)^{-1} V \left|\Psi_{\alpha,\text{in}}\right\rangle.\end{aligned} \tag{1.55}$$

In the second term, we must close the contour for the integral over the energy E_α in the upper half-plane in order to ensure its convergence if $T < 0$. Then, the theorem of residues gives the integral in terms of the singularities of the integrand. Except for the denominator, the other factors generically have their singularities at finite imaginary part, $\text{Im}\,(E_\alpha) > 0$ (because interactions typically lead to a finite lifetime for the particles), and their contribution vanishes when $T \to -\infty$. The denominator leads to poles located at $E_\alpha = E_* - i0^+$, i.e., below the real axis, and therefore it does not contribute to the result. Thus, we have

$$e^{-i\mathcal{H}T} \left|\Psi_{h,\text{in}}\right\rangle \underset{T\to-\infty}{=} e^{-i\mathcal{H}_0 T} \left|\Psi_h^{(0)}\right\rangle, \quad e^{-i\mathcal{H}T} \left|\Psi_{h,\text{out}}\right\rangle \underset{T\to+\infty}{=} e^{-i\mathcal{H}_0 T} \left|\Psi_h^{(0)}\right\rangle.$$

Note that, since the operators $\exp(-i\mathcal{H}T)$ and $\exp(-i\mathcal{H}_0 T)$ are unitary, the states $\left|\Psi_{h,\text{in}}\right\rangle$ and $\left|\Psi_{h,\text{out}}\right\rangle$ are normalized like the states $\left|\Psi_h^{(0)}\right\rangle$. Since this is true for any function $h(\alpha)$, the states $|\Psi_{\alpha,\text{in}}\rangle$ and $|\Psi_{\alpha,\text{out}}\rangle$ each form an orthonormal set of states. In other words, the *in* and *out* states become identical to the eigenstates of the free Hamiltonian in frames translated by a time $T \to -\infty$ or $T \to +\infty$, respectively. Thus, the physical interpretation of the (time-independent) state $|\Psi_{\alpha,\text{in}}\rangle$ is that it encodes all the possible outcomes of an interacting system initialized at $T = -\infty$ in the free state $|\Psi_\alpha^{(0)}\rangle$. Likewise, $|\Psi_{\alpha,\text{out}}\rangle$ encodes all the possible histories that lead to the free state $|\Psi_\alpha^{(0)}\rangle$ at $T = +\infty$. Therefore, the transition amplitude from a state α to a state β should be defined as

$$S_{\beta\alpha} \equiv \langle\Psi_{\beta,\text{out}}|\Psi_{\alpha,\text{in}}\rangle.$$

Note that the definition of the state $|\Psi_{\alpha,\text{in}}\rangle$ can be rewritten as

$$|\Psi_{\alpha,\text{in}}\rangle = \int d\beta\, \underbrace{\left(\delta_{\beta\alpha} + \frac{T_{\beta\alpha}}{E_\alpha - E_\beta + i0^+}\right)}_{\equiv\,\Omega_{\beta\alpha}} |\Psi_\beta^{(0)}\rangle, \quad \text{with } T_{\beta\alpha} \equiv \langle\Psi_\beta^{(0)}|V|\Psi_{\alpha,\text{in}}\rangle$$

($\delta_{\beta\alpha}$ is the combination of delta functions and Kronecker symbols such that $\int d\beta\, f(\beta)\delta_{\beta\alpha} = f(\alpha)$). Inserting the definition of $|\Psi_{\alpha,\text{in}}\rangle$ in $T_{\beta\alpha}$, we obtain an integral equation for T:

$$T_{\beta\alpha} = V_{\beta\alpha} + \int d\gamma\, \frac{V_{\beta\gamma}T_{\gamma\alpha}}{E_\alpha - E_\gamma + i0^+} = \int d\gamma\, V_{\beta\gamma}\,\Omega_{\gamma\alpha}, \tag{1.56}$$

with $V_{\beta\alpha} \equiv \langle\Psi_\beta^{(0)}|V|\Psi_\alpha^{(0)}\rangle$. This equation, known as the *Lipmann–Schwinger equation*, shows how the matrix elements of T are obtained from the matrix elements of the interaction potential in the free basis.

11.a Given the above discussion, we have

$$|\Psi_{\alpha,\text{in}}\rangle \underset{T\to-\infty}{=} e^{i\mathcal{H}T}\, e^{-i\mathcal{H}_0 T}\, |\Psi_\alpha^{(0)}\rangle,$$

which implies that $\Omega_{\beta\alpha} = \lim_{T\to-\infty} \langle\Psi_\beta^{(0)}|e^{i\mathcal{H}T}\, e^{-i\mathcal{H}_0T}|\Psi_\alpha^{(0)}\rangle$. In this form, we see immediately that the $\Omega_{\beta\alpha}$'s are the matrix elements of a unitary operator.

Alternatively, we can reach the same conclusion using only the information provided in the statement of the problem. For this, it is convenient to write

$$\Omega \equiv 1 + \Delta, \quad T_{\beta\alpha} = (E_\alpha - E_\beta + i0^+)\,\Delta_{\beta\alpha},$$
$$|\Psi_{\alpha,\text{in}}\rangle = |\Psi_\alpha^{(0)}\rangle + \int d\gamma\, \Delta^t_{\alpha\gamma}\, |\Psi_\gamma^{(0)}\rangle \quad (\Delta^t_{\alpha\gamma} \equiv \Delta_{\gamma\alpha}).$$

Given the integral equation that relates $V_{\beta\alpha}$ and $T_{\beta\alpha}$ and the definition of $|\Psi_{\alpha,\text{in}}\rangle$, we have also

$$T_{\beta\alpha} = \langle\Psi_\beta^{(0)}|V|\Psi_{\alpha,\text{in}}\rangle.$$

Then, the matrix element $\langle\Psi_{\beta,\text{in}}|V|\Psi_{\alpha,\text{in}}\rangle$ can be written in two ways, depending on whether we expand the state on the left or the state on the right:

$$\begin{aligned}
\langle\Psi_{\beta,\text{in}}|V|\Psi_{\alpha,\text{in}}\rangle &= \underbrace{\langle\Psi_\beta^{(0)}|V|\Psi_{\alpha,\text{in}}\rangle}_{T_{\beta\alpha}} + \int d\gamma\, \Delta^\dagger_{\beta\gamma}\, \underbrace{\langle\Psi_\gamma^{(0)}|V|\Psi_{\alpha,\text{in}}\rangle}_{T_{\gamma\alpha}} \\
&= \underbrace{\langle\Psi_{\beta,\text{in}}|V|\Psi_\alpha^{(0)}\rangle}_{T^*_{\alpha\beta}} + \int d\gamma\, \underbrace{\langle\Psi_{\beta,\text{in}}|V|\Psi_\gamma^{(0)}\rangle}_{T^*_{\gamma\beta}=T^\dagger_{\beta\gamma}}\, \Delta_{\gamma\alpha}.
\end{aligned}$$

Equating these two expressions for the matrix element leads to

$$\begin{aligned}
T^*_{\alpha\beta} - T_{\beta\alpha} &= \int d\gamma\, \left(\Delta^\dagger_{\beta\gamma} T_{\gamma\alpha} - T^\dagger_{\beta\gamma}\Delta_{\gamma\alpha}\right) \\
&= (E_\alpha - E_\beta + i0^+) \int d\gamma\, \Delta^\dagger_{\beta\gamma}\Delta_{\gamma\alpha},
\end{aligned}$$

i.e., $\Delta + \Delta^\dagger + \Delta^\dagger\Delta = 0$. From this relationship, we conclude that $\Omega = 1 + \Delta$ is unitary.

S-matrix: The matrix element $S_{\beta\alpha}$ can also be expressed in terms of $T_{\beta\alpha}$. In order to obtain this relationship, let us start again from (1.55), but this time we take the limit $T \to +\infty$. Since $T > 0$, we must close the integration contour in the lower half-plane. In the limit $T \to +\infty$, the only term that survives when we apply the theorem of residues comes from the zero in the denominator, and we get

$$\begin{aligned}
|\Psi_{h,\text{in}}\rangle_T &\underset{T\to+\infty}{=} e^{-i\mathcal{H}_0 T}\,|\Psi_h^{(0)}\rangle - 2\pi i \int d\alpha d\beta\; h(\alpha)\, e^{-iE_\beta T}\, \delta(E_\alpha - E_\beta)\, T_{\beta\alpha}\,|\Psi_\beta^{(0)}\rangle \\
&\underset{T\to+\infty}{=} \int d\alpha d\beta\; h(\alpha)\, e^{-iE_\beta T} \left(\delta_{\beta\alpha} - 2\pi i\, \delta(E_\alpha - E_\beta)\, T_{\beta\alpha}\right) |\Psi_\beta^{(0)}\rangle.
\end{aligned}$$

Alternatively, we can also write

$$\begin{aligned}
|\Psi_{h,\text{in}}\rangle_T &= \int d\alpha\; h(\alpha)\, e^{-iE_\alpha T}\,|\Psi_{\alpha,\text{in}}\rangle \\
&= \int d\alpha\; h(\alpha)\, e^{-iE_\alpha T} \int d\beta\; |\Psi_{\beta,\text{out}}\rangle\, \underbrace{\langle\Psi_{\beta,\text{out}}|\Psi_{\alpha,\text{in}}\rangle}_{S_{\beta\alpha}} \\
&\underset{T\to+\infty}{=} \int d\alpha d\beta\; h(\alpha)\, e^{-iE_\beta T}\, S_{\beta\alpha}\,|\Psi_\beta^{(0)}\rangle.
\end{aligned}$$

(In the last line, we also used the fact that $S_{\beta\alpha}$ is proportional to a delta function of energy conservation to change E_α into E_β in the exponential.) By comparing the two formulas, we conclude that

$$S_{\beta\alpha} = \delta_{\beta\alpha} - 2\pi i\, \delta(E_\alpha - E_\beta)\, T_{\beta\alpha}.$$

11.b The Lipmann–Schwinger equation (1.56) can be solved iteratively, leading to the following series in powers of the interaction potential:

$$T_{\beta\alpha} = V_{\beta\alpha} + \sum_{n=1}^{+\infty} \int d\gamma_1 \cdots d\gamma_n \frac{V_{\beta\gamma_n} V_{\gamma_n\gamma_{n-1}} \cdots V_{\gamma_1\alpha}}{(E_\alpha - E_{\gamma_1} + i0^+) \cdots (E_\alpha - E_{\gamma_n} + i0^+)}.$$

The physical interpretation of this expression is that the system starts in the state α and evolves to the state β, undergoing a sequence of interactions that take it through the intermediate states $\gamma_1, \gamma_2, \ldots, \gamma_n$.

11.c In this formula, the matrix elements of V are regular, and the only potential singularities come from the poles in the other factors under the integral. However, these poles do not always lead to a divergence in the result. This can be seen, for instance, in the following two toy examples:

$$\int_{-1}^{+1} \frac{dx}{x + i0^+} = -i\pi, \qquad \int_{0}^{+1} \frac{dx}{x + i0^+} = \infty.$$

In these examples, we see that in order to lead to a divergence, the pole should be at one of the endpoints of the integration domain; otherwise we may deform the integration contour to completely avoid the pole. Another configuration leading to an unavoidable divergence is when the integrand has a pair of poles "pinching" the integration contour so that it cannot be deformed.

Two states α and γ may accidentally have the same energy when one varies the momentum of a particle. However, when varying the state γ, these accidental degeneracies occur in the middle of the allowed domain of the variables that parameterize γ. The corresponding pole in $(E_\alpha - E_\gamma + i0^+)^{-1}$ is therefore avoidable by deforming the integration contour, and this does not lead to a divergence.

The state γ also has the same energy as α if it differs from α by one or more ultrasoft particles. In this situation, the resulting pole in $(E_\alpha - E_\gamma + i0^+)^{-1}$ occurs when the energy of these extra particles goes to zero, which is on the boundary of the integration domain for γ, and therefore produces an actual divergence in the result. Another situation where γ has the same energy as α is when a hard massless particle in one of the two states is replaced in the other state by two or more collinear particles with the same total momentum. This case also leads to a divergence, because the pole is reached on the boundary of the angular integration domain.

In order to keep track of these two problematic situations, let us denote by $\mathcal{D}(\alpha)$ the set of states that are degenerate with α, excluding the accidental degeneracies. Consider now the sequence of states

$$\alpha \to \gamma_1 \to \cdots \to \gamma_{i-1} \to \gamma_i \to \gamma_{i+1} \to \cdots \to \gamma_n \to \beta.$$

Each state γ_i in this sequence is related to the preceding and following states by an elementary interaction V. Obviously, in order to produce a pole, the state γ_i must have the same energy as α. But for this pole to lead to a divergence, a stronger condition should be satisfied: the pole must occur on the boundary of the integration domain of γ_i, which is made of the states in $\mathcal{D}(\gamma_{i-1})$ or in $\mathcal{D}(\gamma_{i+1})$.

Assume for instance that $\gamma_i \in \mathcal{D}(\gamma_{i-1})$, and has the same energy as α. But this implies that γ_{i-1} also has the same energy as α, which happens on a harmless zero measure subset of the phase-space of γ_{i-1} unless γ_{i-1} itself belongs to the degeneracy set of γ_{i-2}, $\mathcal{D}(\gamma_{i-2})$. By repeating this inductive argument, we see that this divergence occurs only if all of $\gamma_1, \gamma_2, \ldots, \gamma_i$ belong to $\mathcal{D}(\alpha)$. If instead we make the assumption that $\gamma_i \in \mathcal{D}(\gamma_{i+1})$, then it is the entire chain $\gamma_i, \gamma_{i+1}, \ldots, \gamma_n$ that must be in $\mathcal{D}(\beta)$ in order to have a divergence. This result is consistent with the well-known fact that infrared divergences happen when soft photons are attached to the external legs of a hard process, but not when one attaches the soft photon to a hard internal line of a graph. The same is true of the divergences due to collinear splittings.

11.d We assume now that the initial and final states are not degenerate, i.e., that $\mathcal{D}(\alpha)$ and $\mathcal{D}(\beta)$ do not overlap. Then, we introduce projectors P_α, P_β and P_3 that project respectively on $\mathcal{D}(\alpha)$, on $\mathcal{D}(\beta)$ and on the remaining portion of the Hilbert space. Since the union of these sets is the complete Hilbert space, the sum of these projectors is the identity

$$P_\alpha + P_\beta + P_3 = 1.$$

Moreover, we assume that the volume of $\mathcal{D}(\alpha)$ and $\mathcal{D}(\beta)$ is very small (this is controlled by what is experimentally meant by "soft" or "collinear") compared to the rest, implying that we may neglect these states except when they produce a divergence. In terms of these projectors, the matrix element $T_{\beta\alpha}$ can be written as

$$T_{\beta\alpha} = \sum_{n=0}^{+\infty} \left[V \left(\frac{P_\alpha + P_\beta + P_3}{E_\alpha - \mathcal{H}_0 + i0^+} V \right)^n \right]_{\beta\alpha}.$$

This formula is still exact. The next step is to expand this expression, and to drop P_α and P_β unless they occur in configurations that produce a divergence. Thus we keep the P_α's if they form an uninterrupted chain on the right of the product, and the P_β's if they form a chain on the left of the product:

$$T_{\beta\alpha} \underset{\text{sing.}}{=} \sum_{r,s,n=0}^{+\infty} \left[\left(V \frac{P_\beta}{E_\alpha - \mathcal{H}_0 + i0^+} \right)^s V \left(\frac{P_3}{E_\alpha - \mathcal{H}_0 + i0^+} V \right)^n \left(\frac{P_\alpha}{E_\alpha - \mathcal{H}_0 + i0^+} V \right)^r \right]_{\beta\alpha},$$

which can be written compactly as

$$T_{\beta\alpha} = \left(U^\dagger_{\beta,\text{out}}\, T_3\, U_{\alpha,\text{in}} \right)_{\beta\alpha}, \quad \text{with } T_3 \equiv V \sum_{n=0}^{+\infty} \left(\frac{P_3}{E_\alpha - \mathcal{H}_0 + i0^+} V \right)^n,$$

and

$$U_{\alpha,\text{in}} \equiv \sum_{r=0}^{+\infty} \left(\frac{P_\alpha}{E_\alpha - \mathcal{H}_0 + i0^+} V \right)^r, \quad U_{\beta,\text{out}} \equiv \sum_{s=0}^{+\infty} \left(\frac{P_\beta}{E_\alpha - \mathcal{H}_0 - i0^+} V \right)^s.$$

The factors $U_{\alpha,\text{in}}$ and $U_{\beta,\text{out}}$ contain divergences, while the factor T_3 is divergence-free.

Note now that the operator $U_{\alpha,\text{in}}$ is the restriction to the subspace $\mathcal{D}(\alpha)$ of the operator Ω that we introduced just above (1.56), and that we have shown to be unitary. Therefore, $U_{\alpha,\text{in}}$ is also unitary if restricted to $\mathcal{D}(\alpha)$, which means that

$$U_{\alpha,\text{in}}\, P_\alpha\, U^\dagger_{\alpha,\text{in}} = P_\alpha, \quad U_{\beta,\text{out}}\, P_\beta\, U^\dagger_{\beta,\text{out}} = P_\beta.$$

(We have also written an analogous relationship for $U_{\beta,\text{out}}$.) Then, summing the squared matrix elements over all the states degenerate with α and β, we obtain

$$\begin{aligned}\sum_{a\in\mathcal{D}(\alpha),b\in\mathcal{D}(\beta)} |T_{ba}|^2 &= \sum_{a\in\mathcal{D}(\alpha),b\in\mathcal{D}(\beta)} \left(U^\dagger_{\beta,\text{out}}\, T_3\, U_{\alpha,\text{in}}\right)_{ba} \left(U^\dagger_{\alpha,\text{in}}\, T_3^\dagger\, U_{\beta,\text{out}}\right)_{ab} \\ &= \text{tr}\left(U_{\beta,\text{out}}\, P_\beta\, U^\dagger_{\beta,\text{out}}\, T_3\, U_{\alpha,\text{in}}\, P_\alpha\, U^\dagger_{\alpha,\text{in}}\, T_3^\dagger\right) \\ &= \text{tr}\left(P_\beta\, T_3\, P_\alpha\, T_3^\dagger\right) = \sum_{a\in\mathcal{D}(\alpha),b\in\mathcal{D}(\beta)} \left|(T_3)_{ba}\right|^2,\end{aligned}$$

which is a finite quantity since it contains only the finite operator T_3. In the second line, we have rewritten the sum over the states in $\mathcal{D}(\alpha)$ and $\mathcal{D}(\beta)$ in the form of a trace by introducing the appropriate projectors, and we have used the cyclic invariance of the trace to reorder the factors.

This result, established by Lee and Nauenberg, shows that finite cross-sections are obtained provided one sums over all the initial and final degenerate states. This is rather natural in quantum mechanics since all these states are undistinguishable in practice, given the unavoidable limited resolutions (in energy and in angle) of any detector.

12. Classical External Field Approximation The goal of this problem is to study the scattering between an electron of charge $-e$ and a large atomic nucleus of charge Ze, and to show that, in the limit of a recoilless nucleus at rest, the nucleus can be approximated by its classical Coulomb potential.

12.a Why are multiple scatterings important at large Z?

12.b We are interested in the limit where the momentum exchanges k_i are soft. Why is this the relevant limit for a scattering at large impact parameter? How can we simplify the treatment of the nucleus in this limit?

12.c Express the lower part $\mathcal{A}^{\mu_{\sigma_1}\cdots\mu_{\sigma_n}}_{ss'}(P, P'; \{k_{\sigma_i}\})$ of the graph shown below (surrounded by a box) in this limit, for an arbitrary number of exchanged photons.

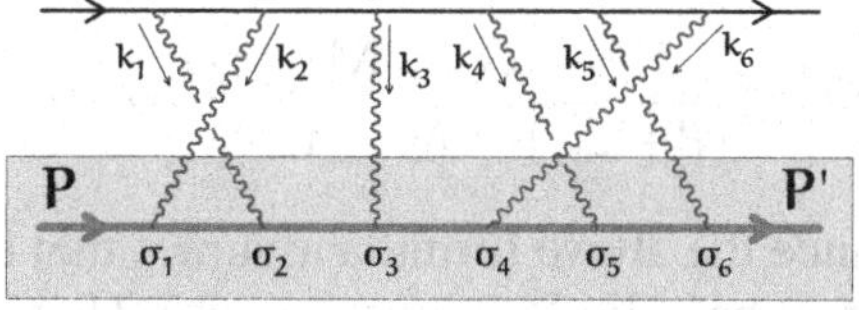

Perform the sum explicitly over all the permutations σ of the attachments of the photons to the nucleus, in order to obtain

$$\sum_{\sigma\in\mathfrak{S}_n} \mathcal{A}^{\mu_{\sigma_1}\cdots\mu_{\sigma_n}}_{ss'}(P, P'; \{k_{\sigma_i}\})$$

$$\underset{k_i\ll P}{\approx} 2\,E_P\,\delta_{ss'}\,(2\pi)^3\delta\big(\mathbf{P} + \textstyle\sum \mathbf{k}_i - \mathbf{P}'\big) \prod_i 2\pi\,(-iZe)\,P^{\mu_i}\,\delta(P\cdot k_i).$$

12.d Show that when the nucleus is at rest, the scattering amplitude of an electron off the nucleus can be obtained by replacing the nucleus by its classical Coulomb potential.

12.a Each photon exchanged between the electron and the nucleus can potentially bring a factor Ze^2. Despite the smallness of the fine-structure constant ($\alpha \sim 137^{-1}$), a large Z can lead to significant multi-photon effects. Here, we must mention an important limitation to this power counting: the coupling of the photon to the nucleus gives a factor Ze only if the photon sees coherently the entire electrical charge of the nucleus. In other words, the wavelength of the photon should be large enough so that it cannot resolve the more elementary charged constituents contained inside the nucleus (protons, or quarks at an even smaller distance scale). Given that the size of a nucleus is of the order of 10 fm, the momentum of the exchanged photon should not exceed 20 MeV in order to benefit from this enhancement.

12.b If we denote by x and y the spacetime coordinates at the endpoints of a photon propagator, the momentum k carried by this photon is Fourier conjugate to the difference $x-y$. In the case of the photons exchanged between the electron and the nucleus propagators, this difference is related to the impact parameter of the scattering. Thus, a small momentum carried by the exchanged photons is equivalent to a large impact parameter.

In fact, the criterion that the photon momentum is small enough for the photon to couple coherently with the entire nucleus is equivalent to having an impact parameter larger than the size of the nucleus. In other words, the electron should pass outside of the nucleus. In this limit, the photon sees only a point-like charge Ze instead of a complicated arrangement of quarks. Moreover, in this limit, the photon momenta ($\lesssim 20$ MeV) are much smaller than the mass of the nucleus, implying that the nucleus suffers a negligible recoil in the scattering.

12.c Using the QED Feynman rules, and the approximation of a point-like nucleus, the lower part of the graph shown in the figure can be written as

$$\begin{aligned}\mathcal{A}_{ss'}^{\mu_{\sigma_1}\cdots\mu_{\sigma_n}}(P,P';\{k_{\sigma_i}\}) &\equiv (2\pi)^4\delta\big(P+\sum_{i=1}^{n}k_i-P'\big)\,\overline{u}_{s'}(P')\\ &\times\Big[(-iZe\gamma^{\mu_{\sigma_n}})\cdots\frac{i(\not{P}+\not{k}_{\sigma_1}+\not{k}_{\sigma_2}+M)}{(P+k_{\sigma_1}+k_{\sigma_2})^2-M^2+i0^+}\\ &\times(-iZe\gamma^{\mu_{\sigma_2}})\frac{i(\not{P}+\not{k}_{\sigma_1}+M)}{(P+k_{\sigma_1})^2-M^2+i0^+}(-iZe\gamma^{\mu_{\sigma_1}})\Big]\,u_s(P).\end{aligned}$$

Note a limitation of our treatment here, since the above formula assumes that the target has spin $1/2$ (although our final result will in fact be valid for arbitrary spin). Using the fact that $P^2=M^2$ and the fact that the exchanged photons are soft, the intermediate propagators can be approximated as

$$\frac{i(\not{P}+\not{k}_{\sigma_1}+\cdots+\not{k}_{\sigma_i}+M)}{(P+k_{\sigma_1}+\cdots+k_{\sigma_i})^2-M^2+i0^+}\approx\frac{i\sum_{s_i=\pm}u_{s_i}(P)\overline{u}_{s_i}(P)}{2P\cdot(k_{\sigma_1}+\cdots+k_{\sigma_i})+i0^+}.$$

Note also that $\overline{u}_{s'}(P')\approx\overline{u}_{s'}(P)$, $\overline{u}_r(P)\gamma^\mu u_s(P)=2\,\delta_{rs}P^\mu$ and

$$P'^0 - P^0 - \sum_i k_i^0 \approx \sqrt{M^2 + \mathbf{P}^2 + 2\,\mathbf{P}\cdot\sum_i \mathbf{k}_i} - E_\mathbf{P} - \sum_i k_i^0 \approx -\frac{P\cdot\sum_i k_i}{E_\mathbf{P}}.$$

Therefore, we have

$$\begin{aligned}\mathcal{A}_{ss'}^{\mu_{\sigma_1}\cdots\mu_{\sigma_n}}(P,P';\{k_{\sigma_i}\}) \underset{k_i\ll P}{=} & \; 2\,E_\mathbf{P}\,\delta_{ss'}\,(-iZe)^n\,P^{\mu_1}P^{\mu_2}\cdots P^{\mu_n}\\ & \times (2\pi)^4\delta(P+\sum k_i - P')\delta(P\cdot\sum k_i)\\ & \times \frac{i}{P\cdot k_{\sigma_1}+i0^+}\,\frac{i}{P\cdot(k_{\sigma_1}+k_{\sigma_2})+i0^+}\cdots\\ & \times \frac{i}{P\cdot(k_{\sigma_1}+\cdots+k_{\sigma_{n-1}})+i0^+}.\end{aligned}$$

The next step is to perform the sum over all the permutations $\sigma \in \mathfrak{S}_n$. One may get a hint about the answer by considering the case $n = 2$, where there are only two permutations:

$$\begin{aligned}&2\pi\delta(P\cdot(k_1+k_2))\left(\frac{i}{P\cdot k_1+i0^+}+\frac{i}{P\cdot k_2+i0^+}\right)\\ &\quad = 2\pi\delta(P\cdot(k_1+k_2))\left(\frac{i}{P\cdot k_1+i0^+}+\frac{i}{-P\cdot k_1+i0^+}\right)\\ &\quad = 2\pi\delta(P\cdot(k_1+k_2))\,2\pi\delta(P\cdot k_1) = 2\pi\delta(P\cdot k_1)\,2\pi\delta(P\cdot k_2).\end{aligned}$$

This can be generalized to any n, by starting from the trivial identity

$$\sum_{\sigma\in\mathfrak{S}_n}\theta(t_{\sigma_1}-t_{\sigma_2})\,\theta(t_{\sigma_2}-t_{\sigma_3})\cdots\theta(t_{\sigma_{n-1}}-t_{\sigma_n}) = 1.$$

Then, we multiply it by $\prod_{i=1}^n \exp\big(i(P\cdot k_i)t_i\big)$ and integrate over all the t_i's. This leads immediately to

$$2\pi\,\delta(P\cdot\sum k_i)\sum_{\sigma\in\mathfrak{S}_n}\frac{i}{P\cdot k_{\sigma_1}+i0^+}\cdots\frac{i}{P\cdot(k_{\sigma_1}+\cdots+k_{\sigma_{n-1}})+i0^+} = \prod_{i=1}^n 2\pi\,\delta(P\cdot k_i).$$

Therefore, we obtain

$$\begin{aligned}&\sum_{\sigma\in\mathfrak{S}_n}\mathcal{A}_{ss'}^{\mu_{\sigma_1}\cdots\mu_{\sigma_n}}(P,P';\{k_{\sigma_i}\})\\ &\qquad \underset{k_i\ll P}{\approx} 2\,E_\mathbf{P}\,\delta_{ss'}\,(2\pi)^3\delta(\mathbf{P}+\sum_i \mathbf{k}_i - \mathbf{P}')\prod_i 2\pi\,(-iZe)\,P^{\mu_i}\,\delta(P\cdot k_i),\end{aligned} \tag{1.57}$$

where the only remaining entanglement among the photon momenta is the delta function of momentum conservation.

12.d In order to go further with the interpretation of this formula, one should recall that with our conventions a state $|\mathbf{P}, s\rangle$ represents a beam of particles of momentum $\mathbf{P}$ and spin s, with a uniform density of $2E_\mathbf{P}$ particles per unit volume, rather than a single particle. (The delta function of momentum conservation in (1.57) is consistent with this: indeed, momentum conservation is equivalent to invariance under spatial translations, which is only possible for states that are themselves translation invariant.) It is easy to construct a normalizable state that represents a single particle, by a linear superposition such as

$$|\Psi\rangle \equiv \sum_s \int \frac{d^3\mathbf{P}}{(2\pi)^3} \frac{e^{i\mathbf{P}\cdot\mathbf{X}_0}}{\sqrt{2E_\mathbf{p}}} \Psi_s(\mathbf{P})\,|\mathbf{P}, s\rangle.$$

Indeed, one may check that

$$\langle\Psi|\Psi\rangle = 1 \quad \text{provided that} \quad \sum_s \int \frac{d^3\mathbf{P}}{(2\pi)^3} |\Psi_s(\mathbf{P})|^2 = 1.$$

In the definition of the state $|\Psi\rangle$, the coordinate $\mathbf{X}_0$ may be viewed as the "center" of the object represented by this state. (For instance, if the momentum-space wavefunction $\Psi_s(\mathbf{P})$ is a Gaussian, the wavefunction of the state in the coordinate representation is a Gaussian centered at X_0.) The analogue of (1.57) for the state $|\Psi\rangle$ is

$$\begin{aligned}
&\mathcal{A}^{\mu_1\cdots\mu_n}(\Psi;\{k_i\}) \\
&\quad \equiv \sum_{s,s'} \int \frac{d^3\mathbf{P}d^3\mathbf{P}'}{(2\pi)^6} \frac{e^{i(\mathbf{P}-\mathbf{P}')\cdot\mathbf{X}_0}}{\sqrt{4E_\mathbf{P}E_{\mathbf{P}'}}} \Psi_s(\mathbf{P})\Psi^*_{s'}(\mathbf{P}') \sum_{\sigma\in\mathfrak{S}_n} \mathcal{A}^{\mu_{\sigma_1}\cdots\mu_{\sigma_n}}_{ss'}(P, P';\{k_{\sigma_i}\}) \\
&\quad \underset{k_i \ll P}{\approx} \sum_s \int \frac{d^3\mathbf{P}}{(2\pi)^3} \Psi_s(\mathbf{P})\Psi^*_{s'}(\mathbf{P}) \prod_i 2\pi(-iZe)\, P^{\mu_i}\, e^{-ik_i\cdot X_0}\, \delta(P\cdot k_i),
\end{aligned}$$

where we have used $\mathbf{P} \approx \mathbf{P}'$ whenever possible. Let us now attach the n photon propagators to the Lorentz indices of this expression:

$$\begin{aligned}
\mathcal{M}^{\nu_1\cdots\nu_n}(\Psi;\{k_i\}) &\equiv \mathcal{A}^{\mu_1\cdots\mu_n}(\Psi;\{k_i\}) \prod_{i=1}^n \frac{-i\, g_{\mu_i}{}^{\nu_i}}{k_i^2 + i0^+} \\
&\underset{k_i \ll P}{=} \sum_s \int \frac{d^3\mathbf{P}}{(2\pi)^3} \Psi_s(\mathbf{P})\Psi^*_{s'}(\mathbf{P}) \prod_i 2\pi\,\delta(P\cdot k_i)\, \frac{-Ze\, P^{\nu_i}\, e^{-ik_i\cdot X_0}}{k_i^2 + i0^+}.
\end{aligned}$$

(Note that the result does not depend on the gauge used to express the photon propagators, because they are contracted into P^{μ_i} and we have $P\cdot k_i = 0$.) Assume now that the nucleus described by the state $|\Psi\rangle$ is at rest. In other words, the wavefunction $\Psi_s(\mathbf{P})$ is a narrow peak centered at $\mathbf{P} = 0$, with a support such that the typical $\mathbf{P}$ is much smaller than the mass M. In this static limit, we have

$$2\pi\,\delta(P\cdot k_i)\,P^{\nu_i} \approx 2\pi\,\delta(Mk_i^0)\,M\,\delta^{\nu_i}{}_0 = 2\pi\,\delta(k_i^0)\,\delta^{\nu_i}{}_0,$$

and

$$\mathcal{M}^{\nu_1\cdots\nu_n}(\Psi;\{k_i\}) \underset{\substack{k_i\ll P\\ \text{static}}}{=} \underbrace{\sum_s\int\frac{d^3P}{(2\pi)^3}\,\Psi_s(P)\Psi^*_{s'}(P)}_{=1}\,\prod_i 2\pi\,\delta(k_i^0)\,\frac{-Ze\,\delta^{\nu_i}{}_0\,e^{-ik_i\cdot X_0}}{k_i^2+i0^+}$$
$$= \prod_i 2\pi\,\delta(k_i^0)\,\frac{Ze\,\delta^{\nu_i}{}_0\,e^{-ik_i\cdot X_0}}{\mathbf{k}_i^2-i0^+}.$$

The interpretation of this expression is more transparent in coordinate space, after a Fourier transformation:

$$\mathcal{M}^{\nu_1\cdots\nu_n}(\Psi;\{x_i\}) \equiv \int\prod_{i=1}^n\frac{d^4k_i}{(2\pi)^4}\,e^{-ik_i\cdot x_i}\,\mathcal{M}^{\nu_1\cdots\nu_n}(\Psi;\{k_i\}) \approx \prod_i \underbrace{\delta^{\nu_i}{}_0\,\frac{Ze}{4\pi\,|\mathbf{x}_i-\mathbf{X}_0|}}_{\equiv A_\Psi^{\nu_i}(x_i)}.$$

In this formula, each factor is nothing but the Coulomb potential produced at $\mathbf{x}_i$ by a point-like electrical charge Ze located at $\mathbf{X}_0$. Therefore, in this limit, the scattering of an electron off the target Ψ is equivalent to a scattering off an external classical field. One would have obtained the same answer simply by adding the Coulomb potential $A^\mu_\Psi(x)$ into the covariant derivative:

$$D^\mu = \partial^\mu - ie\,A^\mu \quad\to\quad \partial^\mu - ie\,(A^\mu + A^\mu_\Psi).$$

This approximation is known as the *external field approximation* or *Weizsäcker–Williams approximation*. Its derivation shows unambiguously that the term of order n in this external field is equivalent to summing over Feynman graphs with n exchanged photons, *summed over all the ways of permuting the attachments of the photons*. Thus, the (single) term of order n in the expansion in powers of the external field contains the contributions from $n!$ Feynman diagrams:

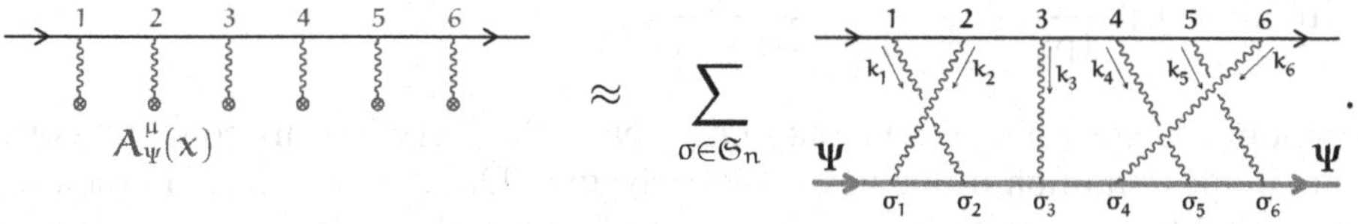

13. Subleading Soft Radiation in Scalar QED The goal of this problem and the next one is to study the first correction to the formula that gives the amplitude for the emission of a soft photon off a hard process. In the present problem, we consider the simpler case where the charged particles are scalar; the result is generalized to spin-1/2 charges in Problem 14.

13.a Show that the formula (1.50) in the statement of Problem 10, which describes the emission of a soft photon of momentum k from an amplitude with hard spin-1/2 charged particles of momenta p_i, is unchanged when the charged particles are scalar, i.e.,

$$\mathcal{M}^{\mu}(k;p_1,\dots,p_n) = \mathcal{M}(p_1,\dots,p_n)\sum_{i=1}^{n}\frac{e_i\,p_i^{\mu}}{p_i\cdot k} + \mathcal{O}(k^0).$$

We recall that, at this order in k, the only contributions are obtained when the soft photon is attached to an external line of the hard process.

13.b Our goal is now to calculate the first subleading correction in k to this formula, in order to obtain the following improved result:

$$\mathcal{M}^{\mu}(k;p_1,\dots,p_n) = \sum_{i=1}^{n}\frac{e_i}{p_i\cdot k}\left(p_i^{\mu} - i\,k_{\nu}\,J_i^{\mu\nu}\right)\mathcal{M}(p_1,\dots,p_n) + \mathcal{O}(k^1),$$

where $J_i^{\mu\nu} \equiv i\left(p_i^{\mu}\frac{\partial}{\partial p_{i\nu}} - p_i^{\nu}\frac{\partial}{\partial p_{i\mu}}\right)$ is the angular momentum operator. *Hints: consider more accurately the insertions of the photon on the external lines. Then, note that there is also a term where the photon is attached to an inner line, but that the leading piece of this contribution is fully constrained by the Ward–Takahashi identities.*

13.a Consider an amplitude $\mathcal{M}(p_1,\dots,p_n)$ in *scalar QED*, with all external particles hard. The emission of a soft photon of momentum k off this amplitude is dominated by Feynman graphs where the soft photon is attached to one of the external lines (the analysis of the denominators is the same in scalar and in spinor QED). The only change comes from the coupling of the photon to a scalar line: the Feynman rule for attaching a photon of momentum k and Lorentz index μ to a scalar of momentum p and charge e is a factor $-ie(2p+k)^{\mu}$. Therefore, when attaching a soft photon to an *outgoing scalar* particle of momentum p, the amplitude is modified by a factor

$$-i\overline{e}\,(2p+k)^{\mu}\,\frac{i}{(p+k)^2-m^2} \underset{k\to 0}{\approx} \frac{\overline{e}\,p^{\mu}}{p\cdot k}.$$

(All the momenta are defined to be outgoing.) Note that here $\overline{e}$ is the coupling constant as it appears in the definition of the covariant derivative, $D_{\mu} \equiv \partial_{\mu} - i\overline{e}A_{\mu}$. In the case of an outgoing scalar, this parameter is also the electrical charge that flows outwards. But consider now an *outgoing antiscalar* of momentum p. In this case, the amplitude is modified by a factor

$$-i\overline{e}\,(-2p-k)^{\mu}\,\frac{i}{(p+k)^2-m^2} \underset{k\to 0}{\approx} \frac{-\overline{e}\,p^{\mu}}{p\cdot k},$$

which differs by a sign from the case of an outgoing scalar. It is possible to make the subsequent formulas look more uniform if we introduce for each external line the corresponding outwards flowing electrical charge, denoted e. The relationship between this charge and the coupling constant $\overline{e}$ is simply

- *Outgoing scalar:* $e = \overline{e}$
- *Outgoing antiscalar:* $e = -\overline{e}$
- *Incoming scalar:* $e = -\overline{e}$
- *Incoming antiscalar:* $e = \overline{e}$.

In terms of this outgoing electrical charge, the modification factor for the emission of a soft photon is $ep^\mu/p \cdot k$ for any kind of external line (in this respect, scalar QED is therefore identical to spinor QED).

Summing over all the charged external lines, we thus obtain

$$\mathcal{M}^\mu(k; p_1, \ldots, p_n) \underset{k \ll p_i}{\approx} \mathcal{M}(p_1, \ldots, p_n) \sum_{i=1}^{n} \frac{e_i\, p_i^\mu}{p_i \cdot k}, \tag{1.58}$$

where the sum runs over all the charged external lines of the hard process. The first neglected terms in this formula are of order zero in k^μ. Recall that, with the conventions used in this problem, conservation of momentum and conservation of electrical charge in the hard process read

$$\sum_{i=1}^{n} p_i^\mu = 0, \quad \sum_{i=1}^{n} e_i = 0.$$

13.b Our task is now to determine the first subleading correction to (1.58), i.e., the next term in the formal expansion in powers of k/p_i. In order to calculate this correction, we need two things:

i. Improve the approximation made when the soft photon is attached to a charged external line of the hard process, so that it remains accurate at least up to the next order in k.

ii. Consider graphs where the soft photon is attached to an internal line of the hard process.

Let us start with **i.** Without doing any approximation, the attachment of a soft on-shell (i.e., $k^2 = 0$) photon to an outgoing scalar line leads to

$$\sum_{i=1}^{n} e_i \frac{(2p_i + k)^\mu}{(p_i + k)^2 - m_i^2} \mathcal{M}(\ldots, p_i + k, \ldots) = \sum_{i=1}^{n} e_i \frac{(2p_i + k)^\mu}{2p_i \cdot k} \mathcal{M}(\ldots, p_i + k, \ldots).$$

Note that this will simplify further when we contract the Lorentz index μ of the soft photon with a physical polarization vector, since $k^\mu \epsilon_{\lambda\mu}(k) = 0$. This contraction will get rid of the term in k^μ, leaving the same eikonal factor $p_i^\mu/(p_i \cdot k)$ as in (1.58). Anticipating this contraction, the only difference from (1.58) is that the hard amplitude must be evaluated at $p_i + k$ instead of p_i when the i-th line emits the soft photon. This is true for all the possible types of external scalar lines (incoming/outgoing, scalar/antiscalar).

Now consider **ii**, i.e., the contributions where the soft photon comes from the interior of the hard process. As we shall show, these extra contributions are constrained by gauge invariance, and these constraints are sufficient to determine the dominant terms in these contributions. Let us write the total amplitude as follows:

$$\mathcal{M}^\mu(k;p_1,\dots,p_n) = \underbrace{\sum_{i=1}^n e_i \frac{(2p_i+k)^\mu}{2p_i\cdot k}\,\mathcal{M}(\dots,p_i+k,\dots)}_{\text{photon attached to external lines}} + \underbrace{\mathcal{N}^\mu(k;p_1,\dots,p_n)}_{\text{photon attached to an internal line}}\ .$$

An important point is that the two terms are in general not gauge invariant separately, but their sum is. In particular, the full amplitude must obey the following Ward–Takahashi identity:

$$\begin{aligned} 0 &= k_\mu\,\mathcal{M}^\mu(k;p_1,\dots,p_n) \\ &= \sum_{i=1}^n e_i\,\mathcal{M}(\dots,p_i+k,\dots) + k_\mu\,\mathcal{N}^\mu(k;p_1,\dots,p_n). \end{aligned} \tag{1.59}$$

(We have used the fact that the photon is on-shell, $k^2 = 0$, in order to simplify the first term.) Note that the term $\mathcal{N}^\mu$ is not singular when $k \to 0$, since the leading k^{-1} behavior of the amplitude is entirely contained in the first term. Thus, the expansion of $\mathcal{N}^\mu$ in powers of k starts with a constant, followed by a term linear in k, etc.:

$$\mathcal{N}^\mu(k;p_1,\dots,p_n) = \mathcal{N}_0^\mu(p_1,\dots,p_n) + k_\nu\,\mathcal{N}_1^{\mu\nu}(p_1,\dots,p_n) + \cdots .$$

We can similarly expand the quantity $\mathcal{M}(\dots,p_i+k,\dots)$:

$$\mathcal{M}(\dots,p_i+k,\dots) = \mathcal{M}(p_1,\dots,p_n) + k_\mu \frac{\partial}{\partial p_{i\mu}}\,\mathcal{M}(p_1,\dots,p_n) + \cdots .$$

The next step is to investigate the consequence of the Ward–Takahashi identity (1.59) order-by-order in k. At order k^{-1}, only the term in $\mathcal{M}$ contributes, and the Ward–Takahashi identity is trivially satisfied thanks to charge conservation in the hard process. At the next order (k^0), the Ward–Takahashi identity reads

$$0 = \sum_{i=1}^n e_i\,k_\mu \frac{\partial}{\partial p_{i\mu}}\,\mathcal{M}(p_1,\dots,p_n) + k_\mu\,\mathcal{N}_0^\mu(p_1,\dots,p_n).$$

Therefore, we have

$$\mathcal{N}_0^\mu(p_1,\dots,p_n) = -\sum_{i=1}^n e_i\,\frac{\partial}{\partial p_{i\mu}}\,\mathcal{M}(p_1,\dots,p_n) + G^\mu(p_1,\dots,p_n),$$

where G^μ is a term that fulfills the Ward–Takahashi identity by itself, $k_\mu G^\mu = 0$. But note that such a G^μ would have to be independent of the photon momentum k, and yet satisfy this identity for any k. The only possibility is therefore $G^\mu(p_1,\dots,p_n) = 0$. Thus, the leading term in $\mathcal{N}^\mu$ is in fact fully determined by gauge invariance, and we have

$$\begin{aligned} &\epsilon_{\lambda\mu}(k)\,\mathcal{M}^\mu(k;p_1,\dots,p_n) \\ &\quad= \sum_{i=1}^n e_i \Bigg(\underbrace{\frac{p_i\cdot\epsilon_\lambda(k)}{p_i\cdot k}}_{\mathcal{O}(k^{-1})} + \underbrace{\Big(\frac{p_i\cdot\epsilon_\lambda(k)}{p_i\cdot k}\,k_\nu - \epsilon_{\lambda\nu}(k)\Big)\frac{\partial}{\partial p_{i\nu}}}_{\mathcal{O}(k^0)} \Bigg)\,\mathcal{M}(p_1,\dots,p_n) + \mathcal{O}(k^1). \end{aligned}$$

Quite remarkably, it is not just the leading term, but also the first subleading correction in the photon momentum, that we can write in this factorized form (provided we allow derivatives with respect to the p_i in the formula). The first non-factorizable term occurs only two orders down, with a suppression $(k/p_i)^2$ relative to the leading term. Note that the above expression can be written in a somewhat more symmetric way:

$$\begin{aligned}&\epsilon_{\lambda\mu}(k)\,\mathcal{M}^{\mu}(k;p_1,\ldots,p_n)\\ &\quad=\sum_{i=1}^{n}\frac{e_i\,p_i^{\mu}}{p_i\cdot k}\Big(\epsilon_{\lambda\mu}(k)+\big(\epsilon_{\lambda\mu}(k)k_\nu-k_\mu\epsilon_{\lambda\nu}(k)\big)\frac{\partial}{\partial p_{i\nu}}\Big)\,\mathcal{M}(p_1,\ldots,p_n)+\mathcal{O}(k^1).\end{aligned}$$

Note also that

$$\begin{aligned}\underbrace{\big(\epsilon_{\lambda\mu}(k)k_\nu-k_\mu\epsilon_{\lambda\nu}(k)\big)}_{\text{antisymmetric in }(\mu,\nu)}p_i^{\mu}\frac{\partial}{\partial p_{i\nu}}&=\frac{\epsilon_{\lambda\mu}(k)k_\nu-k_\mu\epsilon_{\lambda\nu}(k)}{2}\Big(p_i^{\mu}\frac{\partial}{\partial p_{i\nu}}-p_i^{\nu}\frac{\partial}{\partial p_{i\mu}}\Big)\\ &=-i\,\epsilon_{\lambda\mu}(k)k_\nu\underbrace{\Big(i p_i^{\mu}\frac{\partial}{\partial p_{i\nu}}-i p_i^{\nu}\frac{\partial}{\partial p_{i\mu}}\Big)}_{\text{angular momentum op. }J_i^{\mu\nu}}.\end{aligned}$$

Therefore, a very compact way of writing the amplitude for soft emission is

$$\begin{aligned}&\epsilon_{\lambda\mu}(k)\,\mathcal{M}^{\mu}(k;p_1,\ldots,p_n)\\ &\quad=\epsilon_{\lambda\mu}(k)\sum_{i=1}^{n}\frac{e_i}{p_i\cdot k}\Big(p_i^{\mu}-i\,k_\nu\,J_i^{\mu\nu}\Big)\,\mathcal{M}(p_1,\ldots,p_n)+\mathcal{O}(k^1).\end{aligned}$$

This result was first derived by Low (Low, F. (1958), *Phys Rev* 110: 974). It has been subsequently extended to Dirac fermions by Burnett and Kroll (see Problem 14), and it is now known as the *Low–Burnett–Kroll theorem.*

A consequence of this factorization is that, if a hard process is forbidden because of some conflicting quantum numbers, emitting a soft photon will not resolve the conflict (this is true in the first two orders in an expansion in the soft photon energy). For instance, if some decay is forbidden by charge parity (e.g., via Furry's theorem), one may naively think that emitting an extra photon – no matter how soft – lifts the obstruction since the charge parity of a photon is -1. Moreover, since extra soft photons have a probability of order one of being emitted, thanks to the terms in k^{-1}, this extra emission would not lead to a suppressed decay rate, implying that charge parity is in practice not conserved, at odds with experimental evidence. But the above result tells us that this conclusion is not true: it is only via the terms of order k^1 that the extra photon stands a chance of lifting the obstruction, and therefore it cannot be soft (thus, emitting it truly brings a suppression factor $\sim\alpha$). For this reason, another way this theorem is sometimes phrased is by saying that "soft photons do not carry quantum numbers."

---

14. Low–Burnett–Kroll Theorem In this problem, we extend the result of Problem 13 to the case of QED with spin-1/2 charges. As we shall see, the generalization requires that one sum the squared amplitude over the spins of the particles in the hard process, and the resulting formula reads

$$\sum_{\text{spins}}\Big|\epsilon_{\lambda\mu}(k)\mathcal{M}^{\mu}\Big|^2=\sum_{i,j=1}^{n}\frac{e_ie_j(p_i\cdot\epsilon_\lambda)}{(p_i\cdot k)(p_j\cdot k)}\Big(p_j\cdot\epsilon_\lambda-i\,\epsilon_{\lambda\mu}k_\nu J_j^{\mu\nu}\Big)\sum_{\text{spins}}|\mathcal{M}|^2+\mathcal{O}(k^0).$$

This result implies that corrections to a hard process by the emission of an extra soft photon are in general of order $\alpha(\Lambda/k)^2$ relative to the hard process, where Λ is the hard scale. (However, the divergence that would arise when integrating over the energy of the soft photon is canceled when one also includes virtual corrections with a soft photon in a loop.) More importantly, this formula also shows that, when a hard process is forbidden by charge parity, emitting an extra soft photon is not sufficient to lift the obstruction despite the fact that the extra photon flips the charge parity, at least up to a relative order $\alpha(\Lambda/k)^0$. For this reason, this result is often phrased by saying that "soft photons do not carry quantum numbers."

14.a Adapt the QED Feynman rules to the convention where all the external momenta in the amplitude $\mathcal{M}$ are defined to be outgoing.

14.b Consider an outgoing fermion of momentum p_i, spin s_i and charge e_i in the hard amplitude, and single out the corresponding external spinor by writing $\mathcal{M} \equiv \overline{u}_{s_i}(p_i)\mathcal{A}_{s_i}$. Show that attaching a soft photon to this external line can be done by the substitution

$$\overline{u}_{s_i}(p_i)\mathcal{A}_{s_i} \quad \to \quad e_i\overline{u}_{s_i}(p_i)\,\frac{p_i^\mu + p_i^\mu k_\nu \partial_{p_i}^\nu + \frac{1}{2}\gamma^\mu \not{k}}{p_i \cdot k}\,\mathcal{A}_{s_i} + \mathcal{O}(k).$$

Determine the substitution rules for attaching a soft photon to all the other types of external lines.

14.c Use the Ward–Takahashi identity to determine the term where the soft photon is attached to an internal line of the hard process. Write the total amplitude as

$$\mathcal{M}^\mu(k;p_1,\ldots,p_n) = \sum_i e_i \frac{p_i^\mu - ik_\nu\left[J_i^{\mu\nu}\right]_{\mathcal{A},\mathcal{B}}}{p_i \cdot k}\,\mathcal{M}(p_1,\ldots,p_n) + \mathcal{E}^\mu + \mathcal{O}(k^1),$$

$$\mathcal{E}^\mu \equiv \sum_{i\in\{\overline{u},\overline{v}\}} \begin{Bmatrix} \overline{u}_{s_i}(p_i) \\ \overline{v}_{s_i}(-p_i) \end{Bmatrix} \frac{e_i\gamma^\mu \not{k}}{2p_i \cdot k}\mathcal{A}_{s_i} + \sum_{i\in\{u,v\}} \overline{\mathcal{B}}_{s_i}\frac{e_i \not{k}\gamma^\mu}{2p_i \cdot k}\begin{Bmatrix} u_{s_i}(-p_i) \\ v_{s_i}(p_i) \end{Bmatrix},$$

where the subscript $\mathcal{A},\mathcal{B}$ on the angular momentum operators $J^{\mu\nu}$ indicates that they should not act on the external spinors contained in the hard amplitude $\mathcal{M}$.

14.d Square this amplitude and sum over the spins of the hard external fermions in order to prove the announced formula.

14.a In order to provide an easy connection with Problem 13, we use a convention in which all the momenta are outgoing. In spinor quantum electrodynamics, this implies somewhat unusual assignments for the arguments of the spinors that represent the external fermions, and the Dirac equations they satisfy are also modified:

	Feynman rule	Dirac equation
Outgoing fermion	$\overline{u}_s(p)$	$\overline{u}(p)(\not{p} - m) = 0$
Outgoing anti-fermion	$v_s(p)$	$(\not{p} + m)v(p) = 0$
Incoming fermion	$u_s(-p)$	$(\not{p} + m)u(-p) = 0$
Incoming anti-fermion	$\overline{v}_s(-p)$	$\overline{v}(-p)(\not{p} - m) = 0$

14.b Consider an amplitude $\mathcal{M}(p_1, \ldots, p_n)$ with n hard external fermions or anti-fermions. When an extra soft photon is emitted in this process, the leading contributions come from Feynman diagrams where the extra photon is attached to one of these external lines. However, since we now want to calculate the next-to-leading contribution (in an expansion in powers of the photon momentum k), the calculation of the attachments of the photon to the external lines must be refined. Since we have to consider in turn every external line, it is convenient to write the hard amplitude in a factorized form that highlights the spinor corresponding to a given external line. Depending on the type of external line under consideration, we write

$$\mathcal{M} \equiv \bar{u}_{s_i}(\mathbf{p}_i)\, \mathcal{A}_{s_i} \equiv \bar{v}_{s_i}(-\mathbf{p}_i)\, \mathcal{A}_{s_i} \equiv \overline{\mathcal{B}}_{s_i}\, u_{s_i}(-\mathbf{p}_i) \equiv \overline{\mathcal{B}}_{s_i}\, v_{s_i}(\mathbf{p}_i).$$

(In these expressions, there is no sum on the index i.) Let us start with the case of an *outgoing fermion* of momentum $\mathbf{p}_i$. Attaching the soft photon of momentum k to this line gives a contribution that corresponds to the following substitution:

$$\bar{u}_{s_i}(\mathbf{p}_i)\, \mathcal{A}_{s_i}(\ldots, p_i, \ldots)$$

$$\rightarrow \quad \bar{u}_{s_i}(\mathbf{p}_i)\, \bar{e}\gamma^\mu \frac{\slashed{p}_i + m + \slashed{k}}{2\, p_i \cdot k}\, \mathcal{A}_{s_i}(\ldots, p_i + k, \ldots)$$

$$= \frac{\bar{e}}{p_i \cdot k}\, \bar{u}_{s_i}(\mathbf{p}_i) \left\{ p_i^\mu + p_i^\mu k_\nu \partial_{p_i}^\nu + \tfrac{1}{2}\gamma^\mu \slashed{k} \right\} \mathcal{A}_{s_i}(\ldots, p_i, \ldots) + \mathcal{O}(k).$$

In the second line, the denominator has been simplified thanks to the assumption that the external fermion as well as the photon are on-shell. In the last line, we have used the Dirac equation and we have expanded the function $\mathcal{A}_{s_i}(\ldots, p_i + k, \ldots)$ in order to extract the linear order in k. As in Problem 13, $\bar{e}$ denotes the coupling constant that enters in the definition of the covariant derivative, which is equal to the charge of a fermion and is opposite to that of an anti-fermion.

Consider now a soft photon attached to an *outgoing anti-fermion* of momentum $\mathbf{p}_i$. This leads to

$$\overline{\mathcal{B}}_{s_i}(\ldots, p_i, \ldots)\, v_{s_i}(\mathbf{p}_i)$$

$$\rightarrow \quad \overline{\mathcal{B}}_{s_i}(\ldots, p_i + k, \ldots) \frac{-\slashed{p}_i + m - \slashed{k}}{2\, p_i \cdot k}\, \bar{e}\gamma^\mu v_{s_i}(\mathbf{p}_i)$$

$$= \overline{\mathcal{B}}_{s_i}(\ldots, p_i, \ldots) \left\{ p_i^\mu + p_i^\mu k_\nu \overleftarrow{\partial}_{p_i}^\nu + \tfrac{1}{2}\slashed{k}\gamma^\mu \right\} v_{s_i}(\mathbf{p}_i) \frac{-\bar{e}}{p_i \cdot k} + \mathcal{O}(k).$$

Two remarks are in order about this formula. First, note that the derivative with respect to p_i acts on the left, as indicated by an arrow. Second, in the eikonal factor $-\bar{e}/p_i \cdot k$, the numerator $-\bar{e}$ is nothing but the electrical charge *flowing outwards*. In order to further uniformize the notation, it is convenient to express everything in terms of the outgoing electrical charge e, whose relationship with $\bar{e}$ depends on the type of external line under consideration:

- *Outgoing fermion: $e = \bar{e}$*

- *Outgoing anti-fermion: $e = -\bar{e}$*

- *Incoming fermion: $e = -\bar{e}$*

- *Incoming anti-fermion: $e = \bar{e}$.*

With these conventions, it is easy to work out the remaining two cases (incoming fermions and anti-fermions). All the rules for attaching a soft photon, up to terms of order k, to the four types of external fermions lines can be summarized by

$$\begin{Bmatrix} \overline{u}_{s_i}(p_i) \\ \overline{v}_{s_i}(-p_i) \end{Bmatrix} \mathcal{A}_{s_i} \quad \to \quad e_i \begin{Bmatrix} \overline{u}_{s_i}(p_i) \\ \overline{v}_{s_i}(-p_i) \end{Bmatrix} \frac{p_i^\mu + p_i^\mu k_\nu \overrightarrow{\partial}{}^\nu_{p_i} + \frac{1}{2}\gamma^\mu \not{k}}{p_i \cdot k} \mathcal{A}_{s_i},$$

$$\overline{\mathcal{B}}_{s_i} \begin{Bmatrix} u_{s_i}(-p_i) \\ v_{s_i}(p_i) \end{Bmatrix} \quad \to \quad e_i \overline{\mathcal{B}}_{s_i} \frac{p_i^\mu + p_i^\mu k_\nu \overleftarrow{\partial}{}^\nu_{p_i} + \frac{1}{2}\not{k}\gamma^\mu}{p_i \cdot k} \begin{Bmatrix} u_{s_i}(-p_i) \\ v_{s_i}(p_i) \end{Bmatrix}.$$

Recall also that with outgoing momenta and electrical charges, the conservation laws in the hard process take the following simple forms:

$$\sum_i p_i^\mu = 0, \quad \sum_i e_i = 0.$$

14.c The amplitude for the emission of an extra soft photon can be split into a term where the photon is attached to one of the external lines of the hard process and a term where the photon is attached to an internal line:

$$\mathcal{M}^\mu(k; p_1, \ldots, p_n) \equiv \mathcal{M}^\mu_{\text{ext}}(k; p_1, \ldots, p_n) + \mathcal{M}^\mu_{\text{inner}}(k; p_1, \ldots, p_n).$$

The first term is obtained by summing the results of the preceding section over all the external lines:

$$\begin{aligned} \mathcal{M}^\mu_{\text{ext}}(k; p_1, \ldots, p_n) = & \sum_{i\in\{\overline{u},\overline{v}\}} e_i \begin{Bmatrix} \overline{u}_{s_i}(p_i) \\ \overline{v}_{s_i}(-p_i) \end{Bmatrix} \frac{p_i^\mu + p_i^\mu k_\nu \overrightarrow{\partial}{}^\nu_{p_i} + \frac{1}{2}\gamma^\mu \not{k}}{p_i \cdot k} \mathcal{A}_{s_i} \\ & + \sum_{i\in\{u,v\}} e_i \overline{\mathcal{B}}_{s_i} \frac{p_i^\mu + p_i^\mu k_\nu \overleftarrow{\partial}{}^\nu_{p_i} + \frac{1}{2}\not{k}\gamma^\mu}{p_i \cdot k} \begin{Bmatrix} u_{s_i}(-p_i) \\ v_{s_i}(p_i) \end{Bmatrix}, \end{aligned}$$

where the first line contains the contributions from outgoing fermions and incoming anti-fermions, and the second line that of incoming fermions and outgoing anti-fermions (in each case, one should choose the appropriate spinor from the two listed). The above expression can also be rearranged as follows:

$$\begin{aligned} \mathcal{M}^\mu_{\text{ext}}(k; p_1, \ldots, p_n) = & \left(\sum_i e_i \frac{p_i^\mu + p_i^\mu k_\nu \partial^\nu_{p_i}\big|_{\mathcal{A},\mathcal{B}}}{p_i \cdot k} \right) \mathcal{M}(p_1, \ldots, p_n) \\ & + \sum_{i\in\{\overline{u},\overline{v}\}} \begin{Bmatrix} \overline{u}_{s_i}(p_i) \\ \overline{v}_{s_i}(-p_i) \end{Bmatrix} \frac{\frac{e_i}{2}\gamma^\mu \not{k}}{p_i \cdot k} \mathcal{A}_{s_i} + \sum_{i\in\{u,v\}} \overline{\mathcal{B}}_{s_i} \frac{\frac{e_i}{2}\not{k}\gamma^\mu}{p_i \cdot k} \begin{Bmatrix} u_{s_i}(-p_i) \\ v_{s_i}(p_i) \end{Bmatrix}, \end{aligned}$$

where the subscript $\mathcal{A}, \mathcal{B}$ added to the derivative with respect to p_i in the first line indicates that it does not act on the spinors associated with the external lines in the hard amplitude $\mathcal{M}$.

We now need to determine the term $\mathcal{M}^\mu_{\text{inner}}$ where the photon is attached to an inner line of the hard process. This term starts at order k^0, and therefore, to the accuracy of the expansion considered here, it is independent of k. As in the scalar QED case, this term can be constrained by requesting that the full amplitude obey the Ward–Takahashi identity when we contract it with the photon momentum k_μ. Note that the terms in $\not{k}$ vanish identically in this contraction since $\not{k}\not{k} = k^2 = 0$. Using the fact that the total outgoing charge in the hard process is zero, the Ward–Takahashi identity gives

$$\mathcal{M}^\mu_{\text{inner}}(p_1, \dots, p_n) = -\left(\sum_i e_i \left.\frac{\partial}{\partial p_{i\mu}}\right|_{\mathcal{A},\mathcal{B}}\right) \mathcal{M}(p_1, \dots, p_n).$$

(A priori, we could add to this expression a term G^μ such that $k_\mu G^\mu = 0$. But since G^μ should be independent of k, the only way to satisfy this for all k_μ is to have $G^\mu \equiv 0$.)

When we combine this term with $\mathcal{M}^\mu_{\text{ext}}$, the terms with derivatives with respect to the external momenta p_i can be arranged into angular momentum operators $J_i^{\mu\nu} \equiv -i(p_i^\mu \partial_{p_{i\nu}} - p^\nu \partial_{p_{i\mu}})$. This gives

$$\mathcal{M}^\mu(k; p_1, \dots, p_n) = \sum_i e_i \frac{p_i^\mu - i k_\nu \left[J_i^{\mu\nu}\right]_{\mathcal{A},\mathcal{B}}}{p_i \cdot k} \mathcal{M}(p_1, \dots, p_n) + \mathcal{E}^\mu + \mathcal{O}(k^1),$$

$$\mathcal{E}^\mu \equiv \sum_{i \in \{\overline{u}, \overline{v}\}} \begin{Bmatrix} \overline{u}_{s_i}(p_i) \\ \overline{v}_{s_i}(-p_i) \end{Bmatrix} \frac{e_i \gamma^\mu \not{k}}{2 p_i \cdot k} \mathcal{A}_{s_i} + \sum_{i \in \{u, v\}} \overline{\mathcal{B}}_{s_i} \frac{e_i \not{k} \gamma^\mu}{2 p_i \cdot k} \begin{Bmatrix} u_{s_i}(-p_i) \\ v_{s_i}(p_i) \end{Bmatrix}.$$

The first term is identical to the scalar case, except for the fact that the angular momentum operators do not act on the external spinors (as indicated by the subscript $\mathcal{A}, \mathcal{B}$). Another difference from the scalar QED case is that the second term, $\mathcal{E}^\mu$, does not vanish when contracted with a physical polarization vector.

14.d It turns out that, after squaring this amplitude and summing over the spins of the external hard fermions, the terms in $\mathcal{E}^\mu$ provide the missing terms to enlarge the scope of the action of the angular momentum operators to the full hard amplitude. In order to be consistent with the accuracy to which we have determined $\mathcal{M}^\mu$, we should keep only terms of order k^{-2} and k^{-1} in its square, since the higher orders would be incomplete. At this order, the squared amplitude reads

$$\begin{aligned} \sum_{\text{spins}} \left|\epsilon_{\lambda\mu}(k)\mathcal{M}^\mu\right|^2 &= \underbrace{\sum_{i,j} \frac{e_i e_j (p_i \cdot \epsilon_\lambda)}{(p_i \cdot k)(p_j \cdot k)} \left((p_j \cdot \epsilon_\lambda) - i\, \epsilon_{\lambda\mu} k_\nu \left[J_j^{\mu\nu}\right]_{\mathcal{A},\mathcal{B}}\right) \sum_{\text{spins}} |\mathcal{M}|^2}_{\mathcal{O}(k^{-2}) \oplus \mathcal{O}(k^{-1})} \\ &+ \underbrace{\sum_i \frac{e_i (p_i \cdot \epsilon_\lambda)}{p_i \cdot k} \sum_{\text{spins}} \left(\mathcal{M}^* \epsilon_{\lambda\mu} \mathcal{E}^\mu + \mathcal{M} \epsilon_{\lambda\mu} \mathcal{E}^{\mu *}\right)}_{\mathcal{O}(k^{-1})}. \end{aligned} \tag{1.60}$$

Note that in the first line, the derivatives contained in the angular momentum operators act on the two factors of $|\mathcal{M}|^2 = \mathcal{M}^*\mathcal{M}$. In order to calculate the term of the second line, note that

$$\sum_{\text{spins}} \left(\mathcal{M}^* \epsilon_{\lambda\mu}\mathcal{E}^\mu + \mathcal{M}\epsilon_{\lambda\mu}\mathcal{E}^{\mu*}\right) = \sum_{j\in\{\overline{u},\overline{v}\}} e_j\, \overline{\mathcal{A}}_{s_j} \frac{\begin{Bmatrix} \not{p}_j + m \\ -\not{p}_j - m \end{Bmatrix} \not{\epsilon}\not{k} + \not{k}\not{\epsilon} \begin{Bmatrix} \not{p}_j + m \\ -\not{p}_j - m \end{Bmatrix}}{2p_j \cdot k} \mathcal{A}_{s_j}$$

$$+ \sum_{j\in\{u,v\}} e_j\, \overline{\mathcal{B}}_{s_j} \frac{\begin{Bmatrix} -\not{p}_j + m \\ \not{p}_j - m \end{Bmatrix} \not{\epsilon}\not{k} + \not{k}\not{\epsilon} \begin{Bmatrix} -\not{p}_j + m \\ \not{p}_j - m \end{Bmatrix}}{2p_j \cdot k} \mathcal{B}_{s_j}.$$

Using the fact that $k \cdot \epsilon = 0$, we have

$$\frac{(\not{p} \pm m)\not{\epsilon}\not{k} + \not{k}\not{\epsilon}(\not{p} \pm m)}{2p \cdot k} = \frac{(p \cdot \epsilon)\not{k} - (p \cdot k)\not{\epsilon}}{p \cdot k}$$
$$= \frac{1}{p \cdot k}\left(-i\epsilon_\mu k_\nu \left[J_p^{\mu\nu}\right]\right)(\not{p} \pm m).$$

The remarkable feature of this result is that it takes the form of an angular momentum operator acting on spin sums such as $\sum_s u_s(\mathbf{p})\overline{u}_s(\mathbf{p}) = \not{p} + m$. Thanks to this formula, the terms in the second line of (1.60) are precisely what is needed in order to lift the restrictions on the action of the angular momentum operators in the terms of the first line, allowing us to write the complete answer up to (and including) the order k^{-1} in a much more compact fashion:

$$\sum_{\text{spins}} \left|\epsilon_{\lambda\mu}(\mathbf{k})\mathcal{M}^\mu\right|^2 = \sum_{i,j} \frac{e_i e_j (p_i \cdot \epsilon_\lambda)}{(p_i \cdot k)(p_j \cdot k)} \left((p_j \cdot \epsilon_\lambda) - i\,\epsilon_{\lambda\mu}k_\nu J_j^{\mu\nu}\right) \sum_{\text{spins}} |\mathcal{M}|^2.$$

This formula, obtained by Burnett and Kroll (Burnett, T. and Kroll, N. M. (1968), *Phys Rev Lett* 20: 86), generalizes Low's theorem (established in Problem 13) to the case of hard charged external particles of spin 1/2. And the main consequence is the same: if some unpolarized hard process is forbidden by charge parity, emitting extra soft photons is not sufficient to make it possible. (The emission of an extra photon may make this process possible via terms of order $\mathcal{O}(k^0)$, but these terms are not enhanced by inverse powers of the photon energy.)

15. Coherent States in Quantum Field Theory In quantum mechanics, *coherent states* are defined as eigenstates of the annihilation operators. In a certain sense, they are the closest analogues to classical states where the position and momentum are both well defined. The goal of this problem is to generalize the concept of coherent state to a non-interacting scalar quantum field theory, starting from the following definition:

$$|\chi_{\text{in}}\rangle \equiv \mathcal{N}_\chi \exp\left\{\int \frac{d^3\mathbf{k}}{(2\pi)^3 2E_\mathbf{k}}\, \chi(\mathbf{k})\, a^\dagger_{\mathbf{k},\text{in}}\right\} |0_{\text{in}}\rangle,$$

where $\chi(\mathbf{k})$ is a function of 3-momentum and $\mathcal{N}_\chi$ a normalization constant.

15.a Check that this state is an eigenstate of $a_{\mathbf{k},\text{in}}$.

15.b Determine the prefactor $\mathcal{N}_\chi$ so that $\langle\chi_{in}|\chi_{in}\rangle = 1$. *Hint: introduce the operator*

$$T_\chi \equiv \exp\int \frac{d^3k}{(2\pi)^3 2E_k}\left(\chi(k)a^\dagger_{k,in} - \chi^*(k)a_{k,in}\right)$$

and write it in normal-ordered form.

15.c Show that $|\chi_{in}\rangle$ is the ground state of a free scalar field theory with a shifted field:

$$\phi(x) \to \phi(x) - \int \frac{d^3k}{(2\pi)^3 2E_k}\left(\chi^*(k)e^{ik\cdot x} + \chi(k)e^{-ik\cdot x}\right).$$

15.d Consider two coherent states $|\chi, \vartheta_{in}\rangle$, with $\chi(k), \vartheta(k) \propto \delta(k)$, and calculate $\langle\vartheta_{in}|\chi_{in}\rangle$.

15.a Let us denote

$$E_\chi \equiv \int \frac{d^3k}{(2\pi)^3 2E_k}\,\chi(k)\,a^\dagger_{k,in}.$$

The commutation relation between creation and annihilation operators leads to

$$\left[a_{k,in}, E_\chi\right] = \chi(k), \qquad \left[a_{k,in}, f(E_\chi)\right] = \chi(k)\,f'(E_\chi).$$

This implies

$$\begin{aligned} a_{k,in}|\chi_{in}\rangle &= \mathcal{N}_\chi\, a_{k,in}e^{E_\chi}|0_{in}\rangle = \mathcal{N}_\chi\left[a_{k,in}, e^{E_\chi}\right]|0_{in}\rangle \\ &= \mathcal{N}_\chi\,\chi(k)\,e^{E_\chi}|0_{in}\rangle = \chi(k)\,|\chi_{in}\rangle. \end{aligned}$$

Thus, coherent states are eigenstates of the annihilation operators.

15.b Consider now the following operator:

$$T_\chi \equiv \exp\int \frac{d^3k}{(2\pi)^3 2E_k}\left(\chi(k)a^\dagger_{k,in} - \chi^*(k)a_{k,in}\right).$$

First, it is trivial to check that $T^\dagger_\chi = T_{-\chi} = T^{-1}_\chi$. (Therefore, T_χ is unitary.) Then, by using the Baker–Campbell–Hausdorff formula, we can put this operator in normal-ordered form:

$$\begin{aligned} T_\chi = \exp\Big(-\frac{1}{2}\int \frac{d^3k}{(2\pi)^3 2E_k}|\chi(k)|^2\Big)\exp\Big(\int \frac{d^3k}{(2\pi)^3 2E_k}\chi(k)a^\dagger_{k,in}\Big) \\ \times \exp\Big(-\int \frac{d^3k}{(2\pi)^3 2E_k}\chi^*(k)a_{k,in}\Big). \end{aligned}$$

From this expression, we see that $|\chi_{in}\rangle \propto T_\chi|0_{in}\rangle$. Since T_χ is unitary, this implies that the normalization $\langle\chi_{in}|\chi_{in}\rangle = 1$ is obtained by choosing precisely $|\chi_{in}\rangle = T_\chi|0_{in}\rangle$, from which we read the value of the normalization prefactor $\mathcal{N}_\chi$:

$$\mathcal{N}_\chi = \exp\Big(-\frac{1}{2}\int \frac{d^3k}{(2\pi)^3 2E_k}|\chi(k)|^2\Big).$$

15.c If $\mathcal{H}_0$ denotes the Hamiltonian of a free theory, let us now define $\mathcal{H}_\chi \equiv T_\chi \mathcal{H}_0 T_\chi^\dagger$. Note first that $\mathcal{H}_\chi$ is Hermitian. Then, we have

$$\mathcal{H}_\chi |\chi_{\text{in}}\rangle = T_\chi \mathcal{H}_0 \underbrace{T_\chi^\dagger T_\chi}_{=1} |0_{\text{in}}\rangle = 0.$$

Thus, the coherent state is the ground state of the modified Hamiltonian $\mathcal{H}_\chi$.

Using the commutation relations

$$\left[a_{\mathbf{k},\text{in}}, T_\chi\right] = \chi(\mathbf{k})\, T_\chi, \quad \left[a_{\mathbf{k},\text{in}}^\dagger, T_\chi\right] = \chi^*(\mathbf{k})\, T_\chi,$$

we obtain

$$T_\chi a_{\mathbf{k},\text{in}} T_\chi^\dagger = a_{\mathbf{k},\text{in}} - \chi(\mathbf{k}), \quad T_\chi a_{\mathbf{k},\text{in}}^\dagger T_\chi^\dagger = a_{\mathbf{k},\text{in}}^\dagger - \chi^*(\mathbf{k}),$$

and the modified Hamiltonian is more explicitly given by

$$\mathcal{H}_\chi = \int \frac{d^3k}{(2\pi)^3 2E_k}\, E_k \left(a_{\mathbf{k},\text{in}}^\dagger - \chi^*(\mathbf{k})\right)\left(a_{\mathbf{k},\text{in}} - \chi(\mathbf{k})\right).$$

(We have discarded the zero-point energy term, since it is not modified by the action of $T_\chi \cdots T_\chi^\dagger$.) Recalling the formulas that relate a free field operator and the creation and annihilation operators, we see that this is the Hamiltonian of a free theory with a shifted field:

$$\phi(x) \to \phi(x) - \Phi_\chi(x), \quad \Phi_\chi(x) \equiv \int \frac{d^3k}{(2\pi)^3 2E_k} \left(\chi^*(\mathbf{k}) e^{ik\cdot x} + \chi(\mathbf{k}) e^{-ik\cdot x}\right).$$

(Note that $\Phi_\chi(x)$ is an ordinary real-valued function, not an operator.)

15.d Consider two such coherent states, $|\chi_{\text{in}}\rangle$, $|\vartheta_{\text{in}}\rangle$, in the special case where their defining functions only have support at $\mathbf{k} = 0$: $\chi(\mathbf{k}) \equiv (2\pi)^3 \chi_0 \delta(\mathbf{k})$, $\vartheta(\mathbf{k}) \equiv (2\pi)^3 \vartheta_0 \delta(\mathbf{k})$ (we assume $\chi_0, \vartheta_0 \in \mathbb{R}$). The overlap of these two states is given by

$$\begin{aligned}
\langle \vartheta_{\text{in}} | \chi_{\text{in}} \rangle &= \exp\left(-\frac{1}{2}\int \frac{d^3k}{(2\pi)^3 2E_k} \left[|\chi(\mathbf{k})|^2 + |\vartheta(\mathbf{k})|^2\right]\right) \\
&\quad \times \langle 0_{\text{in}}| \exp\left(\int \frac{d^3k}{(2\pi)^3 2E_k}\, \vartheta^*(\mathbf{k})\, a_{\mathbf{k},\text{in}}\right) \exp\left(\int \frac{d^3k}{(2\pi)^3 2E_k}\, \chi(\mathbf{k})\, a_{\mathbf{k},\text{in}}^\dagger\right) |0_{\text{in}}\rangle \\
&= \exp\left(-\frac{1}{2}\int \frac{d^3k}{(2\pi)^3 2E_k} \left(|\chi(\mathbf{k}) - \vartheta(\mathbf{k})|^2 + \chi^*(\mathbf{k})\vartheta(\mathbf{k}) - \vartheta^*(\mathbf{k})\chi(\mathbf{k})\right)\right) \\
&= \exp\left(-\frac{V\,|\chi_0 - \vartheta_0|^2}{4m}\right).
\end{aligned}$$

Therefore, spatially homogeneous coherent states (i.e., ground states of quadratic theories shifted by a uniform field) have an exponentially suppressed overlap, and the argument of the exponential is proportional to the volume. Thus, pairs of coherent states of this type are mutually orthogonal if the volume is infinite.

16. Running Couplings in a Two-Field Scalar Field Theory The purpose of this problem is to study at one loop the scale dependence of the coupling in a scalar field theory with quartic coupling, first with a single field and then with two coupled fields. In the latter case, we show that the theory evolves towards a $\mathrm{U}(1)$ symmetry at large distance.

16.a Draw the graphs that contribute to the β function at lowest order (i.e., one-loop) in a scalar field theory with $\frac{\lambda}{4!}\phi^4$ interaction. Why don't we need to consider self-energy graphs at this order?

16.b Calculate the relevant part of these graphs in dimensional regularization, at the renormalization point where the Mandelstam variables are set to $s = t = u = -\mu^2$. Write their sum as $\frac{3}{2}\lambda^2 L(\mu)$, where $L(\mu)$ is a quantity that should be made explicit. From this, check that the β function is $\beta = 3\lambda^2/(4\pi)^2$.

16.c Next, we consider a theory with two scalar fields ϕ_1 and ϕ_2, with the following Lagrangian:

$$\mathcal{L} \equiv \tfrac{1}{2}(\partial_\mu\phi_1)(\partial^\mu\phi_1) + \tfrac{1}{2}(\partial_\mu\phi_2)(\partial^\mu\phi_2) - \tfrac{\lambda}{4!}(\phi_1^4 + \phi_2^4) - \tfrac{2\rho}{4!}\phi_1^2\phi_2^2.$$

(Note that this theory has a $\mathrm{U}(1)$ invariance in the (ϕ_1, ϕ_2) plane when $\lambda = \rho$.) What are the free Feynman propagators for the fields ϕ_1 and ϕ_2 (in momentum space)? What are the vertices in this theory and the corresponding Feynman rules?

16.d Draw the graphs that give, at lowest order, the β function that controls the scale dependence of the coupling constant λ. Show that the relevant part of these graphs is given by $(\frac{3}{2}\lambda^2 + \frac{1}{6}\rho^2)L(\mu)$, where $L(\mu)$ is the quantity obtained in question 16.b. Check that this β function is $\beta_\lambda = (3\lambda^2 + \rho^2/3)/(4\pi)^2$.

16.e Repeat question 16.d for the β function that controls the scale dependence of the coupling constant ρ. Check that it is $\beta_\rho = (2\lambda\rho + 4\rho^2/3)/(4\pi)^2$.

16.f Derive the renormalization group equation for the ratio ρ/λ. What are the fixed points of this equation? Is $\rho/\lambda = 1$ an attractive fixed point in the ultraviolet? in the infrared?

16.a The calculation of the β function requires the coupling counterterm δ_λ and the wave-function renormalization counterterm δ_z. δ_λ is obtained from the one-loop corrections to the vertex function (i.e., the four-point function in the case of the ϕ^4 theory):

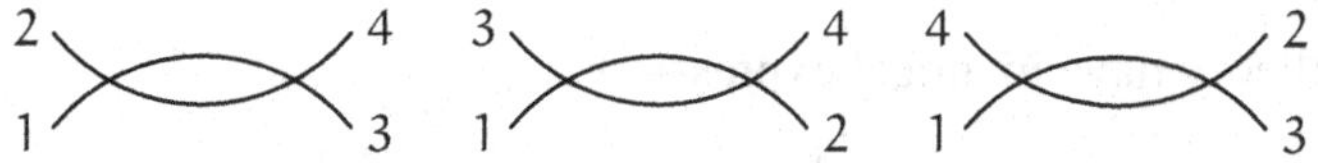

(Note that there are three ways of attaching the four external lines to the loop.) δ_z comes from the momentum dependence of the self-energy. But in the ϕ^4 theory, the self-energy is a constant at one loop, and therefore one has $\delta_z = 0$ at this order.

16.b We need to calculate only one of the above graphs, since the renormalization point we have chosen is symmetric under the exchange of the channels s, t, u. Moreover, since only the ultraviolet divergent part is necessary, we do not need to keep the mass in the propagators. With this in mind, let us consider the first graph. Its contribution to the amputated correlation function Γ_4 reads (we use dimensional regularization, with $d = 4 - 2\epsilon$):

$$\Gamma_4^{(1)}(p_{1,2,3,4}) = i\mu^{2\epsilon}\frac{(-i\lambda)^2}{2}\int \frac{d^dk}{(2\pi)^d}\frac{i}{k^2}\frac{i}{(k+p_1+p_2)^2}$$

$$= i\frac{\lambda^2\mu^{2\epsilon}}{2}\int_0^1 dx \int \frac{d^dk}{(2\pi)^d}\frac{1}{(k^2-\Delta)^2} = -\frac{\lambda^2\mu^{2\epsilon}}{2}\int_0^1 dx \int \frac{d^dk_E}{(2\pi)^d}\frac{1}{(k_E^2+\Delta)^2}$$

$$= -\frac{\lambda^2\mu^{2\epsilon}}{2}\int_0^1 dx \frac{\Delta^{d/2-2}}{(4\pi)^{d/2}}\frac{\Gamma(2-\frac{d}{2})}{\Gamma(2)},$$

with $\Delta \equiv -x(1-x)(p_1+p_2)^2 = x(1-x)\mu^2$. The x dependence of the integrand is of the form $(x(1-x))^\epsilon$ and therefore this integral produces a result of the form $1 + \mathcal{O}(\epsilon)$. Since we are only interested in the ultraviolet divergent part, only the 1 is needed. Thus, we can write

$$\Gamma_4^{(1)}(p_{1,2,3,4}) = -\frac{\lambda^2}{2(4\pi)^2\epsilon}\frac{\mu^{2\epsilon}}{\mu^{2\epsilon}}.$$

Since the renormalization point is symmetric in the s, t, u channels, the above result is simply multiplied by 3 when we include all the graphs:

$$\Gamma_4^{(1+2+3)}(p_{1,2,3,4}) = \frac{3\lambda^2}{2}L(\mu), \quad \text{with } L(\mu) \equiv -\frac{1}{(4\pi)^2\epsilon}\frac{\mu^{2\epsilon}}{\mu^{2\epsilon}}.$$

In this notation, the prefactor $\frac{3}{2}$ is the number of graphs times their symmetry factor, the factor λ^2 is the contribution from the two vertices, and $L(\mu)$ is the value of the loop integral. The counterterm that removes this divergence is

$$\delta_\lambda = -\frac{3\lambda^2}{2}L(\mu),$$

which leads to the following β function:

$$\beta = -\lim_{\epsilon\to 0}\mu\partial_\mu\,\delta_\lambda = \frac{3\lambda^2}{(4\pi)^2}.$$

16.c Consider now an extension of this theory that has two scalar fields ϕ_1 and ϕ_2, with the following Lagrangian:

$$\mathcal{L} \equiv \tfrac{1}{2}(\partial_\mu\phi_1)(\partial^\mu\phi_1) + \tfrac{1}{2}(\partial_\mu\phi_2)(\partial^\mu\phi_2) - \tfrac{\lambda}{4!}(\phi_1^4+\phi_2^4) - \tfrac{2\rho}{4!}\phi_1^2\phi_2^2.$$

The two propagators of this theory have identical expressions:

$$\text{———} = \frac{i}{p^2+i0^+}, \qquad \text{- - - - - -} = \frac{i}{p^2+i0^+}.$$

(Even though the propagators are the same, it is important to use different symbols for the two fields in the diagrams, since there will be various vertices, depending on which fields are attached to them.) This theory has only quartic vertices, ϕ_1^4, ϕ_2^4 and $\phi_1^2\phi_2^2$:

 $= -i\lambda,$  $= -i\lambda,$  $= -i\frac{\rho}{3}.$

For the first two vertices, this is just the Feynman rule for the vertex of the standard ϕ^4 theory, when the interaction term in the Lagrangian is normalized as $\frac{\lambda}{4!}\phi^4$. For the third vertex, the Feynman rule is obtained as

$$-i\underbrace{\frac{2\rho}{4!}}_{\text{coeff. in }\mathcal{L}}\times 2!\times 2! = -i\,\frac{\rho}{3},$$

where the first factor 2! corresponds to the two ways of attaching the fields $\phi_1\phi_1$ to the two solid lines of the vertex, and the second factor 2! counts the number of ways of attaching the fields $\phi_2\phi_2$ to the two dotted lines of the vertex.

16.d In order to determine the β function that controls the scale dependence of the coupling λ, it is sufficient to calculate at one loop the four-point function with only external fields of type 1. The corresponding graphs are:

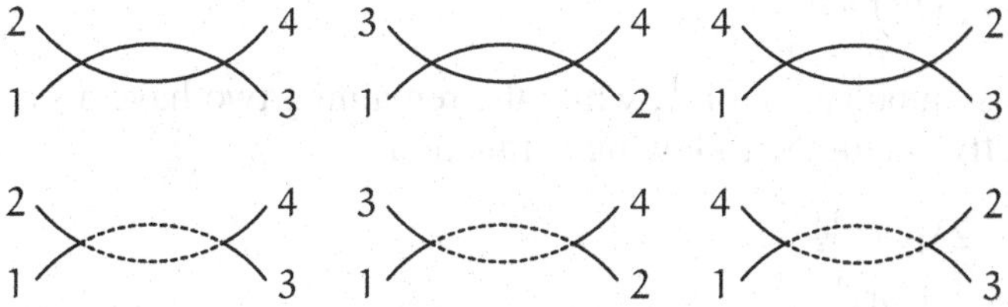

The values of the loops do not depend on the type of field running in them since we neglect masses in this calculation, and are therefore given by the same function $L(\mu)$ in all six graphs. We just need to count the graphs and their symmetry factors ($\frac{1}{2}$ in all cases), and assign them the appropriate vertices to find the correct four-point function:

$$\Gamma_4(p_{1,2,3,4}) = \left(\frac{3}{2}\lambda^2 + \frac{3}{2}\left(\frac{\rho}{3}\right)^2\right) L(\mu) = \left(\frac{3\lambda^2}{2} + \frac{\rho^2}{6}\right) L(\mu).$$

Since $L(\mu)$ is already known, we do not need any new calculation in order to obtain the β function in this case. It is sufficient to appropriately adjust the prefactor, to obtain

$$\beta_\lambda = \frac{3\lambda^2 + \frac{\rho^2}{3}}{(4\pi)^2}.$$

16.e To obtain the β function that drives the evolution of ρ, we need to consider the 4-point function with two external fields ϕ_1 and two external fields ϕ_2. To be definite, let us assume

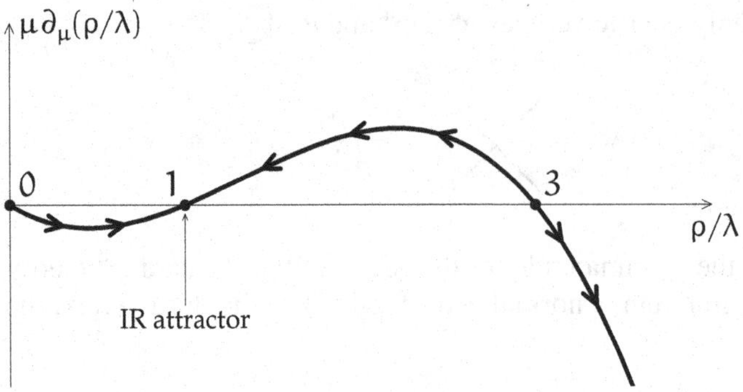

Figure 1.5 The β function that controls the running of the ratio ρ/λ. The arrows indicate the direction of the flow towards long distances.

that the legs 1, 2 carry the field ϕ_1 and the legs 3, 4 are attached to a field ϕ_2. Thus, the one-loop graphs that contribute to this function are the following:

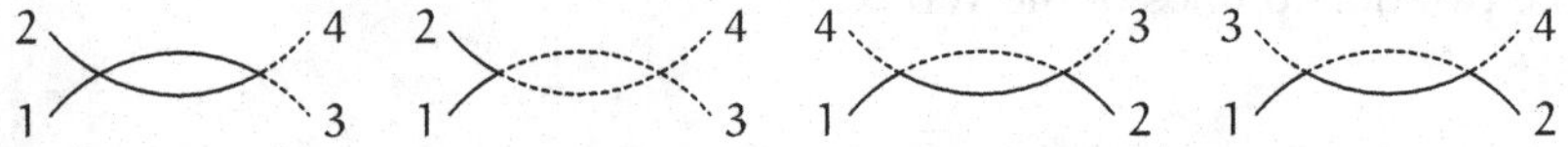

We can directly write the corresponding expression for the four-point function in terms of the integral $L(\mu)$:

$$\Gamma_4(p_{1,2,3,4}) = \left(\frac{2}{2}\frac{\lambda\rho}{3} + 2\left(\frac{\rho}{3}\right)^2\right) L(\mu).$$

(The first two graphs have a symmetry factor $\frac{1}{2}$, while the remaining two have a symmetry factor 1.) From this, we directly obtain the following β function:

$$\beta_\rho = 3 \times \frac{\frac{2\lambda\rho}{3} + \frac{4\rho^2}{9}}{(4\pi)^2} = \frac{2\lambda\rho + \frac{4\rho^2}{3}}{(4\pi)^2}.$$

(On the right-hand side, the prefactor 3 is due to the fact that the four-point function we have calculated is a correction to $\rho/3$, and must therefore be multiplied by 3 in order to obtain a correction to ρ itself.)

16.f From the above results, we have

$$\mu\partial_\mu\lambda = \beta_\lambda, \quad \mu\partial_\mu\rho = \beta_\rho,$$

$$\mu\partial_\mu\left(\frac{\rho}{\lambda}\right) = \frac{\beta_\rho}{\lambda} - \frac{\rho\beta_\lambda}{\lambda^2} = -\frac{\rho}{3}\left(\frac{3 - \frac{4\rho}{\lambda} + \frac{\rho^2}{\lambda^2}}{(4\pi)^2}\right) = -\frac{\rho}{3(4\pi)^2}\left(\frac{\rho}{\lambda} - 1\right)\left(\frac{\rho}{\lambda} - 3\right).$$

From the factorized form of the right-hand side in the preceding equation, the fixed points are

$$\frac{\rho}{\lambda} = 0, 1, 3.$$

To decide whether they are attractive or repulsive, we just need to determine the sign of the right-hand side of this equation in the various intervals between the fixed points. Figure 1.5 shows that $\rho/\lambda = 1$ is an attractive fixed point in the infrared (and repulsive in the ultraviolet).

17. Solution of the Running Equation at Two Loops Consider a theory, such as QCD, in which the β function has the following perturbative expansion up to two loops: $\beta(g) = \beta_0 g^3 + \beta_1 g^5 + \mathcal{O}(g^7)$, with $\beta_0 < 0$. Show that the scale dependence of the coupling is given by

$$g(\mu^2) = -\frac{1}{\beta_0 \ln\left(\frac{\mu^2}{\Lambda^2}\right)}\left(1 + \frac{\beta_1}{\beta_0^2}\frac{\ln\ln\left(\frac{\mu^2}{\Lambda^2}\right)}{\ln\left(\frac{\mu^2}{\Lambda^2}\right)}\right) + \mathcal{O}\left(\ln^{-2}\left(\frac{\mu^2}{\Lambda^2}\right)\right).$$

Hint: recall that Λ is defined as the infrared scale where the coupling constant becomes infinite.

Given the β function, the coupling constant evolves according to $\mu\partial_\mu g = \beta$. This can be turned into a differential equation for the scale dependence of g^2:

$$\mu\frac{\partial g^2}{\partial \mu} = 2\beta_0 g^4(1 + \underbrace{\frac{\beta_1}{\beta_0}}_{\equiv b \sim g^2} g^2),$$

which can be rewritten as

$$\frac{dg^2}{g^4(1+bg^2)} = \beta_0 \frac{d\mu^2}{\mu^2}.$$

To be definite, we integrate this differential equation between a coupling $g^2(\mu^2)$ at the scale μ^2 and a coupling $g^2(\Lambda^2) = \infty$ at the scale Λ^2. In other words, we define Λ to be the scale at which the running coupling becomes infinite. Note that this scale always exists if $\beta_0, \beta_1 < 0$, as is the case in QCD when the number of quark flavors is not too large (see Problem 47). In the following integral, it is legitimate to integrate up to $g^2 = \infty$ since the denominator $1 + bg^2$ does not vanish if $b > 0$. We get

$$\int_{g^2(\mu^2)}^{+\infty} \frac{dg^2}{g^4(1+bg^2)} = \beta_0 \int_{\mu^2}^{\Lambda^2} \frac{d\mu^2}{\mu^2} = -\beta_0 \ln\left(\frac{\mu^2}{\Lambda^2}\right).$$

In order to evaluate the integral on the left-hand side, we use

$$\int_{g^2(\mu^2)}^{+\infty} \frac{dg^2}{g^4(1+\underbrace{bg^2}_{x})} = b\int_{bg^2(\mu^2)}^{+\infty} \frac{dx}{x^2(1+x)} = \frac{1}{g^2(\mu^2)} + b\ln\left(\frac{b\,g^2(\mu^2)}{1+b\,g^2(\mu^2)}\right).$$

Therefore, we have

$$\frac{1}{g^2(\mu^2)} + b\ln\left(\frac{b\,g^2(\mu^2)}{1+b\,g^2(\mu^2)}\right) = -\beta_0 \ln\left(\frac{\mu^2}{\Lambda^2}\right).$$

Now, we must invert this equation in order to express $g^2(\mu^2)$ in terms of the logarithm that appears on the right-hand side. Unfortunately, this cannot be done in closed form, and we must perform some kind of expansion. Let us denote by $g_0^2(\mu^2)$ the solution obtained when keeping only the one-loop β function,

$$g_0^2(\mu^2) = -\frac{1}{\beta_0 \ln\left(\frac{\mu^2}{\Lambda^2}\right)},$$

and then write the full g^2 as

$$g^2(\mu^2) = g_0^2(\mu^2) + g_1^2(\mu^2).$$

When solving for g_1^2, we may replace g^2 by g_0^2 in the term in $b\ln(\cdot)$ because this term is formally of higher order. Therefore, we have

$$\frac{1}{g_0^2(\mu^2)}\left(1 - \frac{g_1^2(\mu^2)}{g_0^2(\mu^2)} + \cdots\right) \approx -\beta_0\,\ell - b\ln\left(\frac{b\,g_0^2(\mu^2)}{1 + b\,g_0^2(\mu^2)}\right),$$

where we denote $\ell \equiv \ln\left(\frac{\mu^2}{\Lambda^2}\right)$ for compactness. This leads to

$$\begin{aligned} g_1^2(\mu^2) &\approx b\,g_0^4(\mu^2)\ln\left(\frac{b\,g_0^2(\mu^2)}{1 + b\,g_0^2(\mu^2)}\right) \\ &= -\frac{\beta_1}{\beta_0^3\ell^2}\Big(\ln(\ell) + \underbrace{\ln\left(\tfrac{\beta_0^2}{\beta_1} + \ell^{-1}\right)}_{\text{const } + \mathcal{O}(\ell^{-1})}\Big) \\ &= -\frac{\beta_1\ln(\ell)}{\beta_0^3\ell^2} + \mathcal{O}(\ell^{-2}). \end{aligned}$$

Therefore, we have

$$g(\mu^2) = -\frac{1}{\beta_0\ln\left(\frac{\mu^2}{\Lambda^2}\right)}\left(1 + \frac{\beta_1}{\beta_0^2}\frac{\ln\ln\left(\frac{\mu^2}{\Lambda^2}\right)}{\ln\left(\frac{\mu^2}{\Lambda^2}\right)}\right) + \mathcal{O}\Big(\ln^{-2}\left(\tfrac{\mu^2}{\Lambda^2}\right)\Big).$$

CHAPTER 2

Functional Methods

Introduction

After reviewing some basic principles of quantum field theory in Chapter 1, we now turn to a series of problems exploring various aspects of functional methods. It lacks a robust mathematical foundation in the case of interacting theories (although the situation in this respect is no better within the canonical formalism), but the formulation of QFT in terms of path integrals considerably simplifies many manipulations that would otherwise be extremely tedious because of the necessity to keep track of the ordering of operators.

In addition to the conventional representation of expectation values of time-ordered products of field operators and their generating functionals in terms of path integrals, we also briefly discus in this chapter the worldline representation for propagators and for one-loop effective actions, which provides an alternative point of view on quantization. Moreover, these ideas are not limited to ordering products of field operators, and can be useful for managing products of other types of non-commuting objects.

Bosonic Path Integral

The vacuum expectation value of a time-ordered product of field operators has the following representation as a path integral:

$$\begin{aligned}\big\langle 0_{\rm out}\big|\mathrm{T}\,\phi(x_1)\cdots\phi(x_n)\big|0_{\rm in}\big\rangle &= \int\big[D\phi(x)D\Pi(x)\big]\;\phi(x_1)\cdots\phi(x_n)\;e^{i\int d^4x\,(\Pi\dot\phi-\mathcal{H})}\\ &= \int\big[D\phi(x)\big]\;\phi(x_1)\cdots\phi(x_n)\;e^{i\,S[\phi]}. \qquad (2.1)\end{aligned}$$

The Hamiltonian version in the first line is always valid, while the Lagrangian version in the second line assumes a Hamiltonian quadratic in the momenta (with a field independent prefactor). Although the same symbol is used for simplicity of notation, ϕ is an operator on the left-hand side, while it is a commuting number on the right-hand side. By infinitesimally shifting the Hamiltonian, $\mathcal{H} \to \mathcal{H}(1 - i0^+)$, one obtains a vacuum expectation value $\langle 0_{\rm out}| \cdots |0_{\rm in}\rangle$ without imposing any ad hoc boundary condition on the integration variables. The path integral representation of the generating functional reads

$$Z[j] = \int [D\phi(x)]\; e^{i(S[\phi] + \int d^4x\, j\phi)}. \tag{2.2}$$

Most of the operations one can perform on ordinary integrals (changes of variable, Fourier transform, etc.) can be extended to path integrals.

Grassmann Variables

In order to extend the path integral representation to the case of fermionic correlation functions, we need anti-commuting "numbers," known as *Grassmann variables.* A set $\{\psi_1, \ldots, \psi_n\}$ of n Grassmann numbers is defined to satisfy $\{\psi_i, \psi_j\} = 0$. This implies that functions of these Grassmann numbers have a finite Taylor series, whose highest-order term reads

$$f(\psi_1, \ldots, \psi_n) = \cdots + f_n\, \psi_1 \cdots \psi_n. \tag{2.3}$$

Differentiation is defined by $\partial_{\psi_j} \psi_i = \delta_{ij}$, with the additional property that derivatives also anti-commute. Integrals of functions of Grassmann variables (also known as *Berezin integrals*) are defined by the following two axioms:

$$\int d\psi\; 1 = 0, \qquad \int d\psi\; \psi = 1, \tag{2.4}$$

in addition to linearity. For the above function, we have

$$\int d\psi_n d\psi_{n-1} \cdots d\psi_1\; f(\psi_1, \ldots, \psi_n) = f_n. \tag{2.5}$$

Fermionic Path Integral

Fermionic correlation functions can be represented as path integrals in terms of Grassmann variables as follows:

$$\begin{aligned} &\langle 0_{\rm out}| T\, \overline{\psi}(x_1) \cdots \overline{\psi}(x_m)\, \psi(y_1) \cdots \psi(y_n) |0_{\rm in}\rangle \\ &\quad = \int [D\psi(x) D\overline{\psi}(x)]\; \overline{\psi}(x_1) \cdots \overline{\psi}(x_m)\, \psi(y_1) \cdots \psi(y_n)\, e^{iS[\overline{\psi}, \psi]}. \end{aligned} \tag{2.6}$$

In this representation, $\psi(x)$ is a complex-valued Grassmann integration variable (in practice, this means that the symbols ψ and $\overline{\psi}$ can be manipulated as if they were independent).

Functional Determinants

Path integrals of Gaussian functionals in general produce a determinant, which can be safely ignored when it is independent of physically relevant quantities (e.g., the other fields of the problem). When we need to keep this determinant, Gaussian integrals are given by

$$\text{Real scalar case (M symmetric):} \qquad \int [D\phi]\, e^{\frac{1}{2}\phi^t M \phi} = \frac{1}{\sqrt{\det M}},$$

$$\text{Complex scalar case (M Hermitian):} \qquad \int [D\phi D\phi^*]\, e^{\phi^\dagger M \phi} = \frac{1}{\det M},$$

$$\text{Real Grassmann case (M antisymmetric):} \qquad \int [D\psi]\, e^{\frac{1}{2}\psi^t M \psi} = \sqrt{\det M},$$

$$\text{Complex Grassmann case (M anti-Hermitian):} \qquad \int [D\psi D\overline{\psi}]\, e^{\overline{\psi}^t M \psi} = \det M. \quad (2.7)$$

Note that the right-hand side of each of these formulas is up to an unwritten (and in general infinite) constant prefactor, which can always be ignored.

The operator M may result from expanding the fields about some external field. In the bosonic case, this corresponds to M of the form $M \equiv \square - V''(\varphi)$, where V is the interaction potential and φ an external field. For such an operator M, the logarithm of the determinant of M is the sum of all the one-loop graphs with an arbitrary number of insertions of the external field.

Since the operator M in the above integrals has in general an unbounded spectrum, its determinant has an ultraviolet divergence. *Zeta function regularization* consists in defining the zeta function of M from its eigenvalues λ_n, $\zeta_M(s) \equiv \sum_n \lambda_n^{-s}$, and analytically continuing it for s in the complex plane. Like the Riemann zeta function, it is in general defined almost everywhere except at a finite number of poles. The determinant of M is obtained as $\ln \det M = -\zeta_M'(0)$.

Quantum Effective Action

The *quantum effective action* $\Gamma[\varphi]$ is the functional of the field that reproduces the full theory solely from tree diagrams (with edges and vertices defined by the derivatives of the functional $\Gamma[\varphi]$, $\delta^n\Gamma/\delta\varphi(x_1)\cdots\delta\varphi(x_n)$). It is related to the generating functional of connected correlation functions, $W[j] \equiv \ln Z[j]$, as follows:

$$i\Gamma[\varphi] = W[j_\varphi] - i\int d^4x\, j_\varphi(x)\varphi(x), \qquad \frac{\delta\Gamma}{\delta\varphi} + j_\varphi = 0. \quad (2.8)$$

(The relationship between φ and j_φ may be viewed as a kind of *quantum equation of motion* that generalizes the classical Euler–Lagrange equation of motion.) The second derivative, $\delta^2\Gamma/\delta\varphi(x_1)\delta\varphi(x_2)$, is the inverse of the exact propagator of the theory. An important property is that the objects $\delta^n\Gamma/\delta\varphi(x_1)\cdots\delta\varphi(x_n)$ are made only of one-particle-irreducible diagrams, i.e., diagrams that remain connected after cutting any single propagator.

The quantum effective action inherits all the linearly realized symmetries of the classical action (i.e., transformations where the infinitesimal variation of the field is linear in the field). This is the case for Lorentz invariance, Abelian gauge symmetry, etc. But non-linearly realized symmetries (e.g., BRST symmetry) are not de facto symmetries of $\Gamma[\varphi]$.

The one-loop contribution to the quantum effective action is obtained by expanding the action to quadratic order about φ and performing the resulting Gaussian integral. In a real scalar field theory, this leads to

$$i\Gamma_{\text{1-loop}}[\varphi] = \ln\left(\left(\det\left(\square + m^2 + V''(\varphi)\right)\right)^{-\frac{1}{2}}\right) = -\tfrac{1}{2}\,\text{tr}\,\ln\left(\square + m^2 + V''(\varphi)\right). \tag{2.9}$$

More generally, the quantum effective action at one loop is given by a determinant (raised to the power $-1, -\frac{1}{2}, +\frac{1}{2}, +1$ depending on the nature of the field running inside the loop; see Eqs. (2.7)) of the second derivative of the action with respect to the field running in the loop, evaluated at the external field φ.

Euclidean Path Integral and Statistical Field Theory

By viewing the canonical density operator, $\rho \equiv \exp(-\beta\mathcal{H})$ (β is the inverse temperature, in units where the Boltzmann constant is equal to 1), as a time evolution operator for an imaginary time shift $-i\beta$, it is possible to express the partition function $Z \equiv \text{tr}\,(\rho)$ as a path integral. The simplest expression involves a Euclidean path integral,

$$Z = \int_{\phi(0,x)=\phi(\beta,x)} [D\phi(x)]\; e^{-S_E[\phi]}, \tag{2.10}$$

where S_E is the Euclidean action. This path integral is restricted to fields that are β-periodic (anti-periodic for fermions) in Euclidean time $\tau \equiv ix^0$. For a scalar field theory, the Euclidean action reads

$$S_E[\phi] = \int_0^\beta d\tau d^3x\, \left\{\tfrac{1}{2}(\partial_\tau\phi)^2 + \tfrac{1}{2}(\nabla\phi)^2 + \tfrac{1}{2}m^2\phi^2 + V(\phi)\right\}. \tag{2.11}$$

This representation leads to a perturbative expansion of $\ln\, Z$ that can be arranged in terms of the same Feynman diagrams as perturbation theory in Minkowski space. The only difference is the substitution of the continuous energy k_0 by a discrete *Matsubara frequency*, $k_0 \to i\omega_n$, with $\omega_n \equiv 2\pi n/\beta\ \ (n \in \mathbb{Z})$. The bare scalar propagator reads

$$G_0(\omega_n, p) = \frac{1}{\omega_n^2 + p^2 + m^2}. \tag{2.12}$$

(The integrals $\int dk_0/2\pi$ are replaced by summations $\beta^{-1}\sum_n$, and the sum of the Matsubara frequencies entering in a vertex should be zero.) From the partition function Z, one obtains various thermodynamical quantities as follows:

$$\begin{aligned} &\text{Energy:} && E = -\frac{\partial \ln(Z)}{\partial\beta}, \\ &\text{Entropy:} && S = \beta E + \ln(Z), \\ &\text{Free energy:} && F \equiv E - \beta^{-1}S = -\beta^{-1}\ln(Z). \end{aligned} \tag{2.13}$$

Worldline Formalism

Another type of path integral representation, known as the *worldline formalism*, is also possible for tree-level propagators in a background field or for the one-loop quantum effective action, in Euclidean spacetime. For a scalar propagator in a background φ (with interaction potential V), one starts from the following representation:

$$G(x,y) = -i \int_0^\infty dT \, \big\langle y \big| e^{-T(\square+m^2+V''(\varphi))} \big| x \big\rangle, \tag{2.14}$$

and then interprets the exponential as an evolution operator in the variable T in order to rewrite it as a path integral,

$$G(x,y) = -i \int_0^\infty dT \int_{\substack{z(0)=x \\ z(T)=y}} [Dz(\tau)] \; e^{-\int_0^T d\tau \, (\frac{1}{4}\dot{z}^2(\tau)+m^2+V''(\varphi(z(\tau))))}. \tag{2.15}$$

In this formula, $z^\mu(\tau)$ is the worldline of a fictitious point particle that starts at the point x at $T = 0$ and ends at the point y at $T = +\infty$. Thus, this representation provides a rather intuitive picture in terms of a particle exploring all the possible worldlines connecting the endpoints of a propagator. A similar representation exists for the one-loop quantum effective action,

$$\Gamma_{\text{1-loop}}[\varphi] = \frac{1}{2} \int_0^\infty \frac{dT}{T} \int_{z(0)=z(T)} [Dz(\tau)] \; e^{-\int_0^T d\tau \, (\frac{1}{4}\dot{z}^2(\tau)+m^2+V''(\varphi(z(\tau))))}. \tag{2.16}$$

For a charged scalar particle moving through an external electromagnetic potential $\mathcal{A}$, this formula becomes

$$\Gamma_{\text{1-loop}}[\mathcal{A}] = \int_0^\infty \frac{dT}{T} \int_{z(0)=z(T)} [Dz(\tau)] \; e^{-\int_0^T d\tau \, (\frac{1}{4}\dot{z}^2(\tau)+ie\dot{z}(\tau)\cdot\mathcal{A}(z(\tau))+m^2)}. \tag{2.17}$$

It is also possible to represent in a similar fashion a loop with a Dirac fermion or a gluon, but this requires an additional ordering for the non-commuting internal degrees of freedom. This can generally be achieved with an additional path integral over Grassmann variables.

About the Problems of this Chapter

- **Problem 18** addresses the quantization of a field theory in which the classical Hamiltonian is not separable, i.e., is not the sum of a function of momenta and a function of the coordinates. *Weyl quantization* is introduced and its main properties are studied, as well as its consequence on the path integral.
- The Legendre transform that relates the quantum effective action Γ to the generating functional of connected correlation functions can be obtained from the request that Γ shall reproduce the full theory solely from trees. In **Problem 19**, we provide a combinatorial interpretation of the Legendre transform that illustrates its action at the level of Feynman graphs.
- **Problem 20** introduces the *coherent state path integral*, by which one may resolve the identity in terms of coherent states instead of states of fixed coordinates or fixed momenta. This formalism is used in **Problem 21** in an alternative, more intuitive, derivation of the bosonic path integral.
- In **Problem 22**, the concept of coherent state is extended to a spin, where it also has the interpretation of a "nearly classical state" (i.e., loosely speaking, an arrow pointing in some direction of three-dimensional space). A path integral expression is derived for the transition amplitude between two such spin coherent states, that exhibits the so-called *Berry phase*, or geometrical phase.
- As an illustration of *zeta function regularization*, **Problem 23** derives a regularized expression of the determinant of the Klein–Gordon operator.
- In **Problem 24**, we derive the *Casimir energy* between two parallel walls in scalar field theory by the traditional approach, which consists in computing the dependence of the zero-point vacuum energy on the separation between the walls. In **Problem 25**, we reconsider this calculation in the spirit of the present chapter, and derive the Casimir energy as the free energy resulting from an explicit modeling of the field–wall interaction. This also provides a concrete application of zeta function regularization.
- The *Gross–Neveu model*, a model of spontaneous chiral symmetry breaking, is introduced in **Problem 26**. In addition to its phenomenological interest, this illustrates a standard technique called *bosonization*, by which a local fermionic interaction may be removed via the introduction of an extra auxiliary bosonic field.
- **Problem 27** derives the *D'Hoker–Gagné formula*, which provides a fermionic path integral representation for the trace of a Wilson line in $\mathrm{SU}(N)$. This illustrates the fact that path integrals may be used for ordering any type of non-commuting objects.
- In **Problem 28**, we introduce the framework of *stochastic quantization*, in which Euclidean correlation functions may be obtained as noise averages of products of fields that obey a Langevin equation with Gaussian white noise. These ideas are explored further in **Problem 29**, in the case of a non-Euclidean toy model, in order to introduce the concept (and problems) of *complex Langevin equations*.

18. Weyl Quantization Weyl's prescription for the quantization of a general classical quantity $f(q,p)$ is

$$F(Q,P) \equiv \int \frac{dp\,dq\,d\mu\,d\nu}{(2\pi)^2}\, f(q,p)\; e^{i(\mu(Q-q)+\nu(P-p))},$$

where Q, P are the position and momentum operators, while q, p are the corresponding classical variables. (This is known as Weyl mapping.)

18.a Show that, for a separable quantity $f(q,p) = a(q) + b(p)$, this definition simply leads to $F(Q,P) = a(Q) + b(P)$.

18.b Show that $(\alpha q + \beta p)^n$ is mapped to $(\alpha Q + \beta P)^n$.

18.c Calculate the Weyl mapping of qp.

18.d Show that the inverse of Weyl mapping is the Wigner transform:

$$f(q,p) \equiv \int ds\; e^{ips}\, \big\langle q - \tfrac{s}{2}\big|F(Q,P)\big|q + \tfrac{s}{2}\big\rangle.$$

18.e Show that, in the path integral, a non-separable Hamiltonian can be handled by evaluating the Hamiltonian at the midpoints of the path discretization.

18.a A first remark that will prove useful in the following is to write the kernel of the Weyl mapping in a factorized form using the Baker–Campbell–Hausdorff formula:

$$e^{i(\mu Q+\nu P)} = e^{i\mu Q}\, e^{i\nu P}\, e^{i\frac{\mu\nu}{2}}.$$

Consider now a function $a(q)$ of the classical coordinate. Its Weyl mapping is given by

$$\begin{aligned} A(Q,P) &= \int \frac{dp\,dq\,d\mu\,d\nu}{(2\pi)^2}\, a(q)\; e^{-i(\mu q+\nu p)} e^{i\mu Q}\, e^{i\nu P}\, e^{i\frac{\mu\nu}{2}} \\ &= \int \frac{d\mu}{2\pi}\, \widetilde{a}(\mu)\; e^{i\mu Q} = a(Q). \end{aligned}$$

In the second line, $\widetilde{a}(\mu)$ is the Fourier transform of $a(q)$. The integral over μ is the inverse Fourier transform, evaluated for an operator argument Q. A heuristic justification of the fact that the result is the operator $a(Q)$ consists in acting on an eigenstate $|q\rangle$ of the position operator (i.e., $Q|q\rangle = q|q\rangle$):

$$\int \frac{d\mu}{2\pi}\, \widetilde{a}(\mu)\; e^{i\mu Q}\, |q\rangle = \int \frac{d\mu}{2\pi}\, \widetilde{a}(\mu)\; e^{i\mu q}\, |q\rangle = a(q)\,|q\rangle = a(Q)\,|q\rangle.$$

Thus, for a function that depends only on the coordinate, the Weyl prescription simply amounts to replacing the classical variable q by the quantum operator Q in the function. A completely analogous result holds for functions that depend only on the momentum p (the proof is identical). Thus, Weyl's prescription amounts to the usual substitution $q \to Q, p \to P$ for separable Hamiltonians.

18.b Let us now see the changes for non-separable Hamiltonians. Consider first the function $f(q,p) \equiv (\alpha q + \beta p)^n$. Its Weyl mapping reads

$$F(Q,P) = \int \frac{dpdqd\mu d\nu}{(2\pi)^2} (\alpha q + \beta p)^n \, e^{-i(\mu q + \nu p)} e^{i(\mu Q + \nu P)}$$
$$= \int \frac{dpdqd\mu d\nu}{(2\pi)^2} \left[i^n (\alpha \partial_\mu + \beta \partial_\nu)^n \, e^{-i(\mu q + \nu p)} \right] e^{i(\mu Q + \nu P)}$$
$$= \int d\mu d\nu \, \delta(\mu)\delta(\nu)(-i)^n (\alpha \partial_\mu + \beta \partial_\nu)^n \, e^{i(\mu Q + \nu P)}.$$

At this stage, the evaluation of the derivatives is complicated by the fact that Q and P do not commute. We can avoid this problem by introducing new variables ρ, σ defined by

$$\mu \equiv \alpha\rho + (1-\alpha)\sigma, \quad \nu \equiv \beta(\rho - \sigma).$$

Under this transformation, we easily check that

$$d\mu d\nu \, \delta(\mu)\delta(\nu) = d\rho d\sigma \, \delta(\rho)\delta(\sigma),$$
$$\alpha\partial_\mu + \beta\partial_\nu = \partial_\rho, \quad \mu Q + \nu P = \rho(\alpha Q + \beta P) + \sigma((1-\alpha)Q - \beta P).$$

In terms of these new integration variables, we can write

$$F(Q,P) = \int d\rho d\sigma \, \delta(\rho)\delta(\sigma)(-i)^n \partial_\rho^n \, e^{i(\rho(\alpha Q + \beta P) + \sigma((1-\alpha)Q - \beta P))}$$
$$= \int d\rho \, \delta(\rho)(-i)^n \partial_\rho^n \, e^{i\rho(\alpha Q + \beta P)} = (\alpha Q + \beta P)^n.$$

18.c In order to obtain the Weyl mapping of qp, we use the fact that $qp = \frac{1}{2}((q+p)^2 - q^2 - p^2)$. The previous results imply the following mappings:

$$q^2 \to Q^2, \quad p^2 \to P^2, \quad (q+p)^2 \to (Q+P)^2 = Q^2 + P^2 + QP + PQ.$$

Therefore, the Weyl mapping of qp is

$$qp \to \tfrac{1}{2}(QP + PQ).$$

This a simple example of a more general result: for any monomial $q^m p^n$, Weyl mapping produces a sum of all the possible orderings of m Q's and n P's, divided by the number of terms. Note that there is not a unique representation of these formulas, since seemingly different but nevertheless equivalent expressions can be obtained by using the canonical commutation relation $[Q, P] = i$.

18.d Let us now consider the inverse problem: find $f(q,p)$ given the quantum operator $F(Q,P)$. Consider the following integral, called the Wigner transform of the operator $F(Q,P)$:

$$\int ds \, e^{ips} \left\langle q - \tfrac{s}{2} \right| F(Q,P) \left| q + \tfrac{s}{2} \right\rangle$$
$$= \int \frac{dsdrdkd\mu d\nu}{(2\pi)^2} \, e^{ips} e^{-i(\mu r + \nu k)} \, f(r,k) \, \underbrace{\left\langle q - \tfrac{s}{2} \right| e^{i\mu Q}}_{e^{i\mu(q-\frac{s}{2})}\left\langle q - \frac{s}{2}\right|} e^{i\nu P} e^{i\frac{\mu\nu}{2}} \left| q + \tfrac{s}{2} \right\rangle.$$

On the right-hand side, we have just replaced F by its expression as the Weyl mapping of f. Then, we insert on the right of the operator $\exp(i\nu P)$ an indentity operator expressed as a complete sum over momentum eigenstates. This leads to

$$\int ds\, e^{ips} \left\langle q-\tfrac{s}{2}\middle|F(Q,P)\middle|q+\tfrac{s}{2}\right\rangle = \int \frac{ds dr d\ell dk d\mu d\nu}{(2\pi)^3}\, f(r,k)$$
$$\times\, e^{ips} e^{-i(\mu r+\nu k)} e^{i\mu(q-\frac{s}{2})} e^{i\frac{\mu\nu}{2}} \underbrace{\left\langle q-\tfrac{s}{2}\middle|e^{i\nu P}\middle|\ell\right\rangle\left\langle \ell\middle|q+\tfrac{s}{2}\right\rangle}_{e^{i\nu\ell}e^{-i\ell s}}.$$

Then, the integral over ν gives a factor $2\pi\,\delta(\ell-k+\frac{\mu}{2})$, and the integral over s gives a factor $2\pi\,\delta(\ell-p+\frac{\mu}{2})$. With these delta functions, we can effortlessly perform the integrals over k and ℓ to obtain

$$\int ds\, e^{ips} \left\langle q-\tfrac{s}{2}\middle|F(Q,P)\middle|q+\tfrac{s}{2}\right\rangle = \int \frac{dr d\mu}{2\pi}\, f(r,p)\, e^{i\mu(q-r)} = f(q,p).$$

This completes the proof of the fact that the Wigner transform is the inverse of Weyl mapping.

18.e Consider now a general classical Hamiltonian $\mathcal{H}(q,p)$, possibly non-separable. In order to derive the path integral representation of a transition amplitude, we need to evaluate an elementary matrix element such as

$$\left\langle q_2\middle|e^{-i\Delta H(Q,P)}\middle|q_1\right\rangle = \left\langle q_2\middle|\Big(1-i\Delta H(Q,P)+\mathcal{O}(\Delta^2)\Big)\middle|q_1\right\rangle.$$

The zeroth-order term is trivial. For our purposes, it is sufficient to calculate the term of order Δ^1. Indeed, since $\Delta \sim N^{-1}$ (N being the number of time slices), the terms of order Δ^2 or higher disappear in the limit $N\to\infty$ when all time slices are combined. Therefore, let us consider

$$\left\langle q_2\middle|H(Q,P)\middle|q_1\right\rangle = \int \frac{dp dq d\mu d\nu}{(2\pi)^2}\, \mathcal{H}(q,p)\, e^{-i(\mu q+\nu p)}\, e^{i\frac{\mu\nu}{2}} \underbrace{\left\langle q_2\middle|e^{i\mu Q}\,e^{i\nu P}\middle|q_1\right\rangle}_{e^{i\mu q_2}\int \frac{dk}{2\pi}\, e^{i\nu k}\, e^{ik(q_2-q_1)}}$$
$$= \int \frac{dp dq d\nu}{2\pi}\, \mathcal{H}(q,p)\, e^{-i\nu p}\, \delta(q_2-q+\tfrac{\nu}{2})\,\delta(\nu+q_2-q_1)$$
$$= \int \frac{dp}{2\pi}\, \mathcal{H}(\tfrac{q_1+q_2}{2},p)\, e^{ip(q_2-q_1)}.$$

All the integrals are trivial, and the final outcome is an expression where the classical Hamiltonian is evaluated at the midpoint coordinate $\frac{q_1+q_2}{2}$ instead of q_1 or q_2.

19. Combinatorial Interpretation of the Legendre Transform This problem provides a combinatorial approach to the Legendre transform, which leads to a simple diagrammatic interpretation. Consider a formal power series $\Gamma(X) \equiv \frac{1}{2}\Gamma^{(2)}_{ab}X_aX_b + \sum_{n>2}\frac{1}{n!}\Gamma^{(n)}_{a_1\cdots a_n}X_{a_1}\cdots X_{a_n}$, with no constant or linear terms and with $\Gamma^{(2)}_{ab}$ invertible. Viewing this series as an "effective action," denote by $W(J)$ the formal series of trees constructed with the "Feynman rules" defined by this action and a source J_a coupled to the field X_a. We wish to prove combinatorially that $\Gamma(X(J)) + J\,X(J) = W(J)$, where $X(J)$ and J are linked by $J_a + \partial\Gamma/\partial X_a = 0$. (The derivative $\partial\Gamma/\partial X$ is a well-defined operation in the ring of formal series, which does not require any notion of convergence.)

19.a Given a tree diagram $\mathcal{G}$, show that $n_V(\mathcal{G}) - n_E(\mathcal{G}) = 1$, where n_V and n_E are the number of vertices (including the sources J) and edges of $\mathcal{G}$, respectively.

19.b Consider the m-th derivative $\partial^m W(J)/\partial J_{a_1} \cdots \partial J_{a_m}\big|_{J=0}$. Show that it is the sum of all the tree graphs with m external points $a_1, a_2, \ldots, a_m$, with edges $\mathcal{E} \equiv -(\Gamma^{(2)})^{-1}$ and vertices $\mathcal{V}_n \equiv \Gamma^{(n)}$. Explain why every graph has a symmetry factor equal to 1.

19.c Consider now the m-th derivative of $\Gamma(X(J)) + JX(J)$. Show that it is represented by the same tree graphs as the derivative of $W(J)$.

19.d Show that each of these graphs occurs $n_V(\mathcal{G}) - n_E(\mathcal{G})$ times. *Hint: show that the derivatives of* $\Gamma(X(J)) + JX(J)$ *can be decomposed as*

$$\frac{\partial^m \Gamma(X(J)) + JX(J)}{\partial J_{a_1} \cdots \partial J_{a_m}}\bigg|_{J=0} = t_1^{(1)} + \sum_{n\geq 3} \frac{1}{n!} [\mathcal{V}_n] t_1^{(n)} \cdots t_n^{(n)} - \frac{1}{2} t_1^{(2)} \mathcal{E}^{-1} t_2^{(2)},$$

where the $t_p^{(n)}$ *are trees. Then, explain why the first two terms produce each tree graph* $n_V(\mathcal{G})$ *times, and the third term produces this graph* $n_E(\mathcal{G})$ *times.*

19.a The functional $W(J)$ is defined as the formal series of *trees* constructed with the rules allowed by the action $\Gamma(X)$, and all the "branches" of these trees are terminated by a source J:

$$W(J) = \sum_{\text{trees}}$$

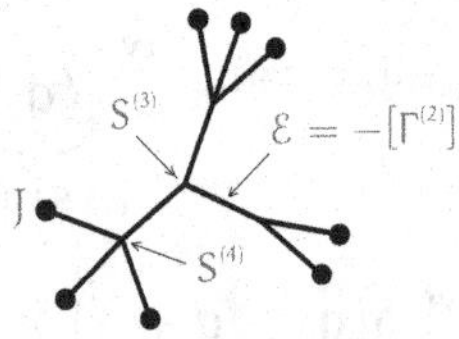

(In quantum field theory, these graphs would give the connected tree-level vacuum-to-vacuum transition amplitude in the presence of an external source J.) Let us denote by $n_V^{(i)}$ the number of vertices of valence i ($i \geq 3$) in such a graph, by n_E its number of edges, and by n_J the number of sources it contains. Since this a tree diagram, we have

$$0 = n_E + 1 - \underbrace{\Big(n_J + \sum_{i\geq 3} n_V^{(i)}\Big)}_{\equiv n_V}.$$

(This is the standard formula for the number of loops in a graph.) Although these graphs are not Feynman graphs but just graphical representations of a formal series, the same type of reasoning applies if we imagine that the lines carry a "momentum." This equation counts the number of unconstrained momenta in the graph, starting from the number of edges and subtracting 1 for each vertex or each source (since the source controls the momentum that flows from it into the graph). As usual, the $+1$ is due to the fact that these constraints are not independent, but obey exactly one linear relationship corresponding to overall momentum conservation.

19.b Each time we differentiate $W(J)$ with respect to a source J_a, we remove a source from the corresponding graphs, as in

$$\frac{\partial W(J)}{\partial J_a} = \sum_{\text{trees}}$$

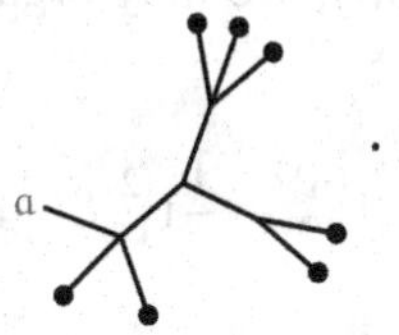

If we differentiate m times, and then set $J \equiv 0$, only the graphs that have exactly m sources contribute, and this operation gives tree graphs with m open external legs (and no source left). Note that the external lines of these graphs carry labels $a_1, \ldots, a_m$ corresponding to the labels of the sources with respect to which the differentiation was performed:

$$\left.\frac{\partial^m W(J)}{\partial J_{a_1} \cdots \partial J_{a_m}}\right|_{J=0} = \sum_{\text{trees}}$$

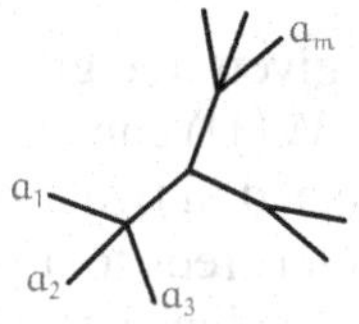

Because of these labels, these graphs have a symmetry factor equal to 1 (i.e., permuting external lines gives a different graph).

19.c The formal series $X(J)$ that satisfies $J_a + \partial\Gamma/\partial X_a = 0$ can be constructed recursively order-by-order in J, starting from the solution of the "lowest-order" equation,

$$\Gamma^{(2)}_{ab} X(J)_b + J_a = 0, \quad \text{i.e.,} \quad X(J)_a = \underbrace{-\left[\Gamma^{(2)}\right]^{-1}_{ab}}_{\equiv \mathcal{E}_{ab}} J_b.$$

We see from this expression that the edges $\mathcal{E}_{ab}$ of the graphs are given by the inverse of $-\Gamma^{(2)}_{ab}$. Moreover, for $n \geq 3$, $\Gamma^{(n)}_{a_1 \cdots a_n}$ is the expression for the vertex $\mathcal{V}_n$ of valence n, and to uniformize the notation we may view the source J_a as a valence-1 vertex $\mathcal{V}_1$. Beyond the lowest order, the series $X(J)$ is given as the sum of tree graphs with one open end and sources attached to all the other external lines:

$X(J) =$ 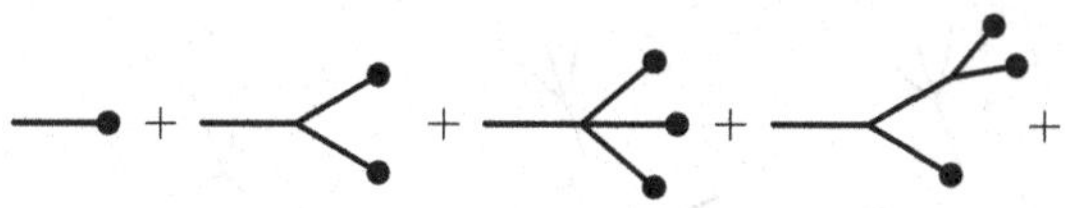 $+ \cdots.$ (2.18)

Consider now the object $A(J) \equiv \Gamma(X(J)) + J\,X(J)$. First, we have

$$\Gamma(X(J)) = \tfrac{1}{2}\Gamma^{(2)}_{ab} X(J)_a X(J)_b + \sum_{n \geq 3} \tfrac{1}{n!}\Gamma^{(n)}_{a_1 \cdots a_n} X(J)_{a_1} \cdots X(J)_{a_n}.$$

Then, $A(J)$ can be written as follows in terms of the edges $\mathcal{E}$ and vertices $\mathcal{V}_n$:

$$A(J) = \left[\mathcal{V}_1\right]_a X(J)_a + \sum_{n \geq 3} \frac{1}{n!}\left[\mathcal{V}_n\right]_{a_1 \cdots a_n} X(J)_{a_1} \cdots X(J)_{a_n} - \frac{1}{2}\left[\mathcal{E}\right]^{-1}_{ab} X(J)_a X(J)_b.$$

Our next task is to differentiate $A(J)$ multiple times with respect to J. Recall that $X(J)_a$ is the sum of trees with one labeled external leg (see (2.18)), and all other legs attached to a source. Obviously, when we differentiate the above expression m times with respect to J and then set $J = 0$, we also obtain tree graphs with exactly m labeled external legs, i.e., a contribution to the m-th derivative of $W(J)$. Formally, we may organize this derivative as follows:

$$\left.\frac{\partial^m A(J)}{\partial J_{a_1} \cdots \partial J_{a_m}}\right|_{J=0} = t_1^{(1)} + \sum_{n\geq 3} \frac{1}{n!}\left[\mathcal{V}_n\right] t_1^{(n)} \cdots t_n^{(n)} - \frac{1}{2} t_1^{(2)} \mathcal{E}^{-1} t_2^{(2)}, \tag{2.19}$$

where the $t_p^{(n)}$ are trees obtained by stripping out the contributions to $X(J)$ of all their J's. (Note that $\mathcal{V}_1$ is no longer present in this expression, since it was equal to a source and therefore disappeared in the differentiation).

19.d Let us now count how many times a given tree graph $\mathcal{G}$ with m labeled external legs (i.e., a contribution to the m-th derivative of $W(J)$) can occur in this expression. First, let us note that in (2.19), the tree graphs are constructed in two different ways. In the first two terms, a tree graph $\mathcal{G}$ is obtained by attaching smaller trees to a vertex $\mathbf{v}$ (which can be one of the external points of $\mathcal{G}$, considered as a vertex of valence 1). Generically, trees built in this way around a vertex $\mathbf{v}$ have the following expression:

$$\mathcal{G} = \frac{1}{n!}\mathbf{v}\prod_{i=1}^{n}\left[\text{Tree}\right]_i.$$

On the right-hand side, $\mathbf{v} = \mathcal{V}_n$ for $n \geq 3$ and $\mathbf{v} = 1$ for $n = 1$, and the prefactor $1/n!$ ensures that we are not overcounting when permuting the trees attached to the exposed vertex. The first two terms of (2.19) provide this kind of representation of $\mathcal{G}$, *for any possible choice of a vertex in* $\mathcal{G}$, as illustrated below:

$t_1^{(1)} =$ 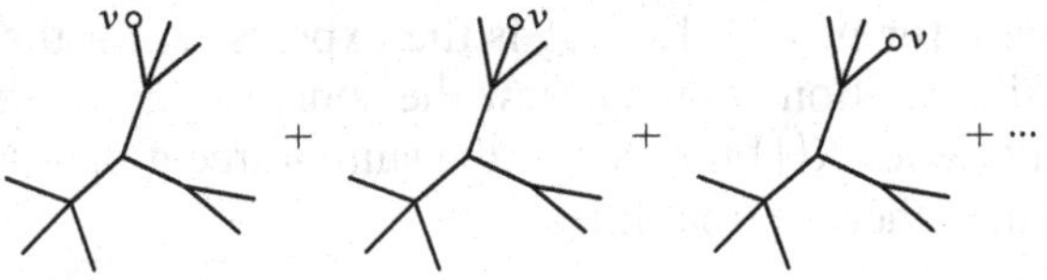

$\sum_{n\geq 3} \frac{1}{n!}\left[\mathcal{V}_n\right] t_1^{(n)} \cdots t_n^{(n)} =$ 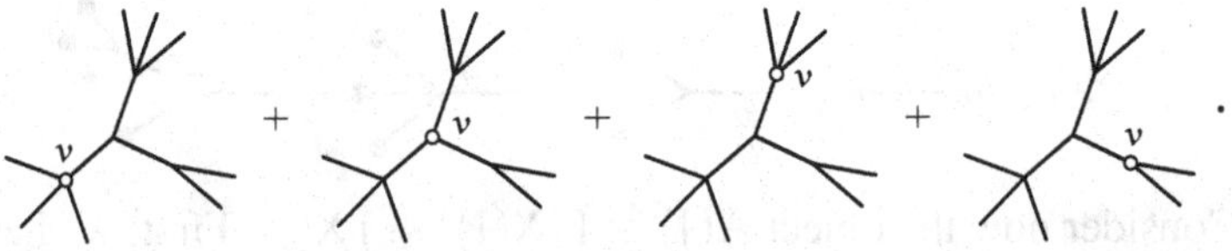

(In the diagrams on the right-hand side, we indicate by an open circle the vertex around which $\mathcal{G}$ is constructed.) Therefore, if the graph has n_v vertices, the first two terms of (2.19) are equal to $n_v \times \mathcal{G}$:

$$t_1^{(1)} + \sum_{n\geq 3} \frac{1}{n!}\left[\mathcal{V}_n\right] t_1^{(n)} \cdots t_n^{(n)} = n_v\, \mathcal{G}.$$

The other way in which a tree graph $\mathcal{G}$ appears in (2.19) is via the third term, in which an edge $\mathbf{e}$ of the graph is singled out and two trees (amputated of one of their external legs) are attached on either side of this edge. In terms of this edge, the graph can be written as

$$\mathcal{G} = \frac{1}{2} \mathrm{L}\,\mathcal{E}\,\mathrm{R} = \frac{1}{2} \underbrace{(\mathrm{L}\mathcal{E})}_{t_1^{(2)}} \mathcal{E}^{-1} \underbrace{(\mathcal{E}\mathrm{R})}_{t_2^{(2)}},$$

where L, R are the sub-trees (amputated of the external leg on the side of the exposed edge) on the left and on the right of the exposed edge. The prefactor $1/2$ is a symmetry factor that compensates for the overcounting due to swapping the two sub-trees. Such a combination occurs in the third term of (2.19), for all possible choices of the edge e:

$$\frac{1}{2} t_1^{(2)} \mathcal{E}^{-1} t_2^{(2)} =$$

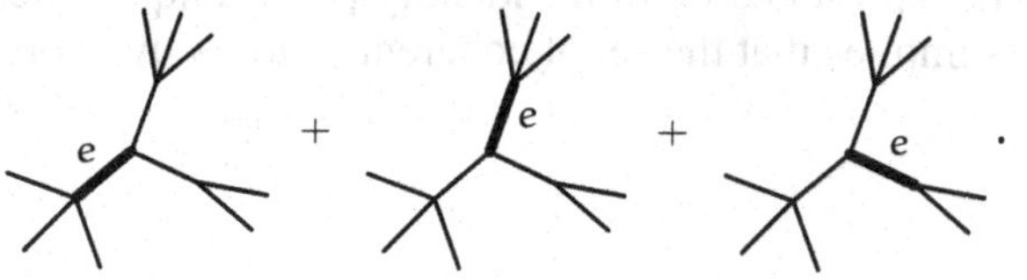

Since the exposed edge e (shown in bold on the right-hand side of the preceding equation) can be arbitrary, the third term of (2.19) is equal to $n_E \times \mathcal{G}$ if the graph $\mathcal{G}$ has n_E edges:

$$\frac{1}{2} t_1^{(2)} \mathcal{E}^{-1} t_2^{(2)} = n_E\, \mathcal{G}.$$

Since this reasoning applies to any tree graph $\mathcal{G}$, we have in fact shown that

$$\left.\frac{\partial^m A(J)}{\partial J_{a_1} \cdots \partial J_{a_m}}\right|_{J=0} = \sum_{\text{trees}} \underbrace{(n_V - n_E)}_{=1} \mathcal{G} = \sum_{\text{trees}} \mathcal{G} = \left.\frac{\partial^m W(J)}{\partial J_{a_1} \cdots \partial J_{a_m}}\right|_{J=0},$$

and the derivatives of $J\,X(J) + \Gamma(X(J))$ and of $W(J)$ are identical, which ensures that the two formal series are identical.
Source: Jackson, D. M., Kempf, A. and Morales, A. H. (2017), *J Phys A* 50: 225201.

20. Coherent State Path Integral This problem builds upon Problem 15, about coherent states. Consider a coherent state defined by

$$\big|\chi_{\text{in}}\big\rangle \equiv \exp\Big(-\frac{1}{2}\int \frac{d^3k}{(2\pi)^3 2E_k}\, |\chi(k)|^2\Big) \exp\Big(\int \frac{d^3k}{(2\pi)^3 2E_k}\, \chi(k)\, a^\dagger_{k,\text{in}}\Big) \big|0_{\text{in}}\big\rangle.$$

20.a Check that

$$a_{p,\text{in}} \big|\chi_{\text{in}}\big\rangle = \chi(p) \big|\chi_{\text{in}}\big\rangle, \quad a^\dagger_{p,\text{in}} \big|\chi_{\text{in}}\big\rangle = \Big(\frac{\chi^*(p)}{2} + (2\pi)^3 2E_p \frac{\delta}{\delta\chi(p)}\Big) \big|\chi_{\text{in}}\big\rangle.$$

20.b Show that the overlap between a coherent state and a Fock state (i.e., a state with a fully specified particle content) is given by

$$\big\langle p_1 \cdots p_{n\,\text{in}} \big| \chi_{\text{in}} \big\rangle = \exp\Big(-\frac{1}{2}\int \frac{d^3k}{(2\pi)^3 2E_k}\, |\chi(k)|^2\Big)\, \chi(p_1) \cdots \chi(p_n).$$

20.c Use this result to show that the identity operator can be written as

$$1 = \int \big[D\chi(\mathbf{k})D\chi^*(\mathbf{k})\big] \; \big|\chi_{\text{in}}\big\rangle\big\langle\chi_{\text{in}}\big|.$$

20.d By using $1 = 1^2$, find another – non-diagonal – representation of the identity in terms of coherent states, of the form

$$1 = \int \big[D\chi D\chi^* D\xi D\xi^*\big] \; F[\chi,\chi^*,\xi,\xi^*] \; \big|\chi_{\text{in}}\big\rangle\big\langle\xi_{\text{in}}\big|.$$

The fact that the representation of the identity is not unique when expressed in terms of coherent states implies that the set of coherent states is *overcomplete.*

20.a In order to determine the action of the annihilation operator $a_{\mathbf{p},\text{in}}$ on a coherent state, we must bring this operator to the right of the exponential containing the creation operators, where it will annihilate the vacuum. We have

$$\begin{aligned}
&\big[a_{\mathbf{p},\text{in}}, a^\dagger_{\mathbf{k},\text{in}}\big] = (2\pi)^3 2E_{\mathbf{p}} \delta(\mathbf{p}-\mathbf{k}),\\
&\Big[a_{\mathbf{p},\text{in}}, \Big(\int \frac{d^3\mathbf{k}}{(2\pi)^3 2E_{\mathbf{k}}}\, \chi(\mathbf{k})\, a^\dagger_{\mathbf{k},\text{in}}\Big)^n\Big] = n\chi(\mathbf{p}) \Big(\int \frac{d^3\mathbf{k}}{(2\pi)^3 2E_{\mathbf{k}}}\, \chi(\mathbf{k})\, a^\dagger_{\mathbf{k},\text{in}}\Big)^{n-1},
\end{aligned}$$

which implies the announced answer after summation on n. To check the formula for the action of a creation operator on a coherent state, we simply differentiate its definition with respect to $\chi(\mathbf{p})$, which gives

$$\frac{\delta}{\delta\chi(\mathbf{p})} \exp\Big(\int \frac{d^3\mathbf{k}}{(2\pi)^3 2E_{\mathbf{k}}}\, \chi(\mathbf{k})\, a^\dagger_{\mathbf{k},\text{in}}\Big) = \frac{a^\dagger_{\mathbf{p},\text{in}}}{(2\pi)^3 2E_{\mathbf{p}}} \exp\Big(\int \frac{d^3\mathbf{k}}{(2\pi)^3 2E_{\mathbf{k}}}\, \chi(\mathbf{k})\, a^\dagger_{\mathbf{k},\text{in}}\Big),$$

$$\frac{\delta}{\delta\chi(\mathbf{p})} \exp\Big(-\int \frac{d^3\mathbf{k}}{(2\pi)^3 4E_{\mathbf{k}}}\, |\chi(\mathbf{k})|^2\Big) = -\frac{\chi^*(\mathbf{p})}{(2\pi)^3 4E_{\mathbf{p}}} \exp\Big(-\int \frac{d^3\mathbf{k}}{(2\pi)^3 4E_{\mathbf{k}}}\, |\chi(\mathbf{k})|^2\Big).$$

20.b Let us now recall the definition of Fock states,

$$\big|\mathbf{p}_1 \cdots \mathbf{p}_{n\text{in}}\big\rangle = a^\dagger_{\mathbf{p}_1,\text{in}} \cdots a^\dagger_{\mathbf{p}_n,\text{in}} \big|0_{\text{in}}\big\rangle,$$

and that they form a complete set of states. From their definition, it is easy to calculate the overlap between two Fock states:

$$\begin{aligned}
&\big\langle \mathbf{q}_1 \cdots \mathbf{q}_{m\text{in}} \big| \mathbf{p}_1 \cdots \mathbf{p}_{n\text{in}}\big\rangle && (2.20)\\
&\quad = \big\langle 0_{\text{in}}\big|\, a_{\mathbf{q}_1,\text{in}} \cdots a_{\mathbf{q}_m,\text{in}}\, a^\dagger_{\mathbf{p}_1,\text{in}} \cdots a^\dagger_{\mathbf{p}_n,\text{in}} \big|0_{\text{in}}\big\rangle\\
&\quad = \delta_{mn} \sum_{\sigma\in\mathfrak{S}_n} (2\pi)^3 2E_{\mathbf{p}_1} \delta(\mathbf{p}_1 - \mathbf{q}_{\sigma(1)}) \cdots (2\pi)^3 2E_{\mathbf{p}_n} \delta(\mathbf{p}_n - \mathbf{q}_{\sigma(n)}). && (2.21)
\end{aligned}$$

(This result is obtained by moving in turn every annihilation operator from the left to the right, where it will eventually annihilate the vacuum state. While doing this, we collect non-zero contributions from their commutators with the creation operators. The final result is in fact very similar to Wick's theorem, and takes the form of a sum over all the possible pairings between the creation and annihilation operators.)

In order to simplify the notation, let us denote by $\mathcal{N}_\chi$ the prefactor that normalizes the coherent state:

$$\mathcal{N}_\chi \equiv \exp\Big(-\frac{1}{2}\int \frac{d^3k}{(2\pi)^3 2E_k}\,|\chi(k)|^2\Big).$$

In order to calculate the overlap of a coherent state with a Fock state, we expand the exponential containing the creation operators:

$$|\chi_{in}\rangle = \mathcal{N}_\chi \left(\sum_{n=0}^{\infty} \frac{1}{n!}\Big(\int \frac{d^3k}{(2\pi)^3 2E_k}\,\chi(k)\, a^\dagger_{k,in}\Big)^n\right)|0_{in}\rangle.$$

When contracting this with an n-particle state, the previous result ensures that only the term of order n in the exponential contributes, because this is the only one that gives matching numbers of creation and annihilation operators:

$$\begin{aligned}\langle p_1 \cdots p_{n\,in}|\chi_{in}\rangle &= \frac{\mathcal{N}_\chi}{n!}\int \frac{d^3k_1}{(2\pi)^3 2E_{k_1}} \cdots \frac{d^3k_n}{(2\pi)^3 2E_{k_n}}\,\chi(k_1)\cdots\chi(k_n)\\ &\quad\times \sum_{\sigma\in\mathfrak{S}_n} (2\pi)^3 2E_{p_1}\delta(p_1 - k_{\sigma(1)})\cdots(2\pi)^3 2E_{p_n}\delta(p_n - k_{\sigma(n)})\\ &= \mathcal{N}_\chi\,\chi(p_1)\cdots\chi(p_n).\end{aligned}$$

(To obtain the final formula, note that the integral over $k_1, \ldots, k_n$ is fully symmetric, implying that all the permutations $\sigma \in \mathfrak{S}_n$ give the same contribution, which cancels the prefactor $1/n!$.)

20.c In order to check the proposed expression for the identity operator, recall that the Fock states are a complete set of states, whose overlap is given in Eq. (2.21). Thus, it is sufficient that the proposed formula gives the same result when sandwiched between two Fock states:

$$\begin{aligned}&\langle q_1 \cdots q_{m\,in}|\Big[\int [D\chi(k)D\chi^*(k)]\;|\chi_{in}\rangle\langle\chi_{in}|\Big]|p_1 \cdots p_{n\,in}\rangle\\ &= \int [D\chi(k)D\chi^*(k)]\;\langle q_1 \cdots q_{m\,in}|\chi_{in}\rangle\langle\chi_{in}|p_1\cdots p_{n\,in}\rangle\\ &= \int [D\chi(k)D\chi^*(k)]\;\exp\Big(-\int\frac{d^3k}{(2\pi)^3 2E_k}|\chi(k)|^2\Big)\chi(q_1)\cdots\chi(q_m)\chi^*(p_1)\cdots\chi^*(p_n).\end{aligned}$$

The final integral is a Gaussian path integral (in momentum space). By means of Wick's theorem, such integrals are fully expressible in terms of the only non-zero two-point correlation function:

$$\int [D\chi(k)D\chi^*(k)]\;\exp\Big(-\int \frac{d^3k}{(2\pi)^3 2E_k}\,|\chi(k)|^2\Big)\,\chi(q)\chi^*(p) = (2\pi)^3 2E_p\delta(p-q) \tag{2.22}$$

(this equation defines the otherwise ambiguous normalization of the path integral). The general result is a sum over all the ways of pairing the χ's and the χ^*'s (thus it is non-zero only if

$m = n$, hence the factor δ_{mn}), which reproduces (2.21) and proves the stated expression for the identity. This result shows that the coherent states form a complete set of states. In particular, any state $|\psi\rangle$ can be written in the coherent state basis as follows:

$$|\psi\rangle = \int [D\chi(\mathbf{k})D\chi^*(\mathbf{k})] \underbrace{\langle\chi_{\text{in}}|\psi\rangle}_{\psi(\chi,\chi^*)} |\chi_{\text{in}}\rangle,$$

where $\psi(\chi, \chi^*)$ can be interpreted as the wavefunction of the state $|\psi\rangle$ in the coherent state representation.

20.d In fact coherent states are an *overcomplete* set of states, in the sense that any state can be obtained as a linear superposition of a strict subset of the coherent states. One way to see this overcompleteness is to consider

$$1 = 1^2 = \int [D\chi D\chi^* D\xi D\xi^*] \langle\chi_{\text{in}}|\xi_{\text{in}}\rangle |\chi_{\text{in}}\rangle\langle\xi_{\text{in}}|. \tag{2.23}$$

Note that the overlap of two coherent states is non-zero:

$$\langle\chi_{\text{in}}|\xi_{\text{in}}\rangle = \exp\Big(-\frac{1}{2}\int \frac{d^3k}{(2\pi)^3 2E_{\mathbf{k}}} \big(|\chi(\mathbf{k}) - \xi(\mathbf{k})|^2 + \xi^*(\mathbf{k})\chi(\mathbf{k}) - \chi^*(\mathbf{k})\xi(\mathbf{k})\big)\Big). \tag{2.24}$$

(This formula was obtained in Problem 15.) Therefore, Eq. (2.23) provides an alternative non-diagonal resolution of the identity operator in terms of coherent states. The existence of several manifestly distinct representations of the identity in terms of a given set of states is an indication that this set is overcomplete.

Additional Note: In the case of one-dimensional *quantum mechanics*, a coherent state is defined by a single complex number χ. In this simpler situation, it is possible to characterize more precisely when a subset of the set of coherent states is complete. Such subsets consist in picking a subset $\{\chi\} \subset \mathbb{C}$. Whether such a subset is complete or not depends crucially on the density of points χ in the complex plane. In particular, for subsets defined as a lattice of points $\{\chi_{mn} \equiv m\chi_1 + n\chi_2 | m, n \in \mathbb{Z}\}$ (with $\chi_{1,2}$ a pair of non-collinear complex numbers, defining a parallelogram of non-zero area), one has the following results:

- The subset is overcomplete if the area of elementary cells of the lattice is $< \pi$.
- The subset is also overcomplete if this area is equal to π, but this time very mildly. Removing any single element of the subset leads to a complete set, and removing any pair of elements gives an incomplete set.
- The subset is not complete if this area is $> \pi$.

Loosely speaking, since coherent states have nearly well-defined position and momentum (corresponding respectively to the real and imaginary parts of χ), a set of coherent states is complete if there is at least "one state per quantum of phase-space, $\Delta q \Delta p \sim 2\pi$". A proof of this result can be found in Bargmann, V., Butera, P., Girardello, L. and Klauder, J. R. (1970), *Rep Math Phys* 2(4): 221. (These properties are part of the more general theory of *Gabor frames*.)

21. Coherent State Path Integral (Continued) This problem follows up on Problem 20.

21.a Show that a coherent state can also be written as

$$|\chi_{in}\rangle = \exp\Big(i\int d^3x\ \big(\Pi(x)\phi_{in}(0,x) - \phi(x)\Pi_{in}(0,x)\big)\Big)\ |0_{in}\rangle,$$

with ϕ, Π two real commuting functions related to the function $\chi(k)$ that defines the coherent state. Thus, the coherent state $|\chi_{in}\rangle$ may also be denoted $|(\phi,\Pi)_{in}\rangle$.

21.b Show that

$$\begin{aligned}&\langle(\phi,\Pi)_{in}|F[\phi_{in}(0,x),\Pi_{in}(0,x)]|(\phi,\Pi)_{in}\rangle\\ &\quad = \langle 0_{in}|F[\phi_{in}(0,x)+\phi(x),\Pi_{in}(0,x)+\Pi(x)]|0_{in}\rangle = F[\phi(x),\Pi(x)].\end{aligned}$$

(The final result is valid under the assumption that F is normal-ordered.) Use this result to obtain the expectation values of $\phi_{in}(0,x)$ and $\Pi_{in}(0,x)$ in a coherent state $|(\phi,\Pi)_{in}\rangle$. *Hint: use Weyl mapping, introduced in Problem 18, in order to express* $F[\phi_{in},\Pi_{in}]$ *in terms of a classical quantity* $f[\phi,\Pi]$.

21.c Show that the identity operator can be expressed as

$$1 = \int \big[D\phi(x)D\Pi(x)\big]\ |(\phi,\Pi)_{in}\rangle\langle(\phi,\Pi)_{in}|.$$

21.d Derive a path integral formula for the transition amplitude between two such coherent states, $\langle(\phi_f,\Pi_f)_{in}|\exp\big(-iH(t_f-t_i)\big)|(\phi_i,\Pi_i)_{in}\rangle$. Specify its boundary conditions.

21.a Recall the Fourier decomposition of the (free) field of the interaction representation and its conjugate momentum,

$$\phi_{in}(x) = \int \frac{d^3k}{(2\pi)^3 2E_k}\Big\{a^\dagger_{k,in}\,e^{+ik\cdot x} + a_{k,in}\,e^{-ik\cdot x}\Big\},$$

$$\Pi_{in}(x) = i\int \frac{d^3k}{(2\pi)^3 2E_k}\ E_k\Big\{a^\dagger_{k,in}\,e^{+ik\cdot x} - a_{k,in}\,e^{-ik\cdot x}\Big\}.$$

The argument of the exponential of the proposed expression for a coherent state can then be written as

$$\begin{aligned}&i\int d^3x\ \big(\Pi(x)\phi_{in}(0,x) - \phi(x)\Pi_{in}(0,x)\big)\\ &= \int\frac{d^3k}{(2\pi)^3 2E_k}\Big\{\underbrace{i\int d^3x\ e^{-ik\cdot x}\big(\Pi(x)-iE_k\phi(x)\big)}_{\chi(k)}\ a^\dagger_{k,in}\\ &\qquad\qquad + \underbrace{i\int d^3x\ e^{+ik\cdot x}\big(\Pi(x)+iE_k\phi(x)\big)}_{-\chi^*(k)}\ a_{k,in}\Big\},\end{aligned}$$

whose exponential is identical to the operator T_χ introduced in Problem 15. Since we have shown there that $|\chi_{in}\rangle = T_\chi\,|0_{in}\rangle$, this proves the announced representation of the coherent state. Note that the relationship between $\chi(\mathbf{k}), \chi^*(\mathbf{k})$ and $\phi(\mathbf{x}), \Pi(\mathbf{x})$ can be inverted, to give

$$\phi(\mathbf{x}) = \int \frac{d^3\mathbf{k}}{(2\pi)^3 2E_{\mathbf{k}}} \left\{\chi^*(\mathbf{k})\, e^{-i\mathbf{k}\cdot\mathbf{x}} + \chi(\mathbf{k})\, e^{+i\mathbf{k}\cdot\mathbf{x}}\right\},$$

$$\Pi(\mathbf{x}) = i\int \frac{d^3\mathbf{k}}{(2\pi)^3 2E_{\mathbf{k}}}\, E_{\mathbf{k}} \left\{\chi^*(\mathbf{k})\, e^{-i\mathbf{k}\cdot\mathbf{x}} - \chi(\mathbf{k})\, e^{+i\mathbf{k}\cdot\mathbf{x}}\right\}.$$

In other words, ϕ is a commuting free field and Π is its canonical conjugate (evaluated at $x^0 = 0$), in which $\chi^*(\mathbf{k})$ and $\chi(\mathbf{k})$ take the place of the creation and annihilation operators, respectively.

21.b Consider now a generic operator built from $\phi_{\text{in}}(0, \mathbf{x})$ and $\Pi_{\text{in}}(0, \mathbf{x})$. In this calculation, it is convenient to view $F[\phi_{\text{in}}, \Pi_{\text{in}}]$ as derived from a corresponding classical quantity $f[\phi, \Pi]$ by means of Weyl mapping (see Problem 18):

$$F[\phi_{\text{in}}, \Pi_{\text{in}}] \equiv \int [D\mu D\nu D\widetilde{\phi} D\widetilde{\Pi}]\ \exp\left(i\int (\mu(\phi_{\text{in}} - \widetilde{\phi}) + \nu(\Pi_{\text{in}} - \widetilde{\Pi}))\right)\ f[\widetilde{\phi}, \widetilde{\Pi}].$$

The expectation value of $F[\phi_{\text{in}}, \Pi_{\text{in}}]$ in the coherent state $\big|(\phi, \Pi)_{\text{in}}\big\rangle$ can be written as

$$\big\langle(\phi, \Pi)_{\text{in}}\big|F[\phi_{\text{in}}, \Pi_{\text{in}}]\big|(\phi, \Pi)_{\text{in}}\big\rangle = \int [D\mu D\nu D\widetilde{\phi} D\widetilde{\Pi}]\ \big\langle 0_{\text{in}}\big|e^{-A} e^{B} e^{A}\ f[\widetilde{\phi}, \widetilde{\Pi}]\ \big|0_{\text{in}}\big\rangle,$$

with the following shorthands:

$$A \equiv i\int (\Pi\phi_{\text{in}} - \phi\Pi_{\text{in}}), \quad B \equiv i\int (\mu(\phi_{\text{in}} - \widetilde{\phi}) + \nu(\Pi_{\text{in}} - \widetilde{\Pi})).$$

Since A and B contain only equal-time field operators and their conjugates, the Baker–Campbell–Hausdorff formula gives

$$e^{B} e^{A} = e^{A} e^{B} e^{[B,A]}, \quad [B, A] = i\int (\mu\phi + \nu\Pi),$$

so that

$$\begin{aligned}
&\big\langle(\phi, \Pi)_{\text{in}}\big|F[\phi_{\text{in}}, \Pi_{\text{in}}]\big|(\phi, \Pi)_{\text{in}}\big\rangle \\
&\quad = \int [D\mu D\nu D\widetilde{\phi} D\widetilde{\Pi}]\big\langle 0_{\text{in}}\big|\exp\left(i\int (\mu(\phi_{\text{in}} + \phi - \widetilde{\phi}) + \nu(\Pi_{\text{in}} + \Pi - \widetilde{\Pi}))\right) f[\widetilde{\phi}, \widetilde{\Pi}]\big|0_{\text{in}}\big\rangle \\
&\quad = \big\langle 0_{\text{in}}\big|F[\phi_{\text{in}} + \phi, \Pi_{\text{in}} + \Pi]\big|0_{\text{in}}\big\rangle.
\end{aligned}$$

In this form, this equation is always valid. (Note that this formula is consistent with the fact that a coherent state is the ground state of a quadratic theory with a shifted field, a property already observed in Problem 15.) To evaluate the right-hand side, we may expand $F[\cdot, \cdot]$ in powers of $\phi_{\text{in}}, \Pi_{\text{in}}$ about $F[\phi, \Pi]$. If all the operators are normal-ordered so that the creation operators are on the left of the annihilation operators, all the terms in this expansion have a null expectation value in the vacuum, except the zeroth-order term, i.e.,

$$\big\langle(\phi,\Pi)_{\rm in}\big|F[\phi_{\rm in},\Pi_{\rm in}]\big|(\phi,\Pi)_{\rm in}\big\rangle = F[\phi,\Pi] = f[\phi,\Pi].$$

($F[\cdot\,,\cdot]$ differs from $f[\cdot\,,\cdot]$ by the ordering of the ϕ's and Π's, but this distinction is irrelevant when the arguments are commuting classical fields.) In particular, we have

$$\big\langle(\phi,\Pi)_{\rm in}\big|\phi_{\rm in}(0,\boldsymbol{x})\big|(\phi,\Pi)_{\rm in}\big\rangle = \phi(\boldsymbol{x}),\quad \big\langle(\phi,\Pi)_{\rm in}\big|\Pi_{\rm in}(0,\boldsymbol{x})\big|(\phi,\Pi)_{\rm in}\big\rangle = \Pi(\boldsymbol{x}).$$

These formulas suggest that the coherent state $\big|(\phi,\Pi)_{\rm in}\big\rangle$ is a close quantum analogue of a classical state of "position" ϕ and "conjugate momentum" Π. Moreover, without normal-ordering, we also have

$$\begin{aligned}
&\big\langle(\phi,\Pi)_{\rm in}\big|\phi_{\rm in}^2\big|(\phi,\Pi)_{\rm in}\big\rangle - \big\langle(\phi,\Pi)_{\rm in}\big|\phi_{\rm in}\big|(\phi,\Pi)_{\rm in}\big\rangle^2\\
&\quad= \big\langle 0_{\rm in}\big|(\phi_{\rm in}+\phi)^2\big|0_{\rm in}\big\rangle - \big\langle 0_{\rm in}\big|\phi_{\rm in}+\phi\big|0_{\rm in}\big\rangle^2\\
&\quad= \big\langle 0_{\rm in}\big|\phi_{\rm in}^2\big|0_{\rm in}\big\rangle - \big\langle 0_{\rm in}\big|\phi_{\rm in}\big|0_{\rm in}\big\rangle^2,
\end{aligned}$$

and likewise

$$\big\langle(\phi,\Pi)_{\rm in}\big|\Pi_{\rm in}^2\big|(\phi,\Pi)_{\rm in}\big\rangle - \big\langle(\phi,\Pi)_{\rm in}\big|\Pi_{\rm in}\big|(\phi,\Pi)_{\rm in}\big\rangle^2 = \big\langle 0_{\rm in}\big|\Pi_{\rm in}^2\big|0_{\rm in}\big\rangle - \big\langle 0_{\rm in}\big|\Pi_{\rm in}\big|0_{\rm in}\big\rangle^2,$$

regardless of ϕ,Π. In other words, the variance of the field fluctuations is the same in a coherent state and in the vacuum (to check this property, we must not normal-order the operators, since normal-ordering suppresses the zero-point vacuum fluctuations).

21.c The integration measure $[D\chi(\boldsymbol{k})D\chi^*(\boldsymbol{k})]$ can be rewritten in terms of the real fields $\phi(\boldsymbol{x}),\Pi(\boldsymbol{x})$:

$$\big[D\chi(\boldsymbol{k})D\chi^*(\boldsymbol{k})\big] = \underbrace{\det\left(\frac{\delta(\chi(\boldsymbol{k}),\chi^*(\boldsymbol{k}))}{\delta(\phi(\boldsymbol{x}),\Pi(\boldsymbol{x}))}\right)}_{\prod_{\boldsymbol{k}}(2iE_{\boldsymbol{k}})}\big[D\phi(\boldsymbol{x})D\Pi(\boldsymbol{x})\big].$$

The determinant is field independent and can thus be ignored. Using the result (2.22) of the previous problem, we thus obtain

$$1 = \int \big[D\phi(\boldsymbol{x})D\Pi(\boldsymbol{x})\big]\;\big|(\phi,\Pi)_{\rm in}\big\rangle\big\langle(\phi,\Pi)_{\rm in}\big|. \tag{2.25}$$

21.d Let us now see how to use this result in order to derive a path integral representation of transition amplitudes between a pair of coherent states. First, we factorize the evolution operator into N factors corresponding to intervals $\epsilon \equiv (t_f - t_i)/N$:

$$e^{-i(t_f-t_i)\mathcal{H}} = \underbrace{e^{-i\epsilon\mathcal{H}}\times\cdots\times e^{-i\epsilon\mathcal{H}}}_{N\ \text{factors}}.$$

Then, we insert between each pair of such factors an identity operator expressed as (2.25). In order to streamline the notation, it is useful to rename the initial state $\big|(\phi_0,\Pi_0)_{\rm in}\big\rangle$ and the final state $\langle(\phi_N, \Pi_N)_{\rm in}|$. We are thus led to evaluating elementary factors such as

$$\begin{aligned}
&\big\langle(\phi_n,\Pi_n)_{\rm in}\big|e^{-i\epsilon\mathcal{H}}\big|(\phi_{n-1},\Pi_{n-1})_{\rm in}\big\rangle\\
&=\big\langle(\phi_n,\Pi_n)_{\rm in}\big|(\phi_{n-1},\Pi_{n-1})_{\rm in}\big\rangle\frac{\big\langle(\phi_n,\Pi_n)_{\rm in}\big|1-i\epsilon\mathcal{H}+\mathcal{O}(\epsilon^2)\big|(\phi_{n-1},\Pi_{n-1})_{\rm in}\big\rangle}{\big\langle(\phi_n,\Pi_n)_{\rm in}\big|(\phi_{n-1},\Pi_{n-1})_{\rm in}\big\rangle}\\
&=\big\langle(\phi_n,\Pi_n)_{\rm in}\big|(\phi_{n-1},\Pi_{n-1})_{\rm in}\big\rangle\left(1-i\epsilon\frac{\big\langle(\phi_n,\Pi_n)_{\rm in}\big|\mathcal{H}\big|(\phi_{n-1},\Pi_{n-1})_{\rm in}\big\rangle}{\big\langle(\phi_n,\Pi_n)_{\rm in}\big|(\phi_{n-1},\Pi_{n-1})_{\rm in}\big\rangle}+\mathcal{O}(\epsilon^2)\right)\\
&=\big\langle(\phi_n,\Pi_n)_{\rm in}\big|(\phi_{n-1},\Pi_{n-1})_{\rm in}\big\rangle\Big(1-i\epsilon\big\langle(\phi_n,\Pi_n)_{\rm in}\big|\mathcal{H}\big|(\phi_{n-1},\Pi_{n-1})_{\rm in}\big\rangle+\mathcal{O}(\epsilon^2)\Big)\\
&=\big\langle(\phi_n,\Pi_n)_{\rm in}\big|(\phi_{n-1},\Pi_{n-1})_{\rm in}\big\rangle\exp\Big(-i\epsilon\big\langle(\phi_n,\Pi_n)_{\rm in}\big|\mathcal{H}\big|(\phi_{n-1},\Pi_{n-1})_{\rm in}\big\rangle+\mathcal{O}(\epsilon^2)\Big)\\
&=\big\langle(\phi_n,\Pi_n)_{\rm in}\big|(\phi_{n-1},\Pi_{n-1})_{\rm in}\big\rangle\exp\Big(-i\epsilon\big\langle(\phi_n,\Pi_n)_{\rm in}\big|\mathcal{H}\big|(\phi_n,\Pi_n)_{\rm in}\big\rangle+\mathcal{O}(\epsilon^2)\Big)\\
&=\big\langle(\phi_n,\Pi_n)_{\rm in}\big|(\phi_{n-1},\Pi_{n-1})_{\rm in}\big\rangle\exp\Big(-i\epsilon\mathcal{H}(\phi_n,\Pi_n)+\mathcal{O}(\epsilon^2)\Big).
\end{aligned}\tag{2.26}$$

In going from the third to the fourth line, we used $\big\langle(\phi_n,\Pi_n)_{\rm in}\big|(\phi_{n-1},\Pi_{n-1})_{\rm in}\big\rangle = 1+\mathcal{O}(\epsilon)$. In the penultimate line, we identified the two coherent states in the second factor, because their difference contributes only at order ϵ^2 and therefore vanishes in the limit $N \to +\infty$.

We must now evaluate the first factor on the right-hand side of (2.26) at order ϵ^1. The overlap between two coherent states is given in Eq. (2.24) in the Problem 20. Here, we need the overlap between two nearby coherent states,

$$\begin{aligned}
\big\langle\chi+\delta\chi_{\rm in}\big|\chi_{\rm in}\big\rangle &= \exp\left(\frac{1}{2}\int\frac{d^3\boldsymbol{k}}{(2\pi)^3 2E_{\boldsymbol{k}}}\,\big((\delta\chi)^*\chi-\chi^*(\delta\chi)\big)+\mathcal{O}(\delta\chi^2)\right)\\
&= \exp\left(\frac{i}{2}\int d^3\boldsymbol{x}\,\big(\Pi(\delta\phi)-\phi(\delta\Pi)\big)+\mathcal{O}(\delta\chi^2)\right),
\end{aligned}$$

where $\delta\phi, \delta\Pi$ are the variations of the functions ϕ, Π corresponding to the variation $\delta\chi$. At this point, we can take the limit $N \to +\infty$ of infinitesimal intervals. The sequence of functions of space $\phi_n(\boldsymbol{x}), \Pi_n(\boldsymbol{x})$ can then be replaced by a pair of spacetime-dependent ones $\phi(t,\boldsymbol{x}), \Pi(t,\boldsymbol{x})$, and the transition amplitude takes the form of a path integral,

$$\begin{aligned}
&\big\langle(\phi_f,\Pi_f)_{\rm in}\big|e^{-i(t_f-t_i)\mathcal{H}}\big|(\phi_i,\Pi_i)_{\rm in}\big\rangle\\
&\quad=\int[\mathcal{D}\phi(\boldsymbol{x})\mathcal{D}\Pi(\boldsymbol{x})]\,\exp\left(i\int_{t_i}^{t_f}d^4x\,\big(\tfrac{1}{2}(\Pi\partial_t\phi-\phi\partial_t\Pi)-\mathcal{H}(\phi,\Pi)\big)\right).
\end{aligned}$$

Up to an integration by parts of the second term inside the exponential, this is the well-known Hamiltonian form of the path integral. Moreover, from the derivation, it should be clear that the boundary values of the functions ϕ, Π are

$$\phi(t_i,\boldsymbol{x})=\phi_i(\boldsymbol{x}),\quad \Pi(t_i,\boldsymbol{x})=\Pi_i(\boldsymbol{x}),\qquad \phi(t_f,\boldsymbol{x})=\phi_f(\boldsymbol{x}),\quad \Pi(t_f,\boldsymbol{x})=\Pi_f(\boldsymbol{x}).$$

22. Spin Coherent States and the Berry Phase Consider a single spin-1/2 object, and denote by $|\pm\rangle$ the usual spin states $|s = \frac{1}{2}, m = \pm\frac{1}{2}\rangle$ (with a quantization axis along the z axis). Let $\widehat{n}$ be a unit vector pointing in the spatial direction corresponding to a polar angle θ and azimuth φ, let $\widehat{n}_0$ be the unit vector in the z direction, and define a *spin coherent state* $|\widehat{n}\rangle$ by

$$|\widehat{n}\rangle \equiv e^{i\theta\widehat{\alpha}_n\cdot s}\,|+\rangle \quad \text{with } \widehat{\alpha}_n \equiv \frac{\widehat{n}\times\widehat{n}_0}{\|\widehat{n}\times\widehat{n}_0\|}, \quad s^i \equiv \frac{\sigma^i}{2}.$$

22.a Check the following properties about $|\widehat{n}\rangle$:

$$|\widehat{n}\rangle = \cos\tfrac{\theta}{2}\,|+\rangle + e^{i\varphi}\sin\tfrac{\theta}{2}\,|-\rangle, \quad \langle\widehat{n}|\widehat{n}\rangle = 1,$$
$$\frac{1}{2\pi}\int_0^{\pi} d\theta\,\sin\theta \int_0^{2\pi} d\varphi\,|\widehat{n}\rangle\langle\widehat{n}| = 1.$$

22.b Show that $\langle\widehat{n}|s|\widehat{n}\rangle = \frac{1}{2}\widehat{n}$. Therefore, a coherent state $|\widehat{n}\rangle$ is a state where the spin points in the definite direction $\widehat{n}$, providing the closest possible analogue to a classical angular momentum.

22.c Show that $\langle\widehat{n}'|\widehat{n}\rangle = \left(\frac{1+\widehat{n}\cdot\widehat{n}'}{2}\right)^{1/2} e^{-i\Phi}$, where Φ is a real phase. For a state $\widehat{n}'$ infinitesimally close to $\widehat{n}$, show that Φ is one-half of the area of the spherical triangle of summits $(\widehat{n}\widehat{n}'\widehat{n}_0)$. Extra: prove that this is also true for arbitrary separations.

22.d Derive a path integral expression for the transition amplitude $\langle\widehat{n}'|e^{-i(t_f-t_i)H}|\widehat{n}\rangle$. Note that the resulting formula contains a purely geometrical term that does not vanish even when $H \to 0$. This term is known as a *Berry phase.*

22.a Let θ, φ be the polar and azimuthal angles in spherical coordinates that represent the vector $\widehat{n}$, respectively. The Cartesian coordinates of $\widehat{n}$, $\widehat{n}_0$ and $\widehat{n}\times\widehat{n}_0$ read

$$\widehat{n} = \begin{pmatrix}\sin\theta\cos\varphi\\ \sin\theta\sin\varphi\\ \cos\theta\end{pmatrix}, \quad \widehat{n}_0 = \begin{pmatrix}0\\0\\1\end{pmatrix}, \quad \widehat{n}\times\widehat{n}_0 = \begin{pmatrix}\sin\theta\sin\varphi\\ -\sin\theta\cos\varphi\\ 0\end{pmatrix}.$$

Since $\sin\theta$ is positive in the interval $0 \le \theta \le \pi$, the unit vector in the direction of $\widehat{n}\times\widehat{n}_0$ reads

$$\widehat{\alpha}_n \equiv \frac{\widehat{n}\times\widehat{n}_0}{\|\widehat{n}\times\widehat{n}_0\|} = \begin{pmatrix}\sin\varphi\\ -\cos\varphi\\ 0\end{pmatrix}.$$

A standard $\mathfrak{su}(2)$ calculation then yields

$$e^{i\theta\widehat{\alpha}_n\cdot s} = \cos\tfrac{\theta}{2} + 2i\sin\tfrac{\theta}{2}\,\widehat{\alpha}_n\cdot s.$$

Note that $\widehat{\alpha}_n\cdot s$ can be expressed in terms of the raising and lowering operators, $s_\pm \equiv s_1 \pm is_2$:

$$\widehat{\alpha}_n\cdot s = \frac{1}{2i}\left(e^{i\varphi}\, s_- - e^{-i\varphi}\, s_+\right).$$

Using $s_+\big|+\big\rangle = 0,\ s_-\big|+\big\rangle = \big|-\big\rangle$, we thus get the following expression for the spin coherent state:

$$\big|\widehat{n}\big\rangle \equiv e^{i\theta\widehat{\alpha}_n\cdot s}\,\big|+\big\rangle = \cos\tfrac{\theta}{2}\,\big|+\big\rangle + e^{i\varphi}\sin\tfrac{\theta}{2}\,\big|-\big\rangle. \tag{2.27}$$

It is immediately clear that this state is normalized to unity, since

$$\big\langle\widehat{n}\big|\widehat{n}\big\rangle = \cos^2\tfrac{\theta}{2} + \sin^2\tfrac{\theta}{2} = 1.$$

Consider now the quantity $\big\langle L\big|\widehat{n}\big\rangle\big\langle\widehat{n}\big|R\big\rangle$, where $\big\langle L\big|, \big|R\big\rangle$ are the basis spin states $\big|\pm\big\rangle$. It is straightforward to obtain the following table for this quantity and its integral over θ, φ:

$\langle L\vert$	$\vert R\rangle$	$\langle L\vert\widehat{n}\rangle\langle\widehat{n}\vert R\rangle$	$\int_0^\pi d\theta\ \sin\theta\int_0^{2\pi} d\varphi\ \langle L\vert\widehat{n}\rangle\langle\widehat{n}\vert R\rangle$
$\langle +\vert$	$\vert +\rangle$	$\cos^2\frac{\theta}{2}$	2π
$\langle +\vert$	$\vert -\rangle$	$e^{-i\varphi}\cos\frac{\theta}{2}\sin\frac{\theta}{2}$	0
$\langle -\vert$	$\vert +\rangle$	$e^{i\varphi}\cos\frac{\theta}{2}\sin\frac{\theta}{2}$	0
$\langle -\vert$	$\vert -\rangle$	$\sin^2\frac{\theta}{2}$	2π

From these results, we conclude that

$$\underbrace{\frac{1}{2\pi}\int_0^\pi d\theta\ \sin\theta\int_0^{2\pi}d\varphi}_{\int\frac{d^2\widehat{n}}{2\pi}}\ \big|\widehat{n}\big\rangle\big\langle\widehat{n}\big| = \big|+\big\rangle\big\langle+\big| + \big|-\big\rangle\big\langle-\big| = 1. \tag{2.28}$$

Therefore, the coherent spin states provide a complete (in fact, overcomplete) set spanning the Hilbert space of spin-1/2 states.

22.b In order to evaluate the expectation value of the spin operator in a coherent state, we need the following:

$$s_1 = \tfrac{s_+ + s_-}{2},\quad s_2 = \tfrac{s_+ - s_-}{2i},\quad s_3\big|\pm\big\rangle = \pm\tfrac{1}{2}\big|\pm\big\rangle.$$

We get

$$\begin{aligned}
\big\langle\widehat{n}\big|s_1\big|\widehat{n}\big\rangle &= \tfrac{1}{2}\big\langle\widehat{n}\big|s_+\big|\widehat{n}\big\rangle + \tfrac{1}{2}\big\langle\widehat{n}\big|s_-\big|\widehat{n}\big\rangle = \tfrac{1}{2}\cos\tfrac{\theta}{2}\sin\tfrac{\theta}{2}\left(e^{i\varphi} + e^{-i\varphi}\right) = \tfrac{1}{2}\sin\theta\cos\varphi,\\
\big\langle\widehat{n}\big|s_2\big|\widehat{n}\big\rangle &= \tfrac{1}{2i}\big\langle\widehat{n}\big|s_+\big|\widehat{n}\big\rangle - \tfrac{1}{2i}\big\langle\widehat{n}\big|s_-\big|\widehat{n}\big\rangle = \tfrac{1}{2i}\cos\tfrac{\theta}{2}\sin\tfrac{\theta}{2}\left(e^{i\varphi} - e^{-i\varphi}\right) = \tfrac{1}{2}\sin\theta\cos\varphi,\\
\big\langle\widehat{n}\big|s_3\big|\widehat{n}\big\rangle &= \tfrac{1}{2}\cos^2\tfrac{\theta}{2} - \tfrac{1}{2}\sin^2\tfrac{\theta}{2} = \tfrac{1}{2}\cos\theta.
\end{aligned}$$

We recognize on the right-hand side of these equations the Cartesian coordinates of the vector $\widehat{n}$, which proves that $\big\langle\widehat{n}\big|s\big|\widehat{n}\big\rangle = \frac{1}{2}\widehat{n}$. In other words, the coherent spin state $\big|\widehat{n}\big\rangle$ is such that the spin operator has an expectation value aligned with the vector $\widehat{n}$ itself, unlike the states $\big|s, m\big\rangle$ in which the spin vector can have a very delocalized orientation.

22.c Given the explicit expression (2.27) for the coherent spin state, the overlap between two such states reads

$$\langle \hat{n}'|\hat{n}\rangle = \cos\tfrac{\theta}{2}\cos\tfrac{\theta'}{2} + e^{i(\varphi-\varphi')}\sin\tfrac{\theta}{2}\sin\tfrac{\theta'}{2},$$

and the scalar product between $\hat{n}$ and $\hat{n}'$ is

$$\begin{aligned}\hat{n}\cdot\hat{n}' &= \sin\theta\sin\theta'(\cos\varphi\cos\varphi' + \sin\varphi\sin\varphi') + \cos\theta\cos\theta'\\ &= \sin\theta\sin\theta'\cos(\varphi-\varphi') + \cos\theta\cos\theta'.\end{aligned}$$

The squared modulus of $\langle \hat{n}'|\hat{n}\rangle$ is given by

$$\begin{aligned}|\langle \hat{n}'|\hat{n}\rangle|^2 &= \Big(\cos\tfrac{\theta}{2}\cos\tfrac{\theta'}{2} + \cos(\varphi-\varphi')\sin\tfrac{\theta}{2}\sin\tfrac{\theta'}{2}\Big)^2 + \Big(\sin(\varphi-\varphi')\sin\tfrac{\theta}{2}\sin\tfrac{\theta'}{2}\Big)^2\\ &= \cos^2\tfrac{\theta}{2}\cos^2\tfrac{\theta'}{2} + \sin^2\tfrac{\theta}{2}\sin^2\tfrac{\theta'}{2} + 2\cos(\varphi-\varphi')\underbrace{\cos\tfrac{\theta}{2}\sin\tfrac{\theta}{2}}_{\frac{1}{2}\sin\theta}\underbrace{\cos\tfrac{\theta'}{2}\sin\tfrac{\theta'}{2}}_{\frac{1}{2}\sin\theta'}\\ &= \tfrac{1}{2} + \tfrac{1}{2}\cos\theta\cos\theta' + \tfrac{1}{2}\cos(\varphi-\varphi')\sin\theta\sin\theta' = \frac{1+\hat{n}\cdot\hat{n}'}{2}.\end{aligned}$$

This proves that $\langle \hat{n}'|\hat{n}\rangle$ is equal to $\left(\frac{1+\hat{n}\cdot\hat{n}'}{2}\right)^{1/2}$ up to a phase.

In order to get a sense of the value of this phase, let us first consider the case where $\hat{n}'$ is very close to $\hat{n}$, i.e., where $\theta' = \theta + d\theta$ and $\varphi' = \varphi + d\varphi$. Expanding to first order the overlap between the two states gives

$$\langle \hat{n}'|\hat{n}\rangle = 1 - i\,d\varphi\,\sin^2\tfrac{\theta}{2} + \cdots = 1 - i\,d\varphi\left(\tfrac{1-\cos\theta}{2}\right) + \cdots.$$

From this expansion, we can read the value of the phase Φ for this infinitesimal separation,

$$\Phi = d\varphi\left(\tfrac{1-\cos\theta}{2}\right).$$

In order to see the relationship between this quantity and the area of a spherical triangle, we may view this area as the flux of a radial field, $\mathbf{B}(x) \equiv \hat{x}/|x|^2$. Then, we note that this field can be obtained as a curl,

$$\mathbf{B} = \boldsymbol{\nabla}\times\mathbf{A} \quad \text{with } \mathbf{A}(x) = \frac{1-\cos\theta}{|x|\sin\theta}\,\mathbf{e}_\phi$$

(with $\mathbf{e}_\phi$ the unit vector along the lines of constant latitude), in order to rewrite this flux as the integral of $\mathbf{A}$ along the boundary of the triangle. Since $\mathbf{A}\parallel \mathbf{e}_\varphi$, the edges $(\hat{n}'\hat{n}_0)$ and $(\hat{n}_0\hat{n})$ of this triangle do not contribute to the contour integral, and only the infinitesimal edge $(\hat{n}\hat{n}')$ contributes, by an amount $\mathbf{A}\cdot(\hat{n}'-\hat{n}) = d\varphi(1-\cos\theta)$ (for a unit radius). Therefore, for an infinitesimal separation, we have indeed shown that

$$\langle \hat{n}'|\hat{n}\rangle = \left(\frac{1+\hat{n}\cdot\hat{n}'}{2}\right)^{1/2}\exp\Big(-\tfrac{i}{2}\,\mathrm{Area}\,(\hat{n}\hat{n}'\hat{n}_0)\Big). \qquad (2.29)$$

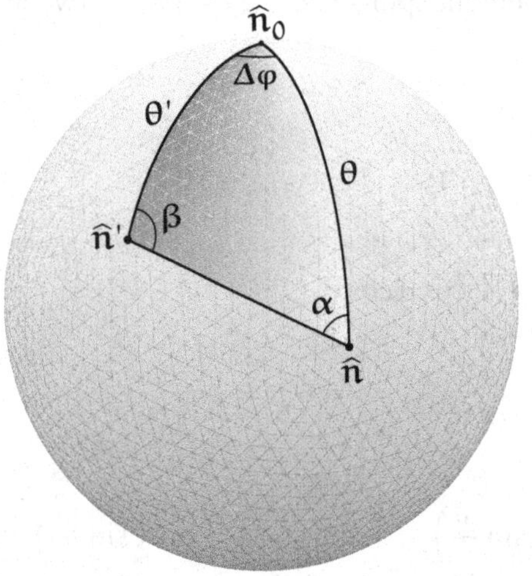

Figure 2.1 Notation for the angles of the spherical triangle $(\widehat{n}\widehat{n}'\widehat{n}_0)$.

General Case: For an arbitrary separation between $\widehat{n}'$ and $\widehat{n}$, the tangent of the phase Φ is given by

$$\tan\Phi = \frac{\tan\frac{\theta}{2}\tan\frac{\theta'}{2}\sin(\Delta\varphi)}{1+\tan\frac{\theta}{2}\tan\frac{\theta'}{2}\cos(\Delta\varphi)},$$

where $\Delta\varphi \equiv \varphi' - \varphi$. If we denote by $\Delta\varphi$, α, β the three angles (see Figure 2.1 for the notation) of the spherical triangle $(\widehat{n}\widehat{n}'\widehat{n}_0)$, we have the following results of spherical trigonometry:

$$\text{Area}\,(\widehat{n}\widehat{n}'\widehat{n}_0) = \alpha+\beta+\Delta\varphi-\pi, \qquad \tan\left(\tfrac{\alpha+\beta}{2}\right)\tan\left(\tfrac{\Delta\varphi}{2}\right) = \frac{\cos\left(\frac{\theta-\theta'}{2}\right)}{\cos\left(\frac{\theta+\theta'}{2}\right)}.$$

The first formula states that the area of a spherical triangle (that is, a triangle whose edges are geodesics on the sphere) on the unit 2-sphere is equal to its *excess angle*, i.e., the deviation of the sum of its angles from that of a planar triangle (this formula is a special case of the *Gauss–Bonnet theorem*; see the extra material at the end of the solution to this problem for an elementary proof). The second formula is known as *Napier's analogy*, or the *tangent rule*. These relations, with the help of some elementary trigonometry, lead to $\Phi = \frac{1}{2}\text{Area}\,(\widehat{n}\widehat{n}'\widehat{n}_0)$, which generalizes (2.29) to coherent states with arbitrary separations.

22.d In order to evaluate the matrix element $\langle\widehat{n}'|e^{-i(t_f-t_i)\mathcal{H}}|\widehat{n}\rangle$, the standard procedure is to slice the time interval $[t_i, t_f]$ into N sub-intervals of size $\epsilon \equiv (t_f - t_i)/N$. Then, we insert between each of the resulting elementary evolution operators an identity operator written in the form of (2.28). We are thus led to evaluating transition amplitudes between nearby coherent

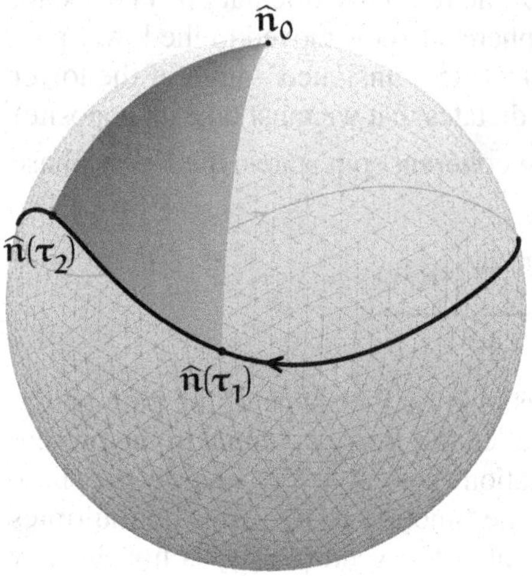

Figure 2.2 Geometrical interpretation of the Berry phase.

states,

$$
\begin{aligned}
&\langle\hat{n}_{i+1}|e^{-i\epsilon\mathcal{H}}|\hat{n}_i\rangle \\
&= \langle\hat{n}_{i+1}|\hat{n}_i\rangle \frac{\langle\hat{n}_{i+1}|e^{-i\epsilon\mathcal{H}}|\hat{n}_i\rangle}{\langle\hat{n}_{i+1}|\hat{n}_i\rangle} = \langle\hat{n}_{i+1}|\hat{n}_i\rangle \frac{\langle\hat{n}_{i+1}|1 - i\epsilon\mathcal{H} + \mathcal{O}(\epsilon^2)|\hat{n}_i\rangle}{\langle\hat{n}_{i+1}|\hat{n}_i\rangle} \\
&= \langle\hat{n}_{i+1}|\hat{n}_i\rangle \Big(1 - i\epsilon\frac{\langle\hat{n}_{i+1}|\mathcal{H}|\hat{n}_i\rangle}{\langle\hat{n}_{i+1}|\hat{n}_i\rangle} + \mathcal{O}(\epsilon^2)\Big) = \langle\hat{n}_{i+1}|\hat{n}_i\rangle \Big(1 - i\epsilon\mathcal{H}(\hat{n}_i) + \mathcal{O}(\epsilon^2)\Big) \\
&= \exp\Big(- i d\Phi_i - i\epsilon\mathcal{H}(\hat{n}_i) + \mathcal{O}(\epsilon^2)\Big),
\end{aligned}
$$

where $d\Phi_i$ is one-half of the area of the spherical triangle $(\hat{n}_i\hat{n}_{i+1}\hat{n}_0)$ (this area is of order ϵ). In taking the limit $N \to +\infty$, the discrete sequence of intermediate states $\hat{n}_i$ becomes a curve $\hat{n}(\tau)$ on the unit sphere, and the transition amplitude can be written as

$$
\langle\hat{n}'|e^{-i(t_f-t_i)\mathcal{H}}|\hat{n}\rangle = \int_{\substack{\hat{n}(t_i)=\hat{n} \\ \hat{n}(t_f)=\hat{n}'}} [D\hat{n}(\tau)] \, \exp\Big(- i\int_{t_i}^{t_f} d\tau \, \big(\tfrac{d\Phi}{d\tau} + \mathcal{H}(\hat{n}(\tau))\big)\Big).
$$

The first term inside the exponential, in $d\Phi/d\tau$, is the spin analogue of the term $p\dot{q}$ in the path integral for the motion of a point particle. This term leads to non-trivial effects, even when the Hamiltonian is zero. In particular, if $\hat{n}' = \hat{n}$ (i.e., the evolution returns to its starting point), this phase is one-half of the area above the closed curve $\hat{n}(\tau)$ on the unit 2-sphere. Loosely speaking, it owes its existence to the fact that parallel transporting a vector around a closed loop on a curved manifold results in a vector that differs from the starting one. This phase, known as a *geometric phase*, or *Berry's phase*, is illustrated in Figure 2.2. The denomination *geometric phase* comes from the fact that this phase depends only on the shape of the path $\hat{n}(\tau)$, but not on the velocity at which it is traveled. Note also that the Berry phase of a closed loop does not depend on the choice of the quantization axis used for defining the states $|s, m\rangle$ and therefore the coherent spin states. Indeed, changing the location of the reference orientation $\hat{n}_0$ does not change the area if $\hat{n}_0$ remains in the upper part of the sphere. If $\hat{n}_0$ is moved to the lower part, the area is transformed according to Area $\to$ Area $- 4\pi$ (the – unsigned – area of the lower part is $4\pi -$ Area, but the orientation of the closed loop dictates that we must take its opposite).

From its origin as the overlap between neighboring coherent spin states, the Berry phase can also be written as

$$\Phi = -i \int d\tau \, \langle \hat{n}(\tau) | \partial_\tau | \hat{n}(\tau) \rangle = -i \int d\hat{n} \cdot \underbrace{\langle \hat{n}(\tau) | \nabla_{\hat{n}} | \hat{n}(\tau) \rangle}_{A(\hat{n})} .$$

The second form makes it clear that the phase depends only on the shape of the path on the 2-sphere. The vector $A(\hat{n})$ acts as a gauge potential, called the *Berry potential* in the present case. This object is not defined uniquely: any transformation $A \to A + \nabla_{\hat{n}}\alpha$ leaves the phase accumulated around a closed loop invariant, provided the function $\alpha(\hat{n})$ varies by multiples of 2π when $\hat{n}$ returns to the starting point. Consequently, the vector $A(\hat{n})$ is not directly observable, while the Berry phase is.

From the gauge potential A, one may define the analogue of a field strength by

$$F_{ab}(\hat{n}) \equiv \nabla^a_{\hat{n}} A^b(\hat{n}) - \nabla^b_{\hat{n}} A^a(\hat{n}),$$

and Stokes's theorem allows us to rewrite the Berry phase of a closed loop γ as an integral over the upper part of the sphere (Σ) of boundary γ:

$$\Phi[\gamma = \partial\Sigma] = -\frac{i}{2} \int_\Sigma d\hat{n}_a \wedge d\hat{n}_b \, F_{ab}(\hat{n}).$$

In the case of the spin considered here, an explicit realization of this is obtained by extending the function $\hat{n}(\tau)$ (shown by the black curve in Figure 2.2) to a function $\hat{n}(\tau, s)$ extended to the part of the sphere above this curve, such that

$$\hat{n}(\tau, s = 0) = \hat{n}(\tau), \quad \hat{n}(\tau, s = 1) = \hat{n}_0.$$

Then, the total area above the curve can be obtained as

$$\text{Area} = \int ds d\tau \, \hat{n} \cdot (\partial_\tau \hat{n} \times \partial_s \hat{n}).$$

This integral is known as a *Wess–Zumino term.*

Elementary Derivation of the Area of a Spherical Triangle (on the Unit 2-Sphere): With the notation of Figure 2.3, we have the following identities:

$$\begin{aligned}
&2\,\text{Area}\,(S) + \text{Area}\,(A) + \text{Area}\,(A') = 4\alpha, \\
&2\,\text{Area}\,(S) + \text{Area}\,(B) + \text{Area}\,(B') = 4\beta, \\
&2\,\text{Area}\,(S) + \text{Area}\,(C) + \text{Area}\,(C') = 4\gamma, \\
&2\,\text{Area}\,(S) + \text{Area}\,(A) + \text{Area}\,(A') \\
&\quad + \text{Area}\,(B) + \text{Area}\,(B') + \text{Area}\,(C) + \text{Area}\,(C') = 4\pi.
\end{aligned}$$

The prefactor 2 accompanying Area (S) arises because there is another domain, antipodal to S, on the hidden side of the sphere. Adding the first three relations and subtracting the last one

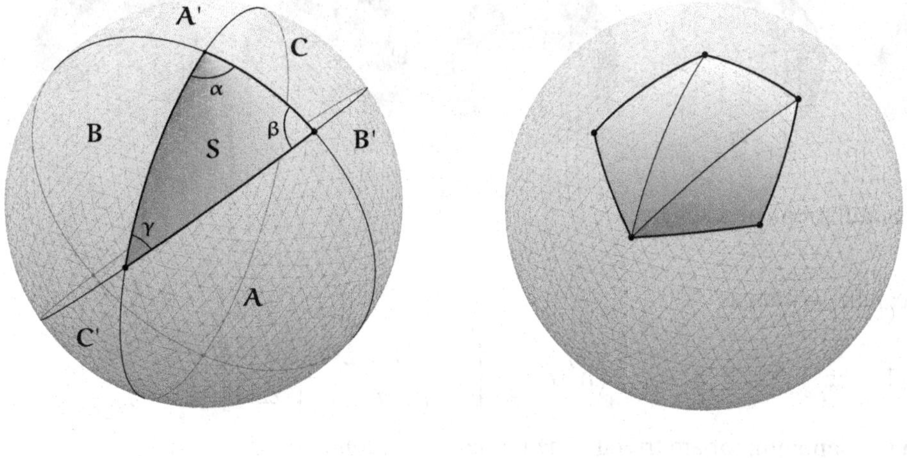

Figure 2.3 Left: Notation for the derivation of the area of a spherical triangle. Right: triangulation of a pentagon by three spherical triangles.

gives

$$\text{Area}\left(S\right) = \alpha + \beta + \gamma - \pi.$$

This formula can be generalized for a spherical p-gon bounded by arcs of great circles,

$$\text{Area}(p\text{-gon}) = \sum_{i=1}^{p} \alpha_i - (p-2)\pi,$$

where the α_i's are the interior angles at the vertices of the p-gon. This formula is obtained by triangulating the p-gon so that it becomes a union of non-overlapping spherical triangles (if one does not introduce any additional vertex, such a triangulation requires $p-2$ triangles, as illustrated in Figure 2.3).

Arbitrary Spins: It is possible to extend this discussion to the case of arbitrary spins S instead of 1/2. The coherent spin states are defined in the same way, starting from the highest-weight state $|S, S\rangle$. In order to more easily extract the spin dependence, one may check that

$$\exp\left(\tfrac{\theta}{2}e^{i\varphi}s_- - \tfrac{\theta}{2}e^{-i\varphi}s_+\right) = e^{zs_-}\,e^{-\ln(1+zz^*)s_3}\,e^{-z^*s_+} \qquad \text{with } z \equiv e^{i\varphi}\tan\tfrac{\theta}{2}.$$

(Although it is tempting to use the Baker–Campbell–Hausdorff formula to prove this, it is much easier to check it by explicit matrix exponentiation in the case of the fundamental representation of $\mathfrak{su}(2)$, where the spin operators are the Pauli matrices divided by 2. Then, one uses the fact that the fundamental representation is faithful, to conclude that this identity is valid in any representation; see Problem 33). Using this identity, the coherent spin states and

Figure 2.4 Triangulations of a sphere.

their overlap can be written as

$$|\widehat{n}\rangle = (1 + zz^*)^{-S}\, e^{zs_-}\,|S, S\rangle, \quad \langle\widehat{n}'|\widehat{n}\rangle = \left[\frac{(1 + z'^*z)^2}{(1 + z'z'^*)(1 + zz^*)}\right]^S.$$

The factors in the denominator are trivial. The numerator is equal to

$$\langle S, S|e^{z'^* s_+} e^{zs_-}|S, S\rangle.$$

Only the terms up to order z^{2S} can contribute in the second exponential (acting once more with s_- after we have reached the state $|S, -S\rangle$ gives zero). Moreover, each term must have a matching number of s_+'s in order to bring the state back to $\langle S, S|$ after its weight has been lowered by the lowering operators. Therefore we have in fact an expansion in powers $(z'^*z)^p$. It is then straightforward to compute the coefficients and check that they are $(2S)!/(p!(2S-p)!)$, giving $(1 + z'^*z)^{2S}$. Thus, for a spin S, the overlap of two coherent spin states is the 2S-th power of its value for $S = \frac{1}{2}$. In particular, the phase is $\Phi = S \cdot \text{Area}\,(\widehat{n}\widehat{n}'\widehat{n}_0)$. Another place where the S dependence enters is the resolution of identity, which becomes

$$\frac{2S+1}{2\pi}\int_0^\pi d\theta\, \sin\theta \int_0^{2\pi} d\varphi\, |\widehat{n}\rangle\langle\widehat{n}| = 1.$$

(The proof consists in inserting the left-hand side between the $(2S + 1)^2$ pairs of states $\langle S, m|\cdots|S, m'\rangle$, in order to check that the result is $\delta_{mm'}$.)

Gauss–Bonnet Theorem: Consider a two-dimensional smooth compact surface S. This surface can be approximated by covering it with a mesh of planar triangles of increasingly smaller areas, as illustrated in Figure 2.4 for the case of a sphere. Let us denote by V the number of vertices, by E the number of edges and by F the number of faces in such a triangulation. The combination $\chi(S) \equiv F + V - E$ is called the *Euler characteristic* of the surface S. It is easy to check that this combination is unchanged as one refines the triangulation, i.e., it is in fact a property of the topology of the surface S, independent of the triangulation by which it is approximated (see the discussion in Problem 42). Since a tetrahedron ($V = 4, E = 6, F = 4$) is the simplest triangulation of a sphere, we have $\chi = 2$ for a sphere. Then, removing one face (i.e., creating a closed boundary in the surface) reduces χ by one unit. If we create two such boundaries and then "glue" them in order to make a handle, χ is reduced by two units. Therefore, we have in general

$$\chi = 2 - 2 \times (\text{number of handles}) - (\text{number of boundaries}).$$

For instance, the Euler characteristic of a torus is 0 (one handle and no boundary), as it is for a finite section of cylinder (no handle and two boundaries).

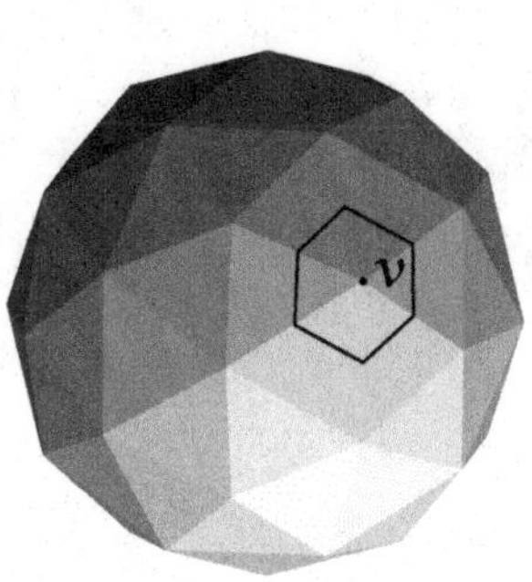

Figure 2.5 Definition of the angle θ_v at a vertex v of a triangulation.

In the case of a general surface S, let us denote by E_B the number of triangle edges that sit on the boundary of S. One may check that

$$3F = 2E - E_B.$$

This formula follows from the fact that internal edges belong to two triangles, while boundary edges belong to a single triangle. Therefore, $3\times$ the number of faces should be $2\times$ the number of internal edges plus $1\times$ the number of boundary edges. Using this relationship, the Euler characteristic can be rewritten as

$$\chi(S) = V - \tfrac{1}{2}F - \tfrac{1}{2}E_B = V - \tfrac{1}{2}F - \tfrac{1}{2}V_B,$$

where V_B is the number of vertices on the boundary of S (since the surface is compact, its boundary is a union of closed curves, and therefore $E_B = V_B$). Let us denote by $\alpha_f, \beta_f, \gamma_f$ the angles at the corners of the face f. Obviously, since this face is a planar triangle, we have $\alpha_f + \beta_f + \gamma_f = \pi$. Thus, we can write

$$\begin{aligned} 2\pi\chi(S) &= 2\pi V - \pi F - \pi V_B = \sum_{v\in\mathcal{I}} 2\pi + \sum_{v\in\partial S} \pi - \sum_{\text{faces}} (\alpha_f + \beta_f + \gamma_f) \\ &= \sum_{v\in\mathcal{I}} (2\pi - \theta_v) + \sum_{v\in\partial S} (\pi - \theta_v), \end{aligned} \tag{2.30}$$

where $\mathcal{I}$ is the set of internal vertices. In the final equality, we have introduced θ_v, the sum of the angles pointing at the vertex v, defined as shown in Figure 2.5. Note that for an internal vertex, $\theta_v = 2\pi$ only if the surface has no curvature at the point v. Likewise, on the boundary, $\theta_v = \pi$ only if the boundary is locally "straight" (which for a curved surface means "along a geodesic"). Eq. (2.30) is a discrete version of the Gauss–Bonnet theorem. We see that it expresses a global topological property of the surface (its Euler characteristic) as a sum of terms that encode the local shape of the surface. In the limit of an infinitely fine triangular mesh, one can show that

$$\sum_{v\in\mathcal{I}} (2\pi - \theta_v) \to \int_S K\, dS, \quad \sum_{v\in\partial S} (\pi - \theta_v) \to \int_{\partial S} k_g\, d\ell + \sum_{\text{cusps}} (\pi - \theta_{\text{cusp}}),$$

where K is the Gaussian curvature of the surface (i.e., the product of its two principal curvatures), and k_g is the geodesic curvature of the boundary (i.e., a measure of how much it deviates

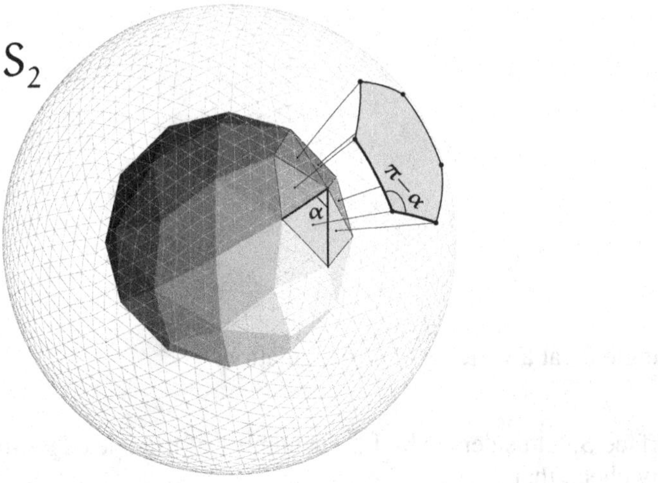

Figure 2.6 Gauss mapping of a vertex of the triangulation of a surface. The area of the spherical polygon image of this vertex is $2\pi - \theta_v$.

from a geodesic). The final term in the second expression is the sum of finite contributions one gets at the cusps of the boundary, assumed to be piecewise smooth. Thus, the continuous version of the Gauss–Bonnet theorem reads

$$2\pi\chi(S) = \int_S K\, dS + \int_{\partial S} k_g\, d\ell + \sum_{\text{cusps}} (\pi - \theta_{\text{cusp}}).$$

The remarkable aspect of this formula is that on the left-hand side the curvatures K and k_g are metric concepts on the manifold S, while the right-hand side is a purely topological property of S.

The relationship between the quantity $2\pi - \theta_v$ and the integrated curvature may be understood as follows. First, observe that in the case of a triangulation, since the faces are flat, the curvature K is concentrated at the vertices in the form of terms proportional to delta functions. The fact that $2\pi - \theta_v$ is the contribution of the vertex v to the integrated curvature can be established by using *Gauss mapping*, which maps a point $x \in S$ of the surface, where the unit normal vector is n_x, to the point n_x of the unit 2-sphere S_2. A crucial property of Gauss mapping is that it preserves the integrated curvature. Under this mapping, the faces, edges and vertices of a triangulation are transformed as follows:

$$\begin{array}{rcl} \text{faces} & \to & \text{points,} \\ \text{edges} & \to & \text{arcs of great circles,} \\ \text{vertices} & \to & \text{geodesic polygons,} \end{array}$$

as illustrated in Figure 2.6. From this figure, we see that the sum of the interior angles of the p-gon which is the image of the vertex v under Gauss mapping is $p\pi - \theta_v$. Therefore, the area of this polygon – which is also the corresponding integrated curvature since $K = 1$ on the unit sphere – is $2\pi - \theta_v$. Since Gauss mapping conserves the integrated curvature, $2\pi - \theta_v$ is the amount of integrated curvature of the triangulation carried by the vertex v.

23. Zeta Function Regularization of the Klein-Gordon Operator The *heat kernel* of a Hermitian operator $\mathcal{O}$ of eigenvalues λ_n and eigenfunctions $\psi_n(x)$ (normalized so that $\sum_n \psi_n(x)\psi_n^*(y) = \delta(x-y)$) is defined as

$$H_{\mathcal{O}}(x, y; \tau) \equiv \sum_n e^{-\lambda_n \tau}\, \psi_n(x)\psi_n^*(y).$$

23.a Find the equation of motion obeyed by $H_{\mathcal{O}}(x, y; \tau)$ and its initial condition at $\tau = 0$.

23.b The zeta function of $\mathcal{O}$ is defined as $\zeta_{\mathcal{O}}(s) \equiv \sum_n \lambda_n^{-s}$. Show that it is given by

$$\zeta_{\mathcal{O}}(s) = \frac{1}{\Gamma(s)} \int_0^\infty \frac{d\tau}{\tau}\, \tau^s \int d^4x\, H_{\mathcal{O}}(x, x; \tau).$$

23.c Use this approach to show that the zeta function of the Euclidean Klein–Gordon operator $\mathcal{O} \equiv m^2 - \partial^2$ in four dimensions is

$$\zeta_{\mathcal{O}}(s) = \frac{m^4 V_4}{16\pi^2} \frac{1}{m^{2s}} \frac{1}{(s-1)(s-2)}$$

(V_4 is the volume of space, assumed to be finite), and deduce from this an expression for its determinant.

23.a From the fact that $\mathcal{O}\psi_n(x) = \lambda_n \psi_n(x)$ and the definition of the heat kernel, we immediately get

$$\partial_\tau H_{\mathcal{O}}(x, y; \tau) = -\mathcal{O}\, H_{\mathcal{O}}(x, y; \tau).$$

When $\mathcal{O}$ is an elliptic differential operator, this is indeed very similar to the heat equation, hence the name *heat kernel* given to the function $H_{\mathcal{O}}$. Moreover, when $\tau = 0$, we have

$$H_{\mathcal{O}}(x, y; 0) = \sum_n \psi_n(x)\psi_n^*(y) = \delta(x-y),$$

where we have used the completeness of the set of eigenfunctions.

23.b From the definition of the heat kernel, we also get

$$\int d^4x\, H_{\mathcal{O}}(x, x; \tau) = \sum_n e^{-\lambda_n \tau} \underbrace{\int d^4x\, \psi_n(x)\psi_n^*(x)}_{\delta_{nn}=1} = \sum_n e^{-\lambda_n \tau}.$$

Recall here the integral that defines the gamma function,

$$\Gamma(s) = \int_0^\infty \frac{d\tau}{\tau}\,\tau^s\, e^{-\tau},$$

from which we obtain

$$\int_0^\infty \frac{d\tau}{\tau}\,\tau^s \sum_n e^{-\lambda_n \tau} = \Gamma(s) \sum_n \lambda_n^{-s} = \Gamma(s)\zeta_{\mathcal{O}}(s),$$

where $\zeta_{\mathcal{O}}(s) \equiv \sum_n \lambda_n^{-s}$ is the zeta function associated with the operator $\mathcal{O}$. Therefore, we have shown that

$$\zeta_{\mathcal{O}}(s) = \frac{1}{\Gamma(s)} \int_0^\infty \frac{d\tau}{\tau}\,\tau^s \int d^4x\, H_{\mathcal{O}}(x, x; \tau).$$

This formula provides, in principle, a way of calculating the zeta function of the operator $\mathcal{O}$.

23.c Consider now the Klein–Gordon operator in d-dimensional Euclidean space, $\mathcal{O} \equiv m^2 - \partial_i\partial_i$. The equation of motion for its heat kernel reads

$$\partial_\tau H_{\mathcal{O}} = (\partial^2 - m^2)\, H_{\mathcal{O}}.$$

The easiest way to solve it is to Fourier transform $H_{\mathcal{O}}$ by defining

$$\widetilde{H}_{\mathcal{O}}(p; \tau) \equiv \int d^dx\, e^{-ip\cdot(x-y)}\, H_{\mathcal{O}}(x, y; \tau),$$

so that

$$\partial_\tau \widetilde{H}_{\mathcal{O}} = -(p^2 + m^2)\, \widetilde{H}_{\mathcal{O}}, \quad \text{with initial condition } \widetilde{H}_{\mathcal{O}}(p; 0) = 1.$$

This is an ordinary differential equation whose solution is trivial:

$$\widetilde{H}_{\mathcal{O}}(p; \tau) = e^{-(m^2+p^2)\tau}.$$

The inverse Fourier transform is a simple Gaussian integral, giving

$$H_{\mathcal{O}}(x, y; \tau) = \frac{e^{-m^2\tau}}{(4\pi\tau)^{d/2}}\, e^{-\frac{(x-y)^2}{4\tau}}. \tag{2.31}$$

From this intermediate result, we have

$$\begin{aligned}\int_0^\infty \frac{d\tau}{\tau}\,\tau^s \int d^dx\, H_{\mathcal{O}}(x, y; \tau) &= \frac{V_d}{(4\pi)^{d/2}} \int_0^\infty \frac{d\tau}{\tau}\,\tau^{s-d/2}\, e^{-(m^2\tau + \frac{(x-y)^2}{4\tau})} \\ &= \frac{2V_d}{(4\pi)^{d/2}} \left(\frac{|x-y|}{2m}\right)^{s-d/2} K_{s-d/2}(m|x-y|).\end{aligned}$$

When doing this calculation, we may view the heat kernel $H_{\mathcal{O}}(x, y; \tau)$ as a function of x and $x - y$. Since (2.31) depends only on the coordinate difference $x - y$, the integration over x at fixed $x - y$ simply produces the volume V_d of d-dimensional Euclidean space (we need this volume to be finite for the integral to converge). We have expressed the final answer in terms of the modified Bessel function of the second kind, whose integral representation is

$$K_\nu(x) \equiv \int_0^\infty \frac{d\tau}{\tau}\,\tau^\nu\,e^{-\frac{x}{2}(\tau+\tau^{-1})}.$$

The only property of this special function that we need is its behavior near $x = 0$,

$$K_\nu(x) \underset{x\approx 0}{=} \frac{\Gamma(\nu)}{2}\left(\frac{2}{x}\right)^\nu + \cdots.$$

Thanks to this formula, we obtain

$$\zeta_{\mathcal{O}}(s) = \frac{V_d}{(4\pi)^{d/2}}\frac{1}{m^{2s-d}}\frac{\Gamma(s-d/2)}{\Gamma(s)} \underset{d=4}{=} \frac{m^4 V_4}{16\pi^2}\frac{1}{m^{2s}}\frac{1}{(s-1)(s-2)}.$$

Here, we can note a slightly unpleasant feature of our definition of the zeta function of an operator, namely the fact that its dimension depends on the dimension of the operator and on the value of the argument s (this is because the eigenvalues λ_n have the same dimension as $\mathcal{O}$ itself). Since in this example $\mathcal{O}$ has mass dimension two, this can be remedied by redefining

$$\zeta_{\mathcal{O}}(s) \to \mu^{2s}\zeta_{\mathcal{O}}(s) \underset{d=4}{=} \frac{m^4 V_4}{16\pi^2}\left(\frac{\mu}{m}\right)^{2s}\frac{1}{(s-1)(s-2)},$$

where μ is some arbitrary mass scale (which one may think of as a renormalization scale). By taking the derivative of this expression at $s = 0$, we get

$$\ln\det\left(m^2 - \partial^2\right) \underset{d=4}{=} \frac{m^4 V_4}{16\pi^2}\left[\ln\left(\frac{m}{\mu}\right) - \frac{3}{4}\right].$$

24. Casimir Zero-Point Energy The goal of this problem is to calculate the Casimir force exerted on a pair of parallel plates in a massless scalar field theory. As shown in Figure 2.7, we consider a box of size $L \times L \times \ell$, containing three parallel plates at $z = 0, a$ and ℓ on which we assume that the field must vanish. $E(r)$ denotes the zero-point energy contained between a pair of such plates separated by r, and the Casimir energy is defined as $\mathcal{E}(a) \equiv E(a) + E(\ell - a) - 2\,E(\frac{\ell}{2})$.

24.a Determine the expression for $E(r)$ (as a combination of integrals and sums that we shall not try to compute at this point). In order to make it finite, introduce a cutoff function on the energy of the modes, $f(\omega/\Lambda)$.

24.b In the case where the separation r is large, approximate the discrete sum on the longitudinal modes by an integral, and show that

$$E(r) \underset{\text{large } r}{=} \frac{\pi^2 L^2 r}{8a^4}\int_0^\infty dn\,\mathcal{F}(n), \quad \text{with } \mathcal{F}(n) \equiv \int_{n^2}^\infty dx\,\sqrt{x}\,f\left(\frac{\pi\sqrt{x}}{a\Lambda}\right).$$

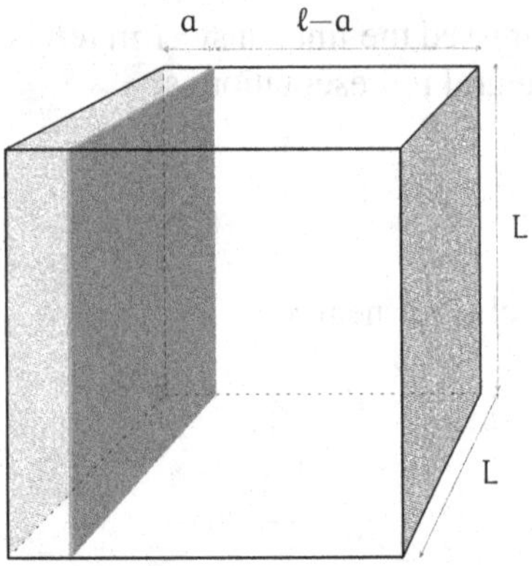

Figure 2.7 Setup for the calculation of the Casimir energy.

24.c When $r = a$, show that

$$E(a) = \frac{\pi^2 L^2}{8a^3} \sum_{n \geq 0} \mathcal{F}(n).$$

24.d Use the Euler–MacLaurin formula in order to obtain the following Casimir energy:

$$\mathcal{E}(a) = \text{const} - \frac{\pi^2 L^2}{1440 a^3}.$$

Calculate the corresponding force.

24.e Determine a constant energy–momentum tensor $T^{\mu\nu}$ such that $\mathcal{E}(a) = T^{00} L^2 a$.

24.a It is common to define the Hamiltonian of quantum field theory as normal-ordered in order to remove the infinite zero-point vacuum energy, on the grounds that it is a constant that does not play a role in physical processes since they are sensitive only to energy differences. However, when a system is subject to non-trivial boundary conditions, it may happen that changes at the boundaries lead to measurable variations of the vacuum energy. This is known as the *Casimir effect*. In this problem, we illustrate this phenomenon in the case of a massless scalar field theory, for a particularly simple geometrical configuration.

Let us denote by $E(r)$ the vacuum zero-point energy in a "slice" of size $L \times L \times r$ (with the field vanishing at $z = 0$ and $z = r$). With the notation of Figure 2.7, the Casimir energy is defined by

$$\mathcal{E}(a) \equiv E(a) + E(\ell - a) - 2\,E(\tfrac{\ell}{2}). \tag{2.32}$$

The subtraction in the third term is not crucial, but removes some infinite a-independent terms (although infinite, these terms are not very important, because they do not contribute to the Casimir force since they are constant). Since we assume that the walls that delimitate the two cavities in the z direction are such that the field vanishes on the walls, while there is no boundary condition in the x, y directions, the allowed Fourier modes $k_{x,y}$ are continuous variables, while k_z is discrete:

$$k_{\|} \equiv (k_x, k_y), \quad k_z = \frac{\pi n_z}{r}, \quad \omega(k_{\|}, n_z) \equiv \sqrt{k_{\|}^2 + k_z^2},$$

where n_z is an integer. Note that changing $n_z \to -n_z$ does not give a distinct eigenmode of the field; therefore, we must restrict it to $n_z \geq 0$. The corresponding zero-point vacuum energy is therefore given by

$$E(r) = L^2 \int \frac{d^2k_{\|}}{(2\pi)^2} \sum_{n_z \geq 0} \frac{\omega(k_{\|}, n_z)}{2} f\left(\tfrac{\omega(k_{\|}, n_z)}{\Lambda}\right),$$

where we have regularized the ultraviolet divergence of this sum/integral by inserting a regulating function $f(\omega/\Lambda)$ chosen such that $f(0) = 1$ and $f(z)$ decreases sufficiently fast when $z \to \infty$.

24.b When r is large, we can also treat the momentum k_z as a continuous variable, since the splitting between two successive modes is inversely proportional to the width of the cavity in this direction. In this limit, we obtain the following expression for $E(\ell)$:

$$\begin{aligned} E(r) \underset{\text{large } r}{=} & \; L^2 \int \frac{d^2k_{\|}}{(2\pi)^2} \int_0^{\infty} dn_z \, \frac{\omega(k_{\|}, n_z)}{2} f\left(\tfrac{\omega(k_{\|}, n_z)}{\Lambda}\right) \\ = & \; \frac{\pi^2 L^2 r}{8a^4} \int_0^{\infty} dn \underbrace{\int_{n^2}^{\infty} dx \, \sqrt{x} \, f\left(\tfrac{\pi\sqrt{x}}{a\Lambda}\right)}_{\equiv \mathcal{J}(n)}, \end{aligned}$$

where we have performed the angular integration in the (k_x, k_y) plane and introduced new integration variables $n \equiv a n_z / r$ and $x \equiv n^2 + \frac{a^2}{\pi^2} k_{\|}^2$ in the final expression.

24.c This approximation of a continuously varying n_z is not applicable to the first term of (2.32), where $r = a$. For this term, we may still perform the angular integration in the (k_x, k_y) plane and introduce the variables $n \equiv n_z$ and $x \equiv n^2 + \frac{a^2}{\pi^2} k_{\|}^2$ in order to obtain

$$\begin{aligned} E(a) &= L^2 \int_0^{\infty} \frac{dk_{\|} k_{\|}}{4\pi} \sum_{n_z \geq 0} \omega(k_{\|}, n_z) f\left(\tfrac{\omega(k_{\|}, n_z)}{\Lambda}\right) \\ &= \frac{\pi^2 L^2}{8a^3} \sum_{n \geq 0} \int_{n^2}^{\infty} dx \sqrt{x} \, f\left(\tfrac{\pi\sqrt{x}}{a\Lambda}\right) = \frac{\pi^2 L^2}{8a^3} \sum_{n \geq 0} \mathcal{J}(n). \end{aligned}$$

24.d Therefore, in the limit $a \ll \ell$, the Casimir energy is given by

$$\mathcal{E}(a) = \frac{\pi^2 L^2}{8a^3}\Big\{\sum_{n\geq 0} \mathcal{J}(n) - \int_0^\infty dn\, \mathcal{J}(n)\Big\}.$$

The most effective way to evaluate this quantity is by using the Euler–MacLaurin formula, which gives

$$\mathcal{E}(a) = \frac{\pi^2 L^2}{8a^3}\Big\{\tfrac{1}{2}\mathcal{J}(0) - \sum_{p=1}^{\infty} \frac{B_{2p}}{(2p)!}\mathcal{J}^{(2p-1)}(0)\Big\},$$

where the B_{2p} are the Bernoulli numbers ($B_2 = \frac{1}{6}, B_4 = -\frac{1}{30}, \ldots$; the odd ones are zero). In writing this expression, we have already used the fact that $\mathcal{J}(n)$ vanishes when $n \to \infty$, as do all its derivatives (this property follows from our assumptions about the behavior of the cutoff function $f(z)$ at large z). By a simple rescaling of the integration variable x, we see that the term in $a^{-3}\mathcal{J}(0)$ is independent of a and therefore only gives a constant (infinite) shift to the Casimir energy. Then, a straightforward calculation gives

$$\mathcal{J}'(0) = 0, \quad \mathcal{J}'''(0) = -4,$$

which leads to the following contribution to the Casimir energy:

$$\mathcal{E}(a) = \text{const} - \frac{\pi^2 L^2}{1440 a^3}.$$

Note that this result does not depend on the choice of the regulator function, aside from the generic properties we have assumed. (However, there are additional terms, suppressed by powers of $1/(a\Lambda)$, that are not universal and depend on the details of how the cutoff is implemented.) The derivative of this energy tells us the force exerted on the plate:

$$\mathcal{F}(a) = -\frac{d\mathcal{E}(a)}{da} = -\frac{\pi^2 L^2}{480 a^4},$$

which is attractive. A similar calculation could have been done in QED, with a result twice as large as the scalar one (because photons have two degrees of polarization per momentum mode). Although predicted in 1948, this force was first observed experimentally only in 1997.

24.e One may also view the Casimir energy as coming from a homogeneous "vacuum energy–momentum tensor," $\mathcal{E}(a) = L^2 a\, T^{00}$. With the geometrical setup of the problem at hand, this energy–momentum tensor must be constructed solely from $g^{\mu\nu}$ and from the unit vector $n^\mu = (0, 0, 0, 1)$ orthogonal to the plates. Moreover, it has to be symmetric, and traceless since we are considering massless fields (a free massless scalar field theory is scale invariant). These constraints lead to

$$T^{\mu\nu} = -\frac{\pi^2}{1440 a^4}(g^{\mu\nu} + 4n^\mu n^\nu).$$

From this, we obtain the pressure in the z direction (i.e., the force per unit of transverse area):

$$P_z \equiv T^{33} = -\frac{\pi^2}{480 a^4}.$$

This negative pressure implies an attraction between the plates, consistent with the force obtained above if multiplied by L^2 (recall that pressure is force per unit area).

25. The Feynman Diagrams behind the Casimir Force In Problem 24, we derived the expression for the Casimir force exerted on two nearby walls by calculating how the zero-point vacuum energy is altered by the presence of the two walls. This derivation may seem somewhat mystifying because nowhere did we talk about the interactions of the scalar field with the walls (these interactions were of course implicit in the boundary conditions imposed on the field). Here, we propose a more explicit derivation of the Casimir energy, achieved by modeling the field–wall interaction and by calculating the corresponding free energy, using the technique developed in Problem 23. This calculation suggests that the Casimir force is an actual wall–wall interaction mediated by a scalar loop.

The walls at $x^3 = 0$ and $x^3 = a$ are modeled by an external field $\sigma(x)$, coupled as follows to the scalar field:

$$\mathcal{L} = \tfrac{1}{2}(\partial_\mu\phi)(\partial^\mu\phi) - \tfrac{g}{2}\sigma\phi^2, \quad \sigma(x) \equiv \delta(x^3) + \delta(x^3 - a).$$

25.a Show that the free energy F of the system is $F = \frac{1}{2}\beta^{-1}\ln\det(g\sigma - \square_E)$, where β is the inverse temperature and $\square_E$ is the Euclidean d'Alembertian operator.

25.b Show that, in the limit $g \to \infty$, the normalized eigenfunctions of $g\sigma - \square_E$ are

$$\psi_{k,n}(x) = \sqrt{2}\, e^{ik\cdot x} \sin\left(\tfrac{\pi n x^3}{a}\right) \quad (k \equiv (k^0, k^1, k^2), x \equiv (x^0, x^1, x^2), n \in \mathbb{N}_+).$$

25.c Show that, when $g \to \infty$, the zeta function of the operator $g\sigma - \square_E$ is equal to

$$\zeta_{g\sigma-\square}(s) = \frac{\beta L^2}{8\pi^{3/2}}\left(\frac{a}{\pi}\right)^{2s-3}\frac{\zeta(2s-3)\Gamma(s-\frac{3}{2})}{\Gamma(s)}.$$

25.d Use this result to recover the expression for the Casimir energy.

25.a The walls are modeled as an external field $\sigma(x)$ localized on the two planes $x^3 = 0$ and $x^3 = a$, which acts as a kind of mass for the scalar field. The corresponding Lagrangian reads

$$\mathcal{L} = \tfrac{1}{2}(\partial_\mu\phi)(\partial^\mu\phi) - \tfrac{g}{2}\sigma\phi^2, \quad \sigma(x) \equiv \delta(x^3) + \delta(x^3 - a),$$

where g is a coupling constant that controls the strength of the interaction of the scalar field with the walls. In the limit $g \to \infty$, the potential of the field ϕ has very steep minima at $x^3 = 0$ and $x^3 = a$, which forces the field to vanish on the walls. Therefore, this limit should be equivalent to the boundary conditions assumed in Problem 24. By going to Euclidean time, we can express the partition function $Z(\beta)$ of the system as a path integral,

$$Z(\beta) = \int_{\phi(0,x)=\phi(\beta,x)} [D\phi]\, e^{-\mathcal{S}_E[\phi]}, \quad \text{with } \mathcal{S}_E[\phi] \equiv \frac{1}{2}\int_0^\beta d\tau d^3x \left(-\phi\square_E\phi + g\sigma\phi^2\right),$$

where β is the inverse temperature and $\square_E$ is the Euclidean d'Alembertian operator. The path integral is over field configurations that are periodic in Euclidean time with a period

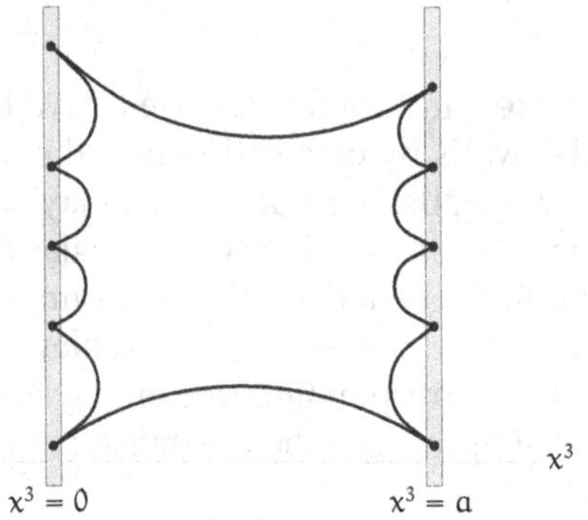

Figure 2.8 Example of a diagram contributing to the free energy in the presence of two walls described as localized external fields. The solid black lines are scalar propagators, and the dots denote insertions of the external field $g\sigma(x)$.

β. The finiteness of the time interval implies that the corresponding frequencies are discrete (they are the Matsubara frequencies). However, in the following, we shall assume that the spacing a between the walls is much smaller than β, which allows us to treat the frequencies in the temporal direction as continuous variables (physically, this assumption means that the temperature is low enough so that the thermal excitations have an energy much smaller than that of the discrete modes in the x^3 direction). From the partition function, we can obtain the free energy of the system as

$$F = -\beta^{-1} \ln(Z).$$

(Note that, since we neglect thermal fluctuations, the entropy is zero, and the energy and the free energy are identical.) Since the above action is quadratic, the path integral is immediate and leads to

$$Z(\beta) = \frac{\text{const}}{\sqrt{\det(g\sigma - \square_E)}}, \quad \text{i.e.,} \quad \ln(Z) = \text{const} - \tfrac{1}{2} \ln\det(g\sigma - \square_E). \tag{2.33}$$

Diagrammatically, the logarithm of the determinant of $g\sigma - \square_E$ is the sum of all the connected one-loop diagrams such as the one shown in Figure 2.8. One may view these graphs as a kind of wall–wall interaction mediated by scalar fields (given the coupling between ϕ and the field σ that describes the walls, this is the only allowed topology).

25.b In order to calculate the logarithm of the determinant that enters in Eq. (2.33), we use the zeta function regularization method developed in Problem 23, which requires first obtaining the eigenfunctions of the operator $g\sigma - \square_E$. Quite generally, these eigenfunctions can be written as

$$\psi_{k,\lambda}(x) \equiv e^{ik\cdot x} f_\lambda(x^3),$$

where we denote $k \equiv (k^0, k^1, k^2), x \equiv (x^0, x^1, x^2)$. The undetermined function $f_\lambda(x^3)$ must satisfy

$$\left(g\sigma(x^3) - \partial_3^2\right) f_\lambda(x^3) = \lambda^2 f_\lambda(x^3). \tag{2.34}$$

Since the function $\sigma(x^3)$ is zero almost everywhere, we look for a function $f_\lambda(x^3)$ of the form

$$f_\lambda(x^3) = \begin{cases} Ae^{i\lambda x^3} + Be^{-i\lambda x^3} & \text{for } x^3 \leq 0, \\ A'e^{i\lambda x^3} + B'e^{-i\lambda x^3} & \text{for } 0 \leq x^3 \leq a, \\ A''e^{i\lambda x^3} + B''e^{-i\lambda x^3} & \text{for } a \leq x^3. \end{cases}$$

Since the differential equation that controls f_λ has only second-order derivatives, the delta functions in $\sigma(x^3)$ produce jumps of the first derivative $f'_\lambda(x^3)$ at $x^3 = 0$ and a, but the function $f(x^3)$ itself is continuous. The continuity of the function leads to two conditions:

$$A + B = A' + B', \quad A'e^{i\lambda a} + B'e^{-i\lambda a} = A''e^{i\lambda a} + B''e^{-i\lambda a}.$$

Two additional conditions are obtained by integrating the differential equation (2.34) over infinitesimal intervals centered at $x^3 = 0$ and $x^3 = a$, respectively. Using the continuity of the function, we obtain a relationship between the values of the function at $x^3 = 0, a$, and the jump of its derivative at these points. This gives

$$\begin{aligned} g(A + B) &= i\lambda\,(A' - A - B' + B), \\ g(A'e^{i\lambda a} + B'e^{-i\lambda a}) &= i\lambda\,(A''e^{i\lambda a} - B''e^{-i\lambda a} - A'e^{i\lambda a} + B'e^{-i\lambda a}). \end{aligned}$$

Although this could in principle be carried out for any value of the coupling g, we restrict ourselves to the limit $g \to \infty$ for simplicity. In this limit, the above four conditions simplify to

$$B = -A, \quad B' = -A', \quad A'(e^{i\lambda a} - e^{-i\lambda a}) = 0, \quad A''e^{i\lambda a} + B''e^{-i\lambda a} = 0.$$

In order to have a non-zero solution in the range $[0, a]$, the frequency λ must be of the form

$$\lambda_n \equiv \frac{\pi n}{a}, \quad n \in \mathbb{N}_+.$$

When this is the case, the last constraint becomes $B'' = -A''$. Therefore, the eigenfunctions in the domain between the two walls are given by

$$\psi_{k,n}(x) = \sqrt{2}\, e^{ik\cdot x} \sin\left(\tfrac{\pi n x^3}{a}\right).$$

The prefactor $\sqrt{2}$ ensures that they are normalized with a unit probability per unit volume, i.e.,

$$\int d^4x\, |\psi_{k,n}(x)|^2 = \beta L^2 a,$$

where β is the length of the Euclidean time interval and L^2 is the transverse section under consideration. The completeness relation satisfied by these eigenfunctions reads

$$a^{-1} \sum_{n \geq 1} \int \frac{d^3k}{(2\pi)^3}\, \psi_{k,n}(x)\psi^*_{k,n}(x') = \delta(x - x').$$

25.c The heat kernel (see Problem 23) of $g\sigma - \Box_E$ is then given by

$$H_{g\sigma-\Box}(x, x'; \tau) \equiv \frac{2}{a}\sum_{n\geq 1}\int \frac{d^3k}{(2\pi)^3}\, e^{-\tau(k^2+\pi^2n^2/a^2)}\, e^{ik\cdot(x-x')}\sin\left(\tfrac{\pi n x^3}{a}\right)\sin\left(\tfrac{\pi n x'^3}{a}\right).$$

Evaluating this at $x = x'$ and integrating over x in a domain of transverse section L^2 between the two walls gives

$$\int d^4x\, H_{g\sigma-\Box}(x, x; \tau) = \beta L^2 \sum_{n\geq 1}\int \frac{d^3k}{(2\pi)^3}\, e^{-\tau(k^2+\pi^2n^2/a^2)},$$

and the corresponding zeta function is therefore

$$\zeta_{g\sigma-\Box}(s) = \frac{\beta L^2}{\Gamma(s)}\int_0^\infty \frac{d\tau}{\tau}\,\tau^s \sum_{n\geq 1}\int \frac{d^3k}{(2\pi)^3}\, e^{-\tau(k^2+\pi^2n^2/a^2)}.$$

The integral over k is an elementary Gaussian integral,

$$\int \frac{d^3k}{(2\pi)^3}\, e^{-\tau k^2} = \frac{1}{8\pi^{3/2}\tau^{3/2}}.$$

Up to a rescaling, the integral over τ gives a gamma function,

$$\int_0^\infty \frac{d\tau}{\tau}\,\tau^{s-\frac{3}{2}}\, e^{-\tau\pi^2n^2/a^2} = \left(\frac{a}{\pi n}\right)^{2s-3}\Gamma(s - \tfrac{3}{2}).$$

Finally, the sum on n produces a Riemann zeta function. Combining everything, we get

$$\zeta_{g\sigma-\Box}(s) = \frac{\beta L^2}{8\pi^{3/2}}\left(\frac{a}{\pi}\right)^{2s-3}\frac{\zeta(2s-3)\Gamma(s-\frac{3}{2})}{\Gamma(s)}.$$

25.d Noting that $\Gamma(s) \approx s^{-1}$ in the vicinity of $s = 0$, its derivative at $s = 0$ is

$$\zeta'_{g\sigma-\Box}(0) = -\ln\det\left(g\sigma - \Box_E\right) = \frac{\pi^{3/2}\beta L^2}{8a^3}\underbrace{\zeta(-3)}_{=\frac{1}{120}}\underbrace{\Gamma(-\tfrac{3}{2})}_{=\frac{4}{3}\pi^{1/2}} = \frac{\pi^2\beta L^2}{720\,a^3}.$$

Finally, we obtain the following expression for the free energy:

$$F = -\frac{1}{2}\beta^{-1}\zeta'_{g\sigma-\Box}(0) = -\frac{\pi^2 L^2}{1440\,a^3},$$

which is identical to the expression for the Casimir vacuum energy obtained in Problem 24. Although the derivation followed here is not technically simpler, it provides a much more satisfactory handle on how the fields interact with the walls. In particular, we see explicitly how the Casimir energy is nothing but the free energy of the two walls (strongly coupled to the fields) interacting by the exchange of scalar particles. Technical challenges aside, there would be no conceptual difficulty in extending the present approach to deal with the case where the field–wall interaction has a finite strength, or walls that have a non-zero thickness, or even to calculate the finite-temperature corrections (neglected here because we have replaced the Matsubara frequencies by a continuous k^0).
Source: this problem was inspired by Jaffe, R. L. (2005), *Phys Rev D* 72: 021301.

26. Gross-Neveu Model The *Gross–Neveu model* is a $(1+1)$-dimensional quantum field theory with N species of massless fermions (indexed by $n = 1, \dots, N$), with a quartic interaction,

$$\mathcal{L} \equiv i \sum_n \overline{\psi}_n \partial\!\!\!/ \psi_n + \frac{g^2}{2}\Big(\sum_n \overline{\psi}_n \psi_n\Big)^2 .$$

This quantum field theory is a toy model for the spontaneous breaking of chiral symmetry, and also provides an example of the technique of *bosonization*.

26.a Check that a valid representation of the Dirac matrices in two dimensions is $\gamma^0 = \sigma^2$, $\gamma^1 = i\sigma^1$ (with $\sigma^{1,2}$ Pauli matrices). Also define $\gamma^5 \equiv \gamma^0\gamma^1$. Check that $(\gamma^5)^2 = 1$, $(\gamma^5)^\dagger = \gamma^5$, $\{\gamma^5, \gamma^\mu\} = 0$ for $\mu = 0, 1$. Show that $\mathcal{L}$ is invariant under $\psi_n \to \gamma^5 \psi_n$. Would a mass term be consistent with this symmetry?
According to the power counting, can this theory be renormalizable in $d = 2$?

26.b Prove the following equality:

$$\int [D\psi D\overline{\psi}]\, e^{i\int \mathcal{L}} = \int [D\psi D\overline{\psi} D\phi]\, e^{i\int \sum_n \overline{\psi}_n (i\partial\!\!\!/ - \phi)\psi_n}\, e^{-\frac{i}{2g^2}\int \phi^2},$$

where ϕ is an auxiliary *scalar* field.

26.c We are now interested in the ground state of this theory, in which ϕ is constant. Perform the integration over the fermions in closed form for a constant ϕ, and show that the result of this integration is a factor

$$\exp\left\{ N \int \frac{d^2p}{(2\pi)^2} \ln(-p^2 + \phi^2) \right\}, \qquad \text{with } p^2 \equiv p_0^2 - p_1^2.$$

26.d Using $\ln(A) = -\partial_\alpha (A^{-\alpha})_{\alpha=0}$, and using dimensional regularization with $d = 2 - 2\epsilon$, show that the integral can be written as

$$\int \frac{d^dp}{(2\pi)^d} \ln(-p^2 + \phi^2) = i\frac{\phi^2}{4\pi}\left(\frac{1}{\epsilon} + \text{const} + 1 - \ln\phi^2 + \mathcal{O}(\epsilon^2)\right).$$

Which operator should one renormalize to eliminate the ultraviolet divergence? After introduction of an appropriate counterterm, show that the effective potential for the field ϕ can be written as

$$V_{\text{eff}}(\phi) = \frac{\phi^2}{2g^2} + \frac{N\phi^2}{4\pi}\left(\ln\left(\frac{\phi^2}{\mu^2}\right) - 1\right),$$

where μ is a constant whose precise value depends on the finite terms included in the counterterm besides the pole in ϵ^{-1}.

26.e Determine analytically the location of the minimum of $V_{\text{eff}}(\phi)$. Could this answer have been found by a perturbative expansion in powers of g^2? What are the consequences for the fermions of having a non-zero expectation value for ϕ?

26.f What are the diagrams that would give the result of the above functional calculation? Why is the answer exact in the limit $N \to \infty$?

26.a In order to verify that the proposed matrices are valid Dirac matrices in two dimensions, we just need to verify that $\{\gamma^\mu, \gamma^\nu\} = 2\,g^{\mu\nu}$:

$$(\gamma^0)^2 = (\sigma^2)^2 = 1, \quad (\gamma^1)^2 = (i\sigma^1)^2 = -1,$$
$$\gamma^0\gamma^1 + \gamma^1\gamma^0 = i\sigma^2\sigma^1 + i\sigma^1\sigma^2 = i(-i\sigma^3) + i(i\sigma^3) = 0.$$

Defining $\gamma^5 \equiv \gamma^0\gamma^1$, we have

$$(\gamma^5)^2 = \gamma^0\gamma^1\gamma^0\gamma^1 = -\gamma^0\underbrace{\gamma^1\gamma^1}_{-1}\gamma^0 = +1,$$
$$(\gamma^5)^\dagger = (i\sigma^2\sigma^1)^\dagger = -i\sigma^1\sigma^2 = +i\sigma^2\sigma^1 = \gamma^5,$$
$$\{\gamma^5, \gamma^\mu\} = \gamma^0\gamma^1\gamma^\mu + \gamma^\mu\gamma^0\gamma^1 = 0.$$

(In the final equality above, since μ is either 0 or 1, γ^μ commutes with one of $\gamma^{0,1}$ and anti-commutes with the other one.) Under the transformation $\psi_n \to \gamma^5\psi_n$, we have

$$\psi_n \to \gamma^5\psi_n,$$
$$\overline{\psi}_n \equiv \psi_n^\dagger\gamma^0 \to \psi_n^\dagger \underbrace{\gamma^{5\dagger}\gamma^0}_{=\gamma^5\gamma^0=-\gamma^0\gamma^5} = -\overline{\psi}_n\gamma^5,$$
$$\overline{\psi}_n\psi_n \to -\overline{\psi}_n\psi_n,$$
$$(\overline{\psi}_n\psi_n)^2 \to (\overline{\psi}_n\psi_n)^2,$$
$$\overline{\psi}_n\gamma^\mu\psi_n \to -\overline{\psi}_n\underbrace{\gamma^5\gamma^\mu\gamma^5}_{-\gamma^\mu}\psi_n = \overline{\psi}_n\gamma^\mu\psi_n,$$

hence the invariance of the Lagrangian. Note that a mass term is not allowed by this symmetry since $\overline{\psi}_n\psi_n \to -\overline{\psi}_n\psi_n$ under this transformation.

The mass dimension of the field ψ_n is $\frac{1}{2}$ in $d = 2$, and therefore the mass dimension of the coupling constant g is zero. This theory should therefore be renormalizable in two dimensions according to the power counting.

26.b In order to check the proposed equality, we proceed by explicitly performing the integration over ϕ on the right-hand side. For this, we need

$$-\tfrac{1}{2g^2}\phi^2 - \phi\sum_n \overline{\psi}_n\psi_n = -\tfrac{1}{2g^2}\Big[\phi^2 + 2g^2\phi\sum_n \overline{\psi}_n\psi_n\Big]$$
$$= -\tfrac{1}{2g^2}\Big[\Big(\phi + g^2\sum_n \overline{\psi}_n\psi_n\Big)^2 - g^4\Big(\sum_n \overline{\psi}_n\psi_n\Big)^2\Big].$$

Then, we define $\phi' \equiv \phi + g^2\sum_n \overline{\psi}_n\psi_n$ and trade the integration variable ϕ in favor of ϕ'. The integral produces a constant prefactor (which we ignore), and leaves a factor $\exp(\frac{g^2}{2}(\sum_n \overline{\psi}_n\psi_n)^2)$, hence the announced equality.

26.c When ϕ is a constant, we can perform the integration over the fermions explicitly. Each pair of integrals $D\psi_n D\overline{\psi}_n$ produces a factor

$$\det\left(i\partial\!\!\!/ - \phi\right).$$

These factors do not depend on the index n, and there are N of them, so the N fermionic integrations give

$$\left[\det\left(i\partial\!\!\!/ - \phi\right)\right]^N = \exp\left\{N \,\mathrm{tr}\,\ln\left(i\partial\!\!\!/ - \phi\right)\right\} = \exp\left\{N\int\frac{d^2p}{(2\pi)^2}\,\mathrm{tr}\,\ln(p\!\!\!/ - \phi)\right\}.$$

Note that the original trace was over both spacetime and the Dirac indices. In the final equality, we have rewritten the trace over spacetime as a trace over momentum and we keep unchanged the trace over the Dirac indices. For the latter, we can write

$$\mathrm{tr}\,\ln(p\!\!\!/ - \phi) = \ln\det\begin{pmatrix} -\phi & i(p^0 - p^1) \\ -i(p^0 + p^1) & -\phi \end{pmatrix} = \ln\left(\phi^2 - p^2\right).$$

(We have used the explicit form of the Pauli matrices in the first equality.) This gives the announced formula.

Alternative Proof: It is possible to do this calculation without resorting to any explicit representation of the Dirac matrices, as follows:

$$\begin{aligned}
\mathrm{tr}\,\ln(p\!\!\!/ - \phi) &= \underbrace{\mathrm{tr}\,\ln(-\phi)}_{=2\ln(-\phi)=\ln(\phi^2)} + \mathrm{tr}\,\ln(1 - p\!\!\!//\phi) \\
&= \ln(\phi^2) - \sum_{n\geq 1}\frac{1}{n}\,\mathrm{tr}\left(\left[\frac{p\!\!\!/}{\phi}\right]^n\right) = \ln(\phi^2) - \sum_{p\geq 1}\frac{1}{2p}\,\mathrm{tr}\left(\left[\frac{p\!\!\!/}{\phi}\right]^{2p}\right) \\
&= \ln(\phi^2) - \sum_{p\geq 1}\frac{1}{p}\left(\frac{p^2}{\phi^2}\right)^p = \ln(\phi^2) + \ln(1 - p^2/\phi^2) = \ln(\phi^2 - p^2).
\end{aligned}$$

(In the third line, we used the fact that $p\!\!\!/^{2p+1} = p^{2p}p\!\!\!/$ has a null trace, and therefore we keep only the terms of even order.)

26.d Then, the integration of the momentum p proceeds as follows:

$$\begin{aligned}
\int\frac{d^dp}{(2\pi)^d}\ln(-p^2 + \phi^2) &= -\partial_\alpha \int\frac{d^dp}{(2\pi)^d}\,\frac{1}{(-p^2+\phi^2)^\alpha}\bigg|_{\alpha=0} \\
&= -i\partial_\alpha \int\frac{d^dp_E}{(2\pi)^d}\,\frac{1}{(p_E^2+\phi^2)^\alpha}\bigg|_{\alpha=0} \\
&= -i\partial_\alpha \frac{\phi^{d-2\alpha}}{(4\pi)^{d/2}}\,\frac{\Gamma(\alpha - \frac{d}{2})}{\Gamma(\alpha)}\bigg|_{\alpha=0}.
\end{aligned}$$

Note that $\Gamma(\alpha) \approx \alpha^{-1}$ when $\alpha \to 0$. Therefore, we have

$$\partial_\alpha \frac{F(\alpha)}{\Gamma(\alpha)}\bigg|_{\alpha=0} = F(0),$$

and we obtain

$$\int \frac{d^dp}{(2\pi)^d} \ln(-p^2+\phi^2) = -i\frac{\phi^d}{(4\pi)^{d/2}}\Gamma(-\tfrac{d}{2}) = -i\frac{\phi^2\phi^{-2\epsilon}}{(4\pi)^{1-\epsilon}} \underbrace{\Gamma(-1+\epsilon)}_{-\frac{1}{\epsilon}+\text{const}+\mathcal{O}(\epsilon)}$$

$$= i\frac{\phi^2}{4\pi}\Big[\frac{1}{\epsilon} + \text{const} + 1 - \ln(\phi^2) + \mathcal{O}(\epsilon^2)\Big]. \tag{2.35}$$

(We could have written const + 1 as some unique constant, but it will be convenient later to pull out this 1.) The pole in ϵ^{-1} can be removed by renormalizing the operator ϕ^2. More precisely, we introduce the counterterm

$$-i\frac{\phi^2}{4\pi}\Big[\frac{1}{\epsilon} + \text{const}'\Big],$$

where const$'$ is a finite constant that may be subtracted along with the pole ϵ^{-1}. Combining the original term in ϕ^2 with the result (2.35) and the counterterm, we get

$$\frac{\phi^2}{2g^2} + \frac{N\phi^2}{4\pi}\Big(\ln\Big(\frac{\phi^2}{\mu^2}\Big) - 1\Big)$$

for the sum of the terms in ϕ^2 inside the exponential.

26.e The derivative of this effective potential is given by

$$\frac{\partial}{\partial\phi}V_{\text{eff}}(\phi) = 2\phi\Big[\frac{1}{2g^2} + \frac{N}{4\pi}\ln\Big(\frac{\phi^2}{\mu^2}\Big)\Big].$$

This derivative vanishes at $\phi = 0$, but this is a local maximum. The other zero of the derivative is at

$$\phi_* \equiv \mu \exp\Big(-\frac{\pi}{g^2N}\Big).$$

One may check that this is a local minimum. Note that the location of this minimum depends on the renormalization scale, but the minimum is always there, no matter which scale we choose. Note also that this minimum is non-perturbative since its value is a non-analytic function of the coupling g (more precisely, it is a function whose Taylor coefficients at $g = 0$ are all zero). Because of the coupling $\overline{\psi}_n\phi\psi_n$, and the non-zero expectation value of the scalar field, the fermions become massive in the ground state of this theory (this is similar to the Higgs mechanism), and the chiral symmetry is spontaneously broken.

26.f Diagrammatically, the determinant resulting from the integration over the fermion fields corresponds to the sum of all the diagrams with one fermion loop, dressed by insertions of the scalar field ϕ. By its position inside the exponential, N acts in a similar way to $\hbar^{-1}$. Taking the limit $N \to +\infty$ has the same effect as taking the limit $\hbar \to 0$, namely eliminating all the higher-order corrections. For this reason, the results we have derived in this problem do not receive any higher-order corrections when N is very large.

27. D'Hoker-Gagné Formula Consider a time-dependent $n \times n$ traceless Hermitian matrix $M(\tau)$, and define the time-ordered exponential $W[M] \equiv T\,\exp(i\int_0^T d\tau\, M(\tau))$. The goal of this problem is to show that its trace may be represented by the following Grassmann path integral with anti-periodic boundary conditions:

$$\mathrm{tr}\,\big(W[M]\big) = 2^n \int_0^{2\pi} \frac{d\phi}{2\pi} \int_{\substack{\text{anti-}\\ \text{periodic}}} \big[D\overline{\chi}D\chi\big]\, e^{i\phi\left(\frac{n}{2}-1+T^{-1}\int \overline{\chi}\chi\right)}\, e^{-\int \overline{\chi}\dot{\chi}+i\int \overline{\chi}M\chi}.$$

27.a Express the Grassmann integral on the right-hand side of the above formula in terms of the determinant of the operator $-\partial_\tau + iM + i\frac{\phi}{T}$. Why should this determinant be evaluated over the subspace of anti-periodic functions?

27.b Show that the eigenvalues of this operator are the numbers $\lambda_{i\ell} = \frac{i}{T}\big(w_i+\phi+\pi(2\ell+1)\big)$, with $1 \le i \le n, \ell \in \mathbb{Z}$, where the w_i are the eigenvalues of $-i\ln W[M]$.

27.c Regularize the determinant by dividing it by a similar determinant defined with $M = 0$ and $\phi = 0$. Use the representation of the cosine as an infinite product to obtain

$$\Big[\det\big(-\partial_\tau + iM + i\tfrac{\phi}{T}\big)\Big]_A = \prod_{i=1}^{n} \cos\big(\tfrac{w_i+\phi}{2}\big).$$

Then, perform the integral over ϕ to obtain the announced formula.

27.a First, note that if M is traceless Hermitian, then the Wilson line $W[M]$ is a unitary $n \times n$ matrix, with a unit determinant. Therefore, it may be diagonalized with eigenvalues on the complex unit circle, whose product is equal to 1. We may thus write $W[M]$ as

$$W[M] = \Omega^\dagger\, \mathrm{diag}\left(e^{iw_1}, \dots, e^{iw_n}\right)\Omega, \quad \text{with} \sum_{i=1}^{n} w_i = 0,$$

and its trace is equal to

$$\mathrm{tr}\,\big(W[M]\big) = \sum_{i=1}^{n} e^{iw_i}.$$

Let us now focus on the evaluation of the right-hand side of the proposed formula. The fermionic path integral leads to

$$\int_{\substack{\text{anti-}\\ \text{periodic}}} \big[D\overline{\chi}D\chi\big]\, e^{i\phi T^{-1}\int \overline{\chi}\chi}\, e^{-\int \overline{\chi}\dot{\chi}+i\int \overline{\chi}M\chi} = \Big[\det\big(-\partial_\tau + iM + i\tfrac{\phi}{T}\big)\Big]_A,$$

where the subscript A indicates that the determinant should be evaluated on the space of anti-periodic functions (since the path integral is restricted to anti-periodic fields).

27.b The equation for the eigenfunctions of the operator inside the determinant,

$$(-\partial_\tau + iM + i\tfrac{\phi}{T})F = \lambda F,$$

is solved by

$$F(\tau) = e^{-\lambda\tau} e^{i\phi\frac{\tau}{T}} \Big(T \exp i \int_0^\tau d\tau' \, M(\tau')\Big) F(0).$$

The condition of anti-periodicity over the interval $[0, T]$ reads

$$\begin{aligned} 0 = F(0) + F(T) &= \Big[1 + e^{-\lambda T} e^{i\phi} \, W[M]\Big] \, F(0) \\ &= \Omega^\dagger \Big[1 + e^{-\lambda T} e^{i\phi} \operatorname{diag}\big(e^{iw_1}, \ldots\big)\Big] \, \Omega \, F(0). \end{aligned}$$

For this equation to have non-null solutions, the $n \times n$ matrix in the square brackets must have a null determinant. Since this matrix is diagonal, this condition is

$$\prod_{i=1}^{n} \big(1 + e^{-\lambda T} e^{i(\phi + w_i)}\big) = 0.$$

This equation for λ has n infinite families of solutions:

$$\lambda_{i\ell} = \tfrac{i}{T} \big(w_i + \phi + \pi(2\ell + 1)\big) \qquad (1 \le i \le n, \ell \in \mathbb{Z}).$$

27.c If we simply multiply these eigenvalues, we obviously obtain an infinite product. We may regularize the determinant by dividing it by the analogous determinant obtained with M and ϕ set to zero, whose eigenvalues are

$$\lambda_{i\ell}^{(0)} = \tfrac{i}{T} \pi(2\ell + 1).$$

Therefore, the regularized determinant can be written as

$$\begin{aligned} \Big[\det\big(-\partial_\tau + iM + i\tfrac{\phi}{T}\big)\Big]_A &= \prod_{i=1}^{n} \prod_{\ell \in \mathbb{Z}} \Big(1 + \frac{w_i + \phi}{\pi(2\ell + 1)}\Big) \\ &= \prod_{i=1}^{n} \prod_{\ell=0}^{+\infty} \Big(1 - \frac{(w_i + \phi)^2}{\pi^2(2\ell + 1)^2}\Big). \end{aligned}$$

In the second equality, we have paired terms with opposite values of the index $2\ell + 1$. To proceed, we need the following representation of the cosine as an infinite product:

$$\cos(x) = \prod_{\ell=0}^{+\infty} \Big(1 - \frac{4x^2}{\pi^2(2\ell+1)^2}\Big).$$

Using this formula, we obtain

$$\Big[\det\big(-\partial_\tau + iM + i\tfrac{\phi}{T}\big)\Big]_A = \prod_{i=1}^{n} \cos\big(\tfrac{w_i+\phi}{2}\big)$$
$$= 2^{-n}\, e^{-i\phi\frac{n}{2}} \prod_{i=1}^{n} \Big(1 + e^{iw_i} e^{i\phi}\Big),$$

where we have used the fact that the sum of the w_i's is zero in order to obtain the final equality. Therefore, using the integral

$$\int_0^{2\pi} \frac{d\phi}{2\pi}\, e^{in\phi} = \delta_{n0} \qquad (n \in \mathbb{Z}),$$

the right-hand side of the proposed formula is

$$2^n \int_0^{2\pi} \frac{d\phi}{2\pi} \int_{\substack{\text{anti-}\\\text{periodic}}} [D\overline{\chi}D\chi]\, e^{i\phi\left(\frac{n}{2}-1+T^{-1}\int \overline{\chi}\chi\right)}\, e^{-\int \overline{\chi}\dot{\chi}+i\int \overline{\chi}M\chi}$$
$$= \int_0^{2\pi} \frac{d\phi}{2\pi}\, e^{-i\phi} \prod_{i=1}^{n} \Big(1 + e^{iw_i} e^{i\phi}\Big) = \sum_{i=1}^{n} e^{iw_i} = \operatorname{tr}\big(W[M]\big).$$

This formula may be used to rewrite the internal degrees of freedom as a Grassmann path integral in the worldline formalism.

Variants: Let us mention a possible modification of this representation. Instead of the term in $T^{-1}\phi \int \overline{\chi}\chi$ in the expononential, we could have substituted

$$T^{-1}\phi \int_0^T d\tau\, \overline{\chi}(\tau)\chi(\tau) \quad \to \quad \phi \int_0^T d\tau\, \delta(\tau-\tau_0)\, \overline{\chi}(\tau)\chi(\tau),$$

where $\tau_0 \in [0, T]$. With this modification, the Grassmann path integral leads to the following determinant:

$$\int_{\substack{\text{anti-}\\\text{periodic}}} [D\overline{\chi}D\chi]\, e^{i\phi \int \delta(\tau-\tau_0)\overline{\chi}\chi}\, e^{-\int \overline{\chi}\dot{\chi}+i\int \overline{\chi}M\chi} = \Big[\det\big(-\partial_\tau + iM + i\phi\delta(\tau-\tau_0)\big)\Big]_A,$$

and the solutions of the differential equation that determines the eigenfunctions read

$$F(\tau) = \begin{cases} e^{-\lambda\tau}\Big(T\exp i\int_0^\tau d\tau'\, M(\tau')\Big) F(0) & \text{for } \tau < \tau_0 \\ e^{-\lambda\tau}\, e^{i\phi}\Big(T\exp i\int_0^\tau d\tau'\, M(\tau')\Big) F(0) & \text{for } \tau > \tau_0 \end{cases}.$$

Since the value of this solution at $\tau = T$ is unchanged, the anti-periodicity condition that determines the eigenvalues remains the same, as does the final result.

Another possible modification is to trade the integral over the continuous variable ϕ for a discrete sum. Indeed, the role of this integration,

$$\int_0^{2\pi} \frac{d\phi}{2\pi}\, e^{-i\phi} \prod_{i=1}^{n} \left(1 + e^{iw_i} e^{i\phi}\right) = \sum_{i=1}^{n} e^{iw_i}, \tag{2.36}$$

is simply to "filter out" all the terms of the integrand that are not linear in the exponentials e^{iw_i}. The same effect can be achieved by using

$$\frac{1}{N} \sum_{\substack{\phi = 2\pi \frac{k}{N} \\ 0 \le k < N}} e^{in\phi} = \delta_{0,\, n \bmod N}.$$

In the integrand of Eq. (2.36), we must cancel all the terms in $e^{i\phi}, e^{2i\phi}, \ldots, e^{i(n-1)\phi}$, which requires us to choose $N \ge n$ in the discrete version. For instance, we may replace (2.36) by

$$\frac{1}{n} \sum_{\substack{\phi = 2\pi \frac{k}{n} \\ 0 \le k < n}} e^{-i\phi} \prod_{i=1}^{n} \left(1 + e^{iw_i} e^{i\phi}\right) = \sum_{i=1}^{n} e^{iw_i}.$$

Source: D'Hoker, E. and Gagné, D. G. (1996), *Nucl Phys B* 467: 297.

28. Stochastic Quantization The goal of this problem is to introduce the framework of *stochastic quantization* and to derive some basic results in this context, in the case of a scalar field theory of action $S_E[\phi]$ in d-dimensional Euclidean space. The starting point is to extend the arguments of the field to include a "fictitious time" τ: $\phi(x) \to \varphi(x, \tau)$, and to let these generalized fields obey the following Langevin equation:

$$\partial_\tau \varphi(x, \tau) = -\frac{\delta S_E[\varphi]}{\delta \varphi(x, \tau)} + \eta(x, \tau), \tag{2.37}$$

where $\eta(x, \tau)$ is a real random Gaussian variable whose distribution is given by

$$\big\langle \eta(x, \tau) \big\rangle_\eta = 0, \quad \big\langle \eta(x, \tau) \eta(x', \tau') \big\rangle_\eta = 2\,\delta(x - x')\delta(\tau - \tau').$$

28.a Write the probability distribution of the noise η.

28.b Denote by $P_\theta[\phi]$ the probability density of an ensemble of field configurations at the fictitious time τ. Assuming each field in this ensemble obeys the Langevin equation, show that the distribution evolves according to the following Fokker–Planck equation:

$$\partial_\tau P_\tau[\phi] = \int d^d x\, \frac{\delta}{\delta \phi(x)} \Big(\frac{\delta S_E}{\delta \phi(x)} + \frac{\delta}{\delta \phi(x)} \Big) P_\tau[\phi].$$

Hint: for the time being, assume that $\delta \varphi(x, \tau)/\delta \eta(x', \tau) = \frac{1}{2} \delta(x - x')$.

28.c Prove that $\delta \varphi(x, \tau)/\delta \eta(x', \tau) = \frac{1}{2} \delta(x - x')$.

28.d Check that $\exp(-\mathcal{S}_E)$ is a fixed point of the Fokker–Planck equation. By defining $Q_\tau \equiv P_\tau e^{\mathcal{S}_E/2}$, show that generic solutions of the Fokker–Planck equation converge towards this fixed point. Conclude that ensemble averages of products of solutions of the Langevin equation converge to the usual time-ordered correlation functions in the limit $\tau \to \infty$.

28.e Sketch the rules of "stochastic perturbation theory," i.e., the rules for the perturbative expansion of φ and noise averages of products of φ. Use them to recover the propagator of the non-interacting theory.

28.a Consider a scalar field theory of action $\mathcal{S}_E[\phi]$ in d-dimensional Euclidean space. The expectation values of time-ordered products of fields have the following path integral representation:

$$\langle T\,\phi(x_1)\cdots\phi(x_n)\rangle = \frac{\int[D\phi]\;\phi(x_1)\cdots\phi(x_n)\;\exp(-\mathcal{S}_E[\phi])}{\int[D\phi]\;\exp(-\mathcal{S}_E[\phi])}.$$

Now, we extend the arguments of the field to include a "fictitious time" τ: $\phi(x) \to \varphi(x,\tau)$ (to reduce the potential for confusion, we denote by ϕ the original field whose support is the Euclidean spacetime $\mathbb{R}^d$ and by φ the extended field whose support is $\mathbb{R}^{d+1}$) and we let this generalized field obey the stochastic equation (2.37) given in the statement of the problem. In this equation, $\eta(x,\tau)$ is a real random Gaussian variable of null mean and variance given by $\langle\eta(x,\tau)\eta(x',\tau')\rangle_\eta = 2\,\delta(x-x')\delta(\tau-\tau')$. (The subscript η indicates an average over the noise. All other correlation functions of the noise are given by Wick's theorem.) This Gaussian distribution can be represented by the following functional density:

$$W[\eta] \equiv W^{-1}\,e^{-\frac{1}{4}\int d^dx d\tau\,\eta^2(x,\tau)}, \quad \text{with } W \equiv \int[D\eta]\;e^{-\frac{1}{4}\int d^dx d\tau\,\eta^2(x,\tau)}.$$

Note that the solution $\varphi(x,\tau)$ of equation (2.37) depends implicitly on the noise function $\eta(x,\tau)$.

28.b Let us start from an ensemble of field configurations at $\tau = 0$, specified by a probability distribution $P_0[\phi]$. Evolving each field in this ensemble with the Langevin equation (2.37) from 0 to τ is equivalent to a change of the distribution $P_0 \to P_\tau$. In order to determine the equation that governs the variations of $P_\tau[\phi]$, consider a functional $F[\phi]$ of the field and express its expectation value at the fictitious time τ in two different ways:

$$\langle F[\varphi(x,\tau)]\rangle_\eta = \int[D\phi]\;P_\tau[\phi]\;F[\phi] = W^{-1}\int[D\eta]\;F[\varphi(x,\tau)]\;e^{-\frac{1}{4}\int\eta^2}.$$

The first expression simply expresses how we could evaluate the expectation value if we knew the probability distribution $P_\tau[\phi]$ at τ. In the second expression, we evaluate F at the solution

φ of the Langevin equation, and then average over all the realizations of the noise. Likewise, the τ derivative of this expectation value can be written in two ways:

$$\partial_\tau \langle F[\varphi(x,\tau)]\rangle_\eta = \int [D\phi]\,(\partial_\tau P_\tau[\phi])\,F[\phi]$$
$$= W^{-1}\int [D\eta]\int d^dx\,(\partial_\tau\varphi(x,\tau))\frac{\delta F[\varphi]}{\delta\varphi(x,\tau)}\,e^{-\frac{1}{4}\int\eta^2}.$$

Using the Langevin equation, the second form of this derivative can be written as

$$W^{-1}\int [D\eta]\int d^dx\Big(-\frac{\delta S_E}{\delta\varphi(x,\tau)}+\eta(x,\tau)\Big)\frac{\delta F[\varphi]}{\delta\varphi(x,\tau)}\,e^{-\frac{1}{4}\int\eta^2}.$$

Let us focus on the second term of this expression. We have

$$\int [D\eta]\int d^dx\,\frac{\delta F[\varphi]}{\delta\varphi(x,\tau)}\eta(x,\tau)\,e^{-\frac{1}{4}\int\eta^2} = -2\int [D\eta]\int d^dx\,\frac{\delta F[\varphi]}{\delta\varphi(x,\tau)}\frac{\delta e^{-\frac{1}{4}\int\eta^2}}{\delta\eta(x,\tau)}$$
$$= 2\int [D\eta]\int d^dx\Big(\frac{\delta}{\delta\eta(x,\tau)}\frac{\delta F[\varphi]}{\delta\varphi(x,\tau)}\Big)\,e^{-\frac{1}{4}\int\eta^2}.$$

Using the chain rule, this becomes

$$2\int [D\eta]\int d^dx\Big(\int d^dy\,\underbrace{\frac{\delta\varphi(y,\tau)}{\delta\eta(x,\tau)}}_{=\frac{1}{2}\delta(x-y)}\frac{\delta^2 F[\varphi]}{\delta\varphi(x,\tau)\delta\varphi(y,\tau)}\Big)\,e^{-\frac{1}{4}\int\eta^2}$$
$$= \int [D\eta]\int d^dx\,\frac{\delta^2 F[\varphi]}{\delta\varphi(x,\tau)^2}\,e^{-\frac{1}{4}\int\eta^2}.$$

Collecting all the terms and rewriting the noise average of a functional of $\varphi(x,\tau)$ as an average weighted by $P_\tau[\phi]$ of the same functional of $\phi(x)$, we have thus shown that

$$\int [D\phi]\,(\partial_\tau P_\tau[\phi])\,F[\phi] = \int [D\phi]\,P_\tau[\phi]\int d^dx\Big(-\frac{\delta S_E}{\delta\phi(x)}+\frac{\delta}{\delta\phi(x)}\Big)\frac{\delta}{\delta\phi(x)}F[\phi]$$
$$= \int [D\phi]\Big\{\int d^dx\,\frac{\delta}{\delta\phi(x)}\Big(\frac{\delta S_E}{\delta\phi(x)}+\frac{\delta}{\delta\phi(x)}\Big)P_\tau[\phi]\Big\}F[\phi].$$

Since this identity must be true for all functional $F[\phi]$, we must in fact have

$$\partial_\tau P_\tau[\phi] = \int d^dx\,\frac{\delta}{\delta\phi(x)}\Big(\frac{\delta S_E}{\delta\phi(x)}+\frac{\delta}{\delta\phi(x)}\Big)P_\tau[\phi],$$

which is a *Fokker–Planck equation* for the probability distribution.

28.c Let us come back to a property used without proof in the derivation of the Fokker–Planck equation, about the value of the derivative $\delta\varphi/\delta\eta$ when the field and the noise are at the same τ. Let us denote

$$X_{x'\tau'}(x,\tau) \equiv \frac{\delta\varphi(x,\tau)}{\delta\eta(x',\tau')}.$$

In order to obtain an evolution equation in τ for this quantity, we differentiate the Langevin equation with respect to $\eta(x',\tau')$. This gives a linear equation of the form

$$\partial_\tau X_{x'\tau'} = \mathcal{D}_x\, X_{x'\tau'} + \delta(x-x')\delta(\tau-\tau'),$$

where $\mathcal{D}_x$ is the second derivative of the action with respect to the field. Since $\mathcal{D}_x$ contains the d'Alembertian operator, this equation is a kind of diffusion equation (with some corrections if the action $\mathcal{S}_E$ has interaction terms) with a source term. Its solution can be formally written as

$$X_{x'\tau'}(x,\tau) = \theta(\tau-\tau')\, e^{(\tau-\tau')\mathcal{D}_x}\, \delta(x-x').$$

(A more correct way of writing this would require a τ-ordering of the exponentials, but this does not change the conclusion.) The main point is that the dependence of φ at τ with respect to the noise at τ' starts at $\tau=\tau'$ in the form of a localized distribution $\delta(x-x')$, and then diffuses as τ increases. Thus, when $\tau=\tau'$ we have

$$X_{x'\tau'}(x,\tau) = \theta(0)\,\delta(x-x') = \tfrac{1}{2}\,\delta(x-x'),$$

where the factor $\frac{1}{2}$ results from a symmetric definition of the step function.

28.d Verifying that $\exp(-\mathcal{S}_E)$ is a fixed point of the Fokker–Planck equation simply requires us to evaluate its right-hand side with this functional as input, which trivially gives zero.

In order to study the convergence of the solutions towards this fixed point, a standard trick is to define a new functional density Q_τ by

$$P_\tau \equiv Q_\tau\, e^{-\mathcal{S}_E/2}.$$

The functional Q_τ satisfies

$$\partial_\tau Q_\tau = e^{\mathcal{S}_E/2}\,\partial_\tau P_\tau = e^{\mathcal{S}_E/2}\int \frac{\delta}{\delta\phi}\Big(\frac{\delta}{\delta\phi}+\frac{\delta\mathcal{S}_E}{\delta\phi}\Big)Q_\tau\, e^{-\mathcal{S}_E/2}$$
$$= -\int\Big(-\frac{\delta}{\delta\phi}+\frac{\delta\mathcal{S}_E}{\delta\phi}\Big)\Big(\frac{\delta}{\delta\phi}+\frac{\delta\mathcal{S}_E}{\delta\phi}\Big)\,Q_\tau.$$

Written in this form, we see that the operator that acts on Q_τ on the right-hand side is self-adjoint and non-positive. Therefore, its spectrum is real and bounded from above by zero. Moreover, the fixed point we have identified provides an eigenfunction with a null eigenvalue, i.e., the largest possible. All the eigenmodes with strictly negative eigenvalues decay as $\tau\to\infty$, suggesting that generic solutions of the Fokker–Planck equation indeed converge towards a distribution proportional to $\exp(-\mathcal{S}_E)$. (The weak point of this argument is that it is in general difficult to prove that the zero mode is unique, and that there is not an accumulation of eigenvalues close to zero, whose corresponding eigenfunctions would decay very slowly.)

The convergence of solutions of the Fokker–Planck equation to $\exp(-\mathcal{S}_E)$ implies the following equality,

$$\lim_{\tau\to\infty} \big\langle \varphi(x_1,\tau)\cdots\varphi(x_n,\tau)\big\rangle_\eta = \big\langle \mathrm{T}\,\phi(x_1)\cdots\phi(x_n)\big\rangle,$$

between noise averages of products of solutions of the Langevin equation (in the limit of large fictitious time τ) and quantum expectation values of time-ordered products of field operators. This approach therefore provides an alternative to the canonical or path integral definitions of a quantum field theory, and for this reason it is called *stochastic quantization.*

28.e The solution of the Langevin equation with a null initial condition can be represented as an infinite sum of tree diagrams:

$\varphi(x,\tau) =$ 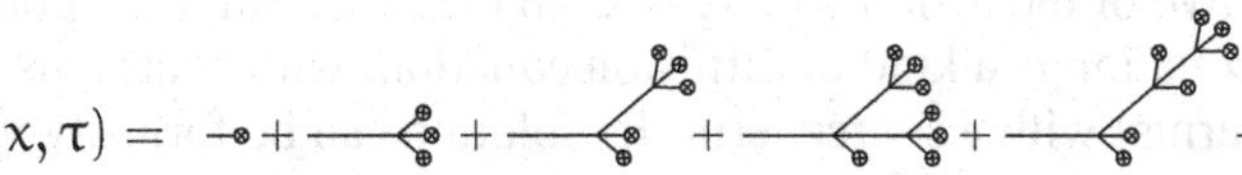 ,

(represented here for a φ^4 interaction term in $\mathcal{S}_E[\varphi]$), where the circled crosses denote the noise η and the lines are Green's functions of the operator $\partial_\tau + m^2 - \square_E$:

$$\big(\partial_\tau + m^2 - \square_E\big)\,\mathbb{G}(x\tau, x'\tau') = \delta(\tau-\tau')\delta(x-x').$$

When an explicit representation of $\mathbb{G}$ is needed, it is convenient to work in the mixed representation (τ, p) (with p the Fourier conjugate of x), in which the propagator is defined by

$$\big(\partial_\tau + m^2 + p^2\big)\,\mathbb{G}(\tau,\tau',p) = \delta(\tau-\tau'),$$

which leads to

$$\mathbb{G}(\tau,\tau',p) = \theta(\tau-\tau')\,e^{-(m^2+p^2)(\tau-\tau')}.$$

(This is the retarded solution, appropriate for solving the Langevin equation from some initial condition at $\tau = 0$.) In this mixed representation, there is an integration over τ at the vertices and at each insertion of the noise, and the momentum p is conserved at the vertices.

The average over the Gaussian noise amounts to connecting the crosses pairwise in the preceding diagrammatic representation. To work in the mixed representation described above, we only need

$$\big\langle \eta(\tau,p)\eta(\tau',p')\big\rangle = 2\,\delta(\tau-\tau')(2\pi)^d\delta(p+p').$$

(Although individual realizations of the noise break translation invariance, the noise distribution is translation invariant. For this reason, momentum conservation at the vertices holds only for noise-averaged quantities.) The simplest of these connections is the following:

$\mathcal{G}(\tau,\tau',p) \equiv$ 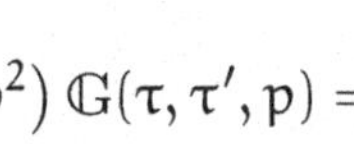

$$= 2\int d\tau''\;\mathbb{G}(\tau,\tau'',p)\mathbb{G}(\tau',\tau'',-p)$$

$$= 2\int_0^{\min(\tau,\tau')} d\tau''\; e^{-(m^2+p^2)(\tau+\tau'-2\tau'')}$$

$$= \frac{e^{-(m^2+p^2)\overbrace{(\tau+\tau'-2\min(\tau,\tau'))}^{|\tau-\tau'|}} - e^{-(m^2+p^2)(\tau+\tau')}}{p^2+m^2}.$$

If we set $\tau = \tau'$ and take the limit $\tau, \tau' \to +\infty$, we indeed obtain the bare (Euclidean) scalar propagator. By combining contributions of orders η and η^3 in two φ's, we can construct the following more elaborate example:

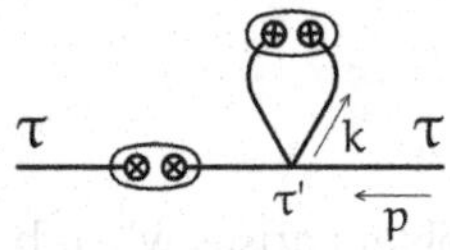

$$
\begin{aligned}
&= \int d\tau' \frac{d^4k}{(2\pi)^4}\, \mathcal{G}(\tau,\tau',p)\mathcal{G}(\tau',\tau',k)\mathbb{G}(\tau,\tau',-p) \\
&= \int_0^\tau d\tau' \frac{d^4k}{(2\pi)^4} \frac{e^{-(m^2+p^2)(\tau-\tau')} - e^{-(m^2+p^2)(\tau+\tau')}}{p^2+m^2} \\
&\quad \times \frac{1-e^{-2(m^2+k^2)\tau'}}{k^2+m^2}\, e^{-(m^2+p^2)(\tau-\tau')} \\
&\underset{\tau\to\infty}{=} \frac{1}{2}\frac{1}{(p^2+m^2)^2}\int \frac{d^4k}{(2\pi)^4}\frac{1}{k^2+m^2}.
\end{aligned}
$$

We again recover the expected perturbative contribution (ultraviolet divergence included). Source: Damgaard, P. H. and Hüffel, H. (1987), *Phys Rep* 152: 227.

---

29. Complex Langevin Equation This problem is a continuation of Problem 28. Its goal is to explore within a toy model what happens when the action is complex-valued, leading to a *complex Langevin equation.* We consider the "action" $S(x) \equiv \frac{\alpha}{2}x^2 + \frac{\beta}{4}x^4$, and write the following Langevin equation:

$$
\partial_\tau x = -S'(x) + \eta, \quad \langle \eta(\tau)\rangle_\eta = 0, \quad \langle \eta(\tau)\eta(\tau')\rangle_\eta = 2\delta(\tau-\tau').
$$

29.a When α, β are complex, why can't the distribution of x defined from solutions of this equation be proportional to $\exp(-S(x))$?

29.b Show that the moments $C_n \equiv \left\langle \frac{z^n(\tau)}{n!}\right\rangle_\eta$ (if they converge) satisfy in the limit $\tau \to \infty$ the same three-term recurrence relation,

$$
C_{n-2} = \alpha n C_n + \beta n(n+1)(n+2)C_{n+2},
$$

as the true moments $\int dx\, \frac{x^n}{n!}\exp(-S(x))/\int dx\, \exp(-S(x))$.

29.c Show that the moments C_n obtained from the complex Langevin equation coincide with the true moments if we have $\lim_{n\to\infty} C_n = 0$.

29.d If $\beta \in \mathbb{R}_+^*$, find a condition on α that guarantees the convergence of the moments C_n (use results from Gibson, P. M. (1971), *Proc Edinb Math Soc* 17(4): 317).

29.e Show that the complex Langevin equation is equivalent to the following Fokker–Planck equation for a distribution $P_\tau(x,y)$ over the complex plane:

$$
\partial_\tau P_\tau(x,y) = \left(\partial_x^2 + \partial_x \mathrm{Re}\, S'(x+iy) + \partial_y \mathrm{Im}\, S'(x+iy)\right) P_\tau(x,y).
$$

29.f For $\beta = 0, \alpha \equiv a + ib$, show that the fixed point of this Fokker–Planck equation is Gaussian,

$$P_\infty(x, y) \propto \exp\Big\{ - a\Big(x^2 + \tfrac{2a}{b}xy + (1 + \tfrac{2a^2}{b})y^2\Big)\Big\}.$$

Although everything looks fine in this case, what practical problem arises when $b \gg a$?

29.a As we have seen in Problem 28, when α, β are positive real integers we can view $\exp(-S(x))$ as a probability distribution on the real axis, and generate an ensemble of x's that samples this distribution by solving the following Langevin equation:

$$\partial_\tau x = -S'(x) + \eta, \quad \big\langle\eta(\tau)\big\rangle = 0, \quad \big\langle\eta(\tau)\eta(\tau')\big\rangle = 2\delta(\tau - \tau'). \tag{2.38}$$

Moreover, the probability distribution $P_\tau(x)$ of an ensemble of x's that obey this Langevin equation evolves according to a Fokker–Planck equation,

$$\partial_\tau P_\tau = \partial_x\big(\partial_x + S'(x)\big)\, P_\tau,$$

that admits $\exp(-S(x))$ as a unique fixed point.

Consider now the case where the coefficients α, β that define $S(x)$ are complex. In this case, the weight $\exp(-S(x))$ in the integral is complex-valued and cannot be interpreted as a probability distribution (moreover, in this case, the integrand has a strong oscillatory behavior that prevents evaluation by Monte Carlo sampling).

On the other hand, when we solve the Langevin equation (2.38) with a complex action, it leads to complex solutions even if the initial condition and the noise are real-valued. For this reason, it is called a *complex Langevin equation*. Thus, the equilibrium distribution of an ensemble of $z \equiv x + iy$'s obeying this Langevin equation is a positive definite distribution *on the complex plane*, i.e., definitely not the complex-valued weight $\exp(-S(x))$ of the original integral.

29.b From the solution of the complex Langevin equation, let us define τ-dependent moments by averaging over the noise, as follows:

$$C_n(\tau) \equiv Z^{-1}\int [D\eta]\, \tfrac{z^n(\tau)}{n!}\, e^{-\frac{1}{4}\int \eta^2}, \quad \text{with } Z \equiv \int [D\eta]\, e^{-\frac{1}{4}\int \eta^2}.$$

From this representation, we obtain the derivative of the moments with respect to τ:

$$
\begin{aligned}
\partial_\tau C_n(\tau) &= Z^{-1}\int [D\eta(\tau)]\, \frac{z^{n-1}(\tau)}{(n-1)!}\partial_\tau z(\tau)\; e^{-\frac{1}{4}\int \eta^2} \\
&= Z^{-1}\int [D\eta(\tau)]\, \frac{z^{n-1}(\tau)}{(n-1)!}\big(-S'(z(\tau))+\eta(\tau)\big)\; e^{-\frac{1}{4}\int \eta^2} \\
&= Z^{-1}\int [D\eta(\tau)]\, \frac{z^{n-1}(\tau)}{(n-1)!}\Big(-S'(z(\tau))-2\frac{\delta}{\delta\eta(\tau)}\Big)\; e^{-\frac{1}{4}\int \eta^2} \\
&= -\Big\langle \frac{x^{n-1}(\tau)}{(n-1)!}\, S'(x(\tau))\Big\rangle_\eta + 2\Big\langle \frac{x^{n-2}(\tau)}{(n-2)!}\, \underbrace{\frac{\delta z(\tau)}{\delta\eta(\tau)}}_{=\frac{1}{2}}\Big\rangle_\eta,
\end{aligned}
$$

i.e.,

$$
\partial_\tau C_n(\tau) = C_{n-2} - \alpha n C_n - \beta n(n+1)(n+2)C_{n+2}. \tag{2.39}
$$

Consider now the true moments of this toy model,

$$
M_n \equiv \frac{\int dx\; e^{-S(x)}\,\frac{x^n}{n!}}{\int dx\; e^{-S(x)}}.
$$

A simple integration by parts shows that

$$
M_{n-2} = \alpha n M_n + \beta n(n+1)(n+2)M_{n+2}.
$$

Interestingly, the fixed points of the equation (2.39) (defined by requiring that $\partial_\tau C_n = 0$) obey exactly the same three-term recurrence relation.

This has led to the conjecture that the probability distribution $P_\tau(x,y)$ in the complex plane (we denote by x and y the real and imaginary parts, respectively) followed by an ensemble of solutions of the complex Langevin equation is such that

$$
\lim_{\tau\to\infty}\int dxdy\; P_\tau(x,y)\, f(x+iy) \underset{?}{=} \int dx\; e^{-S(x)}\, f(x). \tag{2.40}
$$

The remarkable aspect if this equality (if true) is that the left hand side has a positive weight over $\mathbb{C}$ and the test function f is evaluated at complex numbers, while on the right-hand side the function f is evaluated solely on the real axis but weighted with a complex-valued action. This certainly requires at a minimum that the function $f(z)$ be analytic so that the values on the real axis are related to those in the rest of the complex plane (but this limitation should not be a surprise, since we arrived at this conjecture from considerations on the moments). If this conjecture were valid (and generalizable from this toy model to a realistic quantum field theory), it would offer the prospect of turning a path integral that has a sign problem (e.g., a path integral in Minkowski spacetime) into one that has a positive definite weight over a complexified field space (and is thus approachable by Monte Carlo sampling).

The question of the validity of the conjecture (2.40) can in fact be divided into three sub-questions:

1. If the moments $C_n(\tau)$ defined from the complex Langevin equation do converge to a fixed point, does this fixed point actually give the true moments M_n?

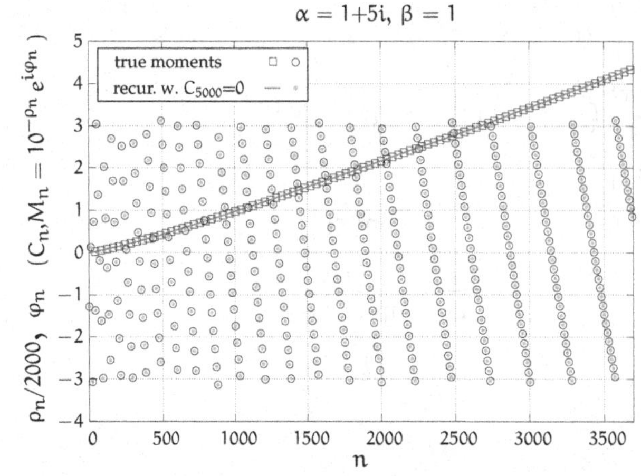

Figure 2.9 Comparison of the true moments M_n (evaluated numerically) and the moments C_n obtained by solving the recurrence with the constraints $C_0 = 1$, $C_{5000} = 0$. The moments are parameterized as $10^{-\rho_n} e^{i\varphi_n}$; the exponent ρ_n (divided by 2000, open squares and solid line) and the phase φ_n (open circles and solid dots) are both shown in the plot.

2. Do they actually converge to a fixed point?

3. In practice one has only a finite number of solutions of the complex Langevin equation, which is equivalent to evaluating the left-hand side of (2.40) by Monte Carlo sampling. Even though the distribution $P_\tau(x, y)$ is positive, are there situations where the integral on x, y cannot be sampled efficiently?

29.c The recurrence relation satisfied both by the fixed point of the C_n and by the true moments M_n is not sufficient to uniquely determine the moments, and therefore cannot guarantee that the C_n and M_n are equal. Since this recurrence is linear and involves three terms, the space of its solutions is a linear space of dimension two. The zeroth-order moment is trivially fixed at $C_0 = M_0 = 1$ in both cases, which reduces to one the dimension of the space of solutions. One may check that the true moments M_n go to zero when $n \to \infty$, thanks to the factorial included in their definition. Therefore, if we also had $C_\infty = 0$, the recurrence would determine a unique sequence of moments, and this sequence would coincide with the true moments. This is illustrated in Figure 2.9 (for practical reasons related to the numerical algorithm used to solve the recurrence, we have set $C_{5000} = 0$ instead of $C_\infty = 0$), where we see a perfect agreement (both in norm and in phase) between the true moments and those obtained by solving the recurrence.

29.d Let us now discuss the convergence of the moments C_n when $\tau \to +\infty$. Since $C_0 = 1$ is independent of τ, and using the fact that the odd moments are zero, we can write the equation of motion of the higher moments as follows:

$$\partial_\tau \mathbf{C} = \mathbf{C}_* + \mathbf{B}\mathbf{C},$$

where we denote

$$\mathbf{C} \equiv \begin{pmatrix} C_2 \\ C_4 \\ \vdots \\ C_{2N} \end{pmatrix}, \quad \mathbf{C}_* \equiv \begin{pmatrix} C_0 \\ 0 \\ \vdots \\ 0 \end{pmatrix}, \quad \mathbf{B} \equiv \begin{pmatrix} -2\alpha & -24\beta & 0 & \cdots & 0 & 0 \\ 1 & -4\alpha & -120\beta & \cdots & 0 & 0 \\ \vdots & \vdots & \vdots & & \vdots & \vdots \\ 0 & 0 & 0 & \cdots & 1 & -2N\alpha \end{pmatrix}.$$

(Here, we have truncated the recurrence by imposing that $C_{2N+2} = 0$. Alternatively, one could view $\mathbf{B}$ as a matrix of infinite size.) The matrix $\mathbf{B}$ is invertible, and the fixed point of the preceding equation is given by

$$\mathbf{C}(\infty) = -\mathbf{B}^{-1}\mathbf{C}_*.$$

Whether the solution actually converges to this fixed point is controlled by the sign of the real part of the eigenvalues of the matrix $\mathbf{B}$. A formal solution of the equation of motion of the moments is

$$\mathbf{C}(\tau) = e^{\tau\mathbf{B}}\left[\mathbf{C}(0) + \int_0^\tau d\tau'\, e^{-\tau'\mathbf{B}}\, \mathbf{C}_*\right],$$

where $\mathbf{C}(0)$ is the initial condition for the moments $C_{2,4,6,\ldots}$. For this solution to forget its initial condition, we must have

$$\lim_{\tau\to+\infty} e^{\tau\mathbf{B}} = 0,$$

which requires that all the eigenvalues of $\mathbf{B}$ have strictly negative real parts. (The matrix $\mathbf{B}$ is non-Hermitian, and therefore not necessarily diagonalizable. However, the property $\exp(\tau\mathbf{B}) \to 0$ if all eigenvalues have negative real parts remains true, as can be seen by using the Jordan form of the matrix.) Moreover, when this is the case, we also have

$$\lim_{\tau\to+\infty} e^{\tau\mathbf{B}} \int_0^\tau d\tau'\, e^{-\tau'\mathbf{B}} = -\mathbf{B}^{-1},$$

which ensures that the solution converges to the expected fixed point.

When $\beta \in \mathbb{R}$ (in this case, β should be positive for the original integral to make sense), we may use the following theorem (see Gibson, P. M. (1971), *Proc Edinb Math Soc* 17(4): 317): given a complex tri-diagonal matrix

$$\Omega \equiv \begin{pmatrix} b_1 & c_1 & 0 & \cdots & 0 & 0 \\ a_1 & b_2 & c_2 & \cdots & 0 & 0 \\ 0 & a_2 & b_3 & \cdots & 0 & 0 \\ \vdots & \vdots & \vdots & & \vdots & \vdots \\ 0 & 0 & 0 & \cdots & b_{n-1} & c_{n-1} \\ 0 & 0 & 0 & \cdots & a_{n-1} & b_n \end{pmatrix},$$

with $a_i, b_i, c_i \in \mathbb{C}$, if $a_i c_i \in \mathbb{R}$ and $a_i c_i \leq 0$ for all $1 \leq i \leq n-1$, then all the eigenvalues λ of Ω satisfy

$$\min\left\{\mathrm{Re}\,(b_i)\right\} \leq \mathrm{Re}\,(\lambda) \leq \max\left\{\mathrm{Re}\,(b_i)\right\}.$$

Therefore, when β is real positive, we may conclude that all the eigenvalues of $\mathbf{B}$ have a real part whose sign is opposite to that of the real part of α:

- $\mathrm{Re}(\alpha) > 0$: $\mathrm{Re}(\lambda) < 0$ for all eigenvalues, and the moments converge to the fixed point, in a fictitious time at most $\tau \sim (2\,\mathrm{Re}(\alpha))^{-1}$ (this is the inverse of the smallest possible real part, among all eigenvalues).

- $\mathrm{Re}(\alpha) < 0$: $\mathrm{Re}(\lambda) > 0$ for all eigenvalues, and the moments diverge (but would converge if we changed the direction of the evolution, $\tau \to -\tau$).

29.e The derivation of the time evolution of the distribution $P_\tau(x,y)$ can be done by the same method as the one used in Problem 28, by considering a generic test function $F(x,y)$ and writing

$$\begin{aligned}\int dxdy\, (\partial_\tau P_\tau)\, F &= Z^{-1}\int [D\eta]\, \Big\{(\partial_\tau x)\partial_x F + (\partial_\tau y)(\partial_y F)\Big\} e^{-\frac{1}{4}\int \eta^2} \\ &= Z^{-1}\int [D\eta]\, \Big\{(-\mathrm{Re}\, S' + \underbrace{\eta}_{-2\frac{\delta}{\delta\eta}})\partial_x F + (-\mathrm{Im}\, S')(\partial_y F)\Big\} e^{-\frac{1}{4}\int \eta^2}.\end{aligned}$$

Then, we integrate the derivative with respect to η by parts to let it act on $\partial_x F$, which gives

$$2\frac{\delta}{\delta\eta(\tau)}\partial_x F = 2\underbrace{\frac{\delta x(\tau)}{\delta\eta(\tau)}}_{=\frac{1}{2}}\partial_x^2 F = \partial_x^2 F.$$

Therefore, we have

$$\begin{aligned}\int dxdy\, (\partial_\tau P_\tau)\, F &= \int dxdy\, P_\tau \Big\{\partial_x^2 F - \mathrm{Re}\, S'(\partial_x F) - \mathrm{Im}\, S'(\partial_y F)\Big\} \\ &= \int dxdy\, F\Big\{\partial_x^2 + \partial_x \mathrm{Re}\, S' + \partial_y \mathrm{Im}\, S'\Big\} P_\tau,\end{aligned}$$

which gives the announced Fokker–Planck equation since this equality must be true for any test function $F(x,y)$.

29.f Let us assume a quadratic action, i.e., $\beta \equiv 0$, $\alpha \equiv a + ib$. The statement of the problem announces an asymptotic solution in the form of the exponential of a polynomial of degree two, of the form

$$-\ln P_\infty(x,y) = \tfrac{1}{2}(ux^2 + vy^2 + 2wxy).$$

Fixed points of the Fokker–Planck equation make its right-hand side vanish. With the above ansatz, this results in the following set of equations for the coefficients u, v, w:

$$u^2 = au + bw, \quad w^2 = av - bw, \quad 2uw = 2aw + b(v-u), \quad u = 2a.$$

(The four equations correspond to equating to zero the coefficients of x^2, y^2, xy and 1, respectively.) Once u is known, it is straightforward to obtain the other coefficients,

$$w = \tfrac{2a^2}{b}, \quad v = 2a\big(1 + \tfrac{2a^2}{b^2}\big).$$

(One should also check that the fourth equation is satisfied by this solution.) Therefore, when $\beta = 0$, we indeed obtain the announced fixed point. This distribution looks completely innocuous, with rapidly falling tails in all directions of the complex plane.

When $b \gg a$ (i.e., when the original integral suffers from a severe sign problem), this distribution becomes nearly isotropic in the x, y plane:

$$P_\infty(x,y) \underset{a \ll b}{=} \frac{a}{\pi} \exp\Big\{ -a(x^2+y^2)\Big\}. \tag{2.41}$$

(We have restored the proper normalization prefactor.) From this isotropic distribution, we would predict $C_0 = 1$, $C_{2,4,6,\ldots} = 0$, while the exact moments are

$$M_{2n} = \frac{1}{2^n n!(a+ib)^n} \approx \frac{1}{2^n n! b^n}.$$

The moment $z^{2n}/(2n)!$ obtained with (2.41) in an angular slice $[\theta, \theta+\epsilon]$ reads

$$\epsilon \int_0^\infty dr\, r\, \tfrac{a}{\pi} e^{-ar^2} \frac{r^{2n} e^{2in\theta}}{(2n)!} = \frac{\epsilon}{\pi} \frac{n! e^{2in\theta}}{a^n (2n)!} \gg M_{2n}.$$

Had we performed the calculation with the exact distribution $P_\infty(x,y)$, we would have

$$\epsilon \int_0^\infty dr\, r\, P_\infty(x,y) \frac{r^{2n} e^{2in\theta}}{(2n)!} = \frac{\epsilon}{\pi} \frac{n! e^{2in\theta}}{a^n (2n)!} \Big\{ 1 \oplus \frac{a}{b} \oplus \cdots \oplus \underbrace{\frac{a^n}{b^n}}_{\substack{\text{order} \\ \text{of } M_{2n}}} \oplus \cdots \Big\}.$$

We see that the correct result for M_{2n} results from cancellations in the angular integration over θ (thanks to the weight $e^{2in\theta}$), such that the result $\sim a^{-n}$ in a small angular slice is reduced to b^{-n} after the angular integration. Thus, the sign problem we sought to circumvent reappears in a (milder) form that renders difficult the evaluation of high-order moments.

CHAPTER 3

Non-Abelian Fields

Introduction

After having explored various basic aspects of quantum field theory in Chapter 1 and functional methods in Chapter 2, we now turn to non-Abelian gauge theories (mostly with an $\mathrm{SU}(N)$ gauge group). The problems in this chapter explore various mathematical questions relevant for non-Abelian gauge theories (in particular, manipulations of $\mathfrak{su}(N)$ generators), some questions related to their perturbative expansion, and some non-perturbative questions related to the rich structure of ground states in Yang–Mills theory.

Lie Groups and Algebras

Non-Abelian gauge theories are quantum field theories endowed with an internal symmetry under a non-Abelian continuous group, i.e., a Lie group $\mathcal{G}$ whose multiplication law is non-commutative. A very important case is that of $\mathcal{G} = \mathrm{SU}(N)$, the group of $N \times N$ unitary matrices of determinant unity.

Given a Lie group $\mathcal{G}$, the associated Lie algebra $\mathfrak{g}$ is the set of matrices X such that $e^{itX} \in \mathcal{G}$ for real t. Geometrically, the Lie algebra is the linear space tangent to the group manifold at the identity element $\Omega = 1$. For every group element Ω and algebra element X, the product $\Omega^{-1}X\Omega$ also belongs to the algebra, which implies that, for two elements X, Y of the algebra, the commutator $i[X, Y]$ is also in the algebra (in this context, the commutator is known as the *Lie bracket*). Given a basis $\{t^a\}$ (the t^a's are called the *generators*) of the algebra, one can therefore write

$$\left[t^a, t^b\right] = i\, f^{abc} t^c, \tag{3.1}$$

with real coefficients f^{abc} called the *structure constants*. The number of linearly independent generators is the dimension of the Lie algebra (equal to $N^2 - 1$ for $\mathfrak{su}(N)$). The sum over cyclic permutations of nested commutators is zero, leading to the *Jacobi identity*,

$$f^{ade}f^{bcd} + f^{bde}f^{cad} + f^{cde}f^{abd} = 0. \tag{3.2}$$

Singlet, Fundamental and Adjoint Representations

A *representation* π of a Lie algebra is a mapping from the Lie algebra into another set (this could also be a linear space of matrices) that preserves the Lie bracket, i.e., $\big[\pi(X), \pi(Y)\big] = \pi\big([X, Y]\big)$. The most frequently encountered representations in non-Abelian gauge theories are:

- *Singlet* (or *trivial*) *representation*: the singlet representation maps every X into the null element. This representation is used to describe non-interacting particles.
- *Fundamental representation*: this is defined as the one-to-one (or faithful) representation of lowest dimension. For the $\mathfrak{su}(N)$ algebra, it is made of $N \times N$ traceless Hermitian matrices. In the case of a compact Lie algebra, one may normalize the corresponding generators by

  $$\mathrm{tr}\,(t_f^a t_f^b) = \tfrac{1}{2}\,\delta^{ab}. \tag{3.3}$$

  For the $\mathfrak{su}(N)$ Lie algebra, the product of two generators is fully specified by

  $$t_f^a t_f^b = \tfrac{1}{2}\left(\tfrac{\delta^{ab}}{n}\,1 + (d^{abf} + i\,f^{abf})\;t_f^f\right) \tag{3.4}$$

  by introducing the *symmetric structure constants* d^{abc}.
- *Adjoint representation*: this representation maps X to the linear operator ad_X on $\mathfrak{g}$, such that $\mathrm{ad}_X(Y) \equiv -i[X, Y]$. The elements of the adjoint representation may be viewed as matrices whose size is the dimension of the Lie algebra, i.e., $N^2 - 1$ for $\mathfrak{su}(N)$. A basis of these matrices is made of the T^a_{adj}'s such that $(T^a_{adj})_{bc} = -if^{cab}$ (thanks to the Jacobi identity, they also fulfill the commutation relation (3.1)). The adjoint representation can be used to represent commutators as $[X, Y] = t^a\big(X_{adj}\big)_{ab} Y_b$.

Fierz identity

The generators of the fundamental representation of $\mathfrak{su}(N)$ obey the following identity:

$$\big(t_f^a\big)_{ij}\big(t_f^a\big)_{kl} = \tfrac{1}{2}\delta_{il}\delta_{jk} - \tfrac{1}{2N}\delta_{ij}\delta_{kl}. \tag{3.5}$$

This formula, known as the *Fierz identity*, also has a useful graphical representation:

$(t_f^a)_{ij}(t_f^a)_{kl} =$ 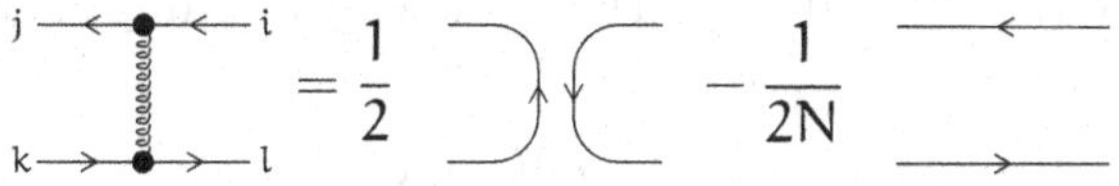 (3.6)

This identity may be used to systematically reduce contractions of fundamental generators of the $\mathfrak{su}(N)$ Lie algebra. It also plays a crucial role in studying the large-N limit of SU(N) Yang–Mills theory.

Yang–Mills Theory

Given a spacetime-dependent Lie group element $\Omega \in \mathcal{G}$, a covariant derivative D^μ is defined to transform into $D^\mu_\Omega \equiv \Omega^{-1}(x) D^\mu \Omega(x)$. By introducing a vector field A^μ in order to write $D^\mu \equiv \partial^\mu - igA^\mu$, this is achieved if A^μ transforms into

$$A^\mu_\Omega \equiv \Omega^{-1} A^\mu \Omega + \frac{i}{g} \Omega^{-1} \partial^\mu \Omega. \tag{3.7}$$

(Note that $A^\mu \in \mathfrak{g}$ for this to be consistent.) If $\Omega \equiv e^{ig\theta}$ and $A^\mu \equiv A^\mu_a t^a$, this can be written more explicitly as

$$A^\mu_{\Omega\,a} = \Omega^{-1}_{ab}\left(A^\mu_b - \partial^\mu \theta_b\right), \tag{3.8}$$

which one may view as an Abelian-like gauge transformation applied independently to each component of the gauge field, followed by a "rotation" that shuffles the group indices. The infinitesimal variation of A^μ_a is $\delta A^\mu_a = -(D^\mu_{\rm adj})_{ab}\theta_b$. A field of the form $A^\mu = (i/g)\Omega^{-1}\partial^\mu\Omega$ is gauge equivalent to the null field, and is called a *pure gauge*.

The commutator of two covariant derivatives defines the field strength $F^{\mu\nu} \equiv ig^{-1}[D^\mu, D^\nu]$, which transforms into $F^{\mu\nu}_\Omega \equiv \Omega^{-1} F^{\mu\nu} \Omega$ under a transformation of $\mathcal{G}$. Note that the field strength, whose explicit expression reads

$$F^{\mu\nu} = \partial^\mu A^\nu - \partial^\nu A^\mu - ig[A^\mu, A^\nu], \tag{3.9}$$

belongs to the algebra, i.e., $F^{\mu\nu} = F^{\mu\nu}_a t^a$. The Lagrangian of Yang–Mills theory is defined as

$$\mathcal{L}_{\rm YM} \equiv -\frac{1}{2}\,{\rm tr}\left(F^{\mu\nu} F_{\mu\nu}\right). \tag{3.10}$$

This is the only gauge and Lorentz invariant local Lagrangian, with at most second derivatives, if one disregards the term $\epsilon_{\mu\nu\rho\sigma} F^{\mu\nu}_a F^{\rho\sigma}_a$ (this term is a total derivative and therefore has no perturbative effect, but it can contribute non-perturbatively). For the Lagrangian (3.10) to give a positive definite energy density, the Lie algebra $\mathfrak{g}$ should be a direct sum of simple compact algebras (for those, the generators in the fundamental representation can be normalized so that ${\rm tr}\,(t^a t^b) = \frac{1}{2}\delta^{ab}$) and $\mathfrak{u}(1)$ algebras.

In contrast to Abelian gauge fields that do not have self-interactions, the gauge bosons represented by the gauge field A^μ_a have self-interactions, because of the cubic and quartic terms in the Lagrangian. Therefore, Yang–Mills theories have a quite rich dynamics, even when they are considered in isolation without any coupling to matter fields. The Lagrangian (3.10) leads to the equation of motion $[D_\mu, F^{\mu\nu}] = 0$, known as the *Yang–Mills equation*.

Fadeev–Popov Ghosts

The Lagrangian (3.10) is invariant under gauge transformations of the fields A^μ_a. This leads to an infinite redundancy of the path integral, which must be tamed by imposing a gauge condition of the form $G^a(A(x)) = \omega^a(x)$. This is done by inserting the following factor under the path integral:

$$1 = \int [D\Omega(x)]\; \delta\left[G^a(A_\Omega) - \omega^a\right] \det\left(\frac{\partial G^a(A)}{\partial A^\mu_b}\left(-D^\mu_{\rm adj}\right)_{bc}\right), \tag{3.11}$$

with $\Omega \equiv e^{ig\theta}$. Unlike in the case of Abelian gauge theories (with linear gauge fixing conditions), the determinant depends on the gauge field and therefore cannot be dropped, but it can be rewritten as a path integral over complex Grassmann fields $\chi_a(x), \overline{\chi}_a(x)$, known as *Fadeev–Popov ghosts*. After gauge fixing and integrating over ω^a with a Gaussian distribution of width ξ^{-1}, the Yang–Mills Lagrangian (3.10) is replaced by the following effective Lagrangian:

$$\mathcal{L}_{\rm eff} \equiv -\frac{1}{2}\,{\rm tr}\,\big(F^{\mu\nu}F_{\mu\nu}\big) - \frac{\xi}{2}\big(G^a(A)\big)^2 + \overline{\chi}_a \frac{\partial G^a(A)}{\partial A^\mu_b}\big(-D^\mu_{\rm adj}\big)_{bc}\chi_c. \tag{3.12}$$

The last term in this Lagrangian in general induces a coupling between the ghosts and the gauge fields (one exception is the strict axial gauge, $n_\mu A^\mu_a = 0$), so that ghosts may appear in the form of loops even in amplitudes with only physical external quanta.

Quantum Chromodynamics

Quantum chromodynamics (QCD) is the quantum field theory that describes strong nuclear interactions, whose Lagrangian reads

$$\mathcal{L}_{\rm QCD} = -\frac{1}{2}\,{\rm tr}\,\big(F^{\mu\nu}F_{\mu\nu}\big) + \sum_f \overline{\psi}_f(i\not{D} - m_f)\psi_f. \tag{3.13}$$

Its field content is made of eight *gluons* (the gauge fields of an $SU(3)$ Yang–Mills theory), and six flavors of *quarks* (up, down, strange, charm, bottom and top), which are spin-$1/2$ fermions minimally coupled to the gluons through the Dirac Lagrangian, and transform according to the fundamental representation of $\mathfrak{su}(3)$. Following are the QCD Feynman rules in the covariant gauge $\partial_\mu A^\mu = \omega$ (ghosts are denoted by a dotted line, the indices $\mu, \nu, \rho, \ldots$ are Lorentz indices, the indices $a, b, c, \ldots$ are adjoint indices ranging from 1 to $N^2 - 1$ and the indices $i, j, k, \ldots$ are fundamental indices ranging from 1 to N):

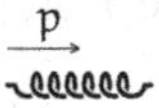

$$= \frac{-i\,g^{\mu\nu}\,\delta_{ab}}{p^2 + i0^+} + \frac{i\,\delta_{ab}}{p^2 + i0^+}\left(1 - \frac{1}{\xi}\right)\frac{p^\mu p^\nu}{p^2}, \tag{3.14}$$

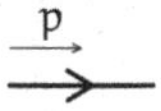

$$= \frac{i\,\delta_{ij}}{\not{p} - m + i0^+},$$

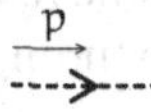

$$= \frac{i\,\delta_{ab}}{p^2 + i0^+}, \tag{3.15}$$

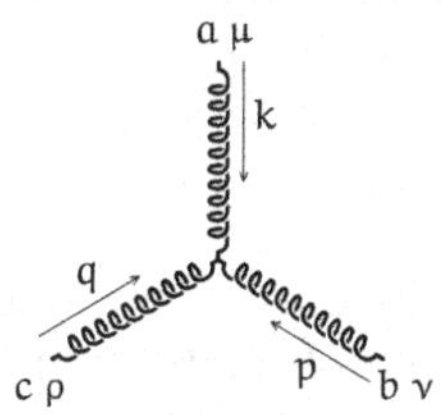

$$= -g\,f^{abc}\big\{g^{\mu\nu}\,(k-p)^\rho + g^{\nu\rho}\,(p-q)^\mu + g^{\rho\mu}\,(q-k)^\nu\big\}, \tag{3.16}$$

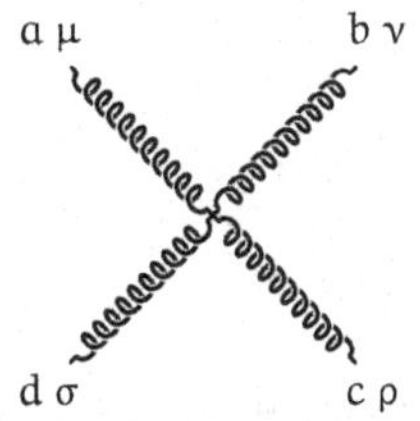

$$\begin{aligned} = -i\,g^2\,\big\{ & f^{abe}f^{cde}\,(g^{\mu\rho}g^{\nu\sigma} - g^{\mu\sigma}g^{\nu\rho}) \\ & + f^{ace}f^{bde}\,(g^{\mu\nu}g^{\rho\sigma} - g^{\mu\sigma}g^{\nu\rho}) \\ & + f^{ade}f^{bce}\,(g^{\mu\nu}g^{\rho\sigma} - g^{\mu\rho}g^{\nu\sigma})\big\}, \end{aligned} \tag{3.17}$$

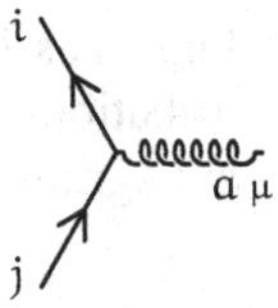 $= -i\, g\, \gamma^{\mu} \left(t_r^a\right)_{ij},$ 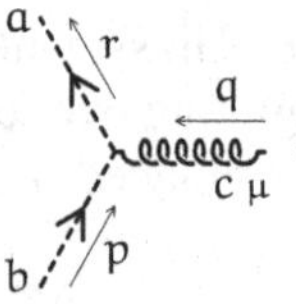 $= g\, f^{abc}\,(p_\mu + q_\mu) = g\, f^{abc}\, r_\mu.$ (3.18)

Since ghosts and quarks obey Fermi statistics, a minus sign comes with every closed ghost or quark loop.

One of the main features of QCD is *asymptotic freedom*, the fact that the running coupling decreases to zero as the energy scale goes to infinity. At one loop, the running coupling reads

$$\alpha_s(\mu) \equiv \frac{g^2(\mu)}{4\pi} = \frac{4\pi}{\left(\frac{11}{3}N - \frac{2}{3}N_f\right)\,\ln\left(\frac{\mu^2}{\Lambda^2_{\text{QCD}}}\right)}, \tag{3.19}$$

where N_f is the number of quark flavors, N the number of quark colors and Λ_{QCD} the infrared scale at which the perturbative coupling becomes infinite.

Non-Abelian Ward–Takahashi Identity

Amplitudes in non-Abelian gauge theories satisfy Ward–Takahashi identities whose application is more restrictive than those of Abelian gauge theories. Given an amplitude with m physical gluons of momenta p_i and n unphysical ones of momenta q_j, we have

$$\underbrace{\epsilon_{\mu_1}(p_1)\cdots\epsilon_{\mu_m}(p_m)}_{\text{physical polarizations}}\;\underbrace{q_{1\nu_1}\cdots q_{n\nu_n}}_{\text{off-shell momenta}}\;\Gamma^{\mu_1\cdots\mu_m;\nu_1\cdots\nu_n}(\{p_i\};\{q_j\};\ldots) = 0. \tag{3.20}$$

The conditions of applicability are the following:

- All the physical gluons must have on-shell momenta, and their Lorentz indices must be contracted with physical polarization vectors.
- The unphysical gluons can have off-shell momenta. All their Lorentz indices must be contracted with the corresponding 4-momenta.
- All other external legs must correspond to physical quanta (fermions) and be on-shell.

BRST Symmetry, Physical States

After gauge fixing, the effective Lagrangian (3.12) is no longer gauge invariant, but it has a remnant of the original gauge symmetry in the form of the *Becchi–Rouet–Stora–Tyutin (BRST) symmetry*, which acts on a field Φ as $\delta\Phi \equiv \vartheta Q_{\text{BRST}}\Phi$, with ϑ a Grassmann constant and

$$\begin{aligned} &Q_{\text{BRST}}A^a_\mu(x) = \left(D^{\text{adj}}_\mu\right)_{ab}\chi_b(x), \quad Q_{\text{BRST}}\psi(x) = i\,g\,\chi_a(x)\,t^a_r\,\psi(x),\\ &Q_{\text{BRST}}\chi_a(x) = -\frac{1}{2}\,g\,f^{abc}\,\chi_b(x)\,\chi_c(x), \quad Q_{\text{BRST}}\overline{\chi}_a(x) = -\xi\,G^a(A). \end{aligned} \tag{3.21}$$

The BRST transformation is nilpotent, i.e., $Q^2_{\text{BRST}} = 0$. By Noether's theorem, the BRST symmetry leads to a conserved charge Ω_{BRST}. The physical states are the states of non-zero norm of the form

$$\big|\psi_{\rm phys}\big\rangle \equiv \big|\psi\big\rangle + \mathcal{Q}_{\rm BRST}\big|\varphi\big\rangle, \quad \text{with } \mathcal{Q}_{\rm BRST}\big|\psi\big\rangle = 0. \tag{3.22}$$

In words, the physical states are elements of the kernel of the BRST charge, up to an element of its image (i.e., they are equivalence classes).

Wilson Lines, Wilson Loops

Given a gauge field $A^\mu(x)$, the *Wilson line* defined on an oriented path γ is

$$W[\gamma, A] \equiv \mathrm{P}\,\exp\Big(ig\int_\gamma dx_\mu\, A^\mu(x)\Big), \tag{3.23}$$

where P is a path-ordering operator, which orders the field operators according to the orientation of the path γ (the operators on the left being those at the endpoint of γ). If the endpoint of the path is varied by $d\gamma^\mu$, we have $d\gamma^\mu D_\mu W[\gamma, A] = 0$ (this property can be turned into a differential equation that defines the Wilson line, if supplemented with an initial condition stating that $W = 1$ for a zero-length path). Under a gauge transformation $\Omega(x)$, a Wilson line transforms as

$$W[\gamma, A] \quad \to \quad \Omega^\dagger(\gamma_f)\, W[\gamma, A]\, \Omega(\gamma_i), \tag{3.24}$$

where $\gamma_{i,f}$ are the initial and final points of γ. Thanks to this property, Wilson lines can be used to construct non-local gauge invariant operators, e.g., $\overline{\psi}(\gamma_f)W[\gamma, A]\psi(\gamma_i)$ (but note that these operators are path-dependent). A *Wilson loop* is a Wilson line defined on a closed contour; its trace does not depend on the starting point of the contour and is gauge invariant.

Instantons

An *instanton* is a gauge field configuration that extremalizes the Euclidean Yang–Mills action. Instantons exist only in four dimensions and their values at infinity are pure gauges, $ig^{-1}\Omega^\dagger\partial^\mu\Omega$. The functions Ω, which map $\mathcal{S}_3 \sim \partial(\mathbb{R}^4)$ into the gauge group $\mathcal{G}$, can be arranged into homotopy classes (field configurations that can be continuously deformed into one another), whose set form a group denoted $\pi_3(\mathcal{G})$ (the third homotopy group of $\mathcal{G}$). The value of the minimal action depends only on the homotopy class of Ω and is given by an integral known as the *Cartan–Maurer invariant*:

$$\begin{aligned} \mathcal{S}_{\min}[A] &= \frac{1}{4}\int d^4x\, \epsilon_{ijkl}\,\mathrm{tr}\,\big(F_{ij}F_{kl}\big) \\ &= \frac{1}{3g^2}\int d\theta_1\, d\theta_2 d\theta_3\, \epsilon_{abc}\,\mathrm{tr}\left\{\Omega^\dagger(\theta)\frac{\partial\Omega(\theta)}{\partial\theta_a}\Omega^\dagger(\theta)\frac{\partial\Omega(\theta)}{\partial\theta_b}\Omega^\dagger(\theta)\frac{\partial\Omega(\theta)}{\partial\theta_c}\right\}, \end{aligned} \tag{3.25}$$

where $\theta_{1,2,3}$ are coordinates on $\mathcal{S}_3$. For $\mathcal{G} = SU(N)$, we have $\pi_3(SU(N)) = \mathbb{Z}$ and $\mathcal{S}_{\min}[A] = 8\pi^2|n|/g^2$, where n is the homotopy index of the function $\Omega(x)$ that defines the boundary behavior of the instanton A^μ.

About the Problems of this Chapter

- **Problem 30** derives some basic identities obeyed by the generators of the fundamental representation of $\mathfrak{su}(n)$, and by the structure constants f^{abc} and d^{abc}. More elaborate identities are obtained in **Problems 31** and **32**.
- In **Problem 33**, we prove an identity obeyed by the generators of $\mathfrak{su}(2)$, which plays a role in problems related to coherent spin states.
- **Problems 34** and **35** explore how Wilson loops and Wilson lines change when the contour on which they are defined is deformed in various ways. This wisdom is then applied in **Problem 35** in order to obtain the *Makeenko–Migdal loop equations.*
- In **Problem 36**, we introduce the concept of a classical particle carrying a non-Abelian charge, and derive the *Wong equations* that describe how such a charge evolves in an external gauge potential. These equations are obtained from the worldline formalism in **Problem 38**.
- The well-known Stokes's theorem relates the integral of a vector along a closed path and the integral of its curl over a surface bounded by this path. The goal of **Problem 39** is to obtain a similar formula that relates a non-Abelian gauge potential and the corresponding field strength.
- In **Problem 40**, we consider the quantization of QED in a non-linear gauge, which leads to complications quite similar to those encountered in non-Abelian gauge theories.
- How gauge invariant scattering amplitudes are obtained from LSZ reduction formulas that contain gauge-dependent fields is discussed in **Problem 41**.
- **Problem 42** is devoted to studying the large-N limit of $SU(N)$ Yang–Mills theory. It is shown that the leading contribution comes from the subset of planar graphs.
- Soft gluon radiation from a quark–antiquark dipole is calculated in **Problem 43**, with emphasis on the antenna patterns of this emission. In **Problem 44**, we derive an auxiliary identity useful in this study. **Problem 45** is devoted to a derivation of the *Low–Burnett–Kroll theorem* in QCD in a simple case.
- In **Problem 46**, we discuss quadratic and cubic operators that are invariant under global gauge transformations.
- **Problem 47** studies some aspects of *Banks–Zaks fixed points*, which are perturbative infrared fixed points in asymptotically free theories.
- Several aspects of instantons are studied in **Problems 48**, **49** and **50**. They are first discussed in the simpler setting of ordinary quantum mechanics in **Problem 48**. Then, it is shown in **Problem 49** that in Yang–Mills theory instantons interpolate between pure gauges of different topological indices, and **Problem 50** is devoted to studying the ground states of Yang–Mills theories and their topological classification.

30. Basic $\mathfrak{su}(n)$ Identities Prove the following $\mathfrak{su}(n)$ identities:

30.a $f^{acd}d^{bcd} = 0$

30.b $\mathrm{tr}\left(t_f^a t_f^b t_f^c\right) = \frac{1}{4}\left(d^{abc} + i\, f^{abc}\right)$

30.c $\mathrm{tr}\left(t_f^a t_f^b t_f^a t_f^c\right) = -\frac{1}{4n}\,\delta_{bc}$

30.d $f^{acd}f^{bcd} = n\,\delta_{ab}$

30.e $d^{acd}d^{bcd} = \left(n - \frac{4}{n}\right)\delta_{ab}$

Hint: for the preceding two questions, write $\mathrm{tr}\left(t_f^a t_f^b t_f^a t_f^c\right)$ *in several ways.*

30.f $f^{ade}f^{bef}f^{cfd} = \frac{n}{2}\,f^{abc}$.

30.a Note that the symmetric structure constants d^{abc} must be real (like the f^{abc}), because the anti-commutator $\left\{t_f^a, t_f^b\right\}$ is Hermitian. Before we start, let us mention that there is a simpler variant of the first equality:

$$f^{abc}d^{abd} = 0,$$

where the contraction is done on the first two indices. This identity follows trivially from the fact that f^{abc} is fully antisymmetric, while d^{abd} is symmetric in the first two indices (this is easily seen from its definition in Eq. (3.4)). The first identity in the statement of the problem is a bit more complicated, because the contraction is on the second and third indices. We may write

$$\begin{aligned} f^{acd}d^{bcd} &= -2i\,\mathrm{tr}\left([t_f^a, t_f^c]\{t_f^b, t_f^c\}\right) \\ &= -2i\,\mathrm{tr}\left(t_f^a t_f^c t_f^b t_f^c - t_f^c t_f^a t_f^b t_f^c + t_f^a t_f^c t_f^c t_f^b - t_f^c t_f^a t_f^c t_f^b\right) = 0. \end{aligned}$$

(We use $t_f^c t_f^c = (n^2-1)/2n$ and the cyclicity of the trace in order to obtain a pairwise cancellation of the four terms.)

30.b In order to obtain the second equality, we start from

$$t_f^a t_f^b = \frac{1}{2}\left(\frac{\delta^{ab}}{n}\,1 + \left(d^{abf} + i\, f^{abf}\right) t_f^f\right), \tag{3.26}$$

multiply on the right by t_f^c and take the trace. This gives

$$\mathrm{tr}\left(t_f^a t_f^b t_f^c\right) = \frac{1}{4}(d^{abc} + i\, f^{abc}).$$

Because d^{abc} and f^{abc} are respectively the real and imaginary parts of this trace, we see that both kinds of structure constants are invariant under cyclic permutations of the indices a, b, c. Since d^{abc} is symmetric under the exchange of the first two indices a, b, its cyclic invariance implies that it is in fact fully symmetric. Likewise, f^{abc} is fully antisymmetric. Therefore the double contraction of a d^{abc} and an f^{abc} on any pair of indices gives zero.

30.c The third identity is easily obtained by multiplying $t_f^a t_f^b t_f^a = -\frac{1}{2N}\, t_f^b$ (derived using the Fierz identity) on the right by t_f^c and taking the trace, which gives

$$\mathrm{tr}\left(t_f^a t_f^b t_f^a t_f^c\right) = -\tfrac{1}{2n}\,\mathrm{tr}\left(t_f^b t_f^c\right) = -\tfrac{1}{4n}\,\delta_{bc}.$$

30.d On the other hand, we could have obtained an alternative expression for this trace as follows:

$$\begin{aligned}\mathrm{tr}\left(t_f^a t_f^b t_f^a t_f^c\right) &= \mathrm{tr}\left(\left([t_f^a, t_f^b] + t_f^b t_f^a\right) t_f^a t_f^c\right) = i f^{abe}\,\mathrm{tr}\left(t_f^e t_f^a t_f^c\right) + \tfrac{n^2-1}{2n}\,\mathrm{tr}\left(t_f^b t_f^c\right)\\ &= \tfrac{i}{4} f^{abe}\left(d^{eac} + i f^{eac}\right) + \tfrac{n^2-1}{4n}\,\delta_{bc} = -\tfrac{1}{4} f^{aeb} f^{aec} + \tfrac{n^2-1}{4n}\,\delta_{bc}.\end{aligned}$$

(We have used the quadratic Casimir operator, $\left(t_f^a t_f^a\right)_{il} = \frac{N^2-1}{2N}\,\delta_{il}$.) Comparing the two forms of this trace gives

$$f^{aeb} f^{aec} = n\,\delta_{bc}.$$

Note that this identity is equivalent to $\mathrm{tr}\left(T_{adj}^b T_{adj}^c\right) = n\,\delta_{bc}$.

30.e Yet another form of the same trace can be obtained by using two copies of (3.26):

$$\begin{aligned}\mathrm{tr}\left(t_f^a t_f^b t_f^a t_f^c\right) &= \tfrac{1}{4}\,\mathrm{tr}\left(\left(\tfrac{\delta^{ab}}{n}\,1 + \left(d^{abe} + i f^{abe}\right) t_f^e\right)\left(\tfrac{\delta^{ac}}{n}\,1 + \left(d^{acf} + i f^{acf}\right) t_f^f\right)\right)\\ &= \tfrac{1}{4}\left(\tfrac{1}{n}\delta_{bc} + \tfrac{1}{2}\left(d^{abe} + i f^{abe}\right)\left(d^{ace} + i f^{ace}\right)\right)\\ &= \tfrac{1}{4}\left(\tfrac{1}{n}\delta_{bc} + \tfrac{1}{2}\left(d^{abe} d^{ace} - f^{abe} f^{ace}\right)\right).\end{aligned}$$

Comparing this expression with the preceding two results leads to

$$d^{abe} d^{ace} = \left(n - \tfrac{4}{n}\right)\delta_{bc}.$$

Since d^{abc} is invariant under cyclic permutations of the three indices, this identity is also equivalent to

$$d^{eab} d^{eac} = d^{bea} d^{cea} = \left(n - \tfrac{4}{n}\right)\delta_{bc}.$$

30.f For the final equality, note first that

$$\begin{aligned}S^{abc} \equiv f^{ade} f^{bef} f^{cfd} &= -i\,\mathrm{tr}\left(T_{adj}^a T_{adj}^b T_{adj}^c\right)\\ &= -i\,\mathrm{tr}\left(\left([T_{adj}^a, T_{adj}^b] + T_{adj}^b T_{adj}^a\right) T_{adj}^c\right) = n\,f^{abc} + S^{bac}.\end{aligned}$$

On the other hand, we can also write

$$\begin{aligned}S^{abc} &= \mathrm{tr}\left(\left(iT_{adj}^a\right)\left(iT_{adj}^b\right)\left(iT_{adj}^c\right)\right) = \mathrm{tr}\left(\left(iT_{adj}^c\right)^{\mathsf{T}}\left(iT_{adj}^b\right)^{\mathsf{T}}\left(iT_{adj}^a\right)^{\mathsf{T}}\right)\\ &= -\mathrm{tr}\left(\left(iT_{adj}^c\right)\left(iT_{adj}^b\right)\left(iT_{adj}^a\right)\right) = -S^{cba} = -S^{bac}.\end{aligned}$$

(We have used the fact that the matrix iT_{adj}^a is antisymmetric, and the cyclic invariance of the trace.) This second relationship proves the antisymmetry of S^{abc}. Combined with the first one, we thus get

$$-i\,\mathrm{tr}\left(T_{adj}^a T_{adj}^b T_{adj}^c\right) = f^{ade} f^{bef} f^{cfd} = \tfrac{n}{2}\,f^{abc}.$$

31. More $\mathfrak{su}(n)$ Identities Consider the generators T^a_{adj} of the adjoint representation of $\mathfrak{su}(n)$, and also denote $(D^a)_{bc} \equiv d^{abc}$. Prove the following identities:

31.a $f^{abc} T^a_{adj} T^b_{adj} T^c_{adj} = \frac{i}{2} n^2 1_{adj}$

31.b $f^{abc} D^a T^b_{adj} T^c_{adj} = 0$

31.c $f^{abc} D^a D^b T^c_{adj} = \frac{i}{2}(n^2 - 4) 1_{adj}$

31.d $f^{abc} D^a D^b D^c = 0.$

Hint: for the last two formulas, one may first prove the following intermediate results:

$$d^{abe} f^{ecd} + d^{bce} f^{ead} + d^{cae} f^{ebd} = 0,$$
$$\tfrac{2}{n}(\delta_{ab}\delta_{cd} - \delta_{ac}\delta_{bd}) + d^{abe} d^{ced} - d^{ace} d^{bed} + f^{bce} f^{aed} = 0.$$

31.a First, we have

$$f^{abc} T^b_{adj} T^c_{adj} = \frac{1}{2} f^{abc} \left[T^b_{adj}, T^c_{adj}\right] = \frac{i}{2} \underbrace{f^{abc} f^{bcd}}_{n\,\delta_{ad}} T^d_{adj} = \frac{in}{2} T^a_{adj}.$$

(The contraction $f^{abc} f^{bcd}$ was derived in Problem 30.) Multiplying by T^a_{adj} on the left, and using $T^a_{adj} T^a_{adj} = n\, 1_{adj}$, we get the first of the listed identities,

$$f^{abc}\, T^a_{adj} T^b_{adj} T^c_{adj} = \frac{i}{2}\, n^2\, 1_{adj}.$$

31.b Instead, we may multiply on the left by D^a to get

$$f^{abc}\, D^a T^b_{adj} T^c_{adj} = \frac{in}{2}\, D^a T^a_{adj}.$$

Recall now that

$$\left(D^a T^a_{adj}\right)_{bc} = -i\, d^{abd} f^{adc} = 0$$

(see Problem 30). Therefore, we have

$$f^{abc}\, D^a T^b_{adj} T^c_{adj} = 0.$$

31.c At this point, the identities proposed in the hint become useful. Let us prove them first. Using the identity

$$[\{A,B\},C]+[\{B,C\},A]+[\{C,A\},B]=0$$

with A, B, C generators of the fundamental representation of $\mathfrak{su}(N)$, we obtain

$$d^{abe}f^{ecd}+d^{bce}f^{ead}+d^{cae}f^{ebd}=0.$$

(This is sometimes called the *second Jacobi identity.*) In order to get the second identity proposed in the hint, the starting point is the following somewhat less-known equality:

$$[A,[B,C]]=\{C,\{A,B\}\}-\{B,\{A,C\}\}.$$

Substituting for A, B, C the three generators t_f^a, t_f^b, t_f^c leads to the announced identity. Recalling that $(T^a_{adj})_{bc}=-if^{abc}$ and $(D^a)_{bc}=d^{abc}$, these identities can be recast as follows:

$$\begin{aligned}
&[D^a,T^b_{adj}]=[T^a_{adj},D^b]=if^{abc}D^c,\\
&T^a_{adj}D^b+T^b_{adj}D^a=D^aT^b_{adj}+D^bT^a_{adj}=d^{abc}T^c_{adj},\\
&[D^a,D^b]=if^{abc}T^c_{adj}+\Delta^{ab},
\end{aligned}$$

with $(\Delta^{ab})_{cd}=\frac{2}{n}(\delta_{cb}\delta_{da}-\delta_{ca}\delta_{db})$. Using these intermediate results, we have

$$\begin{aligned}
&f^{abc}D^bT^c_{adj}=\tfrac{1}{2}f^{abc}\left[D^b,T^c_{adj}\right]=\tfrac{i}{2}\underbrace{f^{abc}f^{bcd}}_{n\,\delta_{ad}}D^d=\tfrac{in}{2}D^a,\\
&f^{abc}\,D^aD^bT^c_{adj}=\tfrac{in}{2}D^aD^a.
\end{aligned}$$

Using the results of Problem 30, we get

$$\left(D^aD^a\right)_{bc}=d^{abd}d^{adc}=\tfrac{n^2-4}{n}\,\delta_{bc},$$

which implies that

$$f^{abc}\,D^aD^bT^c_{adj}=\tfrac{i}{2}\,(n^2-4)\,1_{adj}.$$

31.d We also have

$$f^{abc}D^bD^c=\tfrac{1}{2}f^{abc}\left[D^b,D^c\right]=\tfrac{i}{2}\underbrace{f^{abc}f^{bcd}}_{n\,\delta_{ad}}T^d_{adj}+\tfrac{1}{2}f^{abc}\Delta^{bc}=\tfrac{in}{2}T^a_{adj}+\tfrac{1}{2}f^{abc}\Delta^{bc}.$$

Note now that

$$f^{abc}\left(\Delta^{bc}\right)_{ef}=\tfrac{2i}{n}\left(T^a_{adj}\right)_{bc}\left(\delta_{bf}\delta_{ce}-\delta_{be}\delta_{cf}\right)=\tfrac{2i}{n}\left\{(T^a_{adj})_{fe}-(T^a_{adj})_{ef}\right\}=-\tfrac{4i}{n}(T^a_{adj})_{ef},$$

which implies

$$f^{abc}D^bD^c=i\,\tfrac{n^2-4}{2n}\,T^a_{adj},\quad\text{and } f^{abc}D^aD^bD^c=i\left(\tfrac{n^2-4}{2n}\right)D^aT^a_{adj}=0.$$

32. More $\mathfrak{su}(n)$ Identities (Continued) This problem is a continuation of Problem 31. Prove the following identities:

32.a $d^{abc}T^a_{adj}T^b_{adj}T^c_{adj} = 0$

32.b $d^{abc}D^aT^b_{adj}T^c_{adj} = \frac{1}{2}(n^2-4)\,1_{adj}$

32.c $d^{abc}D^aD^bT^c_{adj} = 0$

32.d $d^{abc}D^aD^bD^c = \frac{(n^2-4)(n^2-12)}{2n^2}\,1_{adj}.$

32.a We start from the following formulas, obtained in Problem 31:

$$\begin{aligned}
&T^a_{adj}D^b + T^b_{adj}D^a = D^aT^b_{adj} + D^bT^a_{adj} = d^{abe}\,T^e_{adj}, \\
&\left[T^a_{adj}, D^b\right] = \left[D^a, T^b_{adj}\right] = i\,f^{abe}\,D^e, \\
&T^a_{adj}D^b + D^aT^b_{adj} = d^{abe}\,T^e_{adj} + i\,f^{abe}\,D^e.
\end{aligned} \tag{3.27}$$

Multiplying (3.27) by T^c_{adj} on the left and taking the trace, we obtain

$$\mathrm{tr}\,(T^c_{adj}D^aT^b_{adj} + T^c_{adj}D^bT^a_{adj}) = d^{abe}\,\underbrace{\mathrm{tr}\,(T^c_{adj}T^e_{adj})}_{n\,\delta_{ce}} = n\,d^{abc}$$

and

$$\mathrm{tr}\,(D^aT^b_{adj}T^c_{adj}) = \frac{n}{2}\,d^{abc}. \tag{3.28}$$

(Here, we assume for the time being that $\mathrm{tr}\,(D^aT^b_{adj}T^c_{adj})$ is symmetric under the permutation of the indices a, b; we shall return to this point after (3.29).)

Writing the trace more explicitly in terms of symmetric and antisymmetric structure constants, one can see that this identity is equivalent to

$$d^{abc}T^b_{adj}T^c_{adj} = \frac{n}{2}\,D^a.$$

Then, multiplying this equation on the left by T^a_{adj} gives

$$d^{abc}T^a_{adj}T^b_{adj}T^c_{adj} = \frac{n}{2}\,T^a_{adj}D^a = 0 \quad (T^a_{adj}D^a = 0 \text{ was proven in Problem 31}).$$

32.b Multiplying instead by D^a on the left gives

$$d^{abc}D^aT^b_{adj}T^c_{adj} = \frac{n}{2}\,D^aD^a = \frac{1}{2}(n^2-4)\,1_{adj}.$$

32.c, 32.d In order to prove the remaining two identities, let us start from

$$[D^a, D^b] = if^{abe}T^e_{adj} + \Delta^{ab} \quad \text{with } (\Delta^{ab})_{cd} = \tfrac{2}{n}(\delta_{cb}\delta_{da} - \delta_{ca}\delta_{db}),$$

also obtained in Problem 31. By explicitly writing the components of all the matrices in this equation, we get $d^{ace}d^{bde} - d^{bce}d^{ade} = f^{abe}f^{ecd} - \frac{2}{n}(\delta_{ac}\delta_{bd} - \delta_{ad}\delta_{bc})$. Then, we subtract from this the same equation with b and c exchanged, which gives

$$d^{ace}d^{bde} - d^{abe}d^{cde} = f^{abe}f^{ecd} - f^{ace}f^{ebd} - \tfrac{2}{n}(\delta_{ac}\delta_{bd} - \delta_{ab}\delta_{cd}),$$

or, equivalently,

$$\left(D^aD^b + T^a_{adj}T^b_{adj}\right)_{df} = d^{abe}\left(D^e\right)_{df} + i\,f^{abe}\left(T^e_{adj}\right)_{df} + \tfrac{2}{n}\left(\delta_{ab}\delta_{df} - \delta_{ad}\delta_{bf}\right).$$

Finally, we multiply this equation on the right by $(D^c)_{fd}$, which gives

$$\begin{aligned}
&\mathrm{tr}\left(D^aD^bD^c + T^a_{adj}T^b_{adj}D^c\right)\\
&= d^{abe}\underbrace{\mathrm{tr}\left(D^eD^c\right)}_{\delta_{ec}(n^2-4)/n} + if^{abe}\underbrace{\mathrm{tr}\left(T^e_{adj}D^c\right)}_{0} + \tfrac{2}{n}\delta_{ab}\underbrace{\mathrm{tr}\left(D^c\right)}_{0} - \tfrac{2}{n}\underbrace{\delta_{ad}\delta_{bf}d^{cfd}}_{d^{cba}=d^{abc}}\\
&= d^{abc}\frac{n^2-6}{n}.
\end{aligned} \tag{3.29}$$

The right-hand side of this equation is fully symmetric in a, b, c. On the left-hand side, it is easy to check (using the commutator $[D^a, D^b]$ derived in Problem 31) that $\mathrm{tr}\left(D^aD^bD^c\right)$ is also fully symmetric, which implies the same property for $\mathrm{tr}\left(T^a_{adj}T^b_{adj}D^c\right)$. This was the missing ingredient in order to complete the proof of (3.28).

Using (3.28), we can then disentangle the two terms on the left-hand side of (3.29), to obtain

$$\mathrm{tr}\left(D^aD^bD^c\right) = d^{abc}\frac{n^2-12}{2n},$$

which is also equivalent to $d^{abc}D^bD^c = \frac{n^2-12}{2n}D^a$. Multiplying this identity on the left respectively by T^a_{adj} and D^a leads to the last two of the announced formulas,

$$d^{abc}T^a_{adj}D^bD^c = \tfrac{n^2-12}{2n}T^a_{adj}D^a = 0,$$
$$d^{abc}D^aD^bD^c = \tfrac{n^2-12}{2n}D^aD^a = \tfrac{(n^2-4)(n^2-12)}{2n^2}1_{adj}.$$

Final Note: In the intermediate steps of the present problem and of Problem 31, a number of identities have been obtained for pairs of T^a_{adj}'s or D^a's contracted with a f^{abc} or a d^{abc}. We collect them here for convenience:

$$f^{abc}T^b_{adj}T^c_{adj} = \tfrac{in}{2}T^a_{adj}, \quad f^{abc}D^bT^c_{adj} = \tfrac{in}{2}D^a, \quad f^{abc}D^bD^c = i\tfrac{n^2-4}{2n}T^a_{adj},$$
$$d^{abc}T^b_{adj}T^c_{adj} = \tfrac{n}{2}D^a, \quad d^{abc}D^bT^c_{adj} = \tfrac{n^2-4}{2n}T^a_{adj}, \quad d^{abc}D^bD^c = \tfrac{n^2-12}{2n}D^a.$$

33. An $\mathfrak{su}(2)$ Identity Let $s_{1,2,3}$ be the spin operators satisfying the $\mathfrak{su}(2)$ algebra, $[s_i, s_j] = i\,\epsilon_{ijk}s_k$, and denote by $s_\pm \equiv s_1 \pm is_2$ the raising and lowering operators. Show that

$$e^{re^{i\varphi}s_- - re^{-i\varphi}s_+} = e^{zs_-}\,e^{-\ln(1+zz^*)s_3}\,e^{-z^*s_+}, \quad \text{with } z \equiv e^{i\varphi}\tan(r).$$

Hint: recall that the fundamental representation is faithful. This identity is very useful for the manipulation of coherent spin states (see Problem 22).

On the surface, it seems that this question could be attacked with the Baker–Campbell–Hausdorff formula. However, this approach would be somewhat involved. It turns out that a direct calculation of the matrix exponentials is quite simple in the *fundamental representation* of $\mathfrak{su}(2)$. Since the fundamental representation is faithful, the validity of an equality in this representation then implies that it is true in any representation. In the fundamental representation, the spin operators are represented by the following 2×2 matrices:

$$s_+ = \begin{pmatrix} 0 & 1 \\ 0 & 0 \end{pmatrix}, \quad s_- = \begin{pmatrix} 0 & 0 \\ 1 & 0 \end{pmatrix}, \quad s_3 = \begin{pmatrix} \frac{1}{2} & 0 \\ 0 & -\frac{1}{2} \end{pmatrix}.$$

Thanks to the fact that $s_\pm^2 = 0$ and $s_3^2 = \frac{1}{4}\cdot\mathbf{1}$, exponentials of s_-, s_+ or s_3 are very simple:

$$e^{\alpha s_-} = \begin{pmatrix} 1 & 0 \\ \alpha & 1 \end{pmatrix}, \quad e^{\gamma s_+} = \begin{pmatrix} 1 & \gamma \\ 0 & 1 \end{pmatrix}, \quad e^{\beta s_3} = \begin{pmatrix} e^{\beta/2} & 0 \\ 0 & e^{-\beta/2} \end{pmatrix}.$$

Therefore, a generic product such as the one on the right-hand side of the formula to be proven reads

$$e^{\alpha s_-}e^{\beta s_3}e^{\gamma s_+} = \begin{pmatrix} 1 & 0 \\ \alpha & 1 \end{pmatrix}\begin{pmatrix} e^{\beta/2} & 0 \\ 0 & e^{-\beta/2} \end{pmatrix}\begin{pmatrix} 1 & \gamma \\ 0 & 1 \end{pmatrix} = \begin{pmatrix} e^{\beta/2} & \gamma\, e^{\beta/2} \\ \alpha\, e^{\beta/2} & \alpha\gamma\, e^{\beta/2} + e^{-\beta/2} \end{pmatrix}. \tag{3.30}$$

Consider now the left-hand side. The generic form of the argument of the exponential is

$$as_- + bs_+ = \begin{pmatrix} 0 & b \\ a & 0 \end{pmatrix}.$$

In order to compute the exponential of this matrix, we need

$$(as_- + bs_+)^{2n} = (ab)^n \begin{pmatrix} 1 & 0 \\ 0 & 1 \end{pmatrix}, \quad (as_- + bs_+)^{2n+1} = (ab)^n \begin{pmatrix} 0 & b \\ a & 0 \end{pmatrix}.$$

Summing the Taylor series that gives the exponential leads to

$$e^{as_- + bs_+} = \begin{pmatrix} \cosh(\sqrt{ab}) & \sqrt{\frac{b}{a}}\sinh(\sqrt{ab}) \\ \sqrt{\frac{a}{b}}\sinh(\sqrt{ab}) & \cosh(\sqrt{ab}) \end{pmatrix}. \tag{3.31}$$

At this point, we simply need to identify (3.30) and (3.31), in the special case where $a = re^{i\varphi}$

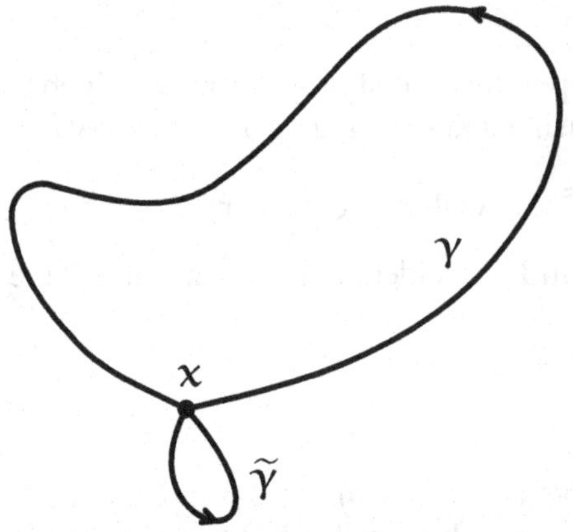

Figure 3.1 Addition of a small appendix $\widetilde{\gamma}$ to a loop γ, at the base point x.

and $b = -re^{-i\varphi}$. This leads to the following set of equations:

$$\cos(r) = e^{\beta/2}, \quad \cos(r) = \alpha\gamma\, e^{\beta/2} + e^{-\beta/2},$$
$$-\,e^{-i\varphi}\sin(r) = \gamma\, e^{\beta/2}, \quad e^{i\varphi}\sin(r) = \alpha\, e^{\beta/2},$$

whose solution is given by

$$\alpha = \tan(r)e^{i\varphi}, \quad \beta = -\ln\left(1 + \tan^2(r)\right), \quad \gamma = -\tan(r)e^{-i\varphi}.$$

(The first, third and last equations directly give β, γ and α, respectively. Then, we must check that the second equation holds with these solutions.)

34. Area Derivative of a Wilson Loop Consider the trace of a Wilson loop $\mathcal{W}[A;\gamma]$ defined on a closed path γ. Pick a point $x \in \gamma$, and at this point add a small closed contour $\widetilde{\gamma}$ that starts and ends at x. Show that

$$\mathrm{tr}\left(\mathcal{W}[A;\gamma\widetilde{\gamma}]\right) - \mathrm{tr}\left(\mathcal{W}[A;\gamma]\right) = ig\,\sigma^{\mu\nu}(\widetilde{\gamma})\,\mathrm{tr}\left(F_{\mu\nu}(x)\,\mathcal{W}[A;\gamma]\right) + \cdots,$$

where $\sigma^{\mu\nu}(x) \equiv \frac{1}{2}\int_{\widetilde{\gamma}} dx^\mu \wedge dx^\nu$ is the integral of the area 2-form on $\widetilde{\gamma}$. The unwritten terms represented by the dots are of higher order in the area enclosed by $\widetilde{\gamma}$.

Consider a Wilson loop $\mathcal{W}[A;\gamma]$ along a closed path γ. This Wilson loop is not gauge invariant, but its trace is. Now, take a point x on the contour γ, and at this point let us add a small closed contour $\widetilde{\gamma}$ that starts and ends at x, as shown Figure 3.1. The Wilson loop on the concatenated contour $\gamma\widetilde{\gamma}$ (this notation means that γ is traversed before $\widetilde{\gamma}$) is the product of the Wilson loops on the individual contours:

$$\mathcal{W}[A;\gamma\widetilde{\gamma}] = \mathcal{W}[A;\widetilde{\gamma}]\,\mathcal{W}[A;\gamma].$$

Note a subtlety here: before we marked a point at x, there was no need to specify the starting point and final point of the path γ in the definition of the *trace* of the Wilson loop, thanks to the cyclicity of the trace. But after we attached the path $\widetilde{\gamma}$ at the point x, the path γ is implicitly assumed to begin and end at the point x, as is $\widetilde{\gamma}$.

When the contour $\widetilde{\gamma}$ is infinitesimal, the corresponding Wilson line can be expanded as follows:

$$\mathcal{W}[A;\widetilde{\gamma}] = 1 + ig\,\sigma^{\mu\nu}(\widetilde{\gamma})\,F_{\mu\nu}(x) + \cdots,$$

where we denote by

$$\sigma^{\mu\nu}(\widetilde{\gamma}) \equiv \frac{1}{2}\int_{\widetilde{\gamma}} dx^\mu \wedge dx^\nu$$

the integrated area 2-form (this measures the oriented area of the projection of $\widetilde{\gamma}$ on the plane $\mu\nu$). The unwritten terms are of higher order in the size of the loop, and therefore become negligible when this size goes to zero. Therefore, we have

$$\mathrm{tr}\left(\mathcal{W}[A;\gamma\widetilde{\gamma}]\right) - \mathrm{tr}\left(\mathcal{W}[A;\gamma]\right) = ig\,\sigma^{\mu\nu}(\widetilde{\gamma})\,\mathrm{tr}\left(F_{\mu\nu}(x)\,\mathcal{W}[A;\gamma]\right) + \cdots.$$

This identity can be formally written as

$$\frac{\delta\,\mathrm{tr}\left(\mathcal{W}[A;\gamma]\right)}{\delta\,\sigma^{\mu\nu}(x)} = ig\,\mathrm{tr}\left(F_{\mu\nu}(x)\,\mathcal{W}[A;\gamma]\right),$$

which is known as the *Mandelstam formula.* It expresses the variation of the trace of a Wilson loop when the contour γ over which it is defined is deformed infinitesimally at a point $x \in \gamma$ (since we have taken the trace, we need not specify in the definition of the area derivative whether the deformation $\widetilde{\gamma}$ of the contour is added "before" or "after" the contour γ). Note an important property of this area derivative: when $x \neq y$, we have

$$\frac{\delta}{\delta\,\sigma^{\mu\nu}(x)}\frac{\delta}{\delta\,\sigma^{\rho\sigma}(y)}\mathrm{tr}\left(\mathcal{W}[A;\gamma]\right) = \frac{\delta}{\delta\,\sigma^{\rho\sigma}(y)}\frac{\delta}{\delta\,\sigma^{\mu\nu}(x)}\mathrm{tr}\left(\mathcal{W}[A;\gamma]\right),$$

which means that when two deformations are applied at distinct points of the contour, the result does not depend on the order in which they are applied. This property is not true if $x = y$: the difference between the two orderings of the area derivative involves the non-zero commutator $[F_{\mu\nu}(x), F_{\rho\sigma}(x)]$.

35. Endpoint Derivative of a Wilson Line This problem is a continuation of Problem 34. Consider a Wilson line defined on an open path γ_{xy} that goes from x to y.

35.a By varying the endpoints of the contour, show that

$$\partial_x^\mu\,\mathcal{W}[A;\gamma_{xy}] = -ig\,\mathcal{W}[A;\gamma_{xy}]\,A^\mu(x), \quad \partial_y^\mu\,\mathcal{W}[A;\gamma_{xy}] = ig\,A^\mu(y)\,\mathcal{W}[A;\gamma_{xy}].$$

35.b Show that $\left[\partial_y^\mu, \partial_y^\nu\right]\mathcal{W}[A;\gamma_{xy}] = ig\,F^{\mu\nu}(y)\,\mathcal{W}[A;\gamma_{xy}]$ and $\left[\partial_\mu^y, \partial_\nu^y\right] = \delta/\delta\sigma^{\mu\nu}(x)$ (see Problem 34 for the definition of the area derivative).

35.c For a gauge field that obeys the classical Yang–Mills equations in the vacuum, check that

$$\partial_x^\mu \frac{\delta \operatorname{tr}\left(W[A;\gamma]\right)}{\delta\sigma^{\mu\nu}(x)} = 0.$$

35.d Is this still zero if we consider the expectation value of the Wilson line, $\langle \operatorname{tr}\left(W[A;\gamma]\right)\rangle$ (i.e., the Wilson line averaged over all configurations of the gauge field) instead of the Wilson line in a fixed A^μ? Determine the non-zero right-hand side in the case of QED.

35.a Consider an open contour γ_{xy} starting at the point x and ending at the point y. Now, let us extend this contour at both endpoints: $\gamma_{xy} \to \delta\gamma_x\,\gamma_{xy}\,\delta\gamma_y$. The Wilson line defined on the new contour can be written as

$$W[A;\delta\gamma_x\,\gamma_{xy}\,\delta\gamma_y] = W[A;\delta\gamma_y]\,W[A;\gamma_{xy}]\,W[A;\delta\gamma_x].$$

(Recall that, in a Wilson line, the fields are ordered from left to right by starting from the endpoint of the contour.) When the added contours are infinitesimal, they may be viewed as variations dx^μ, dy^μ of the endpoints of γ_{xy}. For such small contours, we have

$$W[A;\delta\gamma_y] = 1 + ig\,dy^\mu\,A_\mu(y) + \cdots, \qquad W[A;\delta\gamma_x] = 1 - ig\,dx^\mu\,A_\mu(x) + \cdots,$$

and

$$\begin{aligned} &W[A;\delta\gamma_x\,\gamma_{xy}\,\delta\gamma_y] \\ &\quad = W[A;\gamma_{xy}] + ig\,dy^\mu\,A_\mu(y)\,W[A;\gamma_{xy}] - ig\,dx^\mu\,W[A;\gamma_{xy}]\,A_\mu(x) + \cdots. \end{aligned}$$

The result of this formula can also be written in the form of endpoint derivatives of the Wilson line:

$$\partial_x^\mu\,W[A;\gamma_{xy}] = -ig\,W[A;\gamma_{xy}]\,A^\mu(x), \qquad \partial_y^\mu\,W[A;\gamma_{xy}] = ig\,A^\mu(y)\,W[A;\gamma_{xy}].$$

35.b If we take the derivative with respect to the final point y twice, we obtain

$$\begin{aligned} \partial_y^\nu\partial_y^\mu\,W[A;\gamma_{xy}] &= \partial_y^\nu\Big(ig\,A^\mu(y)\,W[A;\gamma_{xy}]\Big) \\ &= ig\Big(\big(\partial_y^\nu A^\mu(y)\big)\,W[A;\gamma_{xy}] + A^\mu(y)\,\big(\partial_y^\nu\,W[A;\gamma_{xy}]\big)\Big) \\ &= ig\,\Big(\partial_y^\nu A^\mu(y) + ig\,A^\nu(y)A^\mu(y)\Big)\,W[A;\gamma_{xy}]. \end{aligned}$$

From this formula, it is clear that successive endpoint derivatives do not commute. Instead, we have

$$\left[\partial_y^\mu, \partial_y^\nu\right]\,W[A;\gamma_{xy}] = ig\,F^{\mu\nu}(y)\,W[A;\gamma_{xy}].$$

Using the result of Problem 34, we may also interpret this as a relationship between endpoint derivatives and area derivatives:

$$\left[\partial_\mu^y, \partial_\nu^y\right] = \frac{\delta}{\delta\sigma^{\mu\nu}(y)}.$$

35.c Consider now a closed contour γ_{xx} starting and ending at some point x. Upon variation of the point x by dx, the Wilson loop varies according to

$$\mathcal{W}[A;\gamma_{xx}] \to \mathcal{W}[A;\gamma_{xx}] - ig\, dx^\mu\, \mathcal{W}[A;\gamma_{xx}]\, A_\mu(x) + ig\, dx^\mu\, A_\mu(x)\, \mathcal{W}[A;\gamma_{xx}] + \cdots,$$

which we may write as

$$\partial_x^\mu \mathcal{W}[A;\gamma_{xx}] = ig\, \big[A_\mu(x), \mathcal{W}[A;\gamma_{xx}]\big].$$

This implies that $\partial_x^\mu \operatorname{tr}\big(\mathcal{W}[A;\gamma_{xx}]\big) = 0$. This result could have been derived differently. Indeed, varying the common endpoint x of a closed contour by the same amount dx is equivalent to adding a small closed contour $\widetilde{\gamma}$ to γ. As we have seen in Problem 34, the variation of the trace of the Wilson line is proportional to the area of $\widetilde{\gamma}$, and thus should vanish at linear order in dx.

Let us now apply the derivative ∂_x^α to the Mandelstam formula derived in Problem 34. We get

$$\begin{aligned}
\partial_x^\alpha \frac{\delta \operatorname{tr}\big(\mathcal{W}[A;\gamma]\big)}{\delta\, \sigma^{\mu\nu}(x)} &= ig\, \partial_x^\alpha \operatorname{tr}\big(F_{\mu\nu}(x)\, \mathcal{W}[A;\gamma]\big) \\
&= ig \operatorname{tr}\big(F_{\mu\nu}(x)\, (ig\, A^\alpha(x))\mathcal{W}[A;\gamma]\big) \\
&\quad + ig \operatorname{tr}\big(F_{\mu\nu}(x)\, \mathcal{W}[A;\gamma](-ig\, A^\alpha(x))\big) \\
&\quad + ig \operatorname{tr}\big((\partial_x^\alpha F_{\mu\nu}(x))\, \mathcal{W}[A;\gamma]\big) \\
&= ig \operatorname{tr}\big((\partial_x^\alpha F_{\mu\nu}(x) - ig[A^\alpha(x), F_{\mu\nu}(x)])\, \mathcal{W}[A;\gamma]\big) \\
&= ig \operatorname{tr}\big([D^\alpha(x), F_{\mu\nu}(x)]\, \mathcal{W}[A;\gamma]\big).
\end{aligned}$$

If the gauge field A^μ used in the definition of the Wilson loop obeys the vacuum classical Yang–Mills equations, $[D^\mu, F_{\mu\nu}] = 0$, we have

$$\partial_x^\mu \frac{\delta \operatorname{tr}\big(\mathcal{W}[A;\gamma]\big)}{\delta\, \sigma^{\mu\nu}(x)} = 0.$$

35.d Note that if we consider instead the quantum expectation value of the Wilson loop instead of its value in a fixed background, the right-hand side is not zero:

$$\partial_x^\mu \frac{\delta \big\langle \operatorname{tr}\big(\mathcal{W}[A;\gamma]\big)\big\rangle}{\delta\, \sigma^{\mu\nu}(x)} \neq 0.$$

In quantum electrodynamics, one may explicitly calculate the non-zero term that enters on the right-hand side. First, note that

$$\int [DA^\mu]\, \frac{\delta}{\delta A^\mu(x)} \Big(\mathcal{W}[A;\gamma]\, e^{iS[A]}\Big) = 0,$$

where $S[A]$ is the photon action. (The zero follows trivially from the fact that this is the integral of a total derivative.) By explicitly evaluating the functional derivative in the integrand, we get

$$\left\langle\left(\partial_x^\mu F_{\mu\nu}(x) + e\int_\gamma dy_\nu\ \delta(x-y)\right) \mathcal{W}[A;\gamma]\right\rangle = 0,$$

which we may rewrite as

$$\partial_x^\mu \frac{\delta\left\langle \mathcal{W}[A;\gamma]\right\rangle}{\delta\,\sigma^{\mu\nu}(x)} = -ie^2\left\langle \mathcal{W}[A;\gamma]\right\rangle \int_\gamma dy_\nu\ \delta(x-y).$$

(Note that in QED, the trace is superfluous since the value of the Wilson loop is a scalar, not a matrix.)

36. Makeenko–Migdal Loop Equation This is a continuation of Problems 34 and 35. Consider a closed path γ, and the Wilson loop $\mathcal{W}[A;\gamma]$ defined on this contour in Yang–Mills theory.

36.a With the definitions of Problems 34 and 35, show that

$$\partial_x^\mu \frac{\delta\left\langle \mathrm{tr}\left(\mathcal{W}[A;\gamma]\right)\right\rangle}{\delta\,\sigma^{\mu\nu}(x)} + ig^2\int_\gamma dy_\nu\ \delta(y-x)\left\langle \mathrm{tr}\left(t^a\,\mathcal{W}[A;\gamma_{xy}]\,t^a\,\mathcal{W}[A;\gamma_{yx}]\right)\right\rangle = 0,$$

where γ_{xy} and γ_{yx} are the portions of γ that go from x to y, and from y to x, respectively.

36.b Show that the right-hand side of this equation is zero if $x \notin \gamma$, Abelian-like if $x \in \gamma$ and the contour does not have any self-intersection, and contains genuine non-Abelian terms when γ has self-intersections.

36.c In the case of an $\mathrm{SU}(N)$ gauge theory and in the limit of a large number of colors, $N \to +\infty$, show that this equation reduces to

$$\partial_x^\mu \frac{\delta\left\langle \mathrm{tr}\left(\mathcal{W}[A;\gamma]\right)\right\rangle}{\delta\,\sigma^{\mu\nu}(x)} + i\frac{g^2}{2}\int_\gamma dy_\nu\ \delta(y-x)\left\langle \mathrm{tr}\left(\mathcal{W}[A;\gamma_{xy}]\right)\right\rangle\left\langle \mathrm{tr}\left(\mathcal{W}[A;\gamma_{yx}]\right)\right\rangle = 0.$$

36.a In Problem 35, we showed that

$$\partial_x^\mu \frac{\delta\,\mathrm{tr}\left(\mathcal{W}[A;\gamma]\right)}{\delta\,\sigma^{\mu\nu}(x)} = ig\,\mathrm{tr}\left([D^\mu, F_{\mu\nu}(x)]\,\mathcal{W}[A;\gamma]\right). \tag{3.32}$$

Here, since we want an expectation value, we must not assume that the gauge field A^μ obeys

Yang–Mills equations (doing so would only give the classical value of the path integral, not the full quantum result). Therefore, in order to obtain the proposed equation, we must evaluate the expectation value of the right-hand side. Let us start from the following identity (which is trivial because the integrand is a total functional derivative):

$$\int [DA^\mu]\, \frac{\delta}{\delta A_a^\nu(x)} \Big(e^{iS[A]}\, \mathrm{tr}\,\big(\underbrace{P\, e^{ig\int_\gamma dx^\mu A_\mu}}_{W[A;\gamma]} \big)\Big) = 0.$$

(Note that, under the path integral, the path-ordering is still necessary for the final factor. Indeed, even though A^μ is no longer an operator in the path integral, it is a non-commuting matrix-valued quantity.) Let us now apply the derivative to the two factors in turn. First, we have

$$\frac{\delta}{\delta A_a^\nu(x)}\, e^{iS[A]} = i\big(D^\mu_{\mathrm{adj}}\big)_{ab} F^b_{\mu\nu}(x)\, e^{iS[A]} = i\,\big[D^\mu, F_{\mu\nu}(x)\big]_a\, e^{iS[A]}.$$

In order to obtain the derivative of the second factor, note that

$$\frac{\delta}{\delta A_a^\nu(x)}\, P\, e^{ig\int_\gamma dx^\mu A_\mu} = ig \int_\gamma dy_\nu\, \delta(x-y)\, W[A;\gamma_{xy}]\, t^a\, W[A;\gamma_{yx}],$$

where t^a is a generator of the gauge algebra in the representation of interest, and γ_{xy} (resp. γ_{yx}) denotes the portion of the contour γ that goes from the point x to the point y (resp. from y to x). Here, a few more words of explanation are in order. Since this derivative is obviously zero when $x \notin \gamma$, we can assume from now on that the point x lies on γ. Moreover, since at this stage we have not yet taken the trace, we have to choose a starting and a final point on γ (they are identical since γ is a closed loop), in order for the Wilson line to be fully defined. In this case, the simplest choice is to take x as this base point. Thus, the identity we have obtained reads

$$\Big\langle \big[D^\mu, F_{\mu\nu}(x)\big]_a\, W[A;\gamma] \Big\rangle + g \int_\gamma dy_\nu\, \delta(x-y) \Big\langle W[A;\gamma_{xy}]\, t^a\, W[A;\gamma_{yx}] \Big\rangle = 0.$$

Then, we multiply by $ig\, t^a$ on the left and take the trace, and recall (3.32) to get

$$\partial_x^\mu \frac{\delta \big\langle \mathrm{tr}\,(W[A;\gamma]) \big\rangle}{\delta\, \sigma^{\mu\nu}(x)} + ig^2 \int_\gamma dy_\nu\, \delta(x-y)\, \mathrm{tr} \Big\langle t^a\, W[A;\gamma_{xy}]\, t^a\, W[A;\gamma_{yx}] \Big\rangle = 0. \tag{3.33}$$

36.b There are three main cases when discussing the second term:

- The fact that the second term in this equation is zero when $x \notin \gamma$ is trivial, and was in fact already exploited in our form of writing this term.
- When $x \in \gamma$ and the loop γ does not intersect itself, there is a single point in the parameterization of γ such that $y = x$. Therefore, one of the two sub-paths γ_{xy} or γ_{yx}

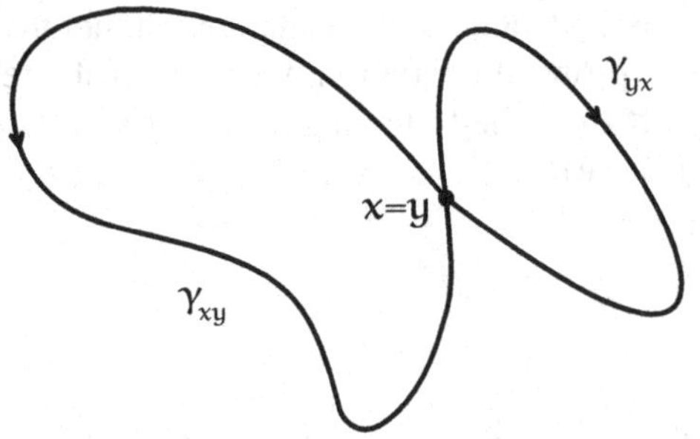

Figure 3.2 Closed path γ with a self-intersection at the point x.

is the trivial path reduced to a single point, while the other one is γ itself. In this case, the remaining Wilson loop is independent of x, y, and the equation (3.33) simplifies to

$$\partial_x^\mu \frac{\delta\left\langle \mathrm{tr}\left(W[A;\gamma]\right)\right\rangle}{\delta\, \sigma^{\mu\nu}(x)} + ig^2 C_{2r}\left\langle \mathrm{tr}\, W[A;\gamma]\right\rangle \int_\gamma dy_\nu\, \delta(x-y) = 0,$$

where C_{2r} is the quadratic Casimir in the representation under consideration. Thus, for non-intersecting loops, we obtain an equation that has exactly the same *linear* structure as in the Abelian case (compare with the final equation obtained in Problem 35).

- When the contour γ possesses self-intersections, and the point x is one of them, then it is possible that the sub-paths γ_{xy} and γ_{yx} are both non-trivial closed contours, as illustrated in Figure 3.2. This is the situation where we have genuine non-Abelian features, since the non-commutativity of the gauge group forbids moving the two Wilson loops next to each other to reconstruct the Wilson loop on the full contour (this is always possible in QED).

36.c Let us focus now on the case of an $SU(N)$ gauge theory, and assume that all the objects in (3.33) are expressed in the fundamental representation. First, the trace in the second term may be written more explicitly as follows:

$$\mathrm{tr}\left\langle t^a\, W[A;\gamma_{xy}]\, t^a\, W[A;\gamma_{yx}]\right\rangle = \left\langle t^a_{ij}\, W_{jk}[A;\gamma_{xy}]\, t^a_{kl}\, W_{li}[A;\gamma_{yx}]\right\rangle.$$

Then, we use the $\mathfrak{su}(N)$ Fierz identity in order to explicitly perform the contraction between the two t^a's,

$$t^a_{ij} t^a_{kl} = \frac{1}{2}\,\delta_{il}\delta_{jk} - \frac{1}{2N}\,\delta_{ij}\delta_{kl}.$$

In the limit where $N \to +\infty$, we can neglect the second term, and we have

$$\mathrm{tr}\left\langle t^a\, W[A;\gamma_{xy}]\, t^a\, W[A;\gamma_{yx}]\right\rangle \underset{N\to+\infty}{=} \frac{1}{2}\left\langle \mathrm{tr}\left(W[A;\gamma_{xy}]\right)\, \mathrm{tr}\left(W[A;\gamma_{yx}]\right)\right\rangle.$$

It turns out this can be simplified further in the large-N limit. In order to see this, note first that, in the fundamental representation, the trace of a Wilson loop scales like N. Recall now that the expectation value of a product of Wilson loops amounts to connecting the fields attached to the loops in all possible ways, according to the rules defined by the Yang–Mills action $S[A]$. In the case of the double loop shown in the preceding figure, there are configurations of these links that do not mix the two loops. These configurations correspond in fact to the product of the expectation values of the two traces:

$$\Big\langle \mathrm{tr}\,(\mathcal{W}[A;\gamma_{xy}])\Big\rangle \Big\langle \mathrm{tr}\,(\mathcal{W}[A;\gamma_{yx}])\Big\rangle = \cdots +$$

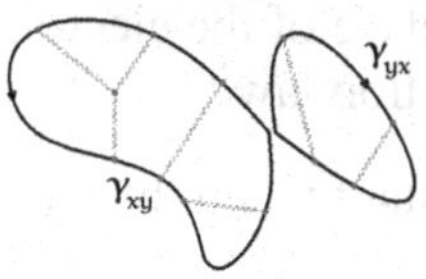

$$+ \cdots = \mathcal{O}(N^2).$$

(In the graphical representation, we have slightly displaced the loops to highlight the fact that there are two separate traces.) In contrast, links that connect the two loops contribute to the "correlated" term:

$$\Big\langle \mathrm{tr}\,(\mathcal{W}[A;\gamma_{xy}])\;\mathrm{tr}\,(\mathcal{W}[A;\gamma_{yx}])\Big\rangle - \Big\langle \mathrm{tr}\,(\mathcal{W}[A;\gamma_{xy}])\Big\rangle \Big\langle \mathrm{tr}\,(\mathcal{W}[A;\gamma_{yx}])\Big\rangle$$

$$=$$

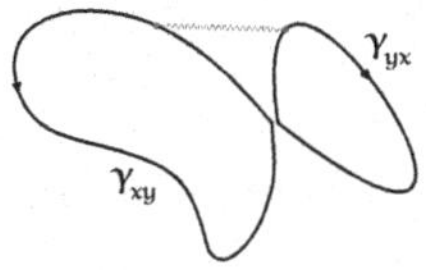

$$+ \cdots = \frac{1}{2}$$

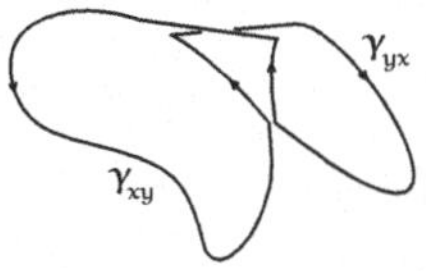

$$- \frac{1}{2N}$$

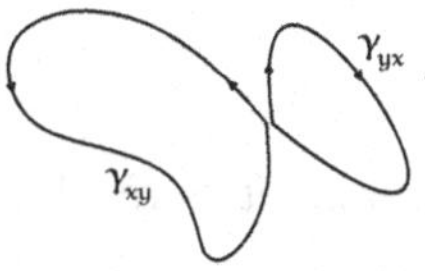

$$+ \cdots .$$

$\mathcal{O}(N)$ $\qquad$ $\mathcal{O}(N)$

Therefore, we see that the correlated part of the double trace is only of order N, and in the large-N limit we have

$$\Big\langle \mathrm{tr}\,(\mathcal{W}[A;\gamma_{xy}])\;\mathrm{tr}\,(\mathcal{W}[A;\gamma_{yx}])\Big\rangle \underset{N\to+\infty}{=} \Big\langle \mathrm{tr}\,(\mathcal{W}[A;\gamma_{xy}])\Big\rangle \Big\langle \mathrm{tr}\,(\mathcal{W}[A;\gamma_{yx}])\Big\rangle .$$

In this limit, the loop equation becomes

$$\partial_x^\mu \frac{\delta\Big\langle \frac{1}{N}\mathrm{tr}\,(\mathcal{W}[A;\gamma])\Big\rangle}{\delta\,\sigma^{\mu\nu}(x)} \underset{N\to+\infty}{=} -i\frac{g^2N}{2}\int_\gamma dy_\nu\,\delta(x-y)\Big\langle \tfrac{1}{N}\mathrm{tr}\,(\mathcal{W}[A;\gamma_{xy}])\Big\rangle\Big\langle \tfrac{1}{N}\mathrm{tr}\,(\mathcal{W}[A;\gamma_{yx}])\Big\rangle .$$

This equation, which is a closed equation for expectation values of traces of Wilson loops, is known as the *Makeenko–Migdal equation.* In this final form, we have normalized each trace by a factor N^{-1} in order to have quantities of order one in the fundamental representation. With this normalization, the prefactor of the second term contains the combination g^2N, called the *'t Hooft coupling.* The limit $N \to +\infty$ we have considered in this derivation in general assumes that the limit is taken while g^2N is kept finite.
Source: Makeenko, Y. M. and Migdal, A. A. (1979), *Phys Lett B* 88: 135.

37. Classical Non-Abelian Particles and the Wong Equations The goal of this problem is to extend to the non-Abelian case the concept of a classical particle that moves along a trajectory $x^\mu(t)$. This is useful in situations where the classical approximation provides the leading answer, for instance in the calculation of *hard thermal loops* in QCD at finite temperature (see Problem 74).

37.a First, consider the Abelian case. Recall the definition of the momentum p^μ of the particle (assume it is massive, with a mass m), and the expression for dp^μ/dt given by the Lorentz force. Write the electromagnetic current created by this classical particle, and check that it is conserved.

37.b Write down a non-Abelian generalization of the Lorentz force, introducing a "color charge" Q^a in place of the electrical charge e. Can Q^a be gauge invariant? If not, what is its transformation law?

37.c Can Q^a be constant when the particle moves through an external color field?

37.d Write down the non-Abelian current generated by the moving particle. What should its conservation equation be? Show that this conservation equation implies the following equation of motion for Q^a:

$$\frac{dx^\mu}{dt}\left[D_\mu, Q\right] = 0 \qquad (Q \equiv Q^a t_f^a).$$

37.a First, let us recall the conservation of the electromagnetic current in QED:

$$J^\mu(x) \equiv e\,\overline{\psi}\gamma^\mu\psi, \qquad \partial_\mu J^\mu = 0,$$

which follows from the equation of motion of the field operator ψ. In this equation, the current is a quantum operator. However, in this case, it is well known how to obtain a classical description of the electromagnetic current in terms of classical charged particles. Let us illustrate this for a single point-like particle of charge e and mass m. We denote by $x^\mu(t)$ the worldline of this particle, parameterized by its proper time. Its momentum is given by

$$p^\mu = m\frac{dx^\mu}{dt},$$

whose equation of motion is driven by the Lorentz force:

$$\frac{dp^\mu}{dt} = e\,F^{\mu\nu}\,\frac{dx_\nu}{dt}. \tag{3.34}$$

The electromagnetic current it produces can be written as

$$J^\mu \equiv e\int dt\,\frac{dx^\mu}{dt}\,\delta(x - x(t)).$$

(In this compact notation, keep in mind that the delta function is in fact a product of four delta functions, one for each dimension of spacetime.) The verification that this current is conserved goes as follows:

$$\partial_\mu J^\mu = e\int dt\,\underbrace{\frac{dx^\mu}{dt}\,\partial_\mu\delta(x - x(t))}_{-\frac{d}{dt}\delta(x-x(t))} = 0.$$

Note an implicit assumption, crucial in obtaining this conservation law: the electrical charge of the particle is time-independent (otherwise, an integration by parts would give a non-zero integrand proportional to the time derivative of e). This is of course very trivial: the electromagnetic current of the particle is conserved precisely because electrical charges are constants.

37.b Now, we would like to generalize the concept of classical particle to particles that carry a non-Abelian charge. The first equation,

$$p^{\mu} = m\frac{dx^{\mu}}{dt},$$

is of course the same. Then, we would like to write an equation for the time derivative of the momentum, involving the non-Abelian field strength. Since dp^{μ}/dt is an $SU(N)$ scalar, we want a scalar on the right-hand side, not a matrix. Then comes the issue of gauge invariance: dp^{μ}/dt describes the trajectory of the particle, and therefore should be gauge invariant, while $F^{\mu\nu}$ is not. Taking the trace of $F^{\mu\nu}$ gives a gauge invariant quantity, but this cannot be correct since the trace is zero. These considerations suggest that we generalize (3.34) as follows in the non-Abelian case:

$$\frac{dp^{\mu}}{dt} = 2\,\mathrm{tr}\left(QF^{\mu\nu}\right)\frac{dx_{\nu}}{dt}, \tag{3.35}$$

where Q is a matrix-valued "charge" (in the same representation of the Lie algebra as $F^{\mu\nu}$). For the right-hand side to be gauge invariant, Q should be gauge-dependent and transform in the same way as the field strength:

$$F^{\mu\nu} \to \Omega^{\dagger}\,F^{\mu\nu}\,\Omega, \qquad Q \to \Omega^{\dagger}\,Q\,\Omega.$$

The prefactor 2 on the right-hand side was chosen with the fundamental representation in mind, so that we also have

$$\frac{dp^{\mu}}{dt} = Q^{a}F^{\mu\nu}_{a}\frac{dx_{\nu}}{dt}.$$

37.c Because of its transformation law and because a gauge transformation Ω can depend on spacetime, Q should be regarded as a spacetime-dependent object. Moreover, since Q describes the charge of a colored particle, and since the fields with which it may interact during its propagation themselves carry color, we expect that Q should evolve according to the background gauge field in which the particle propagates.

37.d In order to derive the equation of motion for Q, let us first define the non-Abelian current by mimicking the QED definition:

$$J^{\mu}(x) \equiv \int dt\; Q(x(t))\,\frac{dx^{\mu}}{dt}\,\delta(x - x(t)). \tag{3.36}$$

The only notable modification is that the charge Q is evaluated at the current point on the trajectory of the particle, to account for its spacetime dependence inherited from gauge transformations. Note also that this current is matrix-valued, and not gauge invariant. It transforms as

$$J^{\mu}(x) \rightarrow \Omega^{\dagger}(x)\, J^{\mu}(x)\, \Omega(x).$$

For this classical current to be a good replacement for the quantum one, it should fulfill the same covariant conservation equation,

$$\left[D_{\mu}, J^{\mu}\right] = 0.$$

(The covariant derivative makes this conservation law gauge invariant.) Therefore, let us evaluate the covariant derivative of the current defined in (3.36):

$$\begin{aligned}\left[D_{\mu}, J^{\mu}\right] &= \int dt\, \Big(-ig\left[A_{\mu}, Q(x(t))\right] \frac{dx^{\mu}}{dt}\, \delta(x - x(t)) \\ &\qquad + \underbrace{Q(x(t))\, \frac{dx^{\mu}}{dt}\, \partial_{\mu}\delta(x - x(t))}_{-\frac{d}{dt}\delta(x-x(t))} \Big) \\ &= \int dt\, \left(\frac{dQ(x(t))}{dt} - ig\frac{dx^{\mu}}{dt}\left[A_{\mu}, Q(x(t))\right] \right) \delta(x - x(t)).\end{aligned}$$

Requesting that this covariant derivative vanish leads to the following equation of motion for the non-Abelian charge Q:

$$\underbrace{\frac{dQ(x(t))}{dt} - ig\frac{dx^{\mu}}{dt}\left[A_{\mu}, Q(x(t))\right]}_{\frac{dx^{\mu}}{dt}\left[D_{\mu}, Q\right]} = 0.$$

This equation, known as the *Wong equation* for Q, describes the variation of the color charge of the particle as it moves through a given background gauge field. Note that even a very soft field (i.e., whose non-zero Fourier modes are much smaller than the momentum of the particle), which does not alter the straight motion of the particle, will change its color. Finally, observe that this equation may also be written as

$$\frac{dx^{\mu}}{dt}\left(D_{\mu}^{adj}\right)_{ab} Q^{b} = 0.$$

In other words, the covariant derivative of Q projected along the trajectory of the particle is vanishing (this equation of motion is reminiscent of the differential equation that defines a Wilson line).
Source: Wong, S. K. (1970), *Nuovo Cimento A* 65: 689.

38. Wong Equations from the Worldline Formalism The goal of this problem is to obtain the Wong equation derived in Problem 37 by using the worldline formalism. Consider a colored complex scalar field (which ultimately we would like to describe as a colored classical particle) embedded in a non-Abelian gauge field A^{μ}.

38.a Write the worldline representation for the contribution of a scalar loop to the quantum effective action of the non-Abelian gauge field. For the time being, use a path ordering to handle the fact that the gauge field is a non-commuting Lie-algebra–valued object.

38.b Use the result of Problem 27 to remove this path-ordering, in order to obtain

$$\Gamma[A] = \text{const} + \int_0^\infty \frac{dT}{T} \int_{z(0)=z(T)} [Dz(\tau)]\; e^{-\int_0^T d\tau\,\left(\frac{1}{4}\dot{z}^2(\tau)+m^2\right)}$$
$$\times\, 2^d \int_0^{2\pi} \frac{d\phi}{2\pi} \int_{\substack{\text{anti-}\\ \text{periodic}}} [D\overline{\chi}D\chi]\; e^{i\phi\left(\frac{d}{2}-1+\overline{\chi}_0\chi_0\right)}\; e^{-\int \overline{\chi}(\dot{z}\cdot D)\chi},$$

where d is the size of the matrices in the representation where the scalar lives, and $\overline{\chi}_0\chi_0$ denotes the value of $\overline{\chi}\chi$ at some fixed $\tau_0 \in [0, T]$.

38.c Introduce a charge defined as $Q^a \equiv g\,\overline{\chi}t^a\chi$, and write the effective Lagrangian of the "worldline particle" and auxiliary Grassmann field χ that appear in this representation.

38.d Derive the equation of motion of χ, and use it to recover the Wong equation for Q^a discussed in Problem 37.

38.a The worldline representation of the one-loop effective action of a *non-Abelian* gauge field due to a scalar loop is almost identical to Eq. (2.17), except that we must keep an explicit path-ordering in order to handle the fact that the gauge field is non-commuting:

$$\Gamma[A] = \text{const} + \int_0^\infty \frac{dT}{T} \int_{z(0)=z(T)} [Dz(\tau)]\; e^{-\int_0^T d\tau\,\left(\frac{1}{4}\dot{z}^2(\tau)+m^2\right)}$$
$$\times\, \text{tr}\, P\left(e^{-ig\int_0^T d\tau\, \dot{z}(\tau)\cdot A(z(\tau))}\right).$$

The other notable difference is the persistence of a trace over the color indices carried by the gauge field (the original trace was over spacetime and color, and the path integral over $z^\mu(\tau)$ only takes care of the spacetime trace).

38.b The trace of the path-ordered exponential in the second line is of the type discussed in Problem 27, with

$$M(\tau) \equiv -g\,\frac{dz^\mu}{d\tau} A_\mu(z(\tau)).$$

Therefore, we can directly write

$$\text{tr}\, P\left(e^{-ig\int_0^T d\tau\, \dot{z}(\tau)\cdot A(z(\tau))}\right) = 2^d \int_0^{2\pi} \frac{d\phi}{2\pi} \int_{\substack{\text{anti-}\\ \text{periodic}}} [D\overline{\chi}D\chi]\; e^{i\phi\left(\frac{d}{2}-1+\overline{\chi}_0\chi_0\right)}\; e^{-\int \overline{\chi}(\dot{\chi}-ig\dot{z}\cdot A\chi)},$$

where d is the dimension of the matrix A^μ in the representation appropriate for the scalar field under consideration. Here, we have adopted the form of this representation wherein the first exponential has a term ϕ is coupled to the value of $\overline{\chi}\chi$ at a fixed τ_0 (which we denote by $\overline{\chi}_0\chi_0$). The argument of the last exponential can be rearranged a bit:

$$\overline{\chi}(\dot{\chi} - ig\dot{z}\cdot A\chi) = \overline{\chi}\Big(\underbrace{\frac{d}{d\tau}}_{\frac{dz^\mu}{d\tau}\frac{\partial}{\partial z^\mu}} - ig\frac{dz^\mu}{d\tau}A_\mu\Big)\chi = \overline{\chi}\Big(\frac{dz^\mu}{d\tau}\underbrace{(\partial_\mu - igA_\mu)}_{D_\mu}\Big)\chi,$$

where D_μ is the covariant derivative in the representation of the scalar field. This establishes the announced formula.

38.c We may also write this formula as

$$\Gamma[A] = \text{const} + \int_0^\infty \frac{dT}{T} \int_{z(0)=z(T)} [Dz(\tau)]$$

$$\times\, 2^d \int_0^{2\pi} \frac{d\phi}{2\pi} \int_{\substack{\text{anti-}\\\text{periodic}}} [D\overline{\chi}D\chi]\; e^{i\phi\left(\frac{d}{2}-1+\overline{\chi}_0\chi_0\right)}\; e^{-\int_0^T d\tau\,\mathcal{L}_{\text{eff}}(z,\overline{\chi},\chi)},$$

with an effective Lagrangian defined as

$$\mathcal{L}_{\text{eff}}(z,\overline{\chi},\chi) \equiv \frac{\dot{z}^2}{4} + m^2 + \overline{\chi}(\dot{z}\cdot D)\chi = \frac{\dot{z}^2}{4} + m^2 - iQ^a\,\dot{z}^\mu A^a_\mu + \overline{\chi}\dot{\chi},$$

where we denote $Q^a \equiv g\,\overline{\chi}t^a\chi$. Note that the first three terms,

$$\frac{\dot{z}^2}{4} + m^2 - iQ^a\,\dot{z}^\mu A^a_\mu,$$

are formally identical to the effective Lagrangian for a scalar particle in a superposition of gauge fields A^μ_a, with charges Q^a controlling the couplings to these various fields. These Q^a play the same role as the non-Abelian charge introduced in Problem 37.

38.d By differentiating the effective Lagrangian with respect to $\overline{\chi}$, we obtain the following equation of motion for χ:

$$(\dot{z}\cdot D)\,\chi = 0, \quad \text{i.e.,}\ \dot{\chi} = ig\,\dot{z}^\mu A_\mu\chi,$$

and by conjugation

$$\dot{\overline{\chi}} = -ig\,\dot{z}^\mu\,\overline{\chi}A_\mu.$$

Thanks to these equations of motion, we can calculate the time derivative of the non-Abelian charge Q^a:

$$\begin{aligned}
\dot{Q}^a &= g\overline{\chi}t^a\dot{\chi} + g\dot{\overline{\chi}}t^a\chi = ig^2\,\dot{z}^\mu\left(\overline{\chi}\,t^aA_\mu\chi - \overline{\chi}\,A_\mu t^a\chi\right)\\
&= ig^2\,\dot{z}^\mu A^b_\mu\overline{\chi}\left[t^a,t^b\right]\chi = -g^2 f^{abc}\,\dot{z}^\mu A^b_\mu\overline{\chi}t^c\chi = ig\,\dot{z}^\mu A^b_\mu(T^b_{\text{adj}})_{ac}\,Q^c\\
&= ig\,\dot{z}^\mu\left(A^{\text{adj}}_\mu\right)_{ac}Q^c.
\end{aligned}$$

These equations are nothing but the Wong equations obtained in Problem 37 by other considerations. Although the result is the same, the present approach provides more solid field

theoretical foundations for the definition and evolution of the non-Abelian charge Q^a.
Source: Jalilian-Marian, J., Jeon, S., Venugopalan, R. and Wirstam J. (2000), *Phys Rev D* 62: 045020.

39. Non-Abelian Stokes's Theorem Consider a closed path γ_{xx} of base point x. Choose a point $y \in \gamma_{xx}$ which divides γ_{xx} into two sub-paths, γ^a_{xy} and γ^b_{yx}, and define a family $\xi^t(x)$ $((t,s) \in [0,1] \times [0,1])$ of interpolating paths between γ^a_{xy} and γ^b_{yx} such that $\xi^0(s) = \gamma^a_{xy}(s)$, $\xi^1(s) = \gamma^b_{xy}(s)$, $\xi^t(0) = x$, $\xi^t(1) = y$. With this notation, for an Abelian gauge field A^μ, Stokes's theorem states that

$$\oint_{\gamma_{xx}} dz^\mu\, A_\mu(z) = \int_0^1 dtds\, \frac{\partial\xi^\mu}{\partial s}\frac{\partial\xi^\nu}{\partial t}\, F_{\mu\nu}(\xi^t(s)).$$

The goal of this problem is to derive a formula, known as the *non-Abelian Stokes's theorem*, which relates a line integral of A^μ and a surface integral of $F^{\mu\nu}$ in the case of non-Abelian fields.

39.a Consider the Wilson loop $W[A;\gamma_{xx}]$ defined on γ_{xx}. Show that it can be written as

$$W[A;\gamma_{xx}] = P_t \exp\left(i\int_0^1 dt\, \mathcal{A}[\xi^t]\right), \quad \text{with } \mathcal{A}[\xi^t] \equiv i\, W^{-1}[A;\xi^t]\,\frac{d}{dt}W[A;\xi^t],$$

where $W[A;\xi^t]$ is the Wilson line defined on the line of constant t.

39.b Show then that

$$i\,\mathcal{A}[\xi^t] = ig\int_0^1 ds\, \frac{\partial\xi^\mu}{\partial s}\frac{\partial\xi^\nu}{\partial t}\, W^{-1}[A;\xi^t_{(0,s)}]\, F_{\mu\nu}(\xi^t(s))\, W[A;\xi^t_{(0,s)}],$$

where $\gamma^t_{(0,s)}$ is the portion of γ^t that extends from $\gamma^t(0) = x$ and $\gamma^t(s)$.

39.c Conclude that the non-Abelian Stokes's theorem reads

$$W[A;\gamma_{xx}] = P_t \exp\left(ig\int_0^1 dtds\, \frac{\partial\xi^\mu}{\partial s}\frac{\partial\xi^\nu}{\partial t} W^{-1}[A;\xi^t_{(0,s)}]F_{\mu\nu}(\xi^t(s))W[A;\xi^t_{(0,s)}]\right).$$

39.a The content of the final formula proposed in the statement of the problem is illustrated in Figure 3.3, with a grid indicating a discretization of the parameters (t, s) to guide the eyes. In this figure, we have shown in gray the integrand inside the exponential for some values of t and s (recall that the field strength weighted by the area form is equal to an infinitesimal closed Wilson loop).

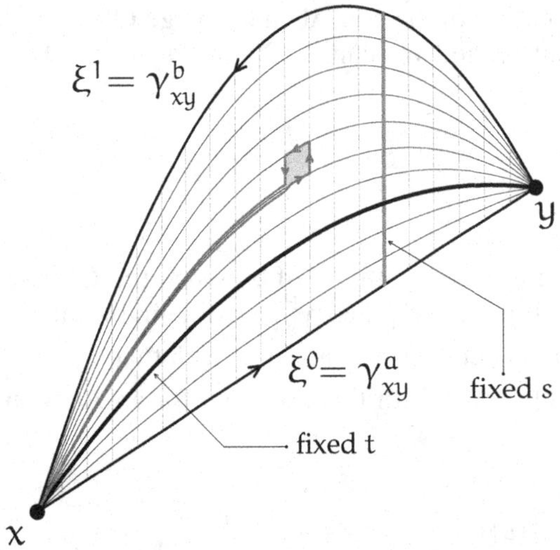

Figure 3.3 Setup for the non-Abelian Stokes's theorem. A grid has been added, in order to indicate lines of constant t and constant s. The shaded area illustrates the integrand in the exponential on the right-hand side of the proposed formula.

As a preliminary remark, let us note that the left- and right-hand sides of the proposed formula transform in the same fashion under a gauge transformation. Indeed, the Wilson line on the left-hand side transforms as

$$W[A;\gamma_{xx}] \quad \to \quad \Omega^\dagger(x)\,W[A;\gamma_{xx}]\,\Omega(x),$$

and the integrand on the right-hand side transforms as

$$\begin{aligned} &W[A;\xi^t_{(0,s)}]F_{\mu\nu}(\xi^t(s))W[A;\xi^t_{(s,0)}] \\ &\quad\to \quad \Omega^\dagger(x)W[A;\xi^t_{(0,s)}]\Omega(\xi^t(s)) \times \Omega^\dagger(\xi^t(s))F_{\mu\nu}(\xi^t(s))\Omega(\xi^t(s)) \\ &\qquad\qquad\qquad \times \Omega^\dagger(\xi^t(s))W[A;\xi^t_{(s,0)}]\Omega(x) \\ &\qquad = \Omega^\dagger(x)\,W[A;\xi^t_{(0,s)}]F_{\mu\nu}(\xi^t(s))W[A;\xi^t_{(s,0)}]\,\Omega(x). \end{aligned}$$

We see that, for this to work, the Wilson lines inserted in the integrand are crucial; otherwise each piece of the integrand would transform differently as we vary t and s.

Let us denote by $W[A;\xi^t]$ the Wilson line defined on a curve of constant t, as shown below.

$W[A;\xi^t] \equiv$ 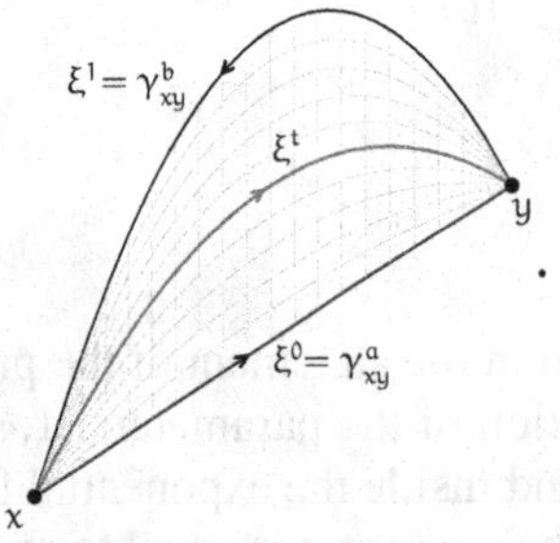 .

From this Wilson line (which belongs to the Lie group), define the following element of the

Lie algebra:

$$\mathcal{A}[\xi^t] \equiv i\, W^{-1}[A;\xi^t]\, \frac{d}{dt} W[A;\xi^t].$$

Note that $\mathcal{A}[\xi^t]$ transforms covariantly at the point x under a gauge transformation, since we have

$$\begin{aligned} W[A;\xi^t] &\to \Omega^\dagger(y)\, W[A;\xi^t]\, \Omega(x), \\ \frac{d}{dt} W[A;\xi^t] &\to \Omega^\dagger(y)\, \frac{d}{dt} W[A;\xi^t]\, \Omega(x), \\ W^{-1}[A;\xi^t] &\to \Omega^\dagger(x)\, W^{-1}[A;\xi^t]\, \Omega(y). \end{aligned}$$

In a way similar to the definition of Wilson lines from a gauge potential, we can define from $\mathcal{A}[\xi^t]$ a group element $\mathcal{S}_t$ that obeys the following differential equation and initial condition:

$$\frac{d}{dt}\mathcal{S}_t = i\,\mathcal{A}[\xi^t]\,\mathcal{S}_t, \quad \mathcal{S}_0 = 1. \tag{3.37}$$

Consider now the following derivative:

$$\begin{aligned} &\frac{d}{dt}\Big(W[A;\xi^t]\,\mathcal{S}_t\, W^{-1}[A;\xi^0]\Big) \\ &\quad = \frac{dW[A;\xi^t]}{dt}\mathcal{S}_t\, W^{-1}[A;\xi^0] + W[A;\xi^t]\frac{d\mathcal{S}_t}{dt}\, W^{-1}[A;\xi^0] \\ &\quad = \Big(-i\, W[A;\xi^t]\mathcal{A}[\xi^t]\Big)\mathcal{S}_t W^{-1}[A;\xi^0] + W[A;\xi^t]\Big(i\mathcal{A}[\xi^t]\mathcal{S}_t\Big)W^{-1}[A;\xi^0] = 0. \end{aligned}$$

Therefore, the quantity being differentiated is constant and equal to its value at $t = 0$:

$$W[A;\xi^t]\,\mathcal{S}_t\, W^{-1}[A;\xi^0] = W[A;\xi^0]\,\mathcal{S}_0\, W^{-1}[A;\xi^0] = 1.$$

Setting $t = 1$, this equality can be recast as

$$\mathcal{S}_1 = W^{-1}[A;\xi^1]\, W[A;\xi^0] = W[A;\gamma_{xx}].$$

Therefore, the Wilson loop we are seeking is equal to the value of $\mathcal{S}_t$ at $t = 1$. We can formally write the solution of (3.37) as a path-ordered exponential, to obtain

$$W[A;\gamma_{xx}] = P_t \exp\Big(i\int_0^1 dt\; \mathcal{A}[\xi^t]\Big). \tag{3.38}$$

39.b Our remaining task is therefore to obtain an expression for $\mathcal{A}[\xi^t]$. From its definition, we can write

$$i\,\mathcal{A}[\xi^t] = \lim_{dt\to 0} \frac{1}{dt}\Big(1 - W^{-1}[A;\xi^t]\,W[A;\xi^{t+dt}]\Big).$$

In order to obtain $\mathcal{A}[\xi^t]$, we thus need to extract the terms linear in dt in the quantity within the parentheses. Let us focus on the second term. Pictorially, it may be represented as follows:

$$W^{-1}[A;\xi^t]\,W[A;\xi^{t+dt}] =$$

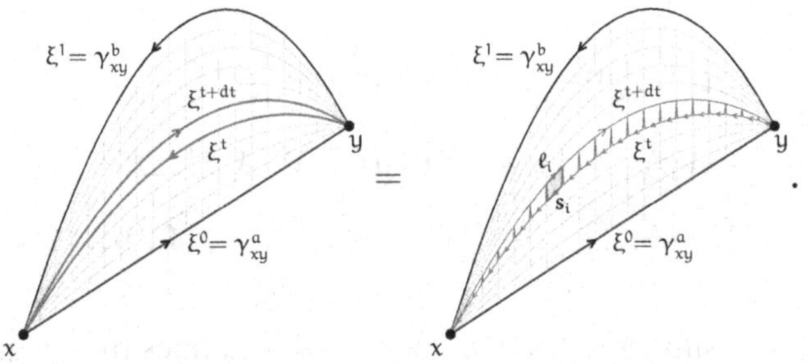

In the second equality, we have broken the path ξ^t into a sequence of "staples" (this does not change the result, since the vertical edges of these staples cancel pairwise). These staples, combined with a small portion of the path ξ^{t+dt}, form a little quadrangle. Let us denote by ℓ_i the Wilson line corresponding to the upper edge of the i-th quadrangle, and by s_i the Wilson line corresponding to the staple formed by the remaining three edges of this quadrangle (oriented according to the arrows in the figure). In terms of these objects, we have

$$W^{-1}[A;\xi^t]\,W[A;\xi^{t+dt}] = \lim_{N\to\infty} s_1 s_2 \cdots s_N\;\ell_N \cdots \ell_2\ell_1. \tag{3.39}$$

The product $s_i\ell_i$ is an infinitesimal closed Wilson loop. Generically, a small Wilson loop can be approximated in terms of the field strength as follows:

$$\begin{aligned}
W[A;\gamma] \equiv P\exp\Big(ig\oint_\gamma dx^\mu A_\mu\Big) &= 1 + ig\oint_\gamma dx^\mu A_\mu + \mathcal{O}(A^2)\\
&= 1 + i\frac{g}{2}\int_\Sigma dx^\mu \wedge dx^\nu\, F^{\text{Abel.}}_{\mu\nu} + \mathcal{O}(A^2)\\
&= 1 + i\frac{g}{2}\,\Sigma^{\mu\nu} F_{\mu\nu} + \text{higher orders in the size of the loop.}
\end{aligned}$$

(In the expansion of the first line, the path-ordering has no effect on the term linear in A_μ. In the second line, we can use the Abelian Stokes's theorem to rewrite this linear term as an integral over the surface Σ whose boundary is γ. In the last line, we reckon that, among the uncalculated terms, some are merely promoting the Abelian field strength into the non-Abelian one. $\Sigma^{\mu\nu}$ denotes the oriented area of the projection of the surface Σ on the $(\mu\nu)$ plane.) Using this result, we may write

$$s_i\ell_i = 1 - ig\,\underbrace{ds\,dt\,\frac{\partial\xi^\mu}{\partial s}\frac{\partial\xi^\nu}{\partial t}}_{\text{area in the }(\mu\nu)\text{ plane}}\,F_{\mu\nu}(\xi^t(s_i)) + \cdots.$$

(Note that the little closed loop formed by $s_i\ell_i$ is traversed clockwise, hence the minus sign. Note also that this formula has no prefactor $\frac{1}{2}$, because the "area tensor" contracted into the field strength is not antisymmetric here.) The terms linear in dt in (3.39) are thus given by

$$W^{-1}[A;\xi^t]\,W[A;\xi^{t+dt}]$$
$$= 1 - ig\,dt \lim_{N\to\infty} ds \sum_{i=1}^{N} \underbrace{\ell_1^{-1}\cdots\ell_{i-1}^{-1}}_{W^{-1}[A;\xi^t_{(0,s_i)}]} \left[\frac{\partial\xi^\mu}{\partial s}\frac{\partial\xi^\nu}{\partial t} F_{\mu\nu}(\xi^t(s_i))\right] \underbrace{\ell_{i-1}\cdots\ell_1}_{W[A;\xi^t_{(0,s_i)}]} + \mathcal{O}(dt^2),$$

where $W[A;\xi^t_{(0,s)}]$ is the Wilson line on the portion of ξ^t that extends from $\xi^t(0) = x$ and $\xi^t(s)$. In order to obtain this formula, we expand each s_i in turn to order dt (expanding more s_i's at once would give higher-order terms). Therefore, all the remaining s_j's are evaluated at zeroth order in dt, where we have simply $s_j = \ell_j^{-1}$. Those located at $j > i$ cancel pairwise with the matching ℓ_j, while those at $j < i$ must be kept in the formula because they cannot be commuted with $F_{\mu\nu}$. Taking the limit of an infinite number of intervals along the s coordinate leads to an integral:

$$W^{-1}[A;\xi^t]\,W[A;\xi^{t+dt}]$$
$$= 1 - ig\,dt \int_0^1 ds\, \frac{\partial\xi^\mu}{\partial s}\frac{\partial\xi^\nu}{\partial t} W^{-1}[A;\xi^t_{(0,s)}]\, F_{\mu\nu}(\xi^t(s))\, W[A;\xi^t_{(0,s)}] + \mathcal{O}(dt^2),$$

and

$$i\mathcal{A}[\xi^t] = ig \int_0^1 ds\, \frac{\partial\xi^\mu}{\partial s}\frac{\partial\xi^\nu}{\partial t} W^{-1}[A;\xi^t_{(0,s)}]\, F_{\mu\nu}(\xi^t(s))\, W[A;\xi^t_{(0,s)}].$$

39.c Inserting this result in (3.38) proves the announced formula,

$$W[A;\gamma_{xx}] = P_t \exp\left\{ig \int_0^1 dtds\, \frac{\partial\xi^\mu}{\partial s}\frac{\partial\xi^\nu}{\partial t} W^{-1}[A;\xi^t_{(0,s)}]F_{\mu\nu}(\xi^t(s))W[A;\xi^t_{(0,s)}]\right\}.$$

This formula is known as the *non-Abelian Stokes's theorem.* It is indeed easy to check that it falls back to the usual Stokes's theorem in the case of an Abelian gauge field. There are several notable points about the non-Abelian version:

- It works only for exponentiated fields, i.e., for elements of the group rather than elements of the algebra.
- It requires a pair of Wilson lines inside the exponential to connect the point where the field strength is evaluated to the base point x, which ensures the proper gauge transformation.
- It requires an ordering of the interpolating paths ξ^t that span the surface (along the s direction, there is also an ordering, hidden in the Wilson lines present inside the exponential). The right-hand side of the non-Abelian Stokes's formula is said to be *surface-ordered.*
- The choice of the family of paths ξ^t is not unique (only the boundary paths ξ^0 and ξ^1 are fixed), which means that the formula is valid for any surface whose boundary is the original contour γ_{xx}.

- Although we have written it in terms of explicit coordinates (t, s), the non-Abelian Stokes's formula is invariant under changes of these coordinates.
- The formula is valid for any choice of the point $y \in \gamma_{xx}$.

Source: Bralić, N. E. (1980), *Phys Rev D* 22: 3090.

40. QED in the 't Hooft–Veltman Gauge Consider the quantization of quantum electrodynamics with the following non-linear gauge condition:

$$\partial_\mu A^\mu + \tfrac{\kappa}{2} A_\mu A^\mu = \omega\,,$$

where κ is a constant. This non-linear gauge is known as the *'t Hooft–Veltman gauge*, and was introduced as a playground to mimic within QED some of the features (ghosts, self-coupling of the gauge bosons) of non-Abelian gauge theories.

40.a Assuming for the time being that the naive quantization procedure works, write the Lagrangian obtained after performing the Gaussian integration over $\omega(x)$. What new couplings does it contain?

40.b Draw the diagrams that contribute to the photon self-energy at one loop and write the corresponding integrals. Do they add up to zero?

40.c Derive the Ward–Takahashi identity that the photon polarization tensor should satisfy in the 't Hooft–Veltman gauge. Is it satisfied by the result of the preceding question?

40.d Reconsider the gauge fixing more carefully. Why is it necessary to introduce ghosts in order to handle the above gauge condition? Determine the ghost term in the gauge-fixed Lagrangian.

40.e Draw and calculate the additional contribution to the photon self-energy due to ghosts. Check that it restores the expected Ward–Takahashi identity.

40.a At a superficial level (i.e., overlooking the fact that the gauge constraint leads to a field-dependent Jacobian), the quantization of QED with this non-linear gauge condition leads to

$$Z_0[j^\mu] \equiv \int [D\omega(x) DA_\mu(x)]\; \exp\Big\{-i\frac{\xi}{2}\int d^4x\; \omega^2(x)\Big\}$$
$$\times\; \delta\big[\partial_\mu A^\mu + \tfrac{\kappa}{2}A_\mu A^\mu - \omega\big]\; \exp\Big\{i\int d^4x\; \big(-\tfrac{1}{4}F^{\mu\nu}F_{\mu\nu} + j^\mu A_\mu\big)\Big\},$$

i.e.,

$$Z_0[j] = \int [DA_\mu(x)]\, \exp\Big\{-i\frac{\xi}{2}\int d^4x\; \big(\partial_\mu A^\mu + \tfrac{\kappa}{2}A_\mu A^\mu\big)^2\Big\}$$
$$\times\, \exp\Big\{i\int d^4x\; \big(-\tfrac{1}{4}F^{\mu\nu}F_{\mu\nu} + j^\mu A_\mu\big)\Big\}.$$

The expansion of the term proportional to ξ, aside from a term in $\xi(\partial_\mu A^\mu)(\partial_\nu A^\nu)$ that goes into the photon propagator (and gives the same propagator as in the linear gauge $\partial_\mu A^\mu = \omega$), produces three-photon and four-photon couplings, respectively of the form $\xi\kappa(\partial_\mu A^\mu)A_\nu A^\nu$ and $\xi\kappa^2 A_\mu A^\mu A_\nu A^\nu$. The corresponding Feynman rules are

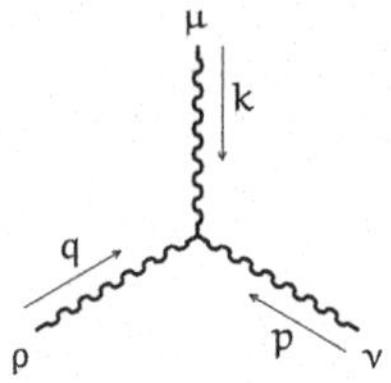

$$= \xi\kappa(k_\mu g_{\nu\rho} + p_\nu g_{\rho\mu} + q_\rho g_{\mu\nu}),$$

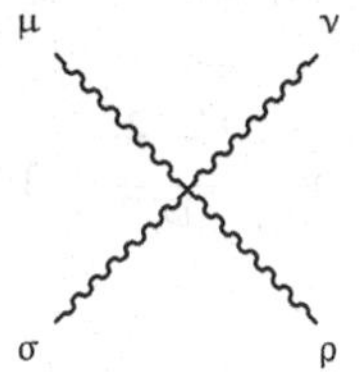

$$= -i\xi\kappa^2(g_{\mu\nu}g_{\rho\sigma} + g_{\mu\rho}g_{\nu\sigma} + g_{\mu\sigma}g_{\nu\rho}).$$

40.b Because of these new vertices, there are photon loop contributions to the self-energy $\Pi_{\mu\nu}(p)$ of the photon. Before we turn to calculating these contributions, let us first make a remark on the gauge dependence of the photon polarization tensor. It is often stated that $\Pi^{\mu\nu}$ is gauge invariant because one can view it as the vacuum expectation value of a current–current correlation function, $\Pi^{\mu\nu}(x,y) = \langle 0|T\,J^\mu(x)J^\nu(y)|0\rangle$ (and since J^μ is itself gauge invariant, since ψ and $\overline{\psi}$ transform with opposite phases). But this identification is only true in gauges where the only coupling of the photon is to a fermion line. In gauges where the photon has self-couplings, such as the non-linear gauge considered here, the photon polarization tensor may contain additional gauge-dependent terms due to these couplings. However, these unusual contributions to the photon polarization should not cause any unphysical effect, such as giving a non-zero mass to transversely polarized photons. With $\xi = 1$, we have

$$= -\frac{\kappa^2}{2}\int \frac{d^D k}{(2\pi)^D}\,\frac{p_\mu g_{\rho\sigma} + k_\rho g_{\mu\sigma} - (p+k)_\sigma g_{\mu\rho}}{k^2} \times \frac{-p_\nu g^{\rho\sigma} - k^\rho g_\nu{}^\sigma + (p+k)^\sigma g_\nu{}^\rho}{(p+k)^2}. \tag{3.40}$$

Let us first rearrange the denominators by introducing a Feynman parameterization,

$$\frac{1}{k^2(k+p)^2} = \int_0^1 \frac{dx}{(\ell^2-\Delta)^2}, \qquad \Delta \equiv -x(1-x)p^2, \quad \ell \equiv k + xp.$$

The next step is to expand the numerator of (3.40), and replace the loop momentum k in terms of ℓ. This leads to

$$\begin{aligned}&[p_\mu g_{\rho\sigma} + k_\rho g_{\mu\sigma} - (p+k)_\sigma g_{\mu\rho}][-p_\nu g^{\rho\sigma} - k^\rho g_\nu{}^\sigma + (p+k)^\sigma g_\nu{}^\rho]\\ &\quad = -2g_{\mu\nu}(\ell^2 + x^2p^2) + (1 - D + 2x^2)p_\mu p_\nu + 2\ell_\mu \ell_\nu.\end{aligned}$$

The integration over the loop momentum ℓ can be performed by using

$$\int \frac{d^D\ell}{(2\pi)^D} \frac{1}{(\ell^2-\Delta)^2} = i\frac{\Delta^{D/2-2}}{(4\pi)^{D/2}} \Gamma(1-\tfrac{D}{2})\,(1-\tfrac{D}{2}),$$

$$\int \frac{d^D\ell}{(2\pi)^D} \frac{\ell^2}{(\ell^2-\Delta)^2} = i\frac{\Delta^{D/2-2}}{(4\pi)^{D/2}} \Gamma(1-\tfrac{D}{2})\,(\tfrac{D}{2}x(1-x)p^2),$$

$$\int \frac{d^D\ell}{(2\pi)^D} \frac{\ell_\mu \ell_\nu}{(\ell^2-\Delta)^2} = i\frac{\Delta^{D/2-2}}{(4\pi)^{D/2}} \Gamma(1-\tfrac{D}{2})\,(\tfrac{1}{2}x(1-x)p^2 g_{\mu\nu}),$$

leading to

$$= -i\frac{\kappa^2}{4}\frac{p^2\,\Gamma(1-\frac{D}{2})}{(4\pi)^{D/2}} \int_0^1 dx\, \Delta^{D/2-2} \Big\{ g_{\mu\nu}\big((1-D)(1-2x^2) + 2(D-2)x^2\big) - \tfrac{p_\mu p_\nu}{p^2}(D-2)(1-D+2x^2) \Big\}. \tag{3.41}$$

In this expression, we have used the fact that Δ is symmetric under the exchange $x \leftrightarrow 1-x$ to eliminate the terms linear in x in the numerator (each such linear term can be replaced by $1/2$ thanks to this symmetry).

In addition, there is a tadpole built with a four-photon vertex,

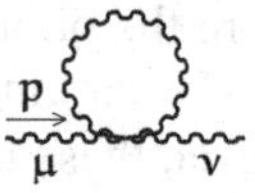

$$= -\frac{(D+2)\kappa^2 g_{\mu\nu}}{2} \int \frac{d^Dk}{(2\pi)^D} \frac{1}{k^2}.$$

Note that this integral is zero in dimensional regularization. Nevertheless, it turns out to be useful to keep this term, after rewriting it as follows:

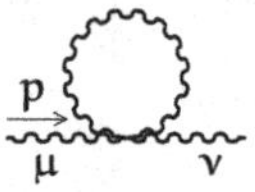

$$\begin{aligned} &= -\frac{(D+2)\kappa^2 g_{\mu\nu}}{4} \int \frac{d^Dk}{(2\pi)^D} \Big(\frac{1}{k^2} + \frac{1}{(p+k)^2}\Big) \\ &= -\frac{(D+2)\kappa^2 g_{\mu\nu}}{2} \int_0^1 dx \int \frac{d^D\ell}{(2\pi)^D} \frac{\ell^2 + x^2p^2}{(\ell^2-\Delta)^2} \\ &= -i\frac{\kappa^2}{4}\frac{p^2\,\Gamma(1-\frac{D}{2})}{(4\pi)^{D/2}} \int_0^1 dx\, \Delta^{D/2-2} \Big\{ 2(D+2) g_{\mu\nu}\big((1-D)x^2 + \tfrac{D}{4}\big) \Big\}. \end{aligned}$$

Even though this expression is nothing but a glorified zero (one can check this by performing the integration over the Feynman parameter x), it will cancel some terms *at the level of the integrand* when combined with the contribution of the first graph. (If we drop this tadpole, these simplifications are difficult to achieve without explicitly performing the integral over x in (3.41).)

There is a third graph made only of photons:

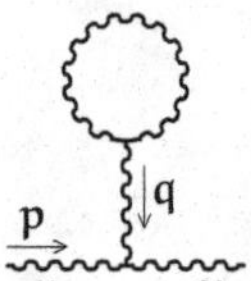

$$= -\frac{\kappa^2}{2} \frac{q_\rho g_{\mu\nu} + p_\mu g_{\nu\rho} - p_\nu g_{\mu\rho}}{q^2} \int \frac{d^Dk}{(2\pi)^D} \frac{k_\sigma g^\rho{}_\sigma - q^\rho g^\sigma{}_\sigma - k^\sigma g_\sigma{}^\rho}{k^2}.$$

As with the tadpole with the four-photon vertex, this is zero in dimensional regularization but it is helpful to transform it as we did for the preceding graph in the hope that it combines nicely with the integrand of the first graph. We get

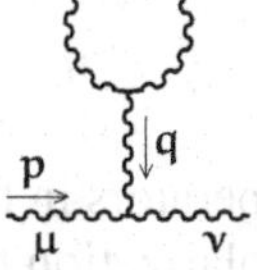

$$= -i\frac{\kappa^2}{4}\frac{p^2\,\Gamma(1-\frac{D}{2})}{(4\pi)^{D/2}}\int_0^1 dx\;\Delta^{D/2-2}\left\{(-2D)g_{\mu\nu}\left((1-D)x^2+\tfrac{D}{4}\right)\right\}.$$

Summing the three contributions, we obtain

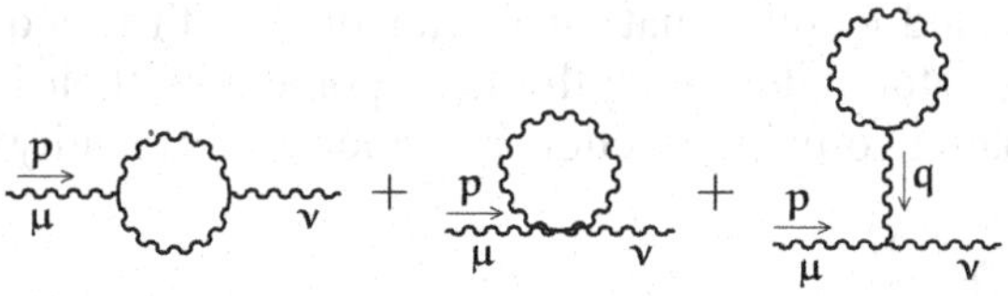

$$= -i\frac{\kappa^2}{4}\frac{p^2\,\Gamma(1-\frac{D}{2})}{(4\pi)^{D/2}}\int_0^1 dx\;\Delta^{D/2-2}\left\{g_{\mu\nu}(1-2x^2)-\tfrac{p_\mu p_\nu}{p^2}(D-2)(1-D+2x^2)\right\},$$

which is a priori not zero.

40.c Before reconsidering the quantization of QED in non-linear gauge with the discussion of Fadeev–Popov ghosts and the calculation of their contribution, it is useful to pause in order to assess how gauge invariance constrains the form of the photon polarization tensor when this non-linear gauge fixing is used. Note that this theory coupled to fermions has a conserved electromagnetic current $J^\mu = \overline{\psi}\gamma^\mu\psi$, due to the $U(1)$ invariance of the Dirac Lagrangian. However, the conservation of this current no longer implies the usual Abelian Ward–Takahashi identities, because an amputated external photon line is not equivalent to an insertion of J^μ in the non-linear gauge under consideration (indeed, the photon can also be attached to another photon – or to a ghost, as we shall soon see – and not only to a fermion). The consequence of this complication is that the standard Abelian Ward–Takahashi identity is not satisfied by the photon polarization tensor,

$$p^\mu\Pi_{\mu\nu}(p) \neq 0,$$

and we should not expect to obtain a transverse photon polarization tensor. In order to derive a constraint obeyed by the polarization tensor, we can instead mimic the derivation of the (weaker) non-Abelian Ward–Takahashi identities. The starting point is to generalize the gauge condition to

$$\partial_\mu A^\mu + \frac{\kappa}{2}A_\mu A^\mu - \zeta = \omega,$$

where $\zeta(x)$ is an arbitrary function. After the Gaussian integration over ω, we obtain additional ζ-dependent terms in the effective Lagrangian,

$$\frac{\xi}{2}\zeta^2 - \xi\zeta\left(\partial_\mu A^\mu + \frac{\kappa}{2}A_\mu A^\mu\right).$$

The term in ζ^2 plays no role since it does not couple to anything else. In the linear terms, ζ acts as an external source coupled to $\partial_\mu A^\mu$ and to $A_\mu A^\mu$. Since ζ is a part of the gauge fixing condition, all physical quantities (e.g., on-shell transition amplitudes with only physical photon polarizations) must be independent of ζ. For this to be true with one insertion of the source ζ, we must have

$$\Big[\partial_x^\mu \langle \mathrm{T}\, A_\mu(x) \cdots \rangle + \frac{\kappa}{2} \langle \mathrm{T}\, A_\mu(x) A^\mu(x) \cdots \rangle\Big] \cdots = 0,$$

where the dots inside the expectation values represent the same set of field operators in the two terms, all appropriately Fourier transformed and contracted into physical polarization tensors or on-shell spinors as prescribed by the LSZ reduction formulas. If $\kappa = 0$ (i.e., when we use a linear covariant gauge), the above identity is just the usual Ward–Takahashi identity. However, when $\kappa \neq 0$, we have a more complicated identity that mixes functions with n and $n+1$ photons (in the latter case, two of the photon fields are at the same point, with their Lorentz indices contracted). Applied to the photon two-point function at one loop, this identity reads

$$\Big[\frac{p^\mu}{p^2}$$ 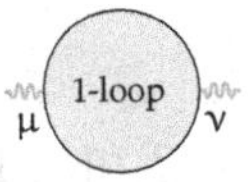

$$+\frac{\kappa}{2}$$ 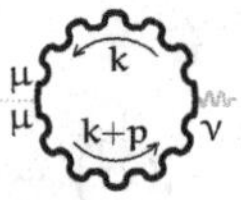

$$\Big]\,\epsilon^\nu(p) = 0.$$

(The short light-gray external photon indicates that the external photon propagator has been amputated. The light-gray dotted line is not attached to a three-photon vertex; instead the endpoints of two photon propagators are Lorentz contracted at this point.) Note that the bracket on the left-hand side can only be proportional to p_ν, and therefore we get zero trivially when contracting with the polarization vector $\epsilon^\nu(p)$ since $p \cdot \epsilon(p) = 0$. If we request independence with respect to ζ at order ζ^2, we get a stronger identity:

$$-\frac{p^\mu p^\nu}{(p^2)^2}$$ 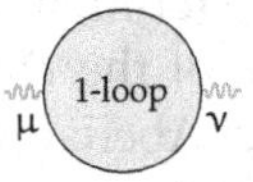

$$-\frac{\kappa p^\nu}{p^2}$$ 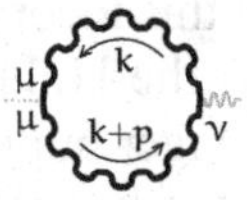

$$+\frac{\kappa^2}{2}$$ 

$$= 0. \tag{3.42}$$

For later reference, let us give here the values in dimensional regularization of the graphs that appear in the second and third terms of the left-hand side:

$$-\frac{\kappa p^\nu}{p^2}$$ 

$$= -i\frac{\kappa^2}{4}\frac{\Gamma(1-\frac{D}{2})}{(4\pi)^{D/2}} \int_0^1 dx\, \Delta^{D/2-2} \big\{2(1-D)(2-D)\big\},$$

$$\frac{\kappa^2}{2}$$ 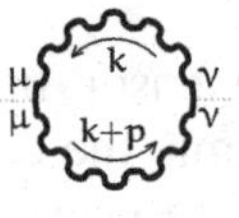

$$= -i\frac{\kappa^2}{4}\frac{\Gamma(1-\frac{D}{2})}{(4\pi)^{D/2}} \int_0^1 dx\, \Delta^{D/2-2} \big\{D(2-D)\big\},$$

and their sum reads

$$-\frac{\kappa p^\nu}{p^2}$$ 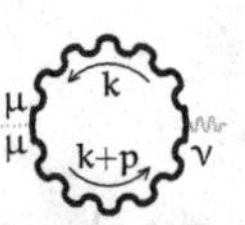

$$+\frac{\kappa^2}{2}$$ 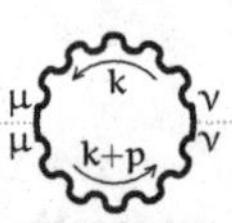

$$= -i\frac{\kappa^2}{4}\frac{\Gamma(1-\frac{D}{2})}{(4\pi)^{D/2}} \int_0^1 dx\, \Delta^{D/2-2} (D-2)^2. \tag{3.43}$$

Note that (3.42) constrains only the longitudinal part of the photon polarization tensor. Then, (3.43) indicates that this longitudinal component must be non-zero and proportional to κ^2 (therefore, it would vanish in linear gauges, i.e., when $\kappa = 0$).

40.d Let us now return to the quantization of QED in the 't Hooft–Veltman gauge. In our naive derivation of the gauge-fixed Lagrangian, we completely ignored the Jacobian associated with the functional delta function that enforces the gauge condition. However, since the gauge condition is non-linear in the photon field, this Jacobian depends on the field and cannot be dropped. As in the non-Abelian case, this Jacobian can be rewritten as a path integral over a pair of ghost fields, $\overline{\chi}, \chi$, and we get an extra term in the gauge-fixed Lagrangian,

$$\mathcal{L}_{\mathrm{FPG}} = \int d^4x\, \overline{\chi}\, \mathcal{M} \chi,$$

where $\mathcal{M}$ is the derivative of the gauge fixing function with respect to the parameter of an infinitesimal gauge transformation. In the present case, we have

$$\mathcal{M} = \frac{\delta\big(\partial_\mu(A^\mu - \partial^\mu\theta) + \frac{\kappa}{2}(A_\mu - \partial_\mu\theta)(A^\mu - \partial^\mu\theta)\big)}{\delta\theta}\bigg|_{\theta=0} = -\square - \kappa A_\mu \partial^\mu.$$

40.e Therefore, the photon self-energy also receives a contribution from a ghost loop, which reads

$$= \kappa^2 \int \frac{d^4k}{(2\pi)^4} \frac{k_\mu (p+k)_\nu}{k^2 (p+k)^2}$$
$$= \kappa^2 \int_0^1 dx \int \frac{d^D\ell}{(2\pi)^D} \frac{\ell_\mu \ell_\nu - x(1-x) p_\mu p_\nu}{(\ell^2 - \Delta)^2}$$
$$= -i \frac{\kappa^2}{4} \frac{p^2\, \Gamma(1-\frac{D}{2})}{(4\pi)^{D/2}} \int_0^1 dx\, \Delta^{D/2-2} \Big\{ -(1-2x^2)\big(g_{\mu\nu} + (D-2)\tfrac{p_\mu p_\nu}{p^2}\big) \Big\}.$$

Combining this with the purely photonic contributions, we get

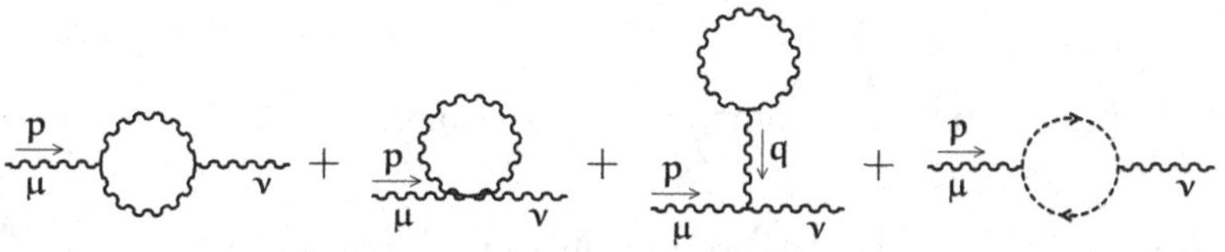

$$= -i \frac{\kappa^2}{4} \frac{\Gamma(1-\frac{D}{2})}{(4\pi)^{D/2}} \int_0^1 dx\, \Delta^{D/2-2} \Big\{ (D-2)^2\, p_\mu p_\nu \Big\}.$$

Note that this result is non-zero, but purely longitudinal. However, by plugging this expression into the first term of (3.42) and using (3.43), we can see that the Ward–Takahashi identity (3.42) is indeed satisfied. Such a purely longitudinal polarization tensor renormalizes the gauge fixing term $\xi(\partial_\mu A^\mu)^2$, but does not affect the transverse part of the photon propagator (hence, it does not introduce unphysical effects such as a photon mass).

41. LSZ Reduction Formula and Gauge Invariance Recall the LSZ reduction formula for scalar particles:

$$\big\langle q_1 \cdots q_n\, {}_{\mathrm{out}} \big| p_1 \cdots p_m\, {}_{\mathrm{in}} \big\rangle = \frac{i^{m+n}}{\sqrt{Z}^{m+n}} \int \prod_{i=1}^{m} d^4x_j\, e^{-ip_i \cdot x_i}\, (\square_{x_i} + m^2)$$
$$\times \int \prod_{j=1}^{n} d^4y_j\, e^{iq_j \cdot y_j}\, (\square_{y_j} + m^2)$$
$$\times \big\langle 0_{\mathrm{out}} \big| T\, \phi(x_1) \cdots \phi(x_m) \phi(y_1) \cdots \phi(y_n) \big| 0_{\mathrm{in}} \big\rangle.$$

41.a Although this formula has integrals over the entire spacetime for each field operator, show that the result depends only on the behavior of the field at $x^0 = +\infty$ for a final-state particle and at $x^0 = -\infty$ for an initial-state one.

41.b Recall the modifications of the LSZ formula for Dirac fermions and gauge fields. How should gauge transformations be restricted for the resulting scattering amplitude to be gauge invariant?

41.c Alternatively, how should the LSZ formula be modified to give an amplitude which is invariant under *any* gauge transformation?

41.a In order to prove the assertion made in the first question, let us consider the following integral, which arises with every initial-state particle:

$$\frac{i}{\sqrt{Z}}\int d^4x\, e^{-ip\cdot x}\,(\square_x + m^2)\,\big\langle 0_{\rm out}\big|T\cdots\phi(x)\cdots\big|0_{\rm in}\big\rangle,$$

where the dots indicate the other field operators present inside the time-ordered product. Since the exponential in the integrand is a solution of the free Klein–Gordon equation, this quantity also reads

$$\begin{aligned}
&\frac{i}{\sqrt{Z}}\int d^4x\, e^{-ip\cdot x}\,\Big(\overrightarrow{\square_x + m^2} - \overleftarrow{\square_x + m^2}\Big)\,\big\langle 0_{\rm out}\big|T\cdots\phi(x)\cdots\big|0_{\rm in}\big\rangle\\
&\quad = \frac{i}{\sqrt{Z}}\int d^4x\, e^{-ip\cdot x}\,\Big(\overrightarrow{\square_x} - \overleftarrow{\square_x}\Big)\,\big\langle 0_{\rm out}\big|T\cdots\phi(x)\cdots\big|0_{\rm in}\big\rangle\\
&\quad = \frac{i}{\sqrt{Z}}\int d^4x\, \partial^\mu\Big(e^{-ip\cdot x}\,\overleftrightarrow{\partial_\mu}\,\big\langle 0_{\rm out}\big|T\cdots\phi(x)\cdots\big|0_{\rm in}\big\rangle\Big)\\
&\quad = \frac{i}{\sqrt{Z}}\int d^3x\,\Big[e^{-ip\cdot x}\,\overleftrightarrow{\partial_0}\,\big\langle 0_{\rm out}\big|T\cdots\phi(x)\cdots\big|0_{\rm in}\big\rangle\Big]_{x^0=-\infty}^{x^0=+\infty}.
\end{aligned}$$

In the final equation, we assume that the boundaries at spatial infinity do not contribute. The next step is to use the time ordering in order to bring the highlighted field to the front or to the end of the product, depending on the limit considered. Using also the fact that $\phi(x)$ goes to $\sqrt{Z}\,\phi_{\rm in}(x)$ or $\sqrt{Z}\,\phi_{\rm out}(x)$ at $x^0 = -\infty$ and $x^0 = +\infty$, respectively, this gives

$$\begin{aligned}
&i\int_{x^0=+\infty} d^3x\, e^{-ip\cdot x}\,\overleftrightarrow{\partial_0}\,\big\langle 0_{\rm out}\big|\phi_{\rm out}(x)\,T\cdots\big|0_{\rm in}\big\rangle - i\int_{x^0=-\infty} d^3x\, e^{-ip\cdot x}\,\overleftrightarrow{\partial_0}\,\big\langle 0_{\rm out}\big|T\cdots\phi_{\rm in}(x)\big|0_{\rm in}\big\rangle\\
&\quad = -\underbrace{\big\langle 0_{\rm out}\big|a^\dagger_{p,\rm out}}_{0}\,T\cdots\big|0_{\rm in}\big\rangle + \big\langle 0_{\rm out}\big|T\cdots\underbrace{a^\dagger_{p,\rm in}\big|0_{\rm in}\big\rangle}_{|p_{\rm in}\rangle}.
\end{aligned}$$

The same sequence of steps may be reproduced for a final-state particle:

$$\frac{i}{\sqrt{Z}}\int d^4x\, e^{+ip\cdot x}\left(\Box_x+m^2\right)\left\langle 0_{out}\middle|T\cdots\phi(x)\cdots\middle|0_{in}\right\rangle$$
$$=\frac{i}{\sqrt{Z}}\int d^3x\left[e^{ip\cdot x}\overleftrightarrow{\partial_0}\left\langle 0_{out}\middle|T\cdots\phi(x)\cdots\middle|0_{in}\right\rangle\right]_{x^0=-\infty}^{x^0=+\infty}$$
$$=\underbrace{\left\langle 0_{out}\middle|a_{p,out}}_{\left\langle p_{out}\right|}T\cdots\left|0_{in}\right\rangle-\left\langle 0_{out}\middle|T\cdots\underbrace{a_{p,in}\middle|0_{in}\right\rangle}_{0}.$$

In addition to providing an a posteriori check of the reduction formulas, this derivation implies the following identities:

$$\frac{i}{\sqrt{Z}}\int d^4x\, e^{-ip\cdot x}\left(\Box_x+m^2\right)\left\langle 0_{out}\middle|T\cdots\phi(x)\cdots\middle|0_{in}\right\rangle$$
$$=-\frac{i}{\sqrt{Z}}\int_{x^0=-\infty} d^3x\, e^{-ip\cdot x}\overleftrightarrow{\partial_0}\left\langle 0_{out}\middle|T\cdots\phi(x)\cdots\middle|0_{in}\right\rangle, \quad (3.44)$$

$$\frac{i}{\sqrt{Z}}\int d^4x\, e^{+ip\cdot x}\left(\Box_x+m^2\right)\left\langle 0_{out}\middle|T\cdots\phi(x)\cdots\middle|0_{in}\right\rangle$$
$$=\frac{i}{\sqrt{Z}}\int_{x^0=+\infty} d^3x\, e^{ip\cdot x}\overleftrightarrow{\partial_0}\left\langle 0_{out}\middle|T\cdots\phi(x)\cdots\middle|0_{in}\right\rangle. \quad (3.45)$$

Therefore, even though the covariant form of the LSZ reduction formula contains integrations over the entire spacetime, the result only depends on the behavior of the field operator at asymptotic times ($x^0=+\infty$ for a final-state particle and $x^0=-\infty$ for an initial-state particle).

41.b For Dirac fermions, the LSZ reduction formulas must contain one of the free spinors $u_s(p), v_s(p), \overline{u}_s(p)$ or $\overline{v}_s(p)$, depending on the case under consideration. For gauge bosons of polarization λ and color a, the LSZ formula must contain the following factors:

$$\frac{i}{\sqrt{Z_3}}\int d^4x\, e^{\pm ip\cdot x}\,\Box_x\,\epsilon_\lambda^\mu(p)\left\langle 0_{out}\middle|T\cdots A_\mu^a(x)\cdots\middle|0_{in}\right\rangle,$$

where Z_3 is the wavefunction renormalization factor for the gauge boson. (In the exponential, the plus sign is for outgoing particles and the minus sign for incoming particles.) The result established for scalar fields remains true for fermions and gauge bosons: the LSZ reduction formula can be rewritten in a way that involves only a spatial integration, in the limit $x^0=-\infty$ for the incoming particles and in the limit $x^0=+\infty$ for the outgoing ones.

Recall now the gauge transformation of scalars, fermions (both assumed to be charged under the non-Abelian interaction under consideration) and gauge fields:

$$A_\mu^a(x)\to\Omega_{ab}^\dagger(x)\left(A_\mu^b(x)-\partial_\mu\theta^b(x)\right),$$
$$\phi_i(x)\to\Omega_{ij}^\dagger(x)\,\phi_j(x),\quad \psi_i(x)\to\Omega_{ij}^\dagger(x)\,\psi_j(x),$$

where $\Omega(x)\equiv\exp(i\theta^a(x)t^a)$. As an example, let us study the effect of a gauge transformation on a scalar field inside the LSZ reduction formulas. Instead of (3.44) and (3.45), we now have

$$\pm\frac{i}{\sqrt{Z}}\int\limits_{x^0=\pm\infty} d^3x\, e^{\pm ip\cdot x}\overleftrightarrow{\partial_0}\,\Omega^\dagger_{ij}(x^0,\mathbf{x})\,\langle 0_{\text{out}}|T\cdots\phi_j(x)\cdots|0_{\text{in}}\rangle.$$

The gauge transformation enters in the same fashion into the LSZ formula for fermions. For gauge bosons, the gauge transformation enters as follows:

$$\pm\frac{i}{\sqrt{Z}}\int\limits_{x^0=\pm\infty} d^3x\, e^{\pm ip\cdot x}\epsilon^\mu_\lambda(\mathbf{p})\overleftrightarrow{\partial_0}\,\Omega^\dagger_{ab}(x^0,\mathbf{x})\,\langle 0_{\text{out}}|T\cdots\left(A^b_\mu(x)-\partial_\mu\theta^b(x)\right)\cdots|0_{\text{in}}\rangle.$$

In the three cases, the factor $\Omega^\dagger(x)$ has no effect provided that the gauge transformation goes to the identity at asymptotic times, $\lim_{x^0\to\pm\infty}\Omega^\dagger(x)=1$. When this is the case, we also have $\lim_{x^0\to\pm\infty}\theta^b(x)=0$ and the term in $\partial_\mu\theta^b$ drops out of the reduction formula for gauge bosons. Therefore, the usual form of the LSZ reduction formula is invariant under gauge transformations that do not affect the fields at asymptotic times.

41.c The reason for this restriction on the allowed gauge transformations is in fact easy to understand. In order to see this, it is useful to rewrite the LSZ formulas in a more compact fashion. For instance, for scalars of color k and momentum $\mathbf{p}$, we write them as

Incoming scalar: $\left(\phi|\Phi_{kp}\right)_{x^0=-\infty}$,

Outgoing scalar: $\left(\Phi_{kp}|\phi\right)_{x^0=+\infty}$,

where $\Phi^i_{kp}(x)\equiv\delta^i_k\,e^{-ip\cdot x}$ is a free solution of the Klein–Gordon equation for color k and momentum $\mathbf{p}$ (in this notation, the subscripts label the quantum numbers of the plane wave – color, polarization, momentum – and the upper indices are the actual indices carried by the field), and where we have introduced the following inner product:

$$\left(\phi_1|\phi_2\right)_{x^0}\equiv\frac{i}{\sqrt{Z}}\int\limits_{\text{fixed }x^0} d^3x\,\phi^\dagger_1(x)\overleftrightarrow{\partial_0}\phi_2(x). \tag{3.46}$$

The LSZ formulas for gauge bosons, fermions and anti-fermions can similarly be written in the form of inner products, by introducing the following free solutions of the Yang–Mills and Dirac equations (linearized, since asymptotic fields are assumed to be non-interacting):

$$A^{\mu a}_{c\lambda p}(x)=\delta^a_c\,\epsilon^\mu_\lambda(\mathbf{p})e^{-ip\cdot x},$$
$$\Psi^i_{+ksp}(x)=\delta^i_k\,u_s(\mathbf{p})e^{-ip\cdot x},\quad\Psi^i_{-ksp}(x)=\delta^i_k\,v_s(\mathbf{p})e^{+ip\cdot x}.$$

When we write the LSZ formula in this way, the gauge invariance of scattering amplitudes is equivalent to the gauge invariance of inner products such as (3.46). The weak form of gauge invariance that we have discussed so far consists simply in restricting ourselves to gauge transformations that go to the identity at times $x^0=\pm\infty$ (i.e., where we need to evaluate the inner products for the LSZ formulas). This can easily be promoted to a more general form of gauge invariance provided that:

- we apply the gauge transformation to both ϕ_1 and ϕ_2 in the inner product. This means that, in the LSZ formulas, we must also use the gauge transformed plane waves, i.e.,

$$\Phi^i_{kp}(x) = \Omega^\dagger_{ij}(x)\delta^j_k\, e^{-ip\cdot x}, \quad \mathbf{A}^{\mu a}_{c\lambda p}(x) = \Omega^\dagger_{ab}(x)\delta^b_c\, \epsilon^\mu_\lambda(p)\, e^{-ip\cdot x},$$
$$\Psi^i_{+ksp}(x) = \Omega^\dagger_{ij}(x)\delta^j_k\, u_s(p)e^{-ip\cdot x}, \quad \Psi^i_{-ksp}(x) = \Omega^\dagger_{ij}(x)\delta^j_k\, v_s(p)e^{+ip\cdot x},$$

  respectively

- we limit ourselves to gauge transformations that become time-independent at large $|x^0|$ (this property is necessary for the Ω to go through the time derivative). In other words, asymptotically, the fields and their conjugate momenta must transform in the same way.

(In the case of gauge bosons, after the Ω's have canceled, the extra term $\partial^\mu\theta^b$ cancels when contracted with physical polarization vectors.)

42. Large-N Expansion in SU(N) Yang–Mills Theory Consider an SU(N) Yang–Mills theory (i.e., a non-Abelian gauge theory without matter fields). The goal of this problem is to study the behavior of graphs when $N \to +\infty$, and to relate this behavior to a topological property of these graphs. (Although we consider the example of vacuum to vacuum graphs in this problem, the result is more general.)

42.a Draw all the one-particle-irreducible vacuum-to-vacuum graphs made only of gluons up to three loops (explain why ghosts and gluons contribute similarly to the leading-N behavior), and determine their leading behavior at large N. Explain why this limit can be obtained by replacing each gluon line by a double quark–antiquark line.

42.b Find a rescaling of the gauge field in the Lagrangian so that the coupling constant g appears as an overall prefactor g^{-2}. For later convenience, introduce $\lambda \equiv g^2 N$. How do the various objects in the Feynman rules scale with λ and N?

42.c Determine the scaling with λ and N of a generic vacuum graph in terms of its numbers of loops, vertices, propagators and faces (we call every closed quark loop in the double line representation a face).

42.d Explain why $\chi \equiv$ (number of vertices) − (number of edges) + (number of faces) depends only on the genus of the closed surface on which the graph is embedded, but not on the other details of the graph.

42.e What is the value of χ for a sphere, for a torus with a single handle, and for a torus with h handles?

42.a Let us first note that the ghost–antighost–gluon vertex has the same color dependence as the three-gluon vertex. Likewise, the ghost propagator is identical to the gluon one as far as the color factors are concerned. Therefore, graphs containing ghosts have the same color structure as graphs of the same topology with the ghosts replaced by gluons. For this reason, we do not need to consider ghosts in this discussion. Up to three loops, the one-particle-irreducible vacuum graphs made only of gluons are the following:

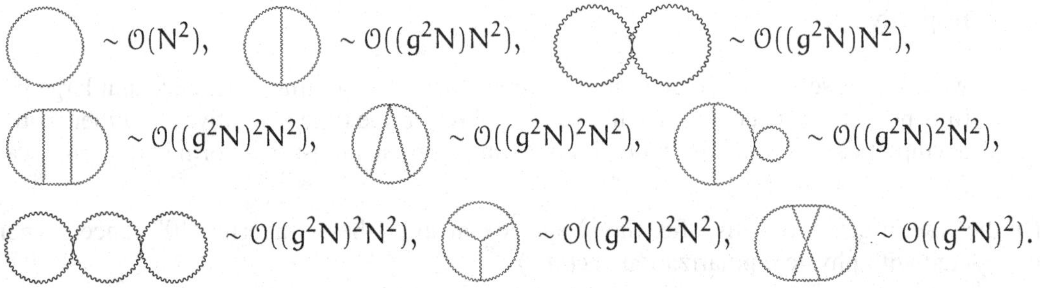

(3.47)

The scaling of the first graph is simply that of a gluon propagator with the two endpoints connected to each other. The sum over the contracted color indices of this propagator gives a factor $\delta_{aa} = N^2 - 1 \sim \mathcal{O}(N^2)$. For the other graphs, we first use the following identity:

$$i\,f^{abc} = 2\,\mathrm{tr}\,(t_f^a t_f^b t_f^c) - 2\,\mathrm{tr}\,(t_f^b t_f^a t_f^c),$$

in order to view the color structure of the three-gluon vertex as that of a quark loop. The next step is to use the $\mathfrak{su}(N)$ Fierz identity in order to perform all the contractions $t_f^a \cdots t_f^a$. Recalling the diagrammatic representation of this identity,

$(t_f^a)_{ij}(t_f^a)_{kl} =$ 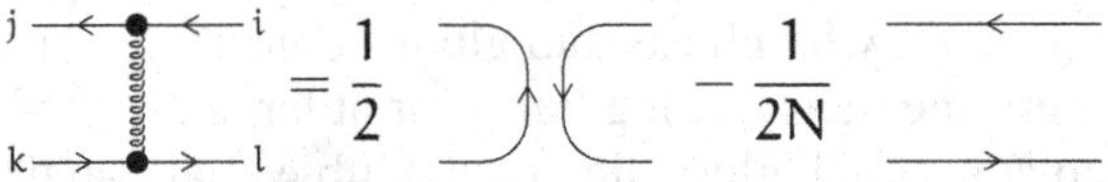 ,

the leading term in N is obtained by replacing the gluon line by a pair of "quark–antiquark" lines (i.e., the first term of the Fierz identity). When we apply this procedure to vacuum diagrams, we obtain only closed "quark" or "antiquark" loops, which in this estimate correspond to the trace of the identity matrix in the fundamental representation, and therefore provide a factor N. For instance, we have

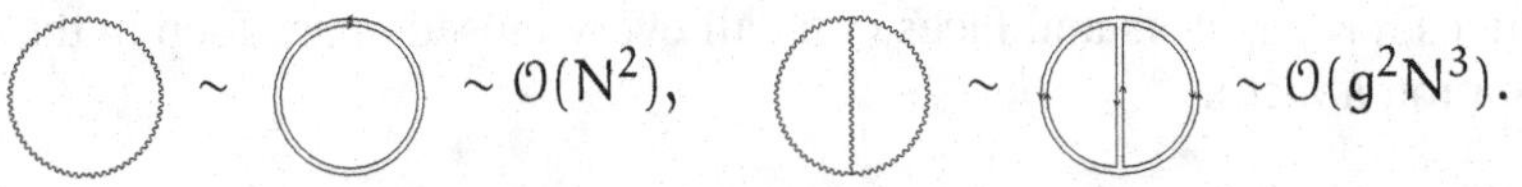

The only missing rule to this recipe is how to handle the four-gluon vertices. For these, it is sufficient to note that the color structure of a four-gluon vertex is equivalent to that of graphs with a pair of three-gluon vertices:

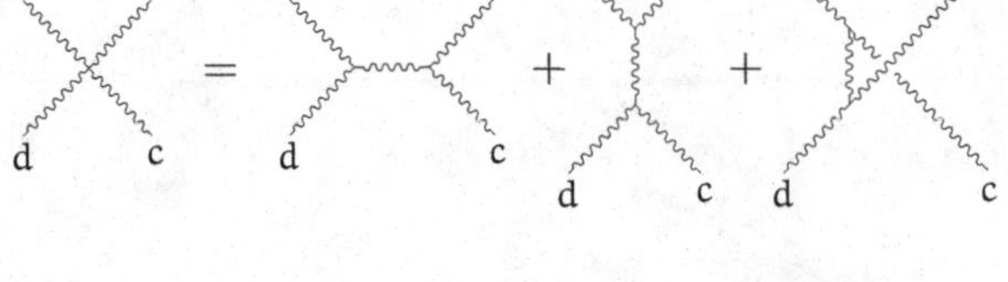

(3.48)

42.b Interestingly, in the examples listed in (3.47), the first eight graphs are of the form $N^2(g^2N)^p$, while the last one is of the form $N^0(g^2N)^p$. Let us now try to understand this pattern more generally. Recall the form of the Yang–Mills action,

$$\mathcal{S}_{YM} = -\frac{1}{2}\int d^4x \,\mathrm{tr}\,(F_{\mu\nu}F^{\mu\nu}), \quad \text{with } F^{\mu\nu} \equiv \partial^\mu A^\nu - \partial^\nu A^\mu - ig\,[A^\mu, A^\nu].$$

Let us define $\widetilde{A}^\mu \equiv g\,A^\mu$. In terms of these rescaled fields, the Yang–Mills action becomes

$$\mathcal{S}_{YM} = -\frac{1}{2g^2}\int d^4x \,\mathrm{tr}\,(\widetilde{F}_{\mu\nu}\widetilde{F}^{\mu\nu}), \quad \text{with } \widetilde{F}^{\mu\nu} \equiv \partial^\mu \widetilde{A}^\nu - \partial^\nu \widetilde{A}^\mu - i\,[\widetilde{A}^\mu, \widetilde{A}^\nu].$$

With this rescaling, the coupling constant g appears only in an overall prefactor of the action, and no longer in the field strength. In view of the observations made on the examples of (3.47), it is useful to rewrite it as

$$\mathcal{S}_{YM} = -\frac{N}{2\lambda}\int d^4x \,\mathrm{tr}\,(\widetilde{F}_{\mu\nu}\widetilde{F}^{\mu\nu}), \quad \text{with } \lambda \equiv g^2N.$$

(λ is known as the *'t Hooft coupling.*) In terms of these rescaled fields and couplings, the various ingredients of the Feynman rules scale as

$$\text{propagator} \quad \longleftrightarrow \quad \lambda/N,$$
$$\text{vertex} \quad \longleftrightarrow \quad N/\lambda.$$

(Note that, after this rescaling, the behaviors of three-gluon and four-gluon vertices are the same.) In addition, in the large-N limit, each gluon line is replaced by a double quark line. Each closed quark line brings a factor N:

$$\text{quark or antiquark loop} \quad \longleftrightarrow \quad N.$$

Each closed quark line corresponds to a face of the diagram, so we may equivalently count the faces of the diagram. In order to properly account for the factor N coming from the outer quark line, we must also count as one face the domain external to the graph (for instance, the first graph of (3.47) has two faces according to this prescription). Topologically, this is equivalent to viewing a vacuum graph as a polytope embedded in three-dimensional space and counting its faces, the propagators being the edges of the polytope.

42.c We are now equipped for counting the powers of λ and N in a generic graph. Let us denote by n_V the number of vertices, by n_I the number of internal propagators, and by n_F the number of faces of a vacuum graph $\mathcal{G}$. We have

$$\mathcal{G} \sim \lambda^{n_I - n_V}\, N^{n_V - n_I + n_F} \sim \lambda^{n_L - 1}\, N^{\chi(\mathcal{G})}, \tag{3.49}$$

where $n_L \equiv n_I - n_V + 1$ is the number of loops of the graph (i.e., the number of unconstrained internal momenta). We have also defined $\chi(\mathcal{G}) \equiv n_V - n_I + n_F$. This quantity is called the *Euler characteristic* of the graph. Before we discuss its properties, note that in (3.49) it was not necessary to introduce a distinction between three-gluon and four-gluon vertices. Indeed, if we replace every four-gluon vertex by a pair of three-gluon vertices according to (3.48), we increase both n_V and n_I by one unit (while n_F remains unchanged), but this does not change n_L, nor $\chi(\mathcal{G})$.

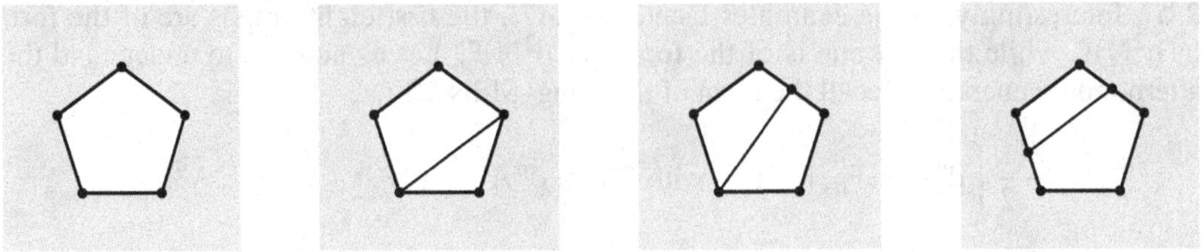

Figure 3.4 Illustration of the invariance of the Euler characteristic when a face of a polytope is split into two faces.

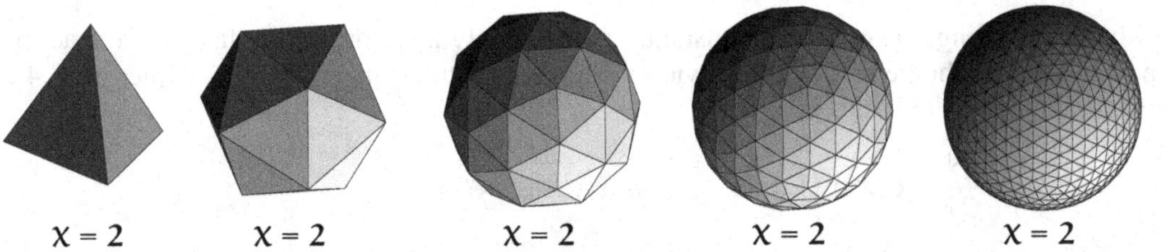

Figure 3.5 Polytopes with the Euler characteristic of a 2-sphere.

42.d The most important property of the Euler characteristic is that it is invariant under the addition of lines to a graph, provided the new lines do not cross the existing ones, as illustrated in Figure 3.4. In the left image, we show one face of a graph (here, a pentagonal face), with the rest of the graph – in the shaded area – not represented. In the second image from the left, we have added one line inside this face, connecting two of the existing vertices. When this is done, the numbers of vertices, lines and faces vary by $\delta n_V = 0, \delta n_I = 1, \delta n_F = 1$, respectively, and the Euler characteristic does not change, $\delta\chi = 0$. In the third image, the new line connects an existing vertex and a newly created vertex. Thus, we have $\delta n_V = 1, \delta n_I = 2, \delta n_F = 1$, but still $\delta\chi = 0$. And in the last image, we have $\delta n_V = 2, \delta n_I = 3, \delta n_F = 1$, and again $\delta\chi = 0$.

This property of being unchanged when lines that do not lead to crossings are added implies that the Euler characteristic does not depend on the details of the graph, but is in fact a topological property of the minimal closed surface on which we can embed the graph without line crossings. For instance, the first five graphs of equation (3.47) can be embedded on a 2-sphere (recall that we must count as a face the domain outside the graph), while the last one requires a torus.

42.e Conversely, we may use this invariance in order to compute the Euler characteristic of all closed surfaces, as follows

- A tetrahedron ($n_V = 4, n_I = 6, n_F = 4$) is topologically equivalent to a two-dimensional sphere. Therefore, $\chi = 2$ for the sphere, and for all the graphs whose minimal embedding is the sphere. Figure 3.5 illustrates the invariance of the Euler characteristic as one refines the triangulation of a surface.
- If we take a polytope and remove one of its faces to create a hole (while keeping the edges and vertices on the boundary of the hole), we decrease the Euler characteristic by one unit, since this operation corresponds to $\delta n_V = 0, \delta n_I = 0, \delta n_F = -1$. If we

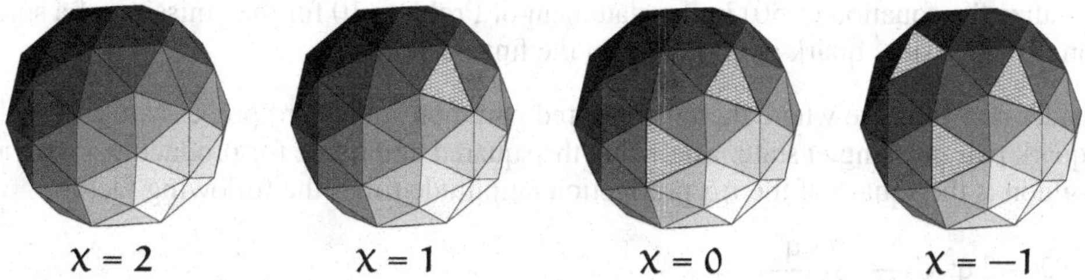

Figure 3.6 Euler characteristics of polytopes with faces removed (indicated as hatched triangles).

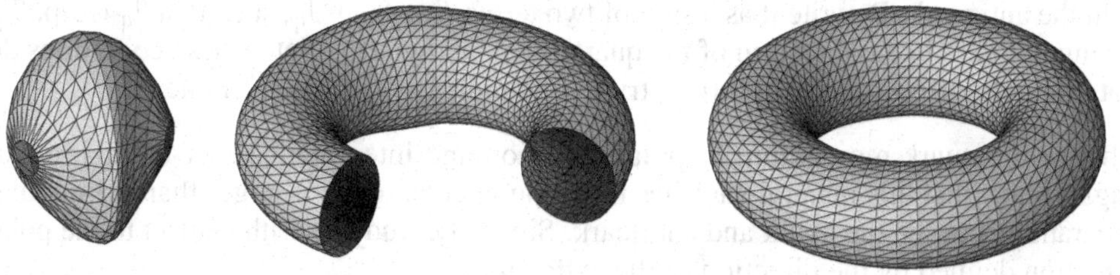

Figure 3.7 Deformation of a polytope with two holes into a torus. The Euler characteristic does not change in the process.

remove two faces in order to create two holes, we have $\delta\chi = -2$, etc. In Figure 3.6, we have indicated the removed faces by hatched areas. Note that removing just the face, or the face as well as the vertices and edges on the boundary, leads to the same χ (indeed, the boundary of a triangle has equal numbers of vertices and edges).

- If we create two holes and then bend the polytope to glue the boundaries of these holes, we also have $\delta\chi = -2$, as illustrated in Figure 3.7 (we show three stages of this process; the Euler characteristic is equal to $\chi = 0$ at all stages). However, this manipulation creates a handle in the polytope: for instance, it transforms a sphere into a torus with one handle, a torus with one handle into a torus with two handles, etc. Therefore, the Euler characteristic of a closed surface with h handles is $\chi = 2 - 2h$.

Thus, the formula (3.49) provides a one-to-one correspondence between the order in N of a vacuum graph (after we have absorbed a factor N in the definition of λ) and the genus of the surface on which the graph may be embedded without line crossings. If we consider the limit of large N at fixed λ, the leading graphs are those that can be embedded on a sphere (the standard QFT terminology for these graphs is to call them *planar graphs*, because the sphere is topologically equivalent to a plane if we remove its north pole; the large-N limit is also called the *planar approximation*), the next-to-leading ones are those that can be embedded on a torus, etc.

43. Soft Radiation and Antenna Effects in QCD The goal of this problem is to study soft gluon radiation from a hard QCD process. The main result is that when a soft gluon is emitted from a quark–antiquark pair in a color-singlet state, its emission at large angles is forbidden.

43.a Generalize the equation (1.50) in the statement of Problem 10 for the emission of a soft gluon off a massless quark or antiquark in the final state.

43.b Consider now the case where the only colored particles in the hard process are a quark–antiquark pair in a singlet state. Show that the squared amplitude for producing an extra soft gluon is the square of the $q\bar{q}$ production amplitude times the following factor:

$$\mathcal{I} \equiv 2\,g^2\,C_f \frac{p \cdot q}{(p \cdot k)(q \cdot k)},$$

where p, q are the momenta of the quark and antiquark, and k that of the gluon.

43.c Check that the above expression is singular when the gluon is aligned with the quark or with the antiquark. Rewrite it as a sum of two terms, $\mathcal{I} \equiv \mathcal{I}_q + \mathcal{I}_{\bar{q}}$, such that $\mathcal{I}_q$ (resp. $\mathcal{I}_{\bar{q}}$) is singular only in the direction of the quark (resp. antiquark). Note: these conditions do not uniquely determine the splitting; try to find the most symmetrical one.

43.d Choose the quark momentum as polar direction and integrate $\mathcal{I}_q$ over the azimuthal angle. Show that this integral vanishes if the gluon polar angle is larger than the angular separation between the quark and antiquark. Similarly, study $\mathcal{I}_{\bar{q}}$ with respect to the polar direction defined by the direction of the antiquark.

43.e Does this result hold in the case where the quark–antiquark pair is produced from a gluon, i.e., is in an octet state?

43.a   Consider a QCD scattering amplitude with a massless quark and a massless antiquark in the final state,

$$\mathcal{M}_{q,\bar{q}}^{ri,sj}(p, q, \dots).$$

(We only explicitly indicate the quark and the antiquark in this notation. The indices i and j are the color indices of the quark and antiquark, r and s their spins, and p, q their momenta.) In order to exhibit the corresponding spinors, we may write

$$\mathcal{M}_{q,\bar{q}}^{ri,sj}(p, q, \dots) = \bar{u}_r(p)\,\mathcal{A}_{q,\bar{q}}^{ij}(p, q, \dots)\,v_s(q).$$

Let us now attach a gluon of momentum k, polarization λ and color index a to the outgoing quark line. This gives the following amplitude:

$$\mathcal{M}_{q+g,\bar{q}}^{ri+\lambda a,sj}(p, q, k, \dots)$$

$$= \bar{u}_r(p)\big(-ig(t_f^a)_{ii'}\gamma^\mu \epsilon_{\lambda\mu}(k)\big)\frac{i(\not{p}+\not{k})}{(p+k)^2}\mathcal{A}_{q,\bar{q}}^{i'j}(p+k, q, \dots)v_s(q)$$

$$\underset{k \ll p,q}{\approx} g(t_f^a)_{ii'}\frac{p \cdot \epsilon_\lambda(k)}{p \cdot k}\underbrace{\bar{u}_r(p)\mathcal{A}_{q,\bar{q}}^{i'j}(p, q, \dots)v_s(q)}_{\mathcal{M}_{q,\bar{q}}^{ri',sj}(p,q,\dots)}.$$

In the last line, we have assumed the gluon to be soft compared to the quark, and we have used $\bar{u}_r(p)\gamma^\mu\not{p} = 2p^\mu\,\bar{u}_r(p)$ (thanks to the Dirac equation satisfied by the massless spinor).

Likewise, we can attach the soft gluon to the outgoing antiquark:

$$\begin{aligned}
&\mathcal{M}_{q,\bar{q}+g}^{ri,sj+\lambda a}(p,q,k,\dots)\\
&\quad = \overline{u}_r(p)\mathcal{A}_{q,\bar{q}}^{ij'}(p,q+k,\dots)\frac{-i(\not{q}+\not{k})}{(q+k)^2}\left(-ig(t_f^a)_{j'j}\gamma^\mu\epsilon_{\lambda\mu}(k)\right)v_s(q)\\
&\quad \underset{k\ll p,q}{\approx} -g(t_f^a)_{j'j}\frac{q\cdot\epsilon_\lambda(k)}{q\cdot k}\underbrace{\overline{u}_r(p)\mathcal{A}_{q,\bar{q}}^{ij'}(p,q,\dots)v_s(q)}_{\mathcal{M}_{q,\bar{q}}^{ri,sj'}(p,q,\dots)}.
\end{aligned}$$

43.b If we assume that the quark–antiquark pair is in a color-singlet state (this is always true if they are the only colored particles in the process), we may write

$$\mathcal{A}_{q,\bar{q}}^{ij}(p,q,\dots)\equiv\delta_{ij}\,\mathcal{A}_{q,\bar{q}}(p,q,\dots).$$

Summing the two terms previously obtained, we obtain the following expression for a $q\bar{q}g$ final state:

$$\mathcal{M}_{q,\bar{q},g}^{ri,sj,\lambda a}(p,q,k,\dots)\approx g(t_f^a)_{ij}\underbrace{\left[\frac{p\cdot\epsilon_\lambda(k)}{p\cdot k}-\frac{q\cdot\epsilon_\lambda(k)}{q\cdot k}\right]}_{J^\mu\epsilon_{\lambda\mu}}\times\overline{u}_r(p)\mathcal{A}_{q,\bar{q}}(p,q,\dots)v_s(q). \tag{3.50}$$

Note that the current J^μ contracted into the gluon polarization vector satisfies an Abelian-like Ward–Takahashi identity,

$$k_\mu\left[\frac{p^\mu}{p\cdot k}-\frac{q^\mu}{q\cdot k}\right]=0,$$

which ensures the gauge invariance of the amplitude (since a gauge transformation modifies the polarization vector by a term proportional to k_μ).

At this stage, squaring the above amplitude and summing over colors and the gluon polarizations leads to

$$\begin{aligned}
&\sum_{ija,\lambda}\left|\mathcal{M}_{q,\bar{q},g}^{ri,sj,\lambda a}(p,q,k,\dots)\right|^2\\
&\quad\approx g^2\underbrace{(t_f^a)_{ij}(t_f^{a*})_{ij}}_{\mathrm{tr}\,(t_f^a t_f^a)=NC_f}J^\mu J^\nu\Big(\underbrace{\sum_\lambda\epsilon_{\lambda\mu}\epsilon_{\lambda\nu}^*}_{-g_{\mu\nu}+\text{terms in }k_{\mu,\nu}}\Big)\left|\overline{u}_r(p)\mathcal{A}_{q,\bar{q}}(p,q,\dots)v_s(q)\right|^2\\
&\quad=\underbrace{2g^2C_f\frac{p\cdot q}{(p\cdot k)(q\cdot k)}}_{\equiv\mathcal{I}}\times\underbrace{N\left|\overline{u}_r(p)\mathcal{A}_{q,\bar{q}}(p,q,\dots)v_s(q)\right|^2}_{\sum_{i,j}\left|\mathcal{M}_{q,\bar{q}}^{ri,sj}(p,q,\dots)\right|^2}.
\end{aligned}$$

43.c Obviously, the factor $\mathcal{J}$ is singular if $p \cdot k = 0$ or $q \cdot k = 0$. Consider the first of these conditions,

$$0 = E_p E_k (1 - \cos\theta_{pk}),$$

where θ_{pk} is the angle between the 3-momenta of the quark and of the gluon. Therefore, this condition is realized when the gluon is emitted collinearly to the quark (note that there would be no singularity if the quark were massive). Likewise, the condition $q \cdot k = 0$ is realized when the gluon is emitted in the direction of the antiquark. These singularities are called *collinear singularities*. The residues at the poles corresponding to the conditions $p \cdot k = 0$ or $q \cdot k = 0$ can be obtained easily:

$$\frac{p \cdot q}{(p \cdot k)(q \cdot k)} = \frac{E_p}{E_k (p \cdot k)} + \text{finite terms when } \theta_{pk} \to 0,$$
$$\frac{p \cdot q}{(p \cdot k)(q \cdot k)} = \frac{E_q}{E_k (q \cdot k)} + \text{finite terms when } \theta_{qk} \to 0.$$

This suggests that we write $\mathcal{J} = \mathcal{J}_q + \mathcal{J}_{\bar{q}}$, with

$$\mathcal{J}_q \equiv g^2 C_f \left[\frac{p \cdot q}{(p \cdot k)(q \cdot k)} - \frac{E_q}{E_k (q \cdot k)} + \frac{E_p}{E_k (p \cdot k)} \right],$$
$$\mathcal{J}_{\bar{q}} \equiv g^2 C_f \left[\frac{p \cdot q}{(p \cdot k)(q \cdot k)} + \frac{E_q}{E_k (q \cdot k)} - \frac{E_p}{E_k (p \cdot k)} \right].$$

(This decomposition is not unique, as we may subtract some finite terms along with the poles.) By construction, $\mathcal{J}_q$ is finite when $\mathbf{k} \parallel \mathbf{q}$ and $\mathcal{J}_{\bar{q}}$ is finite when $\mathbf{k} \parallel \mathbf{p}$, and their sum is trivially equal to $\mathcal{J}$. Note a caveat here: although it is tempting to interpret the factors $\mathcal{J}_{q,\bar{q}}$ as giving the gluon emission off the quark and the antiquark, respectively, this interpretation is not robust because these two quantities are not gauge invariant independently (only their sum is).

43.d Consider now the first quantity, $\mathcal{J}_q$. In this case, it is convenient to use spherical coordinates with a polar axis along $\mathbf{p}$. The various vectors involved in the expression for $\mathcal{J}_q$ have the following Cartesian coordinates:

$$\mathbf{p} = \begin{pmatrix} 0 \\ 0 \\ E_p \end{pmatrix}, \quad \mathbf{q} = \begin{pmatrix} \sin\theta_{qp} \\ 0 \\ \cos\theta_{qp} \end{pmatrix}, \quad \mathbf{k} = \begin{pmatrix} \sin\theta_{pk}\cos\phi \\ \sin\theta_{pk}\sin\phi \\ \cos\theta_{pk} \end{pmatrix},$$

so that

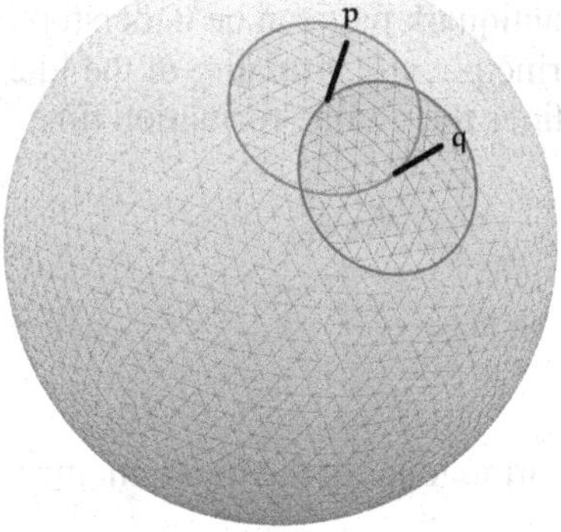

Figure 3.8 Allowed regions (illustrated by the two disks) for the emission of a soft gluon from a quark–antiquark dipole in a singlet color state.

$$\mathcal{I}_q = \frac{1}{E_k^2(1-\cos\theta_{pk})}\left[\frac{\cos\theta_{pk}-\cos\theta_{qp}}{1-\cos\theta_{qp}\cos\theta_{pk}-\sin\theta_{qp}\sin\theta_{pk}\cos\phi}+1\right].$$

Let us now perform the integration over the azimuthal angle ϕ. This integral is of the form

$$\int_0^{2\pi}\frac{d\phi}{a-b\cos\phi} = 2\pi\,\theta(|a|-|b|)\,\frac{\operatorname{sign}(a)}{\sqrt{a^2-b^2}}.$$

This formula, after some trigonometric manipulations, leads to

$$\begin{aligned}\int_0^{2\pi} d\phi\,\mathcal{I}_q &= \frac{2\pi}{E_k^2(1-\cos\theta_{pk})}\left[\frac{\cos\theta_{pk}-\cos\theta_{qp}}{|\cos\theta_{pk}-\cos\theta_{qp}|}+1\right]\\ &= \theta(\theta_{qp}-\theta_{pk})\frac{4\pi}{E_k^2(1-\cos\theta_{pk})}.\end{aligned}$$

Quite remarkably, the azimuthal integral of $\mathcal{I}_q$ vanishes if the emission angle θ_{pk} of the gluon with respect to the quark is larger than the opening of the cone formed by the $q\bar{q}$ pair. The same conclusion holds for the integral of $\mathcal{I}_{\bar{q}}$, but this time with a polar axis along the momentum q of the antiquark. The pattern of allowed soft gluon emission is shown in Figure 3.8. Let us rephrase more precisely the caveat mentioned earlier: $\mathcal{I}_{q,\bar{q}}$ cannot be interpreted as the emission spectra off the quark and antiquark, respectively. Indeed, the emission of a gluon off a quark does not suffer any angular restriction. The angular veto we are observing here is in fact due to a destructive interference between the emissions by the quark and by the antiquark. Let us mention a paradox: the soft radiation factor $\mathcal{J}$ is positive definite, and it is therefore surprising to find regions with seemingly no radiation. But the quantities $\mathcal{J}_q$ and $\mathcal{J}_{\bar{q}}$ into which we decompose $\mathcal{J}$ are not positive definite, which allows their azimuthal integrals to be zero for some polar angles. The paradox is resolved by noting that the polar axis is not the same for the integrals of $\mathcal{J}_q$ and $\mathcal{J}_{\bar{q}}$ (therefore, the sum of these integrals is not the integral of $\mathcal{J}$).

Semi-quantitative Interpretation: The same collimation effect exists in QED for photon emission off an e^+e^- pair, where it is known as the *Chudakov effect.* This suppression of gluon radiation at angles larger than the opening of the quark–antiquark pair can be understood semi-quantitatively as follows. According to the uncertainty principle, the virtuality of the line of momentum $p + k$ just before the gluon emission vertex defines the gluon "formation time,"

$$t_f^{-1} \equiv E_p + E_k - E_{p+k} \approx \frac{(p+k)^2}{2E_p} = E_k(1 - \cos\theta_{pk}) \underset{\theta_{pk}\ll 1}{\approx} \tfrac{1}{2}E_k\theta_{pk}^2.$$

During this time, the quark–antiquark pair grows to a transverse size

$$r_\perp = t_f\theta_{pq}.$$

Moreover, the gluon wavelength (E_k^{-1}), projected on the plane orthogonal to the quark momentum, reads

$$\lambda_\perp = \frac{1}{E_k \sin\theta_{pk}} \underset{\theta_{pk}\ll 1}{\approx} \frac{1}{E_k\theta_{pk}}.$$

For the gluon to be emitted without suppression, its transverse wavelength must be smaller than the size of the quark–antiquark dipole. Otherwise, the gluon cannot resolve the quark–antiquark pair, and sees only its total charge, which is zero in the present case since we have assumed the pair to be in a singlet state. This condition reads

$$\frac{1}{E_k\theta_{pk}} \lesssim r_\perp \quad \Leftrightarrow \quad \theta_{pk} \lesssim \theta_{pq}.$$

This property, known as *angular ordering* in the context of QCD, can be used to formulate gluon emission in the form of "parton cascades" in which each quark, antiquark or gluon has a certain probability of emitting a new gluon. The effect of quantum interference, which would in principle preclude such a probabilistic description, is taken into account by vetoing emission at large angles.

43.e Let us now consider the case where the quark–antiquark pair is produced in an octet state, i.e., from a gluon. In this case, the above qualitative argument fails: even if the produced gluon does not resolve the pair, it sees a non-zero total charge, and there should not be any suppression of its emission (in fact, we expect that the emission at large angle should be equivalent to the emission by the gluon that produced the quark–antiquark pair). First, let us rewrite the analogue of (3.50) in the case where the quark–antiquark pair is not assumed to be singlet:

$$\mathcal{M}^{ri,sj,\lambda a}_{q,\bar{q},g}(p,q,k,\dots) \approx g\,\overline{u}_r(p)\Big[\underbrace{\frac{p\cdot\epsilon_\lambda(k)}{p\cdot k}(t^a_f)_{ii'}\mathcal{A}^{i'j}_{q,\bar{q}}(p,q,\dots) - \frac{q\cdot\epsilon_\lambda(k)}{q\cdot k}\mathcal{A}^{ij'}_{q,\bar{q}}(p,q,\dots)(t^a_f)_{j'j}}_{\equiv J^\mu_{ija}\epsilon_{\lambda\mu}(k)}\Big]v_s(q).$$

Let us check the Ward–Takahashi identity on the current J^μ_{ija}:

$$k_\mu J^\mu_{ija} = \big[t^a_f, \mathcal{A}_{q,\bar{q}}(p,q,\dots)\big]_{ij}.$$

We can see a difficulty here: this current is conserved only when $\mathcal{A}_{q,\bar{q}}$ is singlet (indeed, requesting that the commutator on the right-hand side vanish for all generators t^a_f implies by Schur's lemma that $\mathcal{A}_{q,\bar{q}}$ must be proportional to the identity).

The reason for this problem is that it is impossible for the quark–antiquark pair to be non-singlet if the quark and antiquark are the only colored particles in the hard process. Therefore, we must start from a hard process that has at least one gluon in addition to the $q\bar{q}$ pair:

$$\mathcal{M}^{ri,sj,\lambda a}_{q,\bar{q},g}(p,q,k,\dots) \equiv \epsilon^{\mu}_{\lambda}(\mathbf{k})\overline{u}_r(\mathbf{p})\,\mathcal{A}^{ija}_{\mu}(p,q,k,\dots)v_s(\mathbf{q}).$$

Since this gluon is colored, it also contributes to the soft radiation. Therefore, the next step is to calculate the soft radiation off an external gluon line. To be definite, we take this gluon line to be in the final state. The amplitude corrected by the emission of a soft gluon (of momentum k') off the gluon (of momentum k) line reads

$$\mathcal{M}^{ri,sj,\lambda a+\lambda' a'}_{q,\bar{q},g}(p,q,k,k',\dots) = \epsilon_{\lambda'\mu'}(\mathbf{k}')\frac{-i}{(k+k')^2}(-gf^{aba'})V^{\mu\nu\mu'}(-k,k+k',-k')$$
$$\times\, \epsilon_{\lambda\mu}(\mathbf{k})\overline{u}_r(\mathbf{p})\,\mathcal{A}^{ijb}_{\nu}(p,q,k+k',\dots)v_s(\mathbf{q}),$$

where $V^{\mu\nu\mu'}(-k,k+k',-k')$ is the momentum-dependent factor in the three-gluon vertex:

$$V^{\mu\nu\mu'}(-k,k+k',-k') = g^{\mu\nu}(-2k-k')^{\mu'} + g^{\nu\mu'}(k+2k')^{\mu} + g^{\mu'\mu}(-k'+k)^{\nu}$$
$$\underset{k'\ll k}{\approx} g^{\mu\nu}(-2k^{\mu'}) + \underline{g^{\nu\mu'}k^{\mu}} + \underline{g^{\mu'\mu}k^{\nu}}.$$

The two underlined terms will give zero, since $k^{\mu}\epsilon_{\lambda\mu} = 0$ and $k^{\nu}\mathcal{A}^{ijb}_{\nu}(p,q,k,\dots) = 0$. Using also $(k+k')^2 \approx 2k\cdot k'$, we obtain

$$\mathcal{M}^{ri,sj,\lambda a+\lambda' a'}_{q,\bar{q},g+g}(p,q,k,k'\dots)$$
$$\approx \underbrace{(-igf^{aba'})}_{g(T^{a'}_{\text{adj}})_{ab}}\frac{k\cdot\epsilon_{\lambda'}(\mathbf{k}')}{k\cdot k'}\,\epsilon^{\mu}_{\lambda}(\mathbf{k})\overline{u}_r(\mathbf{p})\,\mathcal{A}^{ijb}_{\mu}(p,q,k,\dots)v_s(\mathbf{q}).$$

(One may check that the soft emission off an incoming gluon is given by the same formula.) Summing the soft emissions off the quark, the antiquark and the gluon, we have

$$\mathcal{M}^{ri,sj,\lambda a,\lambda' a'}_{q,\bar{q},g,g}(p,q,k,k',\dots)$$
$$\approx g\,\epsilon^{\mu}_{\lambda}(\mathbf{k})\overline{u}_r(\mathbf{p})\Big[\frac{k\cdot\epsilon_{\lambda'}(\mathbf{k}')}{k\cdot k'}\big(T^{a'}_{\text{adj}}\big)_{ab}\,\mathcal{A}^{ijb}_{\mu}(p,q,k,\dots)$$
$$\underbrace{+\frac{p\cdot\epsilon_{\lambda'}(\mathbf{k}')}{p\cdot k'}\big(t^{a'}_{f}\big)_{ii'}\mathcal{A}^{i'ja}_{\mu}(p,q,k,\dots) -\frac{q\cdot\epsilon_{\lambda'}(\mathbf{k}')}{q\cdot k'}\mathcal{A}^{ij'a}_{\mu}(p,q,k,\dots)\big(t^{a'}_{f}\big)_{j'j}}_{\equiv\,\epsilon_{\lambda'\mu'}(\mathbf{k})\,J^{\mu'}_{ijaa'}}\Big]v_s(\mathbf{q}). \tag{3.51}$$

This time, we have

$$k'_{\mu'}J^{\mu'}_{ijaa'} = \big(T^{a'}_{\text{adj}}\big)_{ab}\mathcal{A}^{ijb}_{\mu} + \big(t^{a'}_{f}\big)_{ii'}\mathcal{A}^{i'ja}_{\mu} - \mathcal{A}^{ij'a}_{\mu}\big(t^{a'}_{f}\big)_{j'j} = 0. \tag{3.52}$$

(See Problem 44 for a discussion of this identity, which may be viewed as the statement of color conservation in the hard process.) In order to be more specific, consider the case where the quark–antiquark pair is in the octet channel, i.e.,

$$\mathcal{A}^{ija}_{\mu} \equiv \mathcal{A}^{ab}_{\mu}(t^b_f)_{ij}.$$

(In this representation, b is the color of the gluon that produces the pair.) Plugging this representation into Eq. (3.52) leads to

$$\left[T^{a'}_{adj}, \mathcal{A}_\mu\right]_{ac}(t^c_f)_{ij} = 0.$$

For this to be true for all assignments of the color indices, Schur's lemma implies that the object $\mathcal{A}^{ab}_\mu$ must be proportional to δ_{ab},

$$\mathcal{A}^{ab}_\mu \equiv \mathcal{B}_\mu \, \delta_{ab},$$

which is just saying that the gluon that produces the quark–antiquark pair must have the same color as the external gluon in the hard process. We can now plug this specific form of $\mathcal{A}^{ija}_\mu$ into (3.51) and square the amplitude. Because $p^2 = q^2 = k^2 = 0$, the squares of the three terms vanish after summing over the polarizations λ'. Therefore, we need only consider the crossed products,

$$\begin{aligned}
&\sum_{\lambda'} \left|\mathcal{M}^{ri,sj,\lambda a,\lambda' a'}_{q,\bar{q},g,g}(p,q,k,k',\dots)\right|^2 \\
&= \left|\mathcal{M}_{hard}\right|^2 \Big(-\mathcal{I}_{kp}(k')\,(T^{a'}_{adj})_{ab}(t^b_f)_{ij}(t^a_f)_{ji'}(t^{a'}_f)_{i'i} \\
&\qquad + \mathcal{I}_{kq}(k')\,(T^{a'}_{adj})_{ab}(t^b_f)_{ij}(t^{a'}_f)_{jj'}(t^a_f)_{j'i} \\
&\qquad + \mathcal{I}_{pq}(k')\,(t^{a'}_f)_{ii'}(t^a_f)_{i'j}(t^{a'}_f)_{jj'}(t^a_f)_{j'i} + \text{complex conjugate}\Big),
\end{aligned}$$

where we denote

$$\mathcal{I}_{pq}(k') \equiv g^2 \frac{p\cdot q}{(p\cdot k')(q\cdot k')}, \qquad \mathcal{M}_{hard} \equiv \epsilon^\mu_\lambda(\boldsymbol{k})\,\overline{u}_r(\boldsymbol{p})\mathcal{B}_\mu v_s(\boldsymbol{q}).$$

When we sum over the colors, the various color structures that appear in this formula can be simplified easily in the case of $\mathfrak{su}(N)$:

$$\begin{aligned}
(T^{a'}_{adj})_{ab}(t^b_f)_{ij}(t^a_f)_{ji'}(t^{a'}_f)_{i'i} &= -if^{a'ab}\,\mathrm{tr}\,(t^b_f t^a_f t^{a'}_f) = -\tfrac{1}{4}N(N^2-1), \\
(T^{a'}_{adj})_{ab}(t^b_f)_{ij}(t^{a'}_f)_{jj'}(t^a_f)_{j'i} &= -if^{a'ab}\,\mathrm{tr}\,(t^b_f t^{a'}_f t^a_f) = +\tfrac{1}{4}N(N^2-1), \\
(t^{a'}_f)_{ii'}(t^a_f)_{i'j}(t^{a'}_f)_{jj'}(t^a_f)_{j'i} &= \mathrm{tr}\,(t^{a'}_f t^a_f t^{a'}_f t^a_f) = -\tfrac{1}{2}C_f.
\end{aligned}$$

Therefore, we obtain

$$\begin{aligned}
&\sum_{ijaa',\lambda'} \left|\mathcal{M}^{ri,sj,\lambda a,\lambda' a'}_{q,\bar{q},g,g}(p,q,k,k',\dots)\right|^2 \\
&= \tfrac{N^2-1}{2}\left|\mathcal{M}_{hard}\right|^2\Big(N\mathcal{I}_{kp}(k') + N\mathcal{I}_{kq}(k') - \tfrac{1}{N}\mathcal{I}_{pq}(k')\Big).
\end{aligned}$$

Each term in this sum has a specific radiation pattern, as discussed earlier, and the total emission pattern depends on the relative orientations of $\boldsymbol{p}, \boldsymbol{q}, \boldsymbol{k}$. For instance, if these three vectors are all nearly in the same direction, then the soft gluon emission is also collimated in this direction (this happens, for instance, if the gluon was itself radiated from the quark–antiquark pair; this argument indicates that successive emissions of gluons from an initial color-singlet $q\bar{q}$ pair remain collimated). In contrast, if $\mathbf{k} \sim -(\mathbf{p}+\mathbf{q})$, then the soft emission covers most of the solid angle.

From the above formula, one can also see that, when $N \to \infty$, the soft radiation is dominated by the first two terms, i.e., as if the gluon were emitted independently by a $(\mathbf{pk})$ quark–antiquark pair and by a $(\mathbf{kq})$ quark–antiquark pair. This can easily be understood from the fact that we can replace the gluon by a quark–antiquark double line in this limit, which effectively amounts to replacing the $qg\overline{q}$ system by $q\overline{q}q\overline{q}$. This observation is the starting point of the *dipole model* for successive soft gluon emissions in the large-N limit.

44. An Identity Obeyed by On-Shell QCD Amplitudes Consider an amplitude $\mathcal{M}^{a_1\cdots i_1\cdots \bar{\imath}_1\cdots}$, where the indices $a_1, a_2, \ldots$ are the color indices of external gluons, the indices $i_1, i_2, \ldots$ are those of outgoing quarks or incoming antiquarks, and the indices $\bar{\imath}_1, \bar{\imath}_2, \ldots$ those of outgoing antiquarks or incoming quarks. Show that

$$\sum_{\alpha \in \{\text{out. q, in. }\bar{q}\}} (t_f^b)_{i_\alpha i'_\alpha} \mathcal{M}^{\cdots i'_\alpha \cdots} - \sum_{\alpha \in \{\text{in. q, out. }\bar{q}\}} \mathcal{M}^{\cdots \bar{\imath}'_\alpha \cdots} (t_f^b)_{\bar{\imath}'_\alpha \bar{\imath}_\alpha}$$
$$+ \sum_{\alpha \in \{\text{gluons}\}} (T_{\text{adj}}^b)_{a_\alpha a'_\alpha} \mathcal{M}^{\cdots a'_\alpha \cdots} = 0.$$

This identity may be proven in several ways. In the spirit of Problem 43, we can derive it by adding a soft gluon to the amplitude under consideration, in all possible ways. In Problem 43, we worked out the eikonal factors corresponding to the emission of a soft gluon by an outgoing quark or antiquark:

$$\text{outgoing quark:} \qquad \mathcal{M}^{\cdots i \cdots} \to \qquad g\frac{p \cdot \epsilon_\lambda(k)}{p \cdot k}(t_f^b)_{ii'} \mathcal{M}^{\cdots i' \cdots},$$
$$\text{outgoing antiquark:} \qquad \mathcal{M}^{\cdots \bar{\imath} \cdots} \to \qquad -g\frac{p \cdot \epsilon_\lambda(k)}{p \cdot k} \mathcal{M}^{\cdots \bar{\imath}' \cdots} (t_f^b)_{\bar{\imath}'\bar{\imath}}.$$

(Of course, the momentum p should be that of the corresponding outgoing line.) Similar considerations give the following corrective factors for gluon emission off an incoming quark or antiquark:

$$\text{incoming antiquark:}\qquad \mathcal{M}^{\cdots i\cdots} \to g\frac{p\cdot\epsilon_\lambda(k)}{p\cdot k}(t_f^b)_{ii'}\,\mathcal{M}^{\cdots i'\cdots},$$

$$\text{incoming quark:}\qquad \mathcal{M}^{\cdots \bar{\imath}\cdots} \to -g\frac{p\cdot\epsilon_\lambda(k)}{p\cdot k}\,\mathcal{M}^{\cdots \bar{\imath}'\cdots}\,(t_f^b)_{\bar{\imath}'\bar{\imath}}.$$

We can then repeat this calculation with incoming and outgoing gluons, to obtain

$$\text{outgoing gluon:}\qquad \mathcal{M}^{\cdots b\cdots} \to g\frac{p\cdot\epsilon_\lambda(k)}{p\cdot k}(T_{\text{adj}}^b)_{aa'}\,\mathcal{M}^{\cdots a'\cdots},$$

$$\text{incoming gluon:}\qquad \mathcal{M}^{\cdots b\cdots} \to -g\frac{p\cdot\epsilon_\lambda(k)}{p\cdot k}\,\mathcal{M}^{\cdots a'\cdots}\,(T_{\text{adj}}^b)_{a'a}.$$

(Note that the adjoint generators of $\mathfrak{su}(N)$ are antisymmetric, implying that the eikonal factor is the same for incoming and outgoing gluons.)

The amplitude for producing an extra soft gluon is the sum of all these terms, over the various kinds of external lines of the original amplitude. The non-Abelian Ward–Takahashi identity requires that this new amplitude vanish if the polarization vector of the extra gluon, $\epsilon_\lambda^\mu(k)$, is replaced by the momentum k^μ (all the other external lines are on-shell, and contracted with physical polarization vectors in the case of gluons). This condition reads

$$0 = \sum_{\alpha\in\{\text{gluons}\}} \big(T_{\text{adj}}^b\big)_{a_\alpha a'_\alpha}\,\mathcal{M}^{\cdots a'_\alpha\cdots} + \sum_{\alpha\in\{\text{out. q, in. }\bar{q}\}} \big(t_f^b\big)_{i_\alpha i'_\alpha}\,\mathcal{M}^{\cdots i'_\alpha\cdots} - \sum_{\alpha\in\{\text{in. q, out. }\bar{q}\}} \mathcal{M}^{\cdots \bar{\imath}'_\alpha\cdots}\,\big(t_f^b\big)_{\bar{\imath}'_\alpha\bar{\imath}_\alpha}.$$

This identity may be viewed as the statement of color conservation in the original amplitude.

45. Low–Burnett–Kroll Theorem in QCD Consider a color-singlet quark–antiquark bound state, e.g., a J/ψ particle. The *color-octet model* asserts that the production of such a bound state proceeds in two steps: (i) a quark–antiquark pair is first produced in an octet state from a gluon, and (ii) a soft gluon is emitted in order to evaporate the extra color and obtain a singlet state (the idea being that, since the gluon is soft and the strong coupling constant is large at small momentum, there is no suppression of the emission of the soft gluon). In the spirit of Problems 13, 14 and 43, show that this scenario is problematic because the emission of soft gluons is kinematically suppressed in this situation. In other words, "soft gluons cannot bleach color." This calculation can be done by going through the following steps:

45.a Calculate the emissions of the soft gluon from each of the three external lines, at leading order (LO) and next-to-leading order (NLO) in the momentum of the new gluon.

45.b Calculate the emission of the soft gluon from an inner line of the hard amplitude, with the same degree of accuracy.

45.c Simplify the sum of these contributions by projecting the colors of the quark and antiquark on a singlet configuration, since we want to produce a singlet bound state.

45.d Contract the result with the polarization vector $\epsilon^{\nu}_{\sigma}(\ell)$ of the soft gluon and square. Sum over the color of the hard gluon and the spins of the quark and antiquark.

45.a Our starting point is a gluon–quark–antiquark amplitude,

$$\mathcal{M}_{aij} \equiv \overline{u}(p)\,\mathcal{A}_{aij}(p,q,k)\,u(q), \qquad \text{with } \mathcal{A}_{aij}(p,q,k) \equiv \epsilon_{\lambda\mu}(k)\mathcal{A}^{\mu}(p,q,k)\,(t^a_f)_{ij}.$$

Note that $\mathcal{A}^{\mu}$ may be a very complicated object if loops are involved. Obviously, this amplitude gives zero when projected on a color-singlet bound state, since $(t^a_f)_{ij}\delta_{ij} = 0$. Let us now alter this amplitude in order to allow the emission of a gluon of momentum ℓ, color b and polarization σ. This additional gluon can be emitted from one of the external legs, or attached to one of the inner lines in the hard amplitude. Compared to Problem 43, we want to keep an extra order in the expansion in powers of the momentum of the soft gluon, which implies that we must keep the term where the gluon is attached inside the hard amplitude.

Consider first the term where the gluon is attached to the external quark line of momentum p. Using the Dirac equation, this term (before contracting with the polarization vector of the new gluon) reads

$$\mathcal{M}^{\nu}_{abij;q} = g(t^b_f)_{ii'}\,\overline{u}(p)\frac{1}{p\cdot\ell}\Big(\underbrace{p^{\nu}}_{\sim\ell^{-1}} + \underbrace{\tfrac{1}{2}\gamma^{\nu}\not{\ell} + p^{\nu}\ell^{\rho}\frac{\partial}{\partial p^{\rho}}\Big|_{\mathcal{A}}}_{\sim\ell^{0}}\Big)\mathcal{A}_{ai'j}(p,q,k)\,u(q).$$

(We have indicated under each term the order in ℓ at which it contributes.) The derivative in the last term comes from expanding $\mathcal{A}_{aij}(p+\ell,q,k)$ to first order in ℓ, and the subscript $\mathcal{A}$ indicates that the derivative should act only on $\mathcal{A}_{aij}$. Likewise, attaching the gluon to the antiquark line of momentum q gives

$$\mathcal{M}^{\nu}_{abij;\overline{q}} = g(t^b_f)_{j'j}\,\overline{u}(p)\,\mathcal{A}_{aij'}(p,q,k)\Big(-q^{\nu} + \tfrac{1}{2}\not{\ell}\gamma^{\nu} + q^{\nu}\ell^{\rho}\frac{\overleftarrow{\partial}}{\partial q^{\rho}}\Big|_{\mathcal{A}}\Big)\frac{1}{q\cdot\ell}\,u(q).$$

(The arrow indicates that the derivative acts on the factor $\mathcal{A}_{aij'}(p,q,k)$ located to its left.) Finally, the contribution where the new gluon is attached to the external gluon line of momentum k reads

$$\begin{aligned}\mathcal{M}^{\nu}_{abij;g} = {}& g(T^b_{adj})_{aa'}\,\epsilon_{\lambda\mu}(k)\frac{1}{k\cdot\ell}\Big(g^{\mu\rho}k^{\nu} + g^{\nu\mu}\ell^{\rho} - g^{\rho\nu}\ell^{\mu} + g^{\mu\rho}k^{\nu}\ell^{\tau}\frac{\partial}{\partial k^{\tau}}\Big|_{\mathcal{A}}\Big)\\ &\times\overline{u}(p)\,\mathcal{A}_{\rho}(p,q,k)\,(t^{a'}_f)_{ij}\,u(q).\end{aligned}$$

(In this expression, we have discarded a term in ℓ^{ν} by anticipating the fact that it will cancel after contraction with physical polarization vectors. Another term in k^{ρ} has been discarded thanks to the Ward–Takahashi identity satisfied by $\mathcal{A}_{\rho}(p,q,k)$.)

45.b To these three terms, we must add a term $\mathcal{M}^{\nu}_{abij;in}$ coming from attachments of the new gluon inside the hard amplitude. This term does not diverge when $\ell \to 0$. Since we want to go up to order ℓ^0, this term is in fact independent of ℓ, and depends only on p, q, k. As we have seen in Problems 13 and 14, this term is fully determined (at this order in ℓ) by the Ward–Takahashi identity:

$$0 = \ell_\nu\Big(\mathcal{M}^{\nu}_{abij;q} + \mathcal{M}^{\nu}_{abij;\overline{q}} + \mathcal{M}^{\nu}_{abij;g} + \mathcal{M}^{\nu}_{abij;in}\Big)$$
$$= g\,\ell^\rho\Big((t^b_f)_{ii'}\,\frac{\partial\mathcal{M}_{ai'j}}{\partial p^\rho}\Big|_{\mathcal{A}} + (t^b_f)_{j'j}\,\frac{\partial\mathcal{M}_{aij'}}{\partial q^\rho}\Big|_{\mathcal{A}} + (T^b_{adj})_{aa'}\,\frac{\partial\mathcal{M}_{a'ij}}{\partial k^\rho}\Big|_{\mathcal{A}}\Big) + \ell_\nu\mathcal{M}^{\nu}_{abij;in}.$$

(We have used the result of Problem 44 in order to get rid of the terms of order ℓ^0 in this equation.) This constraint determines $\mathcal{M}^{\nu}_{abij;in}$ only up to a term G^ν such that $\ell_\nu G^\nu = 0$. But since we are looking for a function independent of ℓ, this additional term is in fact vanishing. Therefore, we have

$$\mathcal{M}^{\nu}_{abij;in} = -g\,\Big((t^b_f)_{ii'}\,\frac{\partial\mathcal{M}_{ai'j}}{\partial p_\nu}\Big|_{\mathcal{A}} + (t^b_f)_{j'j}\,\frac{\partial\mathcal{M}_{aij'}}{\partial q_\nu}\Big|_{\mathcal{A}} + (T^b_{adj})_{aa'}\,\frac{\partial\mathcal{M}_{a'ij}}{\partial k_\nu}\Big|_{\mathcal{A}}\Big),$$

and the total gluon emission amplitude up to order ℓ^0 reads

$$\mathcal{M}^{\nu}_{abij} = g\frac{(t^b_f)_{ii'}}{p\cdot\ell}(p^\nu - i\ell_\rho\,[J^{\nu\rho}_p]_{\mathcal{A}})\mathcal{M}_{ai'j} + g\frac{(t^b_f)_{j'j}}{q\cdot\ell}(-q^\nu - i\ell_\rho\,[J^{\nu\rho}_q]_{\mathcal{A}})\mathcal{M}_{aij'}$$
$$+\,g\frac{(T^b_{adj})_{aa'}}{k\cdot\ell}(k^\nu - i\ell_\rho\,[J^{\nu\rho}_k]_{\mathcal{A}})\mathcal{M}_{a'ij} + \mathcal{E}^{\nu}_{abij},$$

where $[J_p]^{\nu\rho}_{\mathcal{A}} \equiv i(p^\nu\frac{\partial}{\partial p_\rho} - p^\rho\frac{\partial}{\partial p_\nu})_{\mathcal{A}}$ is the angular momentum operator (with the action of the derivatives restricted to $\mathcal{A}_{aij}$) and where we have defined

$$\mathcal{E}^{\nu}_{abij} \equiv g\Bigg(\frac{(t^b_f)_{ii'}}{p\cdot\ell}\overline{u}(p)\frac{\gamma^\nu\not{\ell}}{2}\mathcal{A}_{ai'j}u(q) + \frac{(t^b_f)_{j'j}}{q\cdot\ell}\overline{u}(p)\mathcal{A}_{aij'}\frac{\gamma^\nu\not{\ell}}{2}u(q)$$
$$+\,\frac{(T^b_{adj})_{aa'}}{k\cdot\ell}(g^{\nu\mu}\ell^\rho - g^{\rho\nu}\ell^\mu)\overline{u}(p)\mathcal{A}_\rho(p,q,k)\,(t^{a'}_f)_{ij}u(q)\epsilon_{\lambda\mu}(k)\Bigg).$$

45.c The next step is to contract this amplitude with the polarization vector $\epsilon_{\sigma\nu}(\ell)$ of the additional gluon, and to square it. In order to address the question of the production of a color-singlet bound state, we can simplify the task considerably by first projecting the colors of the quark and antiquark on a singlet configuration. When we perform this projection, the various color structures that appear in the amplitude simplify as follows:

$$\delta_{ij}\,(t^b_f)_{ii'}(t^a_f)_{i'j} = \mathrm{tr}\,(t^b_f t^a_f) = \tfrac{1}{2}\delta_{ab},$$
$$\delta_{ij}\,(t^b_f)_{j'j}(t^a_f)_{ij'} = \mathrm{tr}\,(t^a_f t^b_f) = \tfrac{1}{2}\delta_{ab},$$
$$\delta_{ij}(T^b_{adj})_{aa'}(t^a_f)_{ij} = 0.$$

The last equation has a simple physical interpretation: the emission of the soft gluon off the external hard gluon does not change the color state of the pair (octet) and leads to a null projection on a singlet. After this projection, the amplitude simplifies to

$$\delta_{ij}\mathcal{M}^{\nu}_{abij} = g\frac{\delta_{ab}}{2}\overline{u}(p)\Bigg(\Big[\underbrace{\frac{p^{\nu}}{p\cdot\ell}-\frac{q^{\nu}}{q\cdot\ell}}_{\equiv R^{\nu}} - i\ell_{\rho}\Big(\frac{[J_p^{\nu\rho}]_{\mathcal{A}}}{p\cdot\ell}+\frac{[J_q^{\nu\rho}]_{\mathcal{A}}}{q\cdot\ell}\Big)\Big]\mathcal{A}^{\mu} + \frac{1}{2}\Big[\frac{\gamma^{\nu}\not{\ell}\mathcal{A}^{\mu}}{p\cdot\ell}+\frac{\mathcal{A}^{\mu}\gamma^{\nu}\not{\ell}}{q\cdot\ell}\Big]\Bigg)u(q)\epsilon_{\lambda\mu}(k).$$

45.d Next, we contract this amplitude with $\epsilon_{\sigma\nu}(\ell)$ and then we square it, keeping only terms up to order ℓ^{-1}. When evaluating this square, we sum over the color of the hard gluon, and over the spins of the quark and antiquark. This leads to

$$\sum_{a,\text{spins}}\Big|\delta_{ij}\mathcal{M}^{\nu}_{abij}\epsilon_{\sigma\nu}(\ell)\Big|^2$$
$$= \frac{g^2}{4}\Bigg((R\cdot\epsilon_\sigma)\Big[(R\cdot\epsilon_\sigma) - i\epsilon_{\sigma\nu}\ell_\rho\Big(\frac{[J_p^{\nu\rho}]_{\mathcal{A}}}{p\cdot\ell}+\frac{[J_q^{\nu\rho}]_{\mathcal{A}}}{q\cdot\ell}\Big)\Big]\sum_{\text{spins}}\Big|\overline{u}(p)[\mathcal{A}\cdot\epsilon_\lambda]u(q)\Big|^2$$
$$+ (R\cdot\epsilon_\sigma)\,\mathrm{tr}\Big((\mathcal{A}\cdot\epsilon_\lambda)(\not{q}+m)(\mathcal{A}^*\cdot\epsilon_\lambda)\frac{(\not{p}+m)\not{\epsilon}_\sigma\not{\ell}+\not{\ell}\not{\epsilon}_\sigma(\not{p}+m)}{2p\cdot\ell}\Big)$$
$$+ (R\cdot\epsilon_\sigma)\,\mathrm{tr}\Big((\mathcal{A}^*\cdot\epsilon_\lambda)(\not{p}+m)(\mathcal{A}\cdot\epsilon_\lambda)\frac{(\not{q}+m)\not{\epsilon}_\sigma\not{\ell}+\not{\ell}\not{\epsilon}_\sigma(\not{q}+m)}{2q\cdot\ell}\Big)\Bigg).$$

In Problem 14, we showed that

$$\frac{(\not{p}+m)\not{\epsilon}_\sigma\not{\ell}+\not{\ell}\not{\epsilon}_\sigma(\not{p}+m)}{2p\cdot\ell} = -\frac{i}{p\cdot\ell}\epsilon_{\sigma\nu}\ell_\rho[J_p^{\nu\rho}]_{u\overline{u}}(\not{p}+m).$$

The subscript $u\overline{u}$ indicates that the angular momentum operator acts only on the p dependence coming from the spinors. Combined with the similar terms that act only on $\mathcal{A}$, we can reconstruct the unrestricted angular momentum operator that acts on the complete squared hard amplitude:

$$\sum_{a,\text{spins}}\Big|\delta_{ij}\mathcal{M}^{\nu}_{abij}\epsilon_{\sigma\nu}(\ell)\Big|^2 = \frac{g^2}{4}(R\cdot\epsilon_\sigma)\Big[(R\cdot\epsilon_\sigma) - i\epsilon_{\sigma\nu}\ell_\rho\Big(\frac{J_p^{\nu\rho}}{p\cdot\ell}+\frac{J_q^{\nu\rho}}{q\cdot\ell}\Big)\Big]\underbrace{\sum_{\text{spins}}\Big|\overline{u}(p)[\mathcal{A}\cdot\epsilon_\lambda]u(q)\Big|^2}_{\text{squared color-stripped hard amplitude}}. \tag{3.53}$$

This formula extends (in a simple case) to QCD the Low–Burnett–Kroll theorem established for QED in Problem 14.

Let us now discuss the significance of (3.53) for the production of color-singlet quark–antiquark bound states. Note first that R^ν is the radiation current of a color-singlet quark–antiquark pair (compare with equation (3.50) in Problem 43). In that problem, we studied in detail the angular pattern of emissions due to such a current, and we saw that emissions are allowed only in cones whose opening is the angular separation between the quark and antiquark. However, in order to merge in a bound state of small size, the quark and antiquark must have nearly parallel momenta, which implies that $R^\nu \approx 0$ (therefore, not only does the leading term of order ℓ^{-2} cancel, but so does the subleading term of order ℓ^{-1}). To gain more intuition on the physical mechanism at work here, one may think of the emission of a soft gluon as being extremely delocalized in time, and happening *after* the color singlet has been formed, hence the suppression. The conclusion is that a soft gluon cannot "bleach" a color-octet quark–antiquark pair in order to produce a color-singlet bound state (this may happen via terms of higher order in ℓ, but these emissions will not be enhanced by inverse powers of ℓ).

46. Globally Gauge Invariant Operators Consider the bilinear and trilinear operators $\alpha_{ab} A_a^\mu A_{b\,\mu}$ and $\beta_{abc}\chi_a\chi_b\chi_c$, where A_a^μ is the gauge field and χ_a is the ghost field. The goal of this problem is to determine the form of the coefficients α_{ab} and β_{abc} such that these operators are invariant under global gauge transformations. A useful tool will be Schur's lemma, which states that any operator C that commutes with all the generators of an irreducible representation of a Lie algebra is proportional to the identity, $C \propto 1$.

46.a Why can we assume that α_{ab} is symmetric and β_{abc} is fully antisymmetric? Why should β vanish when any pair of indices are contracted, e.g., $\beta_{aac} = 0$? How do the gauge field A_a^μ and the ghost χ_a change under a *global* gauge transformation?

46.b Show that the global gauge invariance of $\alpha_{ab} A_a^\mu A_{\mu\, b}$ is equivalent to $\alpha\Omega = \Omega\alpha$, for any Ω in the adjoint representation of the gauge group. Show that this is equivalent to $[\alpha, T^a_{\text{adj}}] = 0$ for all a, and that this implies that one should have $\alpha_{ab} \propto \delta_{ab}$.

46.c Show that $\beta_{abc}\chi_a\chi_b\chi_c$ is invariant under global transformations if $\beta_{abc}\Omega_{ad}\Omega_{be}\Omega_{cf} = \beta_{def}$, for any Ω in the adjoint representation of the gauge group. Introduce matrices B^a defined by $(B^a)_{bc} \equiv i\beta_{abc}$. Show that this condition is equivalent to $[T^a_{\text{adj}}, B^b] = if^{abc}B^c$. From this, conclude that $\beta_{abc} \propto f^{abc}$ is a solution.

46.d Extra: prove that this solution is the only one up to a rescaling.

46.a The operator $A_a^\mu A_{b\,\mu}$ is symmetric in (a, b), while $\chi_a\chi_b\chi_c$ is fully antisymmetric in (a, b, c). Therefore, they must be accompanied by coefficients α_{ab} and β_{abc} that are symmetric and antisymmetric, respectively. Moreover, since $\chi_a\chi_a = 0$, the coefficients β_{abc} should vanish when contracting any pair of indices (to see this for the pair ab, decompose β_{abc} into a term proportional to δ_{ab} and a traceless term; the term in δ_{ab} cancels when contracted with $\chi_a\chi_b$).

Under any gauge transformation Ω, the gauge field A^μ transforms as

$$A^\mu \to \Omega A^\mu \Omega^\dagger + \frac{i}{g}\Omega\partial^\mu\Omega^\dagger.$$

In the case of a global transformation, the second term is zero, and the transformation simplifies to

$$A^\mu \to \Omega A^\mu \Omega^\dagger,$$

or, equivalently, $A^\mu_a \to \Omega_{ab}A^\mu_b$ if written in terms of the color components. For the ghost field, the transformation reads

$$\chi_a \to \Omega_{ab}\chi_b.$$

46.b Under the preceding global gauge transformation, we have

$$\alpha_{ab}A^\mu_a A_{\mu b} \to \alpha_{ab}\Omega_{ac}A^\mu_c\Omega_{bd}A_{\mu d}.$$

Therefore, this operator is invariant if $\alpha_{cd} = \alpha_{ab}\Omega_{ac}\Omega_{bd}$ for all Ω. Noting that the adjoint group elements are real (and therefore obey $\Omega^t = \Omega^{-1}$), this condition can be rewritten as

$$\Omega_{ac}\alpha_{cd} = \alpha_{ab}\Omega_{bd}, \quad \text{i.e.,}\ \Omega\alpha = \alpha\Omega.$$

Writing $\Omega \equiv \exp(i\theta_a T^a_{\text{adj}})$ and requiring that this condition be true for any values of the coefficients θ_a, we obtain the following equivalent condition:

$$\left[\alpha, T^a_{\text{adj}}\right] = 0, \quad \text{for all } a.$$

Then we use Schur's lemma and the fact that the adjoint representation is irreducible to conclude that the matrix of coefficients α must be proportional to the identity.

46.c The operator $\beta_{abc}\chi_a\chi_b\chi_c$ transforms as

$$\beta_{abc}\chi_a\chi_b\chi_c \to \beta_{abc}\Omega_{ad}\chi_d\Omega_{be}\chi_e\Omega_{cf}\chi_f.$$

Therefore, for this operator to be invariant, we must have

$$\beta_{abc}\Omega_{ad}\Omega_{be}\Omega_{cf} = \beta_{def}, \quad \text{for any } \Omega.$$

In terms of the matrices B^a whose components are $(B^a)_{bc} \equiv i\beta_{abc}$, this condition can be transformed into the following equivalent conditions:

$$B^a_{bc}\Omega_{ad}\Omega_{be}\Omega_{cf} = B^d_{ef},$$
$$\Omega_{ad}\left[\Omega^t B^a \Omega\right]_{ef} = B^d_{ef},$$
$$\Omega^t B^a \Omega = B^d \Omega^t_{da} = \Omega_{ad}B^d.$$

(Note also that $\Omega^t = \Omega^\dagger$ for the adjoint representation.) Then, we replace Ω by $\Omega = \exp(i\theta_a T^a_{\text{adj}})$. For the final condition to be true for all values of θ_a, we must have $[T^a_{\text{adj}}, B^b] = if^{abc}B^c$ (to see this, we just need to expand to first order in θ). This condition is obviously true if the B^a are proportional to the adjoint generators T^a_{adj} themselves, which corresponds to β_{abc} being proportional to the structure constant f^{abc}.

46.d Let us now consider the reverse question, namely whether $\beta_{abc} \propto f^{abc}$ is the only solution. Note first that the fact that B^a is an imaginary traceless Hermitian matrix does not imply that it can be written as a linear superposition of the adjoint generators T^a_{adj}, since the dimension of the space of $(n^2-1)\times(n^2-1)$ imaginary traceless Hermitian matrices $(\frac{1}{2}(n^2-1)(n^2-2))$ is in general larger than the dimension of the Lie algebra $\mathfrak{su}(n)$. Therefore, the naive approach that would consist in writing $B^a \equiv b_{ab}T^b_{adj}$ (which would imply $[b, T^a_{adj}] = 0$, i.e., $b_{ab} \sim \delta_{ab}$ by Schur's lemma) is not correct (the only exception is for $n = 2$, where the two dimensions are equal, and therefore this argument is valid).

A proper treatment of this question requires the concept of $SU(n)$ tensors. An $SU(n)$ fundamental vector is an object ψ_i that transforms as

$$\psi_i \to \Omega_i{}^{i'}\psi_{i'}, \quad \psi_i^* \to \psi_{i'}^*\Omega_i^{*i'},$$

where Ω is a matrix in the fundamental representation of $SU(n)$. In this discussion, it is convenient to denote the transformation matrix with a lower and an upper index, in order to easily keep track of which contractions of indices are allowed, and the notation can be made more uniform by also denoting $\psi^i \equiv \psi_i^*, \Omega^i{}_{i'} \equiv \Omega_i^{*i'}$ so that the transformation of the complex conjugate of ψ reads

$$\psi^i \to \Omega^i{}_{i'}\psi^{i'}.$$

A rank-r fundamental $SU(n)$ tensor is an object carrying r indices (up or down), that transforms like an outer product of r vectors (with matching up or down indices):

$$T_{i_1\cdots i_p}{}^{i_{p+1}\cdots i_r} \to \Omega_{i_1}{}^{j_1}\cdots\Omega_{i_p}{}^{j_p}\,\Omega^{i_{p+1}}{}_{j_{p+1}}\cdots\Omega^{i_r}{}_{j_r}\,T_{j_1\cdots j_p}{}^{j_{p+1}\cdots j_r}.$$

Some tensors have the remarkable property of being invariant under $SU(n)$. This is obviously the case for the Kronecker symbol $\delta_i{}^j$ (this follows from the fact that Ω is unitary). It turns out that the Levi–Civita symbol with n indices, all up or all down, is also invariant, since

$$\epsilon_{i_1\cdots i_n} \to \Omega_{i_1}{}^{i'_1}\cdots\Omega_{i_n}{}^{i'_n}\,\epsilon_{i'_1\cdots i'_n} = \epsilon_{i_1\cdots i_n}\underbrace{\det(\Omega)}_{=1}.$$

The most general version of our original problem is to find β_{abc} such that $\beta_{abc}X_aY_bZ_c$ is invariant when X_a, Y_a, Z_a transform according to the adjoint representation of $SU(n)$ (when the fields X, Y, Z are all identical, the solution should be fully symmetric or fully antisymmetric, depending on whether this field is commuting or anti-commuting). Finding such coefficients β_{abc} is equivalent to making a $SU(n)$ scalar out of the following rank-2 fundamental tensors:

$$X_i{}^j \equiv X_a(t_f^a)_i{}^j, \quad Y_i{}^j \equiv Y_a(t_f^a)_i{}^j, \quad Z_i{}^j \equiv Z_a(t_f^a)_i{}^j.$$

The only way to obtain a $SU(n)$ scalar from the outer product of three rank-2 tensors is to contract all the lower indices with the upper ones, possibly also using the $SU(n)$ invariant tensors discussed above, i.e., Kronecker symbols (with mixed indices) and Levi–Civita symbols (with all indices up or all indices down). But since we are only allowed to contract upper indices with lower ones, we must use an even number of Levi–Civita symbols. Therefore, we can always combine them in pairs (one with lower indices and one with upper indices) and use

$$\epsilon_{i_1\cdots i_n}\epsilon^{j_1\cdots j_n} = \det\begin{pmatrix} \delta_{i_1}{}^{j_1} & \dots & \delta_{i_1}{}^{j_n} \\ \delta_{i_2}{}^{j_1} & \dots & \delta_{i_2}{}^{j_n} \\ \vdots & & \vdots \\ \delta_{i_n}{}^{j_1} & \dots & \delta_{i_n}{}^{j_n} \end{pmatrix}.$$

Thanks to this formula, all the Levi–Civita symbols appearing in such a contraction can in fact be replaced by products of Kronecker symbols. This implies that the most general scalar one can make from X, Y, Z is a linear combination of the following terms:

$$\mathrm{tr}\,(t_f^a t_f^b t_f^c)\,X_a Y_b Z_c \quad \text{and permutations},$$
$$\mathrm{tr}\,(t_f^a)\,\mathrm{tr}\,(t_f^b t_f^c)\,X_a Y_b Z_c \quad \text{and permutations}, \qquad \mathrm{tr}\,(t_f^a)\,\mathrm{tr}\,(t_f^b)\,\mathrm{tr}\,(t_f^c)\,X_a Y_b Z_c.$$

In other words, a spanning set for the most general β_{abc} is obtained by forming all the possible products of traces with three fundamental generators (this result generalizes in an obvious manner to scalars constructed with any number of adjoint fields). Since the trace of a single generator is zero, all the terms of the second line are in fact vanishing. Moreover, the trace of three generators is made of only two independent objects, f^{abc} and d^{abc}. In our original problem involving three ghost fields, the only solution is the fully antisymmetric one, i.e., $\beta_{abc} \propto f^{abc}$.

47. Banks–Zaks Fixed Points Consider a theory (e.g., QCD) in which the β function has the following two-loop perturbative expansion: $\beta(g) = \beta_0\, g^3 + \beta_1\, g^5 + \mathcal{O}(g^7)$.

47.a What are the conditions on β_0 and β_1 for the β function to be negative at small coupling and positive at large coupling? How does the theory behave at short distance and at long distance in this situation?

47.b What extra condition on $\beta_{0,1}$ should be satisfied in order to have a weakly coupled theory at the infrared fixed point?

47.c Consider now QCD with N colors and N_f flavors of quarks in the fundamental representation, for which the coefficients $\beta_{0,1}$ are

$$\beta_0 = -\frac{1}{16\pi^2}\,\frac{11N - 2N_f}{3}, \qquad \beta_1 = \frac{1}{(16\pi^2)^2}\left(N_f\left(2C_f + \frac{10N}{3}\right) - \frac{34N^2}{3}\right).$$

What is the range of values of N_f for having a fixed point of the type discussed above?

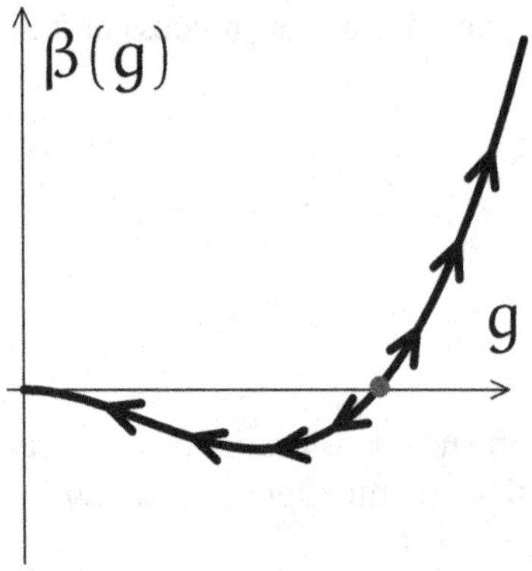

Figure 3.9 Sketch of the β function obtained for $\beta_0 < 0, \beta_1 > 0$. The arrows indicate the renormalization group flow towards the ultraviolet.

47.a With the type of β function proposed in the statement of the problem, the requested behavior of the β function is realized for $\beta_0 < 0, \beta_1 > 0$. Such a β function is illustrated in Figure 3.9. In this case, the β function has a zero at the coupling constant g_* defined by

$$g_*^2 = -\frac{\beta_0}{\beta_1}.$$

Then, if $0 < g < g_*$, the β function is negative. This implies that the coupling decreases as the energy scale rises. Thus the corresponding theory exhibits asymptotic freedom as long as the coupling never exceeds g_*. Conversely, when the energy scale decreases towards the infrared, the coupling constant increases, until it reaches the value g_*. At g_*, the running of the coupling stops no matter how much one decreases the energy scale. g_* is therefore an *infrared fixed point* for this theory. To summarize, a theory with this type of β function has the following fixed points:

i. a non-interacting ultraviolet fixed point,

ii. an interacting infrared fixed point.

At both fixed points, the theory is scale invariant (by virtue of being at a fixed point).

47.b The existence of the infrared fixed point is the main difference between the situation considered here and a situation where the β function would be negative everywhere. In the latter case, the theory becomes strongly coupled in the infrared (the coupling increases indefinitely at large distance), while it stands a chance of being perturbative in the case considered in the present problem. For this, it is not sufficient to have an infrared fixed point, but the value of the coupling at this fixed point should in addition be small enough. To be definite, let us say that perturbation theory is applicable if $g_* < 1$ (this is arguably a bit arbitrary, and may depend on the observable of interest). This additional condition is satisfied if $|\beta_0| < \beta_1$. When it is in the perturbative regime, the infrared fixed point **ii** is known as a *Banks–Zaks fixed point.*

47.c Consider now the specific case of QCD, for which the first two coefficients of the β function are given in the statement of the problem. The condition $\beta_0 < 0$ is equivalent to

$$\beta_0 < 0 \quad \Longleftrightarrow \quad N_f < \frac{11N}{2} \qquad (= 16.5 \text{ for } N = 3).$$

The condition $\beta_1 > 0$ is equivalent to the following condition on N_f:

$$\beta_1 > 0 \quad \Longleftrightarrow \quad N_f > \frac{34N^3}{13N^2 - 3} \qquad (\approx 8 \text{ for } N = 3).$$

Thus, the infrared fixed point cannot be realized with the six families of quarks that are known to exist in nature. Note that the value of the coupling g_* at the infrared fixed point is

$$g_*^2 = \frac{16\pi^2(11N - 2N_f)}{N_f(6C_f + 10N) - 34N^2}.$$

This ratio decreases as N_f increases at fixed N (it goes to zero at the upper end of the allowed range of values of N_f, and to infinity at the lower end of the range of N_f). The value of N_f at which this critical coupling is equal to unity is given by

$$N_f = \frac{88\pi^2 N + 17N^2}{16\pi^2 + 3C_f + 5N} \qquad (\approx 15.6 \text{ for } N = 3).$$

Therefore, the infrared fixed point is perturbative only at the largest allowed value of the number of flavors, very close to the value where the theory ceases to be asymptotically free. Source: Banks, T. and Zaks, A. (1982), *Nucl Phys B* 196: 189.

48. Instantons and Tunneling in Quantum Mechanics This is the first of a series of three problems devoted to the role of instantons in classifying the vacua of Yang–Mills theory. In this first instalment, we start with a warmup in non-relativistic mechanics, and show that instanton-like solutions of the classical equations of motion provide a semi-classical approximation of quantum tunneling amplitudes. Consider a one-dimensional double-well potential $V(x) \equiv \lambda(x^2 - x_*^2)^2$, which has two degenerate minima located at $x = \pm x_*$.

48.a Recall the path integral expression for a transition amplitude from $-x_*$ to $+x_*$, $\mathcal{A} \equiv \big\langle +x_* \big| e^{-iHt} \big| -x_* \big\rangle$. Explain why there is no classical path making the action extremal in this path integral.

48.b Perform a Wick rotation of the time variable by defining $\tau = i\,t$, and write the resulting expression for the transition amplitude.

48.c Show that a saddle point approximation is now possible, and that the leading behavior of the amplitude is given by the quantum mechanical analogue of instantons. Solve the imaginary time equation of motion to obtain this dominant classical path, and calculate the corresponding action. Are there other minima of the action that would give subleading contributions?

48.d The preceding calculation does not give the prefactor in $\mathcal{A}(t)$. How can it be obtained? Without doing an explicit calculation, explain why this prefactor is proportional to the time t. What is the range of t where this semi-classical approximation is valid? What contributions should one include to extend it to larger times?

48.e Let us denote by $|\psi_{0,1,2,\ldots}\rangle$ the eigenstates of the Hamiltonian, with energies $E_{0,1,2,\ldots}$. Express the transition amplitude $\mathcal{A}(t)$ in terms of these eigenstates. Use this alternative representation to show that the tunneling amplitude $\mathcal{A}(t)$ is approximately proportional to the splitting between the energies of the first two levels, $\delta E \equiv E_1 - E_0$.

48.a The transition amplitude from the position $-x_*$ at time 0 to the position $+x_*$ at time t is given by

$$\mathcal{A}(t) \equiv \left\langle +x_* \middle| e^{-iHt} \middle| -x_* \right\rangle = \int_{\substack{x(0)=-x_* \\ x(t)=+x_*}} \big[Dx(t)\big]\; e^{i\mathcal{S}[x]}, \tag{3.54}$$

where the action $\mathcal{S}[x]$ reads (for a mass $m \equiv 1$)

$$\mathcal{S}[x] \equiv \int_0^t dt' \left(\frac{\dot{x}^2(t')}{2} - V(x(t'))\right).$$

Classical trajectories are extrema of the action $\mathcal{S}[x]$ that satisfy $\ddot{x} + V'(x) = 0$. Multiplying by $\dot{x}$ leads to $\frac{1}{2}\dot{x}^2 + V(x) = \text{const}$, i.e., energy conservation. Therefore, a classical particle that starts with no velocity from the bottom of one of the potential wells cannot reach the other well. Because of the absence of a classical solution with the appropriate boundary conditions, the path integral (3.54) is not dominated by any path in particular, and its value is the result of interferences between many paths contributing comparably.

48.b However, we can simplify this problem by defining $\tau \equiv i\,t$. The path integral (3.54) becomes

$$\mathcal{A}(t) = \int \big[Dx(\tau)\big]\; e^{-\mathcal{S}_E[x]},$$

with a Euclidean action equal to

$$\mathcal{S}_E[x] \equiv \int_0^{it} d\tau' \left(\frac{\dot{x}^2(\tau')}{2} \underbrace{+V(x(\tau'))}_{-U(x(\tau'))}\right).$$

(Now the dot denotes a derivative with respect to τ.) This action has the same form as before, if we interpret it as a fictitious particle moving in the inverted potential $U(x) \equiv -V(x)$ (see Figure 3.10).

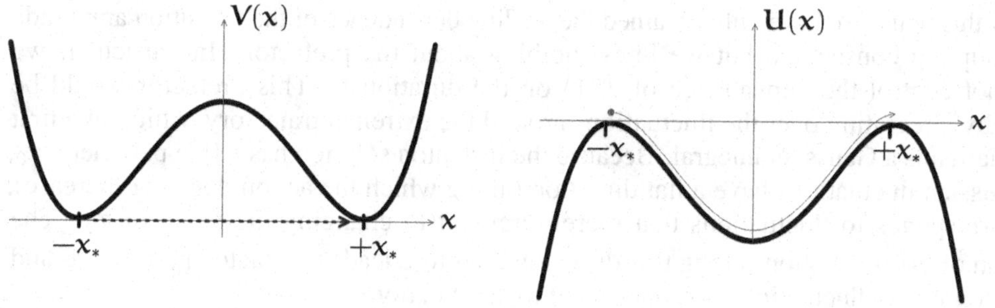

Figure 3.10 Left: double-well potential. Right: classical solution of the Euclidean time equation of motion, that connects the two extrema.

48.c Because the potential U is flipped upside down, the Euclidean action has extremal paths connecting $-x_*$ to $+x_*$, which are the quantum mechanical analogue of instantons (recall that instantons are gauge field configurations that make the Euclidean action extremal). Solving the Euclidean equation of motion is best done by starting from "energy conservation,"

$$\frac{\dot{x}^2}{2} + U(x) = 0,$$

which can be rewritten as

$$\frac{dx}{x_*^2 - x^2} = \sqrt{2\lambda}\, d\tau.$$

This can be solved by

$$x(\tau) = x_* \tanh\left(\tfrac{\omega}{2}(\tau - \tau_0)\right),$$

where $\omega \equiv \sqrt{8\lambda x_*^2}$ and where τ_0 is an integration constant that defines when this solution reaches the position $x = 0$. The value of the Euclidean action for this solution is $S_E[x] = \omega^3/(12\lambda)$. Note that although this solution connects $-x_*$ to $+x_*$ in an infinite Euclidean time, the effective duration of the transition is $\omega^{-1} \sim \lambda^{-1/2}$. Therefore, this path is a good approximation of an extremal path (away from the transition region, this trajectory stays an infinite amount of time at $-x_*$ and at $+x_*$, but these parts of the trajectory do not contribute to the action) for any transition duration $t \gg \omega^{-1}$, and we find

$$\mathcal{A}(t) \underset{t \gg \omega^{-1}}{\sim} e^{-\omega^3/(12\lambda)}.$$

There are other extrema of the action. The classical trajectory that gives the next-to-lowest minimum starts at $-x_*$ and goes to $+x_*$, then back to $-x_*$, and finally returns to $+x_*$. The action of this "triple instanton" is three times the action we have previously obtained, and therefore its contribution to the transition amplitude is considerably suppressed when $\lambda \ll 1$ since it enters in an exponential,

$$e^{-3\omega^3/(12\lambda)} \ll e^{-\omega^3/(12\lambda)}.$$

48.d At this point, we have only obtained the leading dependence of the transition amplitude on the coupling constant λ, but we know nothing about the prefactor. In particular, we also do not control the dependence of $\mathcal{A}(t)$ on the duration t. This prefactor would be obtained by integrating over the fluctuations around the extremal trajectory, which in a first approximation is a Gaussian integral. Because the instanton solution has a free parameter τ_0, these Gaussian fluctuations have a flat direction (along which the action does not increase), which corresponds to fluctuations that merely amount to changing the value of τ_0. The integration of the fluctuations along this direction therefore leads to a factor $\int_0^t d\tau_0 = t$, and the prefactor due to fluctuations is proportional to the duration:

$$\mathcal{A}(t) \underset{t \gg \omega^{-1}}{\sim} \omega t\, e^{-\omega^3/(12\lambda)}.$$

Obviously, the modulus of $\mathcal{A}(t)$ should not be larger than 1 (since its square is a transition probability). Therefore, the validity of this expression is certainly limited by

$$1 \ll \omega t \ll e^{\omega^3/(12\lambda)}.$$

For times comparable to or larger than the upper bound, we need to take the multi-instanton transitions into account, since they will provide corrections in powers of $\omega t e^{-\omega^3/(12\lambda)}$ and their summation to all orders will tame the large time behavior of the single-instanton contribution.

48.e If the eigenstates of the Hamiltonian are $|\psi_{0,1,2,\ldots}\rangle$, we can write the transition amplitude as follows:

$$\begin{aligned}\mathcal{A}(t) &= \sum_{m,n} \left\langle +x_* \middle| \psi_m \right\rangle \underbrace{\left\langle \psi_m \middle| e^{-i\mathcal{H}t} \middle| \psi_n \right\rangle}_{\delta_{mn}\, e^{-iE_m t}} \left\langle \psi_n \middle| -x_* \right\rangle \\ &= \sum_m \left\langle +x_* \middle| \psi_m \right\rangle \left\langle \psi_m \middle| -x_* \right\rangle e^{-iE_m t} \\ &\approx \left\langle +x_* \middle| \psi_0 \right\rangle \left\langle \psi_0 \middle| -x_* \right\rangle e^{-iE_0 t} + \left\langle +x_* \middle| \psi_1 \right\rangle \left\langle \psi_1 \middle| -x_* \right\rangle e^{-iE_1 t},\end{aligned}$$

where we keep only the lowest two eigenstates in the approximation. Recall now that in a symmetric double-well potential (with well-separated wells), the ground state and the next-to-lowest state are the symmetric and antisymmetric combinations of the Gaussian packets $\psi_\pm$ that would give the ground state of each separate well:

$$\psi_0(x) = \frac{1}{\sqrt{2}}\big(\psi_+(x) + \psi_-(x)\big), \quad \psi_1(x) = \frac{1}{\sqrt{2}}\big(\psi_+(x) - \psi_-(x)\big).$$

Moreover, when the wells are well separated, we have $\psi_+(-x_*) \approx 0$ and $\psi_-(+x_*) \approx 0$. Therefore, the above transition amplitude becomes

$$\mathcal{A}(t) \approx \frac{i}{2} \psi_+(+x_*)\psi_-(-x_*)\, e^{-iE_0 t}\, (E_1 - E_0)t.$$

Thus, this formula establishes a direct connection between the tunneling amplitude and the first level splitting of the Hamiltonian. In particular, we see that the symmetric and antisymmetric lowest-energy states are degenerate if the tunneling probability is zero. (This result is not specific to instantons; another area where it is relevant is the discussion of spontaneous symmetry breaking.)

49. Instantons as Transitions between Pure Gauges in Yang–Mills Theory In this problem, we consider time-independent pure gauge fields such that $A^i(x) \to 0$ when $|x| \to \infty$ (in the gauge $A^0 = 0$). We show that these fields can be classified in topological classes whose index is a Cartan–Maurer invariant, and that instantons are field configurations that interpolate between two such pure gauges.

49.a Consider a pure gauge $A^i(x) \equiv ig^{-1}\Omega^\dagger(x)\partial^i\Omega(x)$, with $\Omega(x) \to 1$ at $|x| \to \infty$. Explain why the homotopy classes of these fields are the elements of $\pi_3(SU(N))$, with an index n given by

$$n \equiv \int \frac{d^3x}{24\pi^2}\, \epsilon_{jkl}\, \mathrm{tr}\, \big(\Omega^\dagger(\partial_j\Omega)\Omega^\dagger(\partial_k\Omega)\Omega^\dagger(\partial_l\Omega)\big).$$

49.b We denote by $|n_{\rm in,out}\rangle$ the asymptotic states made of pure gauges in the homotopy class of index n. A representative field configuration of the topological class $n = 0$ is $A^i_a(x) \equiv 0$. How does this field evolve via the classical Yang–Mills equation in the gauge $A^0 = 0$? Conclude from this that the transitions $\langle n_{\rm out}|0_{\rm in}\rangle$ are classically forbidden if $n \neq 0$ and necessarily happen by quantum tunneling.

49.c By analogy with Problem 48, write the path integral representation of this transition amplitude after going to imaginary time, and argue that the amplitude is dominated by Euclidean classical solutions.

49.d Recall that the topological index of the instanton is given by

$$N = g^2 \int \frac{d^4x}{64\pi^2}\, \epsilon_{ijkl} F^a_{ij} F^a_{kl} = \lim_{R\to\infty} g^2 \int_{\mathcal{S}_R} \frac{d^3S_i}{16\pi^2}\, \epsilon_{ijkl}\, \mathrm{tr}\, \big(A_j F_{kl} + \tfrac{2ig}{3} A_j A_k A_l\big),$$

where $\mathcal{S}_R$ is a three-dimensional sphere of radius R. Deform the sphere $\mathcal{S}_\infty$ at infinity into the surface of a hyper-cube, and show that in the gauge $A^0 = 0$ this integral is the difference of two integrals on surfaces at $x^0 = \pm\infty$, respectively, that are the topological indices of two pure gauges. Conclude that the instanton of index N interpolates between pure gauges of index 0 and N.

49.a Since the pure gauge fields $A^i(x)$ under consideration go to a constant when $|x| \to \infty$, we may identify all the points at infinity in $\mathbb{R}^3$, which amounts to identifying $\mathbb{R}^3$ with a 3-sphere $\mathcal{S}_3$ (this mapping can be done explicitly by a stereographic projection). Thus, the group element $\Omega(x)$ that defines the pure gauge can be viewed as a function from $\mathcal{S}_3$ to $SU(N)$, and the homotopy classes are the elements of the third homotopy group of $SU(N)$, $\pi_3(SU(N))$. The topological index of a given $\Omega(x)$ is the appropriately normalized Cartan–Maurer invariant,

$$n \equiv \int \frac{d^3x}{24\pi^2}\, \epsilon_{jkl}\, \mathrm{tr}\, \big(\Omega^\dagger(\partial_j\Omega)\Omega^\dagger(\partial_k\Omega)\Omega^\dagger(\partial_l\Omega)\big).$$

49.b The topological sector $n = 0$ admits as representative the group element $\Omega(x)$ which is constant and equal to the identity everywhere, i.e., the null gauge potential $A^i_a = 0$. In the gauge $A^0 = 0$, the classical Yang–Mills equations leave this field configuration unchanged throughout all the time evolution. Therefore, the transition from a field of index $n = 0$ to a field of index $n \neq 0$ is classically forbidden, and can only be mediated by quantum tunneling.

49.c After a transformation to imaginary time, this transition amplitude may be written as

$$\langle n_{\text{out}} | 0_{\text{in}} \rangle = \int_{[0]}^{[n]} [DA^i]\, e^{-S_E[A]},$$

where S_E is the Euclidean action, and where the notations $[0]$ and $[n]$ on the integral symbol are a reminder of its boundary conditions, namely pure gauge field configurations (which depend only on x since they are at the boundaries in time) of topological indices 0 and n, respectively. This path integral is dominated by the minimal action over the set of fields compatible with these boundary conditions. On the other hand, we know that the fields that give the minima of the Euclidean action are instantons, whose value at the boundary of $\mathbb{R}^4$ is a pure gauge.

49.d The topological index N of an instanton solution is given by the integral

$$N = g^2 \int \frac{d^4x}{64\pi^2}\, \epsilon_{ijkl} F^a_{ij} F^a_{kl} = g^2 \int_{\partial(\mathbb{R}^4)} \frac{d^3 S_i}{32\pi^2}\, K^i, \tag{3.55}$$

with

$$K^i = 2\,\epsilon_{ijkl}\,\mathrm{tr}\left(A_j F_{kl} + \frac{2ig}{3} A_j A_k A_l\right) \underset{|x|\to\infty}{=} \frac{4ig}{3}\,\epsilon_{ijkl}\,\mathrm{tr}\left(A_j A_k A_l\right).$$

(The first term in K^i becomes negligible in the limit $|x| \to \infty$ for fields that go to a pure gauge at infinity.) The second equality in (3.55) comes from the fact that $\epsilon_{ijkl} F^a_{ij} F^a_{kl}$ is equal to the total derivative $2\,\partial_i K^i$. Therefore, by Stokes's theorem, the integral can be rewritten as the flux of K^i through the boundary at infinity. The result does not depend on the precise shape of this boundary, and we can thus take a four-dimensional hyper-cube. Two "faces" of this hyper-cube correspond to the boundaries at infinite time, and all the others are boundaries at infinity in one of the spatial directions.

Let us consider the $A^0 = 0$ gauge. If i is a spatial direction, then in $\epsilon_{ijkl} A^j A^k A^l$, one of j, k, l must be 0 and therefore K^i vanishes at infinity on the two faces of the hyper-cube orthogonal to the spatial direction i. Thus, only the two boundaries at $x^0 = \pm\infty$ may contribute, and the topological index of the instanton can be rewritten as

$$N = \frac{4ig^3}{3}\,\epsilon_{0jkl} \left\{ \int_{x^0=+\infty} \frac{d^3x}{32\pi^2}\,\mathrm{tr}\left(A_j A_k A_l\right) - \int_{x^0=-\infty} \frac{d^3x}{32\pi^2}\,\mathrm{tr}\left(A_j A_k A_l\right) \right\}.$$

In these integrals, the field A^i is a pure gauge $A^i = (i/g)\Omega^\dagger \partial^i \Omega$, with Ω belonging to the topological sector 0 (at $x^0 = -\infty$) or n (at $x^0 = +\infty$), respectively. The integrals are the corresponding Cartan–Maurer invariants, i.e., the topological indices of the initial and final pure gauges. For instance:

$$\frac{4ig^3}{3}\epsilon_{0jkl}\int_{x^0=+\infty}\frac{d^3x}{32\pi^2}\,\mathrm{tr}\,(A_jA_kA_l) = \underbrace{\int\frac{d^3x}{24\pi^2}\,\epsilon_{jkl}\,\mathrm{tr}\left(\Omega^\dagger(\partial_j\Omega)\Omega^\dagger(\partial_k\Omega)\Omega^\dagger(\partial_l\Omega)\right)}_{\text{topological index } n \text{ of } \Omega(x) \text{ at } x^0=+\infty}.$$

Therefore, we have shown that

$$N = n - 0 \quad \text{(more generally, } N = n(+\infty) - n(-\infty)\text{)}.$$

Therefore, the instanton of index N is the Euclidean solution that interpolates between pure gauges of topological indices n and $n + N$.

50. Vacua of Yang–Mills Theory The goal of this problem is to show that classical Yang–Mills theory has a countable infinity of ground states. When quantum effects are taken into account, this leads to a continuous set of ground states known as the θ-vacua. Consider Yang–Mills theory in the temporal gauge $A^0 = 0$, with the coupling set to $g = 1$ for simplicity.

50.a What is the conjugate momentum of A^i? Derive the Hamiltonian, and check that Hamilton's equations are the usual Yang–Mills equations. Why do we need in addition to impose *Gauss's law*, i.e., $G[A] \equiv [D_i, \partial^0 A^i] = 0$, on the classical solutions? (Note that the quantum analogue of Gauss's law is that physical states should be annihilated by $G[A]$.)

50.b Yang–Mills theory in the temporal gauge has a residual invariance under purely spatial gauge transformations $\Omega(x)$. Explain why the gauge transformation $\Omega(x) \equiv e^{i\vartheta_a(x)t^a}$ is generated by

$$Q_\vartheta = -\int d^3x\,(\partial_0 A_a^i)(D_{ab}^i\vartheta_b).$$

If $\Omega(x)$ goes to the identity at spatial infinity, show that this generator becomes proportional to the Gauss's law constraint, implying that physical states are in fact invariant under the action of e^{iQ_ϑ}.

50.c From the above considerations, show that a vacuum state is a superposition of pure gauge fields $i\,U^\dagger\partial^i U$, related to one another by spatial gauge transformations $\Omega(x)$ that go to the identity at infinity.

50.d Consider transition amplitudes between two vacua, the first one corresponding to $U_1 \equiv 1$ and the second one to $U_2(x)$. Show that $U_2(x)$ should go to a constant at infinity for this amplitude to be non-zero. Why can these U's be seen as mappings from the 3-sphere S_3 to the gauge group?

50.e Use the preceding result to conclude that $SU(N)$ Yang–Mills theory has a countable infinity of inequivalent energy minima, indexed by an integer given by the Cartan–Maurer invariant. How do quantum fluctuations alter this classical picture? Determine which linear combinations of these minima are eigenstates of the quantum Hamiltonian. *Hint: introduce $\mathcal{T}$, the "translation operator" that transforms the ground state of index n into the ground state of index $n+1$, and use the fact that $\mathcal{T}$ commutes with the Hamiltonian. Note: there is a close analogy between these superpositions and Bloch states in solid-state physics.*

50.a By using the temporal gauge $A^0 = 0$, we circumvent the problem that A^0_a has a vanishing conjugate momentum, because A^0 is not a dynamical variable in this gauge. Regarding the other components of the gauge potential, the conjugate momentum Π^i of A^i is given by

$$\Pi^i_a \equiv \frac{\partial \mathcal{L}_{\text{YM}}}{\partial \partial_0 A^i_a} = \partial_0 A^i_a.$$

It is convenient to introduce the chromo-electric and chromo-magnetic fields by

$$E^i_a \equiv F^{0i}_a = \Pi^i_a \text{ (in } A^0 = 0 \text{ gauge)}, \quad B^i_a \equiv \tfrac{1}{2}\epsilon_{ijk}\, F^{jk}_a,$$

in terms of which the Lagrangian density can be written as follows:

$$-\frac{1}{4}F^{\mu\nu}_a F^a_{\mu\nu} = -\frac{1}{4}\Big(-F^{0i}_a F^{0i}_a - F^{i0}_a F^{i0}_a + F^{ij}_a F^{ij}_a\Big) = \frac{1}{4}\Big(2E^i_a E^i_a - \epsilon_{ijk}\epsilon_{ijl} B^k_a B^l_a\Big)$$
$$= \frac{1}{4}\Big(2E^i_a E^i_a - (\delta_{jj}\delta_{kl} - \delta_{jk}\delta_{jl}) B^k_a B^l_a\Big) = \frac{1}{2}\left(E^i_a E^i_a - B^i_a B^i_a\right).$$

Then, the Hamiltonian is given by

$$\mathcal{H} = \underbrace{\Pi^i_a(\partial_0 A^i_a)}_{E^i_a E^i_a} - \mathcal{L} = \frac{1}{2}\left(E^i_a E^i_a + B^i_a B^i_a\right) = \frac{1}{2}\Pi^i_a \Pi^i_a + \frac{1}{4}F^{ij}_a F^{ij}_a.$$

(Note that this is also equivalent to the component T^{00} of the energy–momentum tensor in Yang–Mills theory.) The Hamilton equations are

$$\partial_0 A^i_a = \frac{\partial \mathcal{H}}{\partial \Pi^i_a} = \Pi^i_a, \quad \partial_0 \Pi^i_a = -\frac{\partial \mathcal{H}}{\partial A^i_a} = -\partial_j F^{ji}_a - f^{abc} A^b_j F^{ji}_c.$$

The first equation is equivalent to the definition of the conjugate momentum, while the second one reads

$$\partial_0 F^{0i}_a + \left[D_j, F^{ji}\right]_a = 0,$$

which is indeed equivalent to the Yang–Mills equations in the vacuum.

Note, however, that the first Yang–Mills equation,

$$\left[D_\mu, F^{\mu 0}\right] = -\underbrace{\left[D_i, E^i\right]}_{\equiv G[A]} = 0,$$

does not appear among the Hamilton equations. In the Lagrangian formalism, it would appear as the variation of the action with respect to A^0. But when we set $A^0 = 0$ as the gauge condition, this equation has to be imposed separately as a constraint. Note also that $G[A] = 0$ is the non-Abelian generalization of Gauss's law, $\partial_i E^i = 0$. In addition, one may check that the Poisson bracket of $G[A]$ with the Hamiltonian is zero, which implies that the condition $G[A] = 0$ is preserved by the time evolution. In the corresponding quantum theory, we have $\big[\mathcal{H}, G[A]\big] = 0$, and the physical quantum states should be annihilated by the operator $G[A]$ (this is the quantum analogue of $G[A] = 0$ for classical solutions).

50.b The gauge condition $A^0 = 0$ does not completely remove the gauge freedom, since it is still possible to perform purely spatial gauge transformations (they preserve $A^0 = 0$). An infinitesimal gauge transformation of parameter $\vartheta_a(x)$ leads to the following variation of the gauge field:

$$\delta A^i_a(x) = -(D^i_{adj})_{ab}\vartheta_b(x).$$

Therefore, a generic functional $F[A]$ varies as follows:

$$\delta F[A] = \underbrace{-\int d^3x \, \Big((D^i_{adj})_{ab}\vartheta_b(x)\Big)\frac{\delta}{\delta A^i_a(x)}}_{\equiv\, iQ_\vartheta} \, F[A].$$

Note that the derivative $i\delta/\delta A^i_a$ acting on an object in the "coordinate" representation is equivalent to the quantum mechanical operator E^i_a, since it is the canonical conjugate of A^i_a. Therefore, the generator Q_ϑ may also be written as

$$Q_\vartheta = \int d^3x \, \Big((D^i_{adj})_{ab}\vartheta_b(x)\Big)E^i_a(x) = -\int d^3x \, \vartheta_b(x) \, \underbrace{(D^i_{adj})_{ba}E^i_a}_{=\, G[A]_b}.$$

The second equality is valid only for gauge transformations such that $\vartheta_a(x)$ vanishes when $|x| \to \infty$ since it requires an integration by parts. Under this gauge transformation, a quantum state $|\psi\rangle$ transforms as

$$|\psi\rangle \to e^{iQ_\vartheta}\,|\psi\rangle.$$

Therefore, if $|\psi\rangle$ is a physical state annihilated by $G[A]$, it is also invariant under spatial gauge transformations $\Omega(x) \equiv \exp(i\vartheta(x)t^a)$ that go to the identity at infinity.

50.c Now consider the ground states. It is tempting to say that they are the field configurations for which both $E^i_a = B^i_a = 0$ (since this minimizes the classical energy), i.e., pure gauge fields,

$$A^i_U(x) \equiv i\,U^\dagger\partial^i U,$$

with U some element of the gauge group. Let us denote by $|U\rangle$ the state whose wavefunction in the "coordinate representation" is $\delta[A - A_U]$. This state is not gauge invariant (and therefore is not annihilated by $G[A]$, and is not physical), but we may construct a physical state from it by summing over all the possible gauge transformations that go to the identity at infinity:

$$|U\rangle_{phys} \equiv \int \prod_a \,[D\vartheta_a(x)]\; e^{iQ_\vartheta}\,|U\rangle. \tag{3.56}$$

(This state is by construction invariant under gauge transformations that go to the identity at infinity, which also ensures that it is annihilated by Gauss's law.) The set of pure gauge fields can be divided into equivalence classes such that within one class all the fields are related by a gauge transformation that goes to the identity at infinity. Each equivalence class leads to a different physical state defined by the preceding formula.

50.d Now, consider two such states: the state obtained from $U \equiv 1$, $|1\rangle_{\text{phys}}$, and another state $|U\rangle_{\text{phys}}$. We are interested in the transition amplitude between these two states,

$$_{\text{phys}}\langle U|1\rangle_{\text{phys}}.$$

Let us first assume that U does not go to a constant at infinity (this implies that the state $|U\rangle_{\text{phys}}$ has non-zero A^i fields at spatial infinity). To connect the two states, the trajectories that enter into the path integral representation of this transition amplitude must have $E^i = \partial_0 A^i \neq 0$ in an infinite volume for at least a finite amount of time. Therefore, there is an infinite energy barrier between the two states in this case, and the transition amplitude is zero. From this, we conclude that the only allowed transitions from $|1\rangle_{\text{phys}}$ are to states $|U\rangle_{\text{phys}}$ defined by a U that goes to a constant at infinity. When a group element U goes to a constant (this constant can be chosen to be the identity, by applying a global gauge transformation) at $|\boldsymbol{x}| \to +\infty$, we may identify all the points at spatial infinity, which in topological terms means replacing $\mathbb{R}^3$ by a 3-sphere $\mathcal{S}_3$ (this may be done by a stereographic projection). Consequently, the reachable states $|U\rangle_{\text{phys}}$ are defined by U's that map $\mathcal{S}_3$ into the gauge group $\mathcal{G}$. These mappings may be divided into equivalence classes that are the elements of the homotopy group $\pi_3(\mathcal{G})$. In the case of $SU(N)$ Yang–Mills theory, this homotopy group is $\mathbb{Z}$, and the topological index is obtained from the Cartan–Maurer invariant calculated with a representative $U(x)$. The averaging over all gauge transformations e^{iQ_ϑ} that go to the identity at infinity in (3.56) involves only gauge fields that belong to the same topological sector (e^{iQ_ϑ} can be continuously deformed into the identity by taking the limit $\vartheta_a \to 0$; therefore this transformation does not change the topological index). Thus, each value of the topological index corresponds to a distinct ground state.

50.e So far, our discussion has been essentially classical, leading to the conclusion that the Yang–Mills Hamiltonian has a discrete infinity of degenerate minima. A first quantum effect is that these minima, which correspond to a well-defined field configuration (up to a gauge transformation), are "spread out" by the uncertainty principle into states that we shall denote by $|0_n\rangle$ (the integer n is the topological index), in the same way that the quantum ground state in the harmonic potential x^2 is not the state of wavefunction $\delta(x)$ but a Gaussian packet centered at $x = 0$. Let us now discuss how quantum effects lead to the fact that the true eigenstates of the Hamiltonian are certain linear superpositions of the $|0_n\rangle$'s rather than the $|0_n\rangle$'s themselves. Let us introduce the "shift operator" $\mathcal{T}$ that increments the topological index by one unit:

$$\mathcal{T}|0_n\rangle = |0_{n+1}\rangle.$$

Because all the topological sectors are equivalent, $\mathcal{T}$ commutes with the Hamiltonian, $[\mathcal{T}, \mathcal{H}] = 0$, and therefore the true eigenstates $|\psi\rangle$ of the Hamiltonian must also be eigenstates of $\mathcal{T}$:

$$\mathcal{T}|\psi\rangle = \lambda_\psi|\psi\rangle.$$

If we look for $|\psi\rangle$ in the form of a linear superposition of the $|0_n\rangle$'s, it must therefore be of the form

$$\big|\theta\big\rangle \equiv \sum_{n\in\mathbb{Z}} e^{in\theta}\,\big|0_n\big\rangle.$$

Note that these states – called the θ-vacua – are completely analogous to the Bloch states encountered in solid-state physics, i.e., the energy eigenstates of an electron in an infinite periodic crystal lattice. Instead of having discrete electron states confined around each atomic site, the eigenstates of the Hamiltonian form a continuous spectrum of delocalized states (these extended states are responsible for electrical conduction in metals: any electron whose energy lies in this continuous band is able to move over macroscopic distances). In the Yang–Mills case, the eigenstates of the Hamiltonian also form a continuum indexed by the variable θ.

Let us state a few properties of the θ-vacua. Note first that their definition implies that they are 2π-periodic: $\big|\theta+2\pi\big\rangle = \big|\theta\big\rangle$. Then, the overlap of two such states reads

$$\big\langle\theta'\big|\theta\big\rangle = \sum_{m,n\in\mathbb{Z}} e^{i(n\theta-m\theta')}\underbrace{\big\langle 0_m\big|0_n\big\rangle}_{\delta_{mn}} = \sum_{n\in\mathbb{Z}} \delta(\theta-\theta'+2\pi n).$$

Likewise, the Euclidean transition amplitude between two θ-vacua is given by

$$\begin{aligned}\big\langle\theta'\big|e^{-T\mathcal{H}}\big|\theta\big\rangle &= \sum_{m,n\in\mathbb{Z}} e^{i(n\theta-m\theta')}\underbrace{\big\langle 0_m\big|e^{-T\mathcal{H}}\big|0_n\big\rangle}_{f(m-n)} = \sum_{m,p\in\mathbb{Z}} e^{in(\theta-\theta')}e^{-ip\theta'}\,f(p)\\ &= \widetilde{f}(\theta)\sum_{n\in\mathbb{Z}}\delta(\theta-\theta'+2\pi n).\end{aligned}$$

(In the semi-classical approximation, $\big\langle 0_m\big|e^{-T\mathcal{H}}\big|0_n\big\rangle$ depends only on the difference $m-n$ because the transition between the states m and n is due to the instanton of index $m-n$, as shown in Problem 49.) Since transitions between distinct (modulo 2π) θ-vacua are forbidden, they should be viewed as defining distinct theories (i.e., θ is not a dynamical quantity but merely a parameter that defines the ground state of the theory). In the semi-classical approximation, using the results of Problems 48 and 49, the energy $E(\theta)$ of θ-vacua may be estimated as follows:

$$\frac{\big\langle\theta\big|e^{-T\mathcal{H}}\big|\theta\big\rangle}{\big\langle\theta\big|\theta\big\rangle} = e^{-TE(\theta)} = \sum_{n\in\mathbb{Z}} e^{-in\theta}\underbrace{\big\langle 0_n\big|e^{-T\mathcal{H}}\big|0_0\big\rangle}_{\approx\, c_n T^n\, e^{-|n|S[A_1]}} \approx c_0 + 2c_1 T\cos(\theta)e^{-S[A_1]}+\cdots,$$

where $S[A_1]$ is the action of the instanton of index $n=1$. (The dots denote suppressed terms involving multi-instanton contributions.) This gives

$$E(\theta) \approx \text{const} - \frac{2c_1}{c_0}\cos(\theta)\,e^{-S[A_1]}+\cdots.$$

From their origin, the constants $c_{0,1}$ are positive, implying that the minimal energy is obtained for $\theta = 0$. The splitting between the minimal and maximal energy is proportional to the tunneling amplitude between successive vacua, p and $p+1$, which is controlled by the instanton action.

Selecting a particular vacuum $|\theta\rangle$ is done by adding a θ-dependent term,

$$\mathcal{L}_\theta \equiv \frac{g^2\theta}{64\pi^2}\epsilon_{\mu\nu\rho\sigma}F_a^{\mu\nu}F_a^{\rho\sigma}$$

(called the θ-term), to the Yang–Mills Lagrangian. In the path integral, the effect of this term is to weight gauge field configurations of topological index n by a factor $e^{in\theta}$, but it has no perturbative effects since it is a total derivative. The θ angle has non-perturbative effects on some physical observables such as the neutron electric dipole moment, whose present experimental upper limit indicates that $|\theta| \leq 10^{-10}$. The smallness of this parameter, combined with the lack of a symmetry principle that would force it to be zero, is known as the *strong CP problem*. One of the most popular ideas for avoiding this problem is the so-called *axion mechanism*, in which θ is promoted to a dynamical field, whose natural trend is to relax towards the minimum of $E(\theta)$ located at $\theta = 0$. Excitations of θ about $\theta = 0$ imply the existence of a new particle, called the axion.

CHAPTER 4

Scattering Amplitudes

Introduction

This chapter is devoted to novel methods (spinor-helicity formalism, on-shell recursion, generalized unitarity) developed in the past 20 years for computing scattering amplitudes, which bypass the traditional workflow (Lagrangian $\to$ Feynman rules $\to$ diagrams $\to$ amplitudes via LSZ formulas). The recurring theme of this approach is to avoid, as much as possible, direct references to the underlying Lagrangian, whose gauge invariance is a source of unnecessary redundancies. This chapter covers both the case of tree-level amplitudes (with some extensions to treat a few examples with scalar particles or fermions) and a few problems about one-loop amplitudes (this is limited to some rather simple examples, since calculating loop amplitudes by hand remains a challenging task, even with modern tools).

Color Ordering

The first step in organizing the calculation of scattering amplitudes in Yang–Mills theories is to separate the dependence on the color indices from the kinematical dependence. For purely gluonic amplitudes at tree level, this decomposition reads

$$\mathcal{M}_n^{\text{tree}}(1\cdots n) \equiv 2 \sum_{\sigma\in\mathfrak{S}_n/\mathbb{Z}_n} \text{tr}\,(t_f^{a_{\sigma_1}}\cdots t_f^{a_{\sigma_n}})\,\mathcal{A}_n^{\text{tree}}(\sigma_1\cdots\sigma_n), \tag{4.1}$$

where σ is a permutation of $[1, n]$ (only permutations that are not equivalent up to a cyclic permutation need to be considered) and $\mathcal{A}_n^{\text{tree}}$ is a tree-level *cyclic-ordered partial amplitude*. Color-ordered amplitudes can be obtained from color-stripped Feynman rules (see below), and contain only planar graphs with a specific ordering of the external gluon lines.

At one loop, a similar but more involved decomposition exists:

$$\begin{aligned}\mathcal{M}_n^{1-\text{loop}}(1\cdots n) &= N \sum_{\sigma\in\mathfrak{S}_n/Z_n} \operatorname{tr}\left(t_f^{a_{\sigma_1}}\cdots t_f^{a_{\sigma_n}}\right) \mathcal{A}_{n;0}^{1-\text{loop}}(\sigma_1\cdots\sigma_n)\\ &+ \sum_{\substack{c=2\\ \sigma\in\mathfrak{S}_n/Z_{c;n-c}}}^{\lfloor\frac{n}{2}\rfloor} \operatorname{tr}\left(t_f^{a_{\sigma_1}}\cdots t_f^{a_{\sigma_c}}\right)\operatorname{tr}\left(t_f^{a_{\sigma_{c+1}}}\cdots t_f^{a_{\sigma_n}}\right) \mathcal{A}_{n;c}^{1-\text{loop}}(\sigma_1\cdots\sigma_n),\end{aligned} \tag{4.2}$$

where $\lfloor\frac{n}{2}\rfloor$ is the integral part of $n/2$ and $Z_{c;n-c}$ is the set of products of cyclic permutations of c elements by cyclic permutations of $n-c$ elements. The second sum is over all permutations of n elements that do not leave the two traces invariant. The $\mathcal{A}_{n;0}^{1-\text{loop}}$, called the *leading-color partial amplitude*, are the sum of the corresponding color-ordered one-loop planar diagrams. The subleading partial amplitudes $\mathcal{A}_{n;c}$ with $c \geq 2$ can be expressed as sums of permutations of the $\mathcal{A}_{n;0}$.

Color-Stripped Feynman Rules

Color-ordered partial amplitudes (only the leading-color ones at one loop) are obtained as sums of planar graphs with the following Feynman rules (the propagator below is written in covariant gauge):

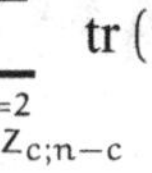

$$= \frac{-i\,g^{\mu\nu}}{p^2+i0^+} + \frac{i}{p^2+i0^+}\left(1-\frac{1}{\xi}\right)\frac{p^\mu p^\nu}{p^2}, \tag{4.3}$$

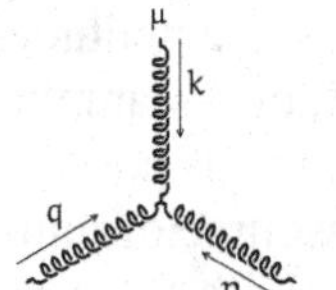

$$= \begin{aligned} g\,\{&g^{\mu\nu}(k-p)^\rho\\ &+g^{\nu\rho}(p-q)^\mu\\ &+g^{\rho\mu}(q-k)^\nu\},\end{aligned}$$

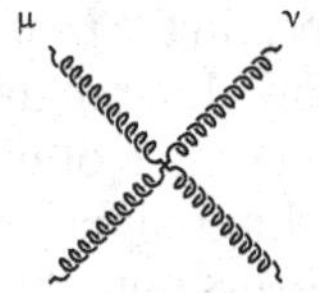

$$= \begin{aligned} -i\,g^2\,(&2\,g^{\mu\rho}g^{\nu\sigma}\\ &-g^{\mu\sigma}g^{\nu\rho}\\ &-g^{\mu\nu}g^{\rho\sigma}).\end{aligned} \tag{4.4}$$

In addition to having simpler Feynman rules, the color ordering also considerably reduces the number of contributing graphs.

Spinor-Helicity Formalism

The complexity of the expression for scattering amplitudes can be considerably reduced by using the *spinor-helicity formalism*. Every 4-momentum can be mapped into a 2×2 matrix as follows:

$$p_\mu \quad\rightarrow\quad P_{ab} \equiv p_\mu \sigma^\mu_{ab}, \quad \text{with } \sigma^\mu \equiv (1,\sigma^i), \quad \sigma^{1,2,3} = \text{Pauli matrices}. \tag{4.5}$$

For on-shell massless momenta, $p_\mu p^\mu = 0$, this matrix has rank 1 and factorizes as

$$P_{ab} = {}_a\big|p\big]\big\langle p\big|_b, \tag{4.6}$$

where $\langle p|_b$ and ${}_a|p]$ are two-component spinors that admit the following explicit representation:

$$|p] = \begin{pmatrix} \sqrt{p_0 + p_3} \\ \frac{p_1 + ip_2}{\sqrt{p_0+p_3}} \end{pmatrix}, \qquad \langle p| = \begin{pmatrix} \sqrt{p_0 + p_3} & \frac{p_1 - ip_2}{\sqrt{p_0+p_3}} \end{pmatrix}. \tag{4.7}$$

Note also the alternative matrix representation:

$$p_\mu \to \overline{p}^{ab} = {}^a|p\rangle [p|^b = -p_\mu (\overline{\sigma}^\mu)^{ab}, \tag{4.8}$$

with $\overline{\sigma}^\mu \equiv (1, -\sigma^i)$, $[p|^a = {}_b|p]\epsilon^{ba}$ and ${}^a|p\rangle = \epsilon^{ab}\langle p|_b$. For a real momentum, the above spinors satisfy $|p]^\dagger = \langle p|$ and $|p\rangle^\dagger = -[p|$ (these identities do not hold for complex momenta). Given two on-shell momenta p, q, we have $(p+q)^2 = \langle pq\rangle [pq]$.

The expression for momentum conservation $\sum_i p_i^\mu = 0$ in this formalism reads

$$\sum_i |p_i\rangle [p_i| = 0. \tag{4.9}$$

The *Schouten identities*,

$$|p\rangle\langle qr\rangle + |q\rangle\langle rp\rangle + |r\rangle\langle pq\rangle = 0, \quad |p][qr] + |q][rp] + |r][pq] = 0, \tag{4.10}$$

are also often useful for simplifying expressions.

In the spinor-helicity formalism, one usually works with explicit polarization vectors in the helicity basis, which read

$$\epsilon_+^\mu(p;q) \equiv \frac{\langle q|\overline{\sigma}^\mu|p]}{\sqrt{2}\langle qp\rangle}, \quad \epsilon_-^\mu(p;q) \equiv \frac{\langle p|\overline{\sigma}^\mu|q]}{\sqrt{2}[pq]}, \tag{4.11}$$

where q is an auxiliary vector (changing q alters the polarization vectors by unphysical terms proportional to p^μ).

Three-Point Amplitudes, Little-Group Scaling

Three-point amplitudes cannot exist with on-shell massless real momenta, but do exist with complex momenta. Moreover, one may show that they are expressed solely in terms of angle brackets, or solely in terms of square brackets.

The representation (4.6) is invariant under the rescaling $|p] \to \lambda^{-1}|p]$, $\langle p| \to \lambda\langle p|$, reminiscent of the little-group transformations that leave p^μ invariant. Likewise, the polarization vectors transform as $\epsilon_+^\mu(p;q) \to \lambda^{-2}\epsilon_+^\mu(p;q)$, $\epsilon_-^\mu(p;q) \to \lambda^2\epsilon_-^\mu(p;q)$. This implies one of the following two forms for a three-point amplitude (for massless particles):

$$\mathcal{A}_3(1^{h_1}2^{h_2}3^{h_3}) = \begin{cases} \text{const} \times \langle 12\rangle^{h_3-h_1-h_2}\langle 23\rangle^{h_1-h_2-h_3}\langle 31\rangle^{h_2-h_3-h_1} \\ \text{const} \times [12]^{-h_3+h_1+h_2}[23]^{-h_1+h_2+h_3}[31]^{-h_2+h_3+h_1} \end{cases}. \tag{4.12}$$

The expected dimension of $\mathcal{A}_3$ allows us to disambiguate between the two forms, and the prefactor is determined by inspection of the relevant terms in the Lagrangian. For three-gluon amplitudes in Yang–Mills theory, we have

$$\mathcal{A}_3(1^-2^-3^+) = \sqrt{2}\,g\,\frac{\langle 12\rangle^3}{\langle 23\rangle\langle 31\rangle}, \quad \mathcal{A}_3(1^+2^+3^-) = \sqrt{2}\,g\,\frac{[12]^3}{[23]\,[31]},$$
$$\mathcal{A}_3(1^-2^-3^-) = \mathcal{A}_3(1^+2^+3^+) = 0. \tag{4.13}$$

Similarly simple expressions exist for three-point amplitudes with photons and/or gravitons. The vanishing amplitudes in the second line are a special case of

$$\mathcal{A}_n(1^+\cdots n^+) = \mathcal{A}_n(1^-\cdots n^-) = 0 \quad \text{at tree level.} \tag{4.14}$$

For $n \geq 4$, we also have

$$\mathcal{A}_n(1^-2^+\cdots n^+) = \mathcal{A}_n(1^+2^-\cdots n^-) = 0 \quad \text{at tree level.} \tag{4.15}$$

The first non-zero amplitudes are the *maximally helicity-violating* (MHV) amplitudes, which have two helicities of one kind, and all other helicities of the other kind. The tree MHV amplitudes are given explicitly by the *Parke–Taylor formula*:

$$\mathcal{A}_n\underbrace{(\cdots i^-\cdots j^-\cdots)}_{\text{all unwritten hel. are }+} = (\sqrt{2}\,g)^{n-2}\,i^{n-3}\,\frac{\langle ij\rangle^4}{\langle 12\rangle\langle 23\rangle\langle 34\rangle\cdots\langle (n-1)n\rangle\langle n1\rangle}. \tag{4.16}$$

(There is a similar formula, with square brackets, for amplitudes with 2 positive helicities and $n-2$ negative ones.)

Britto-Cachazo-Feng-Witten On-Shell Recursion

The Britto–Cachazo–Feng–Witten (BCFW) method consists in shifting the momenta of two external lines i, j as follows:

$$p_i^\mu \to \widehat{p}_i^\mu \equiv p_i^\mu + zk^\mu, \quad p_j^\mu \to \widehat{p}_j^\mu \equiv p_j^\mu - zk^\mu, \quad \text{with } \mathbf{K} \equiv \big|j\big]\big\langle i\big|,\ z \in \mathbb{C}. \tag{4.17}$$

The shifted momenta fulfill momentum conservation and are on-shell for all z. By integrating the z-dependent tree amplitude $\mathcal{A}_n$ resulting from this shift, divided by z, over a circle of large radius in the complex plane, one obtains

$$\mathcal{A}_n(1\cdots n) = \sum_{\text{poles } z_*}\sum_{h=\pm} \mathcal{A}_{\text{L}}(\cdots\widehat{\imath}\cdots\widehat{K}_*^h)\,\frac{i}{K_*^2}\,\mathcal{A}_{\text{R}}(-\widehat{K}_*^{-h}\cdots\widehat{\jmath}\cdots). \tag{4.18}$$

In this formula, the poles z_* of the shifted amplitude result from the vanishing denominators of the propagators that connect the external lines i and j (there is a single path from i to j in a tree diagram). Each singular propagator divides the original amplitude into a left part $\mathcal{A}_{\text{L}}$ (which contains the external leg i) and a right part $\mathcal{A}_{\text{R}}$ (which contains the external leg j). K_* is the momentum of the singular propagator before the shift, and all the shifted lines (indicated by a hat) must be evaluated at the pole z_* (they are thus on-shell). The left and right amplitudes on the right-hand side of Eq. (4.18) are on-shell amplitudes with a number of external legs strictly less than n. Therefore, this formula allows a recursive construction of on-shell amplitudes, starting from the three-point ones.

There is an important condition for the validity of Eq. (4.18): the shifted amplitude must go to zero when $|z| \to \infty$ in order to avoid a contribution from the circle at infinity. In Yang–Mills theory, with the shift defined in Eqs. (4.17), this is realized for the helicity assignments $(h_i, h_j) \in \{(++), (--), (-+)\}$ but not for $(+-)$.

When applying BCFW recursion in order to calculate a tree amplitude, one may choose any pair of external lines i, j to which to apply the shift (provided the above condition is fulfilled). However, some choices may be better than others in that they produce results with fewer terms.

Reduction of One-Loop Integrals

At one loop, after the color decomposition discussed above, one is left with the evaluation of leading-color partial amplitudes $\mathcal{A}_{n;0}^{1-\text{loop}}$, which can be written as loop integrals of rational functions (each denominator being made up of the denominators of the propagators around the loop, and each numerator containing the momentum dependence coming from three-gluon vertices and ghost–gluon couplings). Any loop integral of this kind can be decomposed into a sum of *scalar integrals* (i.e., with a constant numerator) with *one, two, three, four and five denominators* (the last of these appears only when the dimension of the loop momentum is extended to $d \neq 4$, to cope with dimensional regularization), called *master integrals.*

These master integrals are all well known, and the main task is therefore to obtain the coefficients in the decomposition. There are well-known algorithms (*Passarino–Veltman, Van Neerven–Vermaseren*) for this purpose, but applying them to all the Feynman graphs contributing to the amplitude of interest is not practical except in the simplest cases. *Generalized unitarity* (see below) is an alternative approach that gives these coefficients in terms of tree-level on-shell amplitudes, for which there are powerful methods that avoid evaluating Feynman graphs.

Ossola–Papadopoulos–Pittau Decomposition

The reduction described above works at the level of the integral, i.e., for the amplitude itself. A similar reduction also exists for the *integrand* $\mathcal{I}(\ell)$, which takes the form

$$\mathcal{I}(\ell) = \sum_{i_1} \frac{P_{i_1}(\ell)}{D_{i_1}} + \sum_{i_1,i_2} \frac{P_{i_1 i_2}(\ell)}{D_{i_1} D_{i_2}} + \cdots + \sum_{i_1,\cdots,i_5} \frac{P_{i_1 \ldots i_5}(\ell)}{D_{i_1} \cdots D_{i_5}}, \tag{4.19}$$

where $D_i \equiv (\ell + q_i)^2$ (with ℓ the loop momentum and q_i a combination of external momenta) is a denominator, and the $P_{\ldots}(\ell)$ are polynomials with a known structure (but coefficients to be determined, which depend on the amplitude under consideration).

Generalized Unitarity

Fully determining a decomposition such as (4.19) can be done by evaluating the equality at a set of values of the loop momentum ℓ (the number of which should equal the number of unknown coefficients in the numerators on the right-hand side). The method of *generalized unitarity* consists in choosing values of the loop momentum that make certain denominators vanish. In four-dimensional spacetime, the loop momentum has five independent components if one uses dimensional regularization (because all external momenta and polarization vectors are kept four-dimensional, the integrand depends on the extra dimensional components $\ell_\perp^\mu$ only via $\ell_\perp^2$). Therefore, one can make at most five denominators vanish simultaneously. By choosing a loop momentum ℓ_*^μ such that $D_{i_1} = D_{i_2} = D_{i_3} = D_{i_4} = D_{i_5} = 0$, one obtains

$$P_{i_1 \ldots i_5}(\ell_*) = D_{i_1} \cdots D_{i_5} \mathcal{I}(\ell_*). \tag{4.20}$$

The right-hand side of this equality is finite (the five denominators multiplying $\mathcal{I}(\ell_*)$ precisely cancel its singularities when evaluated at ℓ_*). Moreover, this combination is a product of five tree-level on-shell amplitudes, which can be determined without having to evaluate the corresponding Feynman diagrams. After all the terms with five denominators have been evaluated, one chooses loop momenta at which four denominators vanish, which gives the polynomials in the numerators of terms with four denominators from products of four on-shell tree amplitudes, etc. (Algebraic geometry indicates that the number of unknown coefficients in these polynomials is related to the number of independent ℓ_* at which the denominators vanish, ensuring that there are always enough solutions to determine all the coefficients.)

Berends–Giele Recursion

In some implementations of generalized unitarity, one needs to evaluate tree amplitudes in dimensions that differ from four. For this, the BCFW recursion is not well suited since the spinor-helicity formalism is very much tied to four dimensions. An alternative is the *Berends–Giele recursion*, which applies in any dimension to color-ordered tree-level amplitudes $J^\mu(1, \ldots, n)$ with one off-shell leg and n on-shell legs:

$J^\mu(1, \ldots, n) =$ 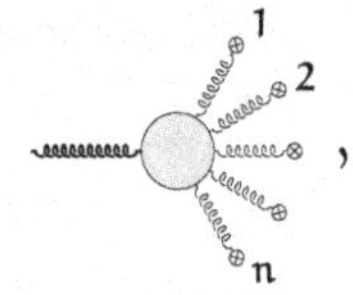 , (4.21)

where each gluon terminated by a circled cross indicates an on-shell line contracted with a physical polarization vector. Diagrammatically, the Berends–Giele recursion reads

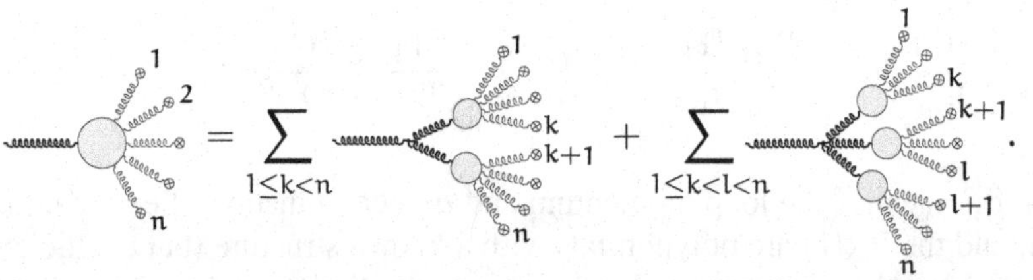

(4.22)

The starting point of the recursion is $J^\mu(1)$, i.e., a single polarization vector $\epsilon^\mu(p_1)$.

About the Problems of this Chapter

- **Problem 51** is devoted to counting the tree graphs that contribute to n-point amplitudes in various quantum field theories (scalar field theories, Yang–Mills theory). In the latter case, the method is extended to counting color-ordered graphs. **Problem 64** generalizes this to one-loop graphs.
- In **Problem 52**, we apply the BCFW recursion in order to obtain an expression with two terms only for the six-point $1^-2^-3^-4^+5^+6^+$ tree amplitude.
- Some frequently used identities for traces in the spinor-helicity formalism are derived in **Problem 53**.
- Various properties of polarization vectors in the spinor-helicity formalism are established in **Problems 54** and **55**.
- **Problem 56** discusses the important *photon decoupling identity*.
- In **Problem 57**, we calculate some tree amplitudes with gluons and a pair of colored scalar and antiscalar particles. These results will be used to derived some one-loop amplitudes by the generalized unitarity method in **Problems 67** and **68**.
- The color decomposition of tree amplitudes with gluons and a quark–antiquark pair is discussed in **Problem 58**. In **Problem 59**, we explain how to deal with massless quarks in the spinor-helicity formalism, and determine the expression for quark–antiquark–gluon amplitudes. Four- and five-point amplitudes containing a quark–antiquark pair are calculated in **Problem 60**.
- In **Problem 61**, we apply the spinor-helicity formalism and the Parke–Taylor formula in order to derive the expression for the $gg \to gg$ tree-level scattering amplitude. Along the way, we obtain the *Kleiss–Kuijf relation* for a simple case.
- **Problem 62** studies the limit where a gluon becomes soft in an amplitude (this is the spinor-helicity analogue of the eikonal approximation), and in **Problem 63** the collinear limit is considered.
- One-loop amplitudes are complicated to calculate, even with modern methods. For this reason, we restrict ourselves to particularly simple examples. In **Problems 65** and **66**, we show that one-loop amplitudes with only gluons of positive helicity are ultraviolet finite and rational.
- In **Problem 67**, we use the generalized unitarity method to obtain a very simple expression for the $1^+2^+3^+4^+$ amplitude with a scalar loop. This is extended in **Problem 68** to the one-loop $1^+2^+3^+4^+5^+$ amplitude.

51. Enumeration of Tree Graphs The goal of this problem is to count the Feynman graphs contributing to tree amplitudes in scalar field theory and in Yang–Mills theory. The basic observation is that counting tree graphs can be achieved with dummy Feynman rules in which all the propagators and vertices are equal to 1, so that each graph contributes the number 1.

51.a Consider first a scalar ϕ^3 theory. Explain why the number of tree graphs contributing to the n-point amplitude is the coefficient of $j^{n-1}/(n-1)!$ in the solution of $X = j + X^2/2$. Solve this equation and calculate the first few coefficients. Show that the number of n-point tree graphs in this theory is $(2n-5)!! = 1 \cdot 3 \cdots (2n-5)$.

51.b Now consider Yang–Mills theory, and write the equation whose solution gives the number of graphs contributing to the tree-level n-gluon amplitude. Perform a Taylor expansion of the solution of this equation to obtain the numbers of graphs contributing to tree amplitudes in Yang–Mills theory.

51.c Likewise, explain how the numbers of *color-ordered* tree graphs in Yang–Mills theory can be obtained from the Taylor coefficients of the solution of the equation $X = j + X^2 + X^3$.

51.a Let us start with ϕ^3 theory at tree level. The solution $X(j)$ of $X = j + X^2/2$ can be represented graphically as the sum of all the tree graphs with trivalent vertices, where X stands at the root of the graph, and a j is attached to each leaf. The vertices and edges of these graphs carry no structure (i.e., they are assigned the trivial "Feynman rule" 1). Therefore, the prefactor of the term of order j^n in this diagrammatic expansion is the number of tree graphs with $n+1$ *labeled* external legs, divided by $n!$ because the n sources are identical:

$$X(j) = \sum_{n=1}^{\infty} \begin{pmatrix} \text{number of tree graphs} \\ \text{with } 1+n \text{ labeled legs} \end{pmatrix} \frac{j^n}{n!}. \tag{4.23}$$

The external legs of scattering amplitudes are labeled, since a different momentum is attached to each of them. Therefore, the number of graphs contributing to the n-point tree amplitude in a scalar theory with cubic interactions is the coefficient of $j^{n-1}/(n-1)!$ in $X(j)$. In the present case, the equation $X = j + X^2/2$ can be solved analytically:

$$X(j) = 1 \pm \sqrt{1-2j}.$$

We must take the solution $1 - \sqrt{1-2j}$ since this is the one that has positive Taylor coefficients:

$$1 - \sqrt{1-2j} = \frac{j}{1!} + \frac{j^2}{2!} + \frac{3\,j^3}{3!} + \frac{15\,j^4}{4!} + \frac{105\,j^5}{5!} + \frac{945\,j^6}{6!} + \frac{10,395\,j^7}{7!} + \cdots.$$

From this expansion, we read the number of tree graphs with n external points in this theory, listed in Table 4.1. In fact, we can even obtain an analytic formula for the Taylor coefficients,

$$X(j) = \sum_{n=1}^{\infty} \frac{(2n-2)!}{2^{n-1}(n-1)!}\frac{j^n}{n!},$$

so that the number of n-point tree graphs in a theory with cubic interactions is $(2n-4)!/(2^{n-2}(n-2)!) = (2n-5)!!$.

Number of external legs	Tree graphs in ϕ^3 theory	Tree graphs in Yang–Mills theory
2	1	1
3	1	1
4	3	4
5	15	25
6	105	220
7	945	2485
8	10,395	34,300
9	135,135	559,405
10	2,027,025	10,525,900

Table 4.1 Number of graphs contributing to n-point tree amplitudes in ϕ^3 scalar field theory and in Yang–Mills theory.

51.b For Yang–Mills theory, which has both cubic and quartic couplings, the reasoning is the same, but the equation one needs to solve is now

$$X = j + \frac{X^2}{2} + \frac{X^3}{6}. \tag{4.24}$$

This is a cubic algebraic equation, which could in principle be solved in closed analytical form by using Cardano's formulas. But this approach leads to very complicated expressions. Alternatively, one may simply solve the equation iteratively, starting from $X_{(1)} \equiv j$ and injecting the result of order $n-1$ into the right-hand side of the equation that gives the order n:

$$X_{(n)} = j + \frac{X^2_{(n-1)}}{2} + \frac{X^3_{(n-1)}}{6}\bigg|_{\text{order} \leq n}.$$

(The subscript means that we truncate the right-hand side to keep only terms up to order j^n; higher-order terms could be kept, but would not be correct.) Going up to the 9th order, this gives

$$X_{(9)} = j + \frac{j^2}{2!} + \frac{4\,j^3}{3!} + \frac{25\,j^4}{4!} + \frac{220\,j^5}{5!} + \frac{2485\,j^6}{6!} + \frac{34,300\,j^7}{7!} + \frac{559,405\,j^8}{8!} + \frac{10,525,900\,j^9}{9!},$$

from which we can read the number of tree graphs up to $n = 10$ external points (see the last column of Table 4.1).

Additional Note: The above results can be justified in a different way as follows. First, we define a zero-dimensional "quantum field theory" by its action:

$$S(X) \equiv \frac{X^2}{2} - \frac{X^3}{3!} \quad (\phi^3 \text{ theory}), \qquad S(X) \equiv \frac{X^2}{2} - \frac{X^3}{3!} - \frac{X^4}{4!} \quad (\text{Yang–Mills theory}).$$

Note that the propagators and vertices associated with this action are all equal to 1. Then, we define the generating function

$$Z(j) \equiv e^{W(j)} \equiv \int dX\, e^{S(X)-jX}.$$

Because the Feynman rules are all equal to 1, the Taylor coefficients of $W(j)$ merely count the number of connected graphs (both tree-level graphs and graphs with loops). In order to isolate the tree-level graphs, it is useful to introduce the quantum effective action $\Gamma(X)$ by

$$\Gamma(X) \equiv W(j) + jX(j), \quad \text{with } \Gamma'(X(j)) = j.$$

At tree level, we have $\Gamma(X) = S(X)$ and therefore

$$W(j) \underset{\text{tree}}{=} S(X(j)) - jX(j), \quad \text{with } S'(X(j)) = j.$$

Since in our earlier discussion one of the legs of the tree graphs was singled out, the corresponding generating series is in fact the first derivative of $W(j)$. At tree level, we have

$$\partial_j W(j) \underset{\text{tree}}{=} \partial_j(X(j))\underbrace{\big(S'(X(j)) - j\big)}_{=0} + X(j) = X(j).$$

51.c Let us now consider the case of tree-level color-ordered graphs in Yang–Mills theory. The only difference from the non-ordered case is that we are not allowed to reshuffle the labels on the external legs, since we must preserve their ordering, which has the effect of removing the symmetry factors. Thus, by replacing Eq. (4.24) with

$$Y = j + Y^2 + Y^3,$$

we should have

$$Y(j) = \sum_{n=1}^{\infty} \begin{pmatrix} \text{number of color-ordered} \\ \text{tree graphs with } 1+n \text{ legs} \end{pmatrix} j^n.$$

Note that, unlike in (4.23), there is no factor $1/n!$ here since the order of the n factors j matters (one may view j as a vector, and j^n as the product of its first n components: $j^n \to j_1 j_2 \cdots j_n$). As before, we solve this equation iteratively:

$$Y_{(n)} = j + Y_{(n-1)}^2 + Y_{(n-1)}^3 \Big|_{\text{order} \le n},$$

which gives $Y_{(9)} = j + j^2 + 3j^3 + 10j^4 + 38j^5 + 154j^6 + 654j^7 + 2871j^8 + 12{,}925j^9$. The coefficients of this power series give directly the number of tree-level color-ordered graphs in Yang–Mills theory.

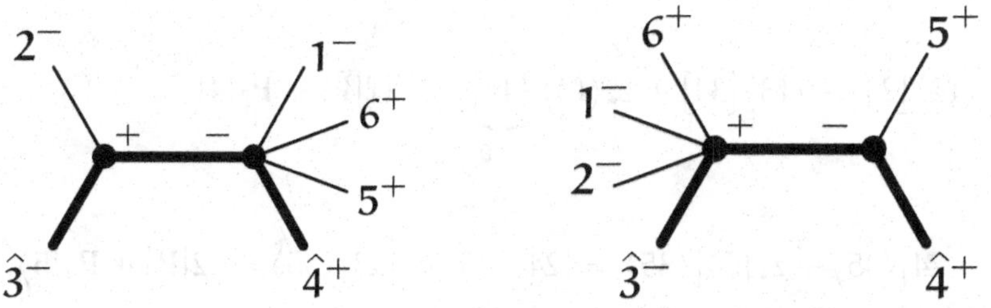

Figure 4.1 Contributions to the six-point color-ordered amplitude $1^-2^-3^-4^+5^+6^+$, with a shift applied to the lines 3 and 4 (the lines carrying the shift are shown in boldface).

52. Six-Point $---+++$ Amplitude Using a BCFW shift on lines 3 and 4, derive an expression for the color-ordered amplitude $1^-2^-3^-4^+5^+6^+$ that has only two terms:

$$\mathcal{A}_6(1^-2^-3^-4^+5^+6^+) = -\frac{4ig^4}{[2|P_3+P_4|5\rangle}\left\{\frac{\langle 1|\overline{P}_2+\overline{P}_3|4]^3}{[23]\,[34]\,\langle 56\rangle\langle 61\rangle(p_2+p_3+p_4)^2} + \frac{\langle 3|\overline{P}_4+\overline{P}_5|6]^3}{[61]\,[12]\,\langle 34\rangle\langle 45\rangle(p_3+p_4+p_5)^2}\right\}.$$

When the shift is applied to lines 3 and 4, there are only two ways to arrange the remaining lines and to choose the internal helicities, as shown in Figure 4.1. Note that the graph on the right contains an anti-MHV amplitude with two positive helicities and three negative ones. For this sub-amplitude, we will accept without proof that the Parke–Taylor formula with square brackets instead of angle brackets is valid. The shift that has good behavior when $|z| \to \infty$ is the one with $|\widehat{4}] = |4]$ and $|\widehat{3}\rangle = |3\rangle$ (and the direction of the shift is a complex momentum k^μ such that $|k\rangle = |3\rangle$ and $|k] = |4]$). Therefore, the sum of these two graphs gives the following expression for the six-point amplitude:

$$\mathcal{A}_6(1^-2^-3^-4^+5^+6^+) = -4ig^4\left\{\frac{1}{s_{23}}\frac{\langle 23\rangle^3}{\langle 3\widehat{K}_I\rangle\langle\widehat{K}_I 2\rangle}\frac{\langle 1\widehat{K}_I\rangle^3}{\langle\widehat{K}_I\widehat{4}\rangle\langle\widehat{4}5\rangle\langle 56\rangle\langle 61\rangle} + \frac{1}{s_{45}}\frac{[45]^3}{[5\widehat{L}_I]\,[\widehat{L}_I 4]}\frac{[\widehat{L}_I 6]^3}{[61]\,[12]\,[2\widehat{3}]\,[\widehat{3}\widehat{L}_I]}\right\},$$

where we denote $s_{23} \equiv (p_2+p_3)^2$, $s_{45} \equiv (p_4+p_5)^2$. We have defined $K_I \equiv -(p_2+p_3)$ in the first term and $L_I = -(p_4+p_5)$ in the second one. Consider the first term, for which the value of z at the pole is given by $0 = \widehat{K}_I^2 = K_I^2 + 2z_* K_I\cdot k = s_{23} - z_*\langle kK_I\rangle[K_I k] = s_{23} + z_*\langle 32\rangle[24]$, which leads to

$$z_* = -\frac{s_{23}}{\langle 32\rangle[24]} = -\frac{\langle 23\rangle[23]}{\langle 32\rangle[24]} = \frac{[23]}{[24]}, \quad |\widehat{4}\rangle = |4\rangle + \frac{[23]}{[24]}|3\rangle.$$

The angle brackets containing $\widehat{K}_I$ can be evaluated as follows. For a generic on-shell momentum

ℓ^μ, we have

$$\langle \ell \widehat{K}_I \rangle [\widehat{K}_I 4] = -\langle \ell 2 \rangle [24] - \langle \ell 3 \rangle [34] + z_* \langle \ell 3 \rangle \underbrace{[44]}_{=0} = -\langle \ell | \overline{P}_2 + \overline{P}_3 | 4].$$

Moreover,

$$\begin{aligned}
[24]\langle \widehat{4}5 \rangle &= [24]\langle 45 \rangle + z_* [24]\langle 35 \rangle = [24]\langle 45 \rangle + [23]\langle 35 \rangle = [2|P_3 + P_4|5\rangle, \\
\langle \widehat{4}|\overline{P}_2 + \overline{P}_3|4] &= \langle 4|\overline{P}_2 + \overline{P}_3|4] + z_* \langle 3|\overline{P}_2 + \overline{P}_3|4] \\
&= \langle 42 \rangle [24] + \langle 43 \rangle [34] - s_{23} = -2\,(p_2 \cdot p_4 + p_3 \cdot p_4 + p_2 \cdot p_3) \\
&= -(p_2 + p_3 + p_4)^2.
\end{aligned}$$

Therefore, the first term can be rewritten as

$$\mathcal{A}_6^{(1)}(1^-2^-3^-4^+5^+6^+) = -\frac{4ig^4}{(p_2+p_3+p_4)^2} \frac{\langle 1|\overline{P}_2 + \overline{P}_3|4]^3}{[23]\,[34]\langle 56 \rangle \langle 61 \rangle [2|P_3 + P_4|5\rangle}.$$

The same method can be used to simplify the second term, leading to the following final result:

$$\mathcal{A}_6(1^-2^-3^-4^+5^+6^+) = -\frac{4ig^4}{[2|P_3+P_4|5\rangle} \left\{ \frac{\langle 1|\overline{P}_2 + \overline{P}_3|4]^3}{[23]\,[34]\langle 56 \rangle \langle 61 \rangle (p_2+p_3+p_4)^2} + \frac{\langle 3|\overline{P}_4 + \overline{P}_5|6]^3}{[61]\,[12]\langle 34 \rangle \langle 45 \rangle (p_3+p_4+p_5)^2} \right\}.$$

This is a remarkably simple expression given the fact that it would be obtained as the sum of 38 color-ordered Feynman graphs in the traditional approach. Non-adjacent configurations of the three negative helicities and the three positive ones lead to more complicated formulas, but it is easy to see that BCFW recursion produces at most six terms in the worst case.

53. Two Spinor-Helicity Identities Define the following four 4×4 matrices:

$$\gamma^\mu \equiv \begin{pmatrix} 0 & \sigma^\mu \\ \overline{\sigma}^\mu & 0 \end{pmatrix}, \quad \text{with } \sigma^\mu \equiv (1, \sigma^i), \quad \overline{\sigma}^\mu \equiv (1, -\sigma^i).$$

53.a Check that these γ^μ are a valid representation of the Dirac matrices in four dimensions.

53.b Show that

$$\begin{aligned}
\mathrm{tr}\,(\overline{\sigma}^\mu \sigma^\nu \overline{\sigma}^\rho \sigma^\sigma) &= 2\left(g^{\mu\nu}g^{\rho\sigma} - g^{\mu\rho}g^{\nu\sigma} + g^{\mu\sigma}g^{\nu\rho} + i\,\epsilon^{\mu\nu\rho\sigma}\right), \\
\mathrm{tr}\,(\sigma^\mu \overline{\sigma}^\nu \sigma^\rho \overline{\sigma}^\sigma) &= 2\left(g^{\mu\nu}g^{\rho\sigma} - g^{\mu\rho}g^{\nu\sigma} + g^{\mu\sigma}g^{\nu\rho} - i\,\epsilon^{\mu\nu\rho\sigma}\right).
\end{aligned}$$

53.c Use this result to prove the following identities (with $s_{ij} \equiv (p_i + p_j)^2$):

$$\begin{aligned}
&[12]\langle 23 \rangle [34] \langle 41 \rangle - \langle 12 \rangle [23] \langle 34 \rangle [41] = 4\,i\,p_{1\mu}p_{2\nu}p_{3\rho}p_{4\sigma}\,\epsilon^{\mu\nu\rho\sigma}, \\
&\frac{\langle 12 \rangle [23]}{\langle 14 \rangle [43]} = \frac{s_{12}s_{34} - s_{13}s_{24} + s_{14}s_{23} - 4\,i\,p_{1\mu}p_{2\nu}p_{3\rho}p_{4\sigma}\,\epsilon^{\mu\nu\rho\sigma}}{2\,s_{14}s_{34}}.
\end{aligned}$$

53.a Let us start from

$$\gamma^\mu \equiv \begin{pmatrix} 0 & \sigma^\mu \\ \overline{\sigma}^\mu & 0 \end{pmatrix}.$$

Using $\sigma^\mu = (1, \sigma^i)$, $\overline{\sigma}^\mu = (1, -\sigma^i)$, an exhaustive check shows that

$$\gamma^\mu\gamma^\nu + \gamma^\nu\gamma^\mu = 2\,g^{\mu\nu}.$$

Therefore, the proposed γ^μ is a valid representation of the Dirac matrices in four dimensions.

53.b Then, one may also check that

$$\gamma^\mu\gamma^\nu\gamma^\rho\gamma^\sigma = \begin{pmatrix} \sigma^\mu\overline{\sigma}^\nu\sigma^\rho\overline{\sigma}^\sigma & 0 \\ 0 & \overline{\sigma}^\mu\sigma^\nu\overline{\sigma}^\rho\sigma^\sigma \end{pmatrix}.$$

Taking the trace of this equation provides a first identity:

$$\begin{aligned} \mathrm{tr}\,(\sigma^\mu\overline{\sigma}^\nu\sigma^\rho\overline{\sigma}^\sigma) + \mathrm{tr}\,(\overline{\sigma}^\mu\sigma^\nu\overline{\sigma}^\rho\sigma^\sigma) &= \mathrm{tr}\,(\gamma^\mu\gamma^\nu\gamma^\rho\gamma^\sigma) \\ &= 4\,(g^{\mu\nu}g^{\rho\sigma} - g^{\mu\rho}g^{\nu\sigma} + g^{\mu\sigma}g^{\nu\rho}). \end{aligned} \tag{4.25}$$

Consider now $\gamma^5 \equiv i\,\gamma^0\gamma^1\gamma^2\gamma^3$. In the above representation of the Dirac matrices, an explicit calculation shows that

$$\gamma^5 = \begin{pmatrix} -1 & 0 \\ 0 & 1 \end{pmatrix}, \quad \gamma^5\gamma^\mu\gamma^\nu\gamma^\rho\gamma^\sigma = \begin{pmatrix} -\sigma^\mu\overline{\sigma}^\nu\sigma^\rho\overline{\sigma}^\sigma & 0 \\ 0 & \overline{\sigma}^\mu\sigma^\nu\overline{\sigma}^\rho\sigma^\sigma \end{pmatrix}.$$

Taking the trace of this equation yields a second identity:

$$\mathrm{tr}\,(\overline{\sigma}^\mu\sigma^\nu\overline{\sigma}^\rho\sigma^\sigma) - \mathrm{tr}\,(\sigma^\mu\overline{\sigma}^\nu\sigma^\rho\overline{\sigma}^\sigma) = \mathrm{tr}\,(\gamma^5\gamma^\mu\gamma^\nu\gamma^\rho\gamma^\sigma) = 4\,i\,\epsilon^{\mu\nu\rho\sigma}. \tag{4.26}$$

(Recall that we use the convention $\epsilon_{0123} = +1$.) The sum and difference of (4.25) and (4.26) lead to

$$\begin{aligned} \mathrm{tr}\,(\overline{\sigma}^\mu\sigma^\nu\overline{\sigma}^\rho\sigma^\sigma) &= 2\,\big(g^{\mu\nu}g^{\rho\sigma} - g^{\mu\rho}g^{\nu\sigma} + g^{\mu\sigma}g^{\nu\rho} + i\,\epsilon^{\mu\nu\rho\sigma}\big), \\ \mathrm{tr}\,(\sigma^\mu\overline{\sigma}^\nu\sigma^\rho\overline{\sigma}^\sigma) &= 2\,\big(g^{\mu\nu}g^{\rho\sigma} - g^{\mu\rho}g^{\nu\sigma} + g^{\mu\sigma}g^{\nu\rho} - i\,\epsilon^{\mu\nu\rho\sigma}\big). \end{aligned}$$

53.c Consider now

$$\begin{aligned} [12]\langle 23\rangle[34]\langle 41\rangle &= \mathrm{tr}\,(\overline{P}_1 P_2 \overline{P}_3 P_4) = p_{1\mu}p_{2\nu}p_{3\rho}p_{4\sigma}\;\mathrm{tr}\,(\overline{\sigma}^\mu\sigma^\nu\overline{\sigma}^\rho\sigma^\sigma) \\ &= 2\,p_{1\mu}p_{2\nu}p_{3\rho}p_{4\sigma}\,\big(g^{\mu\nu}g^{\rho\sigma} - g^{\mu\rho}g^{\nu\sigma} + g^{\mu\sigma}g^{\nu\rho} + i\,\epsilon^{\mu\nu\rho\sigma}\big) \\ &= \frac{s_{12}s_{34} - s_{13}s_{24} + s_{14}s_{23}}{2} + 2\,i\,p_{1\mu}p_{2\nu}p_{3\rho}p_{4\sigma}\,\epsilon^{\mu\nu\rho\sigma}. \end{aligned}$$

We have used the fact that all momenta are on-shell and massless to write $2p_1 \cdot p_2 =$

$(p_1+p_2)^2 \equiv s_{12}$, etc. Likewise, we have

$$\langle 12\rangle[23]\langle 34\rangle[41] = \frac{s_{12}s_{34} - s_{13}s_{24} + s_{14}s_{23}}{2} - 2i\,p_{1\mu}p_{2\nu}p_{3\rho}p_{4\sigma}\,\epsilon^{\mu\nu\rho\sigma}. \tag{4.27}$$

By taking the difference of the last two equations, we get

$$[12]\langle 23\rangle[34]\langle 41\rangle - \langle 12\rangle[23]\langle 34\rangle[41] = 4i\,p_{1\mu}p_{2\nu}p_{3\rho}p_{4\sigma}\,\epsilon^{\mu\nu\rho\sigma}.$$

Dividing (4.27) by $s_{14}s_{34} = \langle 14\rangle[14]\langle 34\rangle[34]$, we obtain

$$\frac{s_{12}s_{34} - s_{13}s_{24} + s_{14}s_{23} - 4i\,p_{1\mu}p_{2\nu}p_{3\rho}p_{4\sigma}\,\epsilon^{\mu\nu\rho\sigma}}{2\,s_{14}s_{34}} = \frac{\langle 12\rangle[23]\langle 34\rangle[41]}{\langle 14\rangle[14]\langle 34\rangle[34]} = \frac{\langle 12\rangle[23]}{\langle 14\rangle[43]},$$

which proves the second of the announced identities.

---

54. Auxiliary Vector Dependence of the Polarization Vectors The goal of this problem is to check that varying the auxiliary vector q in the definition of the polarization vector $\epsilon_+^\mu(p;q)$ leads to a change of the polarization vector by a term proportional to p^μ that does not affect any physical quantity. More precisely, show that

$$\epsilon_+^\mu(p;q') - \epsilon_+^\mu(p;q) = \frac{\sqrt{2}\,\langle qq'\rangle}{\langle qp\rangle\langle q'p\rangle}\,p^\mu.$$

Hint: it is useful to rewrite the Schouten identity for square spinors as follows:

$$1 = -\frac{|q]\,[q'|}{[qq']} - \frac{|q']\,[q|}{[q'q]}.$$

From the definition of the polarization vectors, we have

$$\begin{aligned}
\epsilon_+^\mu(p;q') - \epsilon_+^\mu(p;q) &= \frac{\langle q'|\overline{\sigma}^\mu|p]}{\sqrt{2}\,\langle q'p\rangle} - \frac{\langle q|\overline{\sigma}^\mu|p]}{\sqrt{2}\,\langle qp\rangle} \\
&= \frac{\langle qp\rangle\langle q'|\overline{\sigma}^\mu|p] - \langle q'p\rangle\langle q|\overline{\sigma}^\mu|p]}{\sqrt{2}\,\langle q'p\rangle\langle qp\rangle} \\
&= \frac{\langle q|\overline{\sigma}^\mu|p]\langle pq'\rangle - \langle q'|\overline{\sigma}^\mu|p]\langle pq\rangle}{\sqrt{2}\,\langle q'p\rangle\langle qp\rangle} \\
&= p_\nu\,\frac{\langle q|\overline{\sigma}^\mu\sigma^\nu|q'\rangle - \langle q'|\overline{\sigma}^\mu\sigma^\nu|q\rangle}{\sqrt{2}\,\langle q'p\rangle\langle qp\rangle}.
\end{aligned} \tag{4.28}$$

The Schouten identity for square spinors may be rewritten as

$$\big|r\big] = -\frac{\big|q\big]\big[q'r\big]}{\big[qq'\big]} - \frac{\big|q'\big]\big[qr\big]}{\big[q'q\big]}.$$

Since this must be true for any $\big|r\big]$, we must in fact have

$$1 = -\frac{\big|q\big]\big[q'\big|}{\big[qq'\big]} - \frac{\big|q'\big]\big[q\big|}{\big[q'q\big]}.$$

(This identity is valid over the space of square spinors, but it does not make sense to contract it with angle spinors.) The next step is to insert this representation of the identity between $\overline{\sigma}^\mu$ and σ^ν in (4.28), which gives

$$\begin{aligned}\big\langle q\big|\overline{\sigma}^\mu\sigma^\nu\big|q'\big\rangle - \big\langle q'\big|\overline{\sigma}^\mu\sigma^\nu\big|q\big\rangle &= \frac{\big\langle q\big|\overline{\sigma}^\mu\big|q'\big]\big[q\big|\sigma^\nu\big|q'\big\rangle}{\big[qq'\big]} + \frac{\big\langle q'\big|\overline{\sigma}^\mu\big|q\big]\big[q'\big|\sigma^\nu\big|q\big\rangle}{\big[qq'\big]}\\ &\quad - \frac{\big\langle q\big|\overline{\sigma}^\mu\big|q\big]\big[q'\big|\sigma^\nu\big|q'\big\rangle}{\big[qq'\big]} - \frac{\big\langle q'\big|\overline{\sigma}^\mu\big|q'\big]\big[q\big|\sigma^\nu\big|q\big\rangle}{\big[qq'\big]}.\end{aligned}$$

The third and final terms on the right-hand side are the easiest to calculate, since we have

$$\big\langle q\big|\overline{\sigma}^\mu\big|q\big] = \mathrm{tr}\,(\overline{\sigma}^\mu Q) = q_\alpha\,\underbrace{\mathrm{tr}\,(\overline{\sigma}^\mu\sigma^\alpha)}_{2\,g^{\mu\alpha}} = 2\,q^\mu,$$

$$\big[q\big|\sigma^\nu\big|q\big\rangle = \mathrm{tr}\,(\sigma^\nu\overline{Q}) = -q_\alpha\,\mathrm{tr}\,(\sigma^\nu\overline{\sigma}^\alpha) = -2\,q^\nu.$$

Therefore the last two terms become

$$-\frac{\big\langle q\big|\overline{\sigma}^\mu\big|q\big]\big[q'\big|\sigma^\nu\big|q'\big\rangle}{\big[qq'\big]} - \frac{\big\langle q'\big|\overline{\sigma}^\mu\big|q'\big]\big[q\big|\sigma^\nu\big|q\big\rangle}{\big[qq'\big]} = \frac{4\,(q^\mu q'^\nu + q^\nu q'^\mu)}{\big[qq'\big]}.$$

To evaluate the first two terms, we use

$$\begin{aligned}\big[q\big|\sigma^\nu\big|q'\big\rangle = \big[q\big|^c(\sigma^\nu)_{cd}\big|q'\big\rangle^d &= {}_e\big|q\big]\epsilon^{ec}(\sigma^\nu)_{cd}\epsilon^{df}\big\langle q'\big|_f\\ &= \big\langle q'\big|\underbrace{(\epsilon\sigma^\nu\epsilon)^t}_{-\overline{\sigma}^\nu}\big|q\big] = -\big\langle q'\big|\overline{\sigma}^\nu\big|q\big].\end{aligned}$$

Therefore, we have

$$\begin{aligned}&\big\langle q\big|\overline{\sigma}^\mu\big|q'\big]\big[q\big|\sigma^\nu\big|q'\big\rangle + \big\langle q'\big|\overline{\sigma}^\mu\big|q\big]\big[q'\big|\sigma^\nu\big|q\big\rangle\\ &= -\big\langle q\big|\overline{\sigma}^\mu\big|q'\big]\big\langle q'\big|\overline{\sigma}^\nu\big|q\big] - \big\langle q'\big|\overline{\sigma}^\mu\big|q\big]\big\langle q\big|\overline{\sigma}^\nu\big|q'\big]\\ &= -\mathrm{tr}\,(Q\overline{\sigma}^\mu Q'\overline{\sigma}^\nu) - \mathrm{tr}\,(Q'\overline{\sigma}^\mu Q\overline{\sigma}^\nu).\end{aligned}$$

Then, we use the results of Problem 53, to obtain

$$\frac{\big\langle q\big|\overline{\sigma}^\mu\big|q'\big]\big[q\big|\sigma^\nu\big|q'\big\rangle}{\big[qq'\big]} + \frac{\big\langle q'\big|\overline{\sigma}^\mu\big|q\big]\big[q'\big|\sigma^\nu\big|q\big\rangle}{\big[qq'\big]} = -4\frac{q^\mu q'^\nu + q^\nu q'^\mu - g^{\mu\nu}(q\cdot q')}{\big[qq'\big]},$$

and finally

$$\epsilon_+^\mu(p;q') - \epsilon_+^\mu(p;q) = p_\nu\,\frac{4\,g^{\mu\nu}\,(q\cdot q')}{\sqrt{2}\,\big\langle qp\big\rangle\big\langle q'p\big\rangle\big[qq'\big]} = \sqrt{2}\,\frac{\big\langle qq'\big\rangle}{\big\langle qp\big\rangle\big\langle q'p\big\rangle}\,p^\mu.$$

In other words, changing the auxiliary vector q in the definition of the polarization vectors amounts to varying them by a term proportional to the 4-momentum of the gluon, which has no effect, thanks to gauge invariance (the verification of this statement for negative-helicity polarization vectors works in the same way).

55. Polarization Sum in the Spinor-Helicity Formalism The goal of this problem is to show that the sum over the physical polarizations of the bilinear $\epsilon_h^\mu(p;q)\epsilon_h^{\nu*}(p;q)$ is what one would expect in the strict axial gauge $q \cdot A = 0$.

55.a Check that $\epsilon_+^{\mu*}(p;q) = -\epsilon_-^\mu(p;q) = \epsilon_-^\mu(-p;q)$.

55.b Show that

$$\sum_{h=\pm} \epsilon_h^\mu(p;q)\epsilon_h^{\nu*}(p;q) = -g^{\mu\nu} + \frac{p^\mu q^\nu + p^\nu q^\mu}{p \cdot q}.$$

55.a Let us first recall the expression for the polarization vectors:

$$\epsilon_+^\mu(p;q) \equiv \frac{\langle q|\overline{\sigma}^\mu|p]}{\sqrt{2}\langle qp\rangle}, \quad \epsilon_-^\mu(p;q) \equiv \frac{\langle p|\overline{\sigma}^\mu|q]}{\sqrt{2}\,[pq]}.$$

Then, we have

$$\epsilon_+^{\mu*}(p;q) = \frac{\langle q|\overline{\sigma}^\mu|p]^*}{\sqrt{2}\langle qp\rangle^*} = \frac{\langle p|\overline{\sigma}^\mu|q]}{\sqrt{2}\,[qp]} = -\epsilon_-^\mu(p;q).$$

Changing $p \to -p$ requires a bit of care, since the definition of $|p]$ and $\langle p|$ contains a square root ($\sqrt{p_0 + p_3}$), which leads to a sign ambiguity. The ambiguity is resolved by making the choice that ensures that the two spinors are mutually complex conjugate. This prescription leads to

$$|-p] = i\,|p], \quad \langle -p| = -i\,\langle p|.$$

Therefore, we have

$$\epsilon_-^\mu(-p;q) = \frac{\langle -p|\overline{\sigma}^\mu|q]}{\sqrt{2}\,[-pq]} = -\frac{\langle p|\overline{\sigma}^\mu|q]}{\sqrt{2}\,[pq]} = -\epsilon_-^\mu(p;q) = \epsilon_+^{\mu*}(p;q).$$

55.b Then, we can write

$$\sum_{h=\pm} \epsilon_h^\mu(\mathbf{p}; \mathbf{q})\epsilon_h^{\nu*}(\mathbf{p}; \mathbf{q}) = -\epsilon_+^\mu(\mathbf{p}; \mathbf{q})\epsilon_-^\nu(\mathbf{p}; \mathbf{q}) - \epsilon_-^\mu(\mathbf{p}; \mathbf{q})\epsilon_+^\nu(\mathbf{p}; \mathbf{q})$$

$$= -\frac{\langle q|\overline{\sigma}^\mu|p]\langle p|\overline{\sigma}^\nu|q] + \langle p|\overline{\sigma}^\mu|q]\langle q|\overline{\sigma}^\nu|p]}{2\langle qp\rangle[pq]}$$

$$= -\frac{\mathrm{tr}\,(\overline{\sigma}^\mu P\overline{\sigma}^\nu Q) + \mathrm{tr}\,(\overline{\sigma}^\mu Q\overline{\sigma}^\nu P)}{2\langle qp\rangle[pq]}$$

$$= -\frac{4(p^\mu q^\nu + p^\nu q^\mu - g^{\mu\nu}(p\cdot q))}{2\langle qp\rangle[pq]} = -g^{\mu\nu} + \frac{p^\mu q^\nu + p^\nu q^\mu}{p\cdot q}.$$

(We have used the results of Problem 53 for the traces, and $2p\cdot q = \langle pq\rangle[pq]$ in the last line). Interestingly, this bilinear combination of polarization vectors is the same as the numerator of the gauge propagator in the strict axial gauge defined by $q\cdot A = 0$. This is another way of seeing that changing the auxiliary vector $\mathbf{q}$ amounts to a change of gauge. Note that another way of writing the identity we have derived is

$$\sum_{h=\pm} \epsilon_{+h}^\mu(\mathbf{p}; \mathbf{q})\epsilon_{-h}^\nu(-\mathbf{p}; \mathbf{q}) = -g^{\mu\nu} + \frac{p^\mu q^\nu + p^\nu q^\mu}{p\cdot q}.$$

This form is more intuitive since it highlights the fact that a positive helicity along $+\mathbf{p}$ is equivalent to a negative helicity along $-\mathbf{p}$.

---

56. Photon Decoupling Identity The goal of this problem is to prove a relationship between various n-point color-ordered amplitudes, known as the *photon decoupling identity*. Recall from the discussion of the color decomposition of a tree amplitude that the color ordering has the same form for $SU(N)$ and $U(N)$ gauge theories.

56.a Why are the color-ordered amplitudes the same in $SU(N)$ and $U(N)$?

56.b By considering an n-point amplitude in which the first leg is the $U(1)$ gauge boson of a $U(N)$ gauge theory and the remaining $n-1$ legs are $SU(N)$ gauge bosons, show that

$$A_n(1234\cdots n) + A_n(2134\cdots n)$$
$$+ A_n(2314\cdots n) + \cdots + A_n(234\cdots(n-1)1n) = 0.$$

56.a Recall the color ordering of an n-point tree amplitude:

$$\mathcal{M}_n(1\cdots n) \equiv 2 \sum_{\sigma\in\mathfrak{S}_n/\mathbb{Z}_n} \mathrm{tr}\,(t_f^{a_{\sigma(1)}}\cdots t_f^{a_{\sigma(n)}})\,A_n(\sigma(1)\cdots\sigma(n)). \tag{4.29}$$

This formula was obtained by rewriting the structure constants f^{abc} (which enter into the color structure of the three-gluon and four-gluon vertices) in terms of traces of three fundamental generators t_f^a. Then, we rewrite all the contractions of the form $t_f^a \cdots t_f^a$ by using the Fierz identity. In the case of $\mathfrak{su}(N)$, this formula contains a term of order N^0 and a term of order N^{-1}:

$$(t_f^a)_{ij}(t_f^a)_{kl} =$$

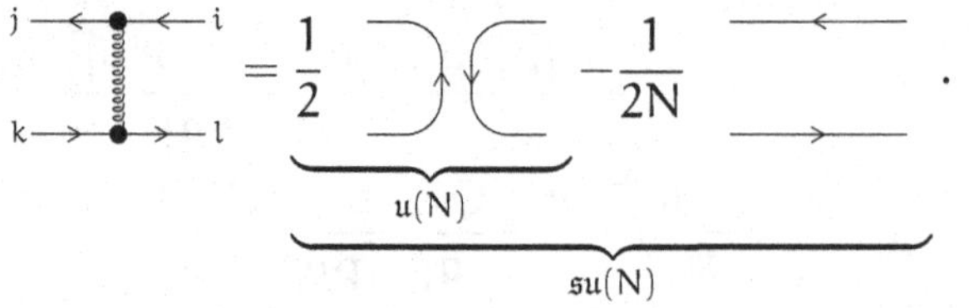

A noteworthy aspect of (4.29) is that all the terms of order N^{-p} with $p > 0$ have canceled. In fact, the formula (4.29) is also the expression one would obtain for the $\mathfrak{u}(N)$ algebra: the $\mathfrak{u}(N)$ and $\mathfrak{su}(N)$ results are identical because the extra gauge boson in $\mathfrak{u}(N) = \mathfrak{su}(N) \oplus \mathfrak{u}(1)$ is an Abelian gauge boson that does not couple to the gluons.

Let us now discuss more specifically the $U(N)$ theory. The structure constants that control the couplings of its gauge bosons differ in a trivial way from those of the $SU(N)$ theory:

$$f^{abc}_{\mathfrak{u}(N)} = \begin{cases} f^{abc}_{\mathfrak{su}(N)} & \text{if } a, b, c < N^2, \\ 0 & \text{if } a, b \text{ or } c = N^2. \end{cases}$$

(The adjoint indices a, b, c for $\mathfrak{u}(N)$ take N^2 different values instead of $N^2 - 1$.) Aside from this, everything else is the same. In particular, the kinematical dependence in the color-ordered Feynman rules is the same, implying that the color-ordered amplitudes $\mathcal{A}_n(\sigma(1) \cdots \sigma(n))$ are the same in $U(N)$ and $SU(N)$. Therefore, in (4.29), the only difference between $U(N)$ and $SU(N)$ is the trace $\mathrm{tr}\,(t_f^{a_{\sigma(1)}} \cdots t_f^{a_{\sigma(n)}})$ since the fundamental representation of $\mathfrak{u}(N)$ contains one more generator, associated with the extra $U(1)$ factor. Since this factor is Abelian, this additional generator is simply the identity.

56.b Consider now an amplitude $\mathcal{M}_n(1 \cdots n)$ in which leg 1 is a $U(1)$ gauge boson, and legs 2 to n are $SU(N)$ ones. This amplitude must be zero, since the Abelian boson does not couple to the non-Abelian ones:

$$\mathcal{M}_n(\underbrace{1}_{U(1)}, \underbrace{2, \ldots, n}_{SU(N)}) = 0.$$

Therefore, when the first generator is $t_f^{a_1} = 1$, we have

$$0 \underset{t_f^{a_1}=1}{=} \sum_{\sigma \in \mathfrak{S}_n/\mathbb{Z}_n} \mathrm{tr}\,(t_f^{a_{\sigma(1)}} \cdots t_f^{a_{\sigma(n)}})\, \mathcal{A}_n(\sigma(1) \cdots \sigma(n)).$$

The sum over the permutations up to a cyclic permutation can be performed by imposing that $\sigma(1) = 1$. Therefore, we need only sum over the set $\mathfrak{S}_{[2,n]}$ of the permutations of $[2, n]$. Using the fact that $t_f^{a_1} = 1$, the above identity becomes

$$0 = \sum_{\sigma \in \mathfrak{S}_{[2,n]}} \mathrm{tr}\,(t_f^{a_{\sigma(2)}} t_f^{a_{\sigma(3)}} \cdots t_f^{a_{\sigma(n)}})\, \mathcal{A}_n(1\sigma(2)\sigma(3) \cdots \sigma(n)).$$

We cannot conclude that the color-ordered amplitudes under the sum are individually zero, however, since the color traces are not all independent because of their cyclic invariance. Thus, the preceding equation should be first rewritten as

$$0 = \sum_{\sigma \in \mathfrak{S}_{[2,n]}/\mathbb{Z}_{n-1}} \mathrm{tr}\,(t_f^{a_{\sigma(2)}} t_f^{a_{\sigma(3)}} \cdots t_f^{a_{\sigma(n)}}) \underbrace{\sum_{\text{cycles } \rho} \mathcal{A}_n(1\sigma(\rho(2))\sigma(\rho(3))\cdots\sigma(\rho(n)))}_{0}.$$

Taking σ to be the identity, we get the following relationship:

$$\mathcal{A}_n(1234\cdots n) + \mathcal{A}_n(134\cdots n2) + \mathcal{A}_n(14\cdots n23) + \cdots + \mathcal{A}_n(1n234\cdots(n-1)) = 0.$$

Using the invariance of color-ordered amplitudes under cyclic permutations, this identity can be rewritten in such a way that the leg n is always in the last position:

$$\mathcal{A}_n(\mathbf{1}234\cdots n) + \mathcal{A}_n(2\mathbf{1}34\cdots n) + \mathcal{A}_n(23\mathbf{1}4\cdots n) + \cdots + \mathcal{A}_n(234\cdots(n-1)\mathbf{1}n) = 0,$$

where we have highlighted in boldface the index 1, which is sweeping from left to right in the $n-1$ terms of this formula while all the other indices do not move. This identity is known as the *photon decoupling identity*.

57. Scalar-Antiscalar-Gluon Amplitudes Consider a Yang–Mills theory coupled to a complex scalar field of mass m. The goal of this problem is to calculate color-ordered amplitudes, with a scalar–antiscalar pair of respective incoming momenta k and k', and n gluons of positive helicity of momenta $p_{1,2,\ldots,n}$. (We will use these results in Problems 67 and 68 in order to calculate some one-loop amplitudes via generalized unitarity.) We denote these amplitudes by $\mathcal{A}_{2+n}(k, 1^+2^+\cdots n^+, k'^*)$.

57.a Check that $\mathcal{A}_3(k, 1^+, k'^*) = ig\sqrt{2}\,\frac{\langle q|\overline{K}|1]}{\langle q1\rangle}$, where q is the auxiliary vector in the polarization vector of the gluon. Extra: check that this does not depend on q.

57.b Verify that

$$\mathcal{A}_3(k, 1^+, -k'^*)\mathcal{A}_3(k', 2^+, k''^*) = -2g^2m^2\frac{[12]}{\langle 12\rangle}.$$

57.c Show that

$$\mathcal{A}_4(k, 1^+2^+, k'^*) = -\frac{2ig^2m^2}{(k+p_1)^2 - m^2}\frac{[12]}{\langle 12\rangle}.$$

Hint: use a BCFW shift on the two gluon lines.

57.d Show that

$$\mathcal{A}_5(k, 1^+2^+3^+, k'^*) = \frac{2i\sqrt{2}g^3m^2}{((k+p_1) - m^2)((k'+p_3)^2 - m^2)}\frac{[3|(P_1+P_2)\overline{K}|1]}{\langle 12\rangle\langle 23\rangle}.$$

Hint: use a BCFW shift on the first two gluon lines.

57.a First, recall the kinematical part of the Feynman rules for a complex scalar coupled to a gauge field. After the color factors have been stripped out in the process of color decomposition, they are identical to the Feynman rules of scalar QED:

$$= \frac{i}{p^2 - m^2 + i0^+},$$

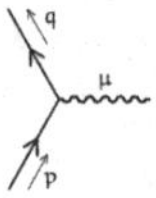

$$= -ig(p+q)^\mu.$$

(There is in addition a $\phi^*\phi A_\mu A^\mu$ four-point vertex, but this vertex is not necessary when applying BCFW recursion, since one can bootstrap everything from the three-point vertex only). The amplitude $\mathcal{A}_3(k, 1^+, k'^*)$ is obtained by simply contracting the three-point vertex into a positive-helicity polarization vector:

$$\mathcal{A}_3(k, 1^+, k'^*) = -ig\underbrace{(k-k')_\mu}_{(2k+p_1)_\mu}\frac{\langle q|\overline{\sigma}^\mu|1]}{\sqrt{2}\langle q1\rangle} = ig\frac{\langle q|(2\overline{K} + \overbrace{|1\rangle[1|)|1]}^{=0}}{\sqrt{2}\langle q1\rangle}$$

$$= ig\sqrt{2}\,\frac{\langle q|\overline{K}|1]}{\langle q1\rangle}.$$

Varying q changes the polarization vector by an amount proportional to the gluon momentum p_1^μ (see Problem 54). Therefore, under the change $q \to q'$, the above three-point amplitude varies by an amount proportional to

$$p_1 \cdot (k-k') = p_1 \cdot (2k+p_1) = (p_1+k)^2 - k^2 = m^2 - m^2 = 0.$$

57.b Next, let us consider

$$\mathcal{A}_3(k, 1^+, -k'^*)\mathcal{A}_3(k', 2^+, k''^*) = -2g^2\frac{\langle q_1|\overline{K}|1]}{\langle q_1 1\rangle}\frac{\langle q_2|\overline{K}'|2]}{\langle q_2 2\rangle}.$$

To simplify this expression, it is convenient to rely on its independence with respect to the choice of q_1 and q_2, to take $q_1 = p_2$ and $q_2 = p_1$. This leads to

$$\mathcal{A}_3(k, 1^+, -k'^*)\mathcal{A}_3(k', 2^+, k''^*) = -2g^2\,\frac{\langle 2|\overline{K}|1]}{\langle 21\rangle}\frac{\langle 1|\overline{K}'|2]}{\langle 12\rangle} = -2g^2\,\frac{\langle 2|\overline{K}|1]}{\langle 21\rangle}\frac{\langle 1|\overline{K}|2]}{\langle 12\rangle}$$

$$= -2g^2\,\frac{\mathrm{tr}\,(\overline{K}P_1\overline{K}P_2)}{\langle 21\rangle\langle 12\rangle} = 4g^2\,\frac{2(k\cdot p_1)(k\cdot p_2) - (p_1\cdot p_2)k^2 + i\,k_\mu p_{1\nu}k_\rho p_{2\sigma}\epsilon^{\mu\nu\rho\sigma}}{\langle 12\rangle^2}.$$

Then, we use

$$2k\cdot p_1 = \underbrace{(k+p_1)^2}_{m^2} - \underbrace{k^2}_{m^2} - \underbrace{p_1^2}_{0} = 0, \quad k_\mu p_{1\nu}k_\rho p_{2\sigma}\epsilon^{\mu\nu\rho\sigma} = 0,$$

in order to obtain

$$\mathcal{A}_3(k, 1^+, -k'^*)\mathcal{A}_3(k', 2^+, k''^*) = -4g^2\,\frac{m^2\,(p_1\cdot p_2)}{\langle 12\rangle^2} = -2g^2m^2\,\frac{[12]}{\langle 12\rangle}. \qquad (4.30)$$

A remarkable feature of this formula is that the right-hand side depends only on the momenta of the two gluons, despite the fact that their couplings to the scalar line depend on the momentum of the scalar particle.

57.c In order to calculate the amplitude $\mathcal{A}_4(k, 1^+2^+, k'^*)$, we can apply a BCFW shift to the two gluon lines 1^+ and 2^+:

$$\widehat{p}_1 = p_1 + z\eta, \quad \widehat{p}_2 = p_2 - z\eta, \quad \eta \equiv |2]\langle 1|,$$
$$\text{i.e.,} \quad |\widehat{1}\rangle = |1\rangle, \quad |\widehat{1}] = |1] + z|2], \quad |\widehat{2}\rangle = |2\rangle - z|1\rangle, \quad |\widehat{2}] = |2].$$

With this shift, we have

$$\mathcal{A}_4(k, 1^+2^+, k'^*) = \underbrace{\mathcal{A}_3(k, \widehat{1}^+, -\widehat{K}_I^*)\mathcal{A}_3(\widehat{K}_I, \widehat{2}^+, k'^*)}_{=-2g^2m^2\frac{[\widehat{1}\widehat{2}]}{\langle\widehat{1}\widehat{2}\rangle}=-2g^2m^2\frac{[12]}{\langle 12\rangle}} \frac{i}{K_I^2 - m^2}, \quad K_I \equiv k + p_1$$
$$= -2g^2m^2 \frac{[12]}{\langle 12\rangle} \frac{i}{(k+p_1)^2 - m^2}.$$

(We have reused the result (4.30) on the product of two three-point amplitudes.) Interestingly, this four-point amplitude factorizes into a product of two *on-shell and unshifted* three-point amplitudes, times an intermediate scalar propagator.

57.d In the case of three external gluons, we can again use the same shift on the first two gluons, 1^+ and 2^+:

$$\widehat{p}_1 = p_1 + z\eta, \quad \widehat{p}_2 = p_2 - z\eta, \quad \eta \equiv |2]\langle 1|,$$
$$\text{i.e.,} \quad |\widehat{1}\rangle = |1\rangle, \quad |\widehat{1}] = |1] + z|2], \quad |\widehat{2}\rangle = |2\rangle - z|1\rangle, \quad |\widehat{2}] = |2].$$

However, the BCFW recursion now produces two terms,

$$\mathcal{A}_5(k, 1^+2^+3^+, k'^*) = \mathcal{A}_3(k, \widehat{1}^+, -\widehat{K}_I^*)\frac{i}{(k+p_1)^2 - m^2}\mathcal{A}_4(\widehat{K}_I, \widehat{2}^+3^+, k'^*)$$
$$+ \mathcal{A}_4(k, \widehat{1}^+ - \widehat{L}_I^+, k'^*)\frac{i}{(k+k'+p_1)^2}\mathcal{A}_3(\widehat{L}_I^-\widehat{2}^+3^+),$$

with respectively an intermediate scalar and an intermediate gluon. Note that only one of the helicity assignments is possible for the intermediate gluon in the second term, because of the three-gluon amplitude on the right:

$$\mathcal{A}_3(\widehat{L}_I^-\widehat{2}^+3^+) = g\sqrt{2}\,\frac{[\widehat{2}3]^3}{[3\widehat{L}_I][\widehat{L}_I\widehat{2}]}.$$

Moreover, the pole condition for this term is

$$0 = \widehat{L}_I^2 = (\widehat{p}_2 + p_3)^2 = 2\widehat{p}_2 \cdot p_3 = \langle\widehat{2}3\rangle[\widehat{2}3] = \langle 23\rangle[\widehat{2}3],$$

from which we conclude that $[\widehat{2}3] = 0$ at the pole. Then, we have

$$|\widehat{L}_I\rangle[\widehat{L}_I 3] = (|\widehat{2}\rangle[\widehat{2}| + |3\rangle[3|)|3] = |2\rangle[\widehat{2}3] = 0.$$

Therefore, $[\widehat{L}_I 3] = 0$ at the pole. Likewise, $[\widehat{2}\widehat{L}_I] = 0$. Therefore, the three-gluon amplitude $\mathcal{A}_3(\widehat{L}_I^-\widehat{2}^+3^+)$ that appears in the second term is zero (it behaves as $(z-z_*)^3/(z-z_*)^2$ when

z approaches the pole), and we have

$$\mathcal{A}_5(k, 1^+2^+3^+, k'^*) = \mathcal{A}_3(k, \hat{1}^+, -\hat{K}_I^*) \frac{i}{(k+p_1)^2 - m^2} \mathcal{A}_4(\hat{K}_I, \hat{2}^+3^+, k'^*)$$
$$= -2i\sqrt{2}g^3m^2 \frac{\langle 2|\overline{K}|1]}{\langle 21\rangle} \frac{[\hat{2}3]}{\langle 23\rangle} \frac{i}{(k+p_1)^2 - m^2} \frac{i}{(k'+p_3)^2 - m^2},$$

where we have made the choice of auxiliary vector $q_1 = p_2$ in the second line.

Note now that

$$\langle 2|\overline{K}|1] = \mathrm{tr}\,(\overline{K}\eta) = -2\,k \cdot \eta,$$

and

$$[\hat{2}3] = [23] - z_*\,[13], \quad z_* = -\frac{k \cdot p_1}{k \cdot \eta} = \frac{\langle 1|\overline{K}|1]}{2\,k \cdot \eta}.$$

Therefore, we have

$$\langle 2|\overline{K}|1]\,[\hat{2}3] = \underbrace{\langle 2|\overline{K}|1]}_{-2\,k\cdot\eta} \left([23] + \frac{k \cdot p_1}{k \cdot \eta}\,[13]\right)$$
$$= \langle 2|\overline{K}|1]\,[23] - \underbrace{2\,(k \cdot p_1)}_{-\langle 1|\overline{K}|1]}\,[13]$$
$$= \langle 2|\overline{K}|1]\,[23] + \langle 1|\overline{K}|1]\,[13] = -[3|(P_1 + P_2)\overline{K}|1].$$

Therefore, the five-point amplitude reads

$$\mathcal{A}_5(k, 1^+2^+3^+, k'^*) = \frac{2i\sqrt{2}g^3m^2}{((k+p_1) - m^2)((k'+p_3)^2 - m^2)} \frac{[3|(P_1 + P_2)\overline{K}|1]}{\langle 12\rangle\langle 23\rangle}.$$

Source: Badger, S. D., Glover, E. W. N., Khoze, V. V. and Svrcek, P. (2005), *J High Energy Phys* 07: 025.

58. Color Decomposition with a Quark-Antiquark Pair The goal of this problem is to work out the color decomposition of tree amplitudes (in an $SU(N)$ gauge theory) with a quark–antiquark pair and n gluons, which we shall denote by $\mathcal{M}_{n+2}(\overline{q}12\cdots nq)$. This is the next-to-simplest case after the purely gluonic amplitudes (amplitudes with more than one quark–antiquark pair have a more complicated color decomposition).

58.a Recall the color decomposition for a gluon tree. Then write the color factors for a gluon tree attached to a quark line.

58.b Use the Fierz identity to simplify the color structure. Explain why the terms in N^{-1} cancel.

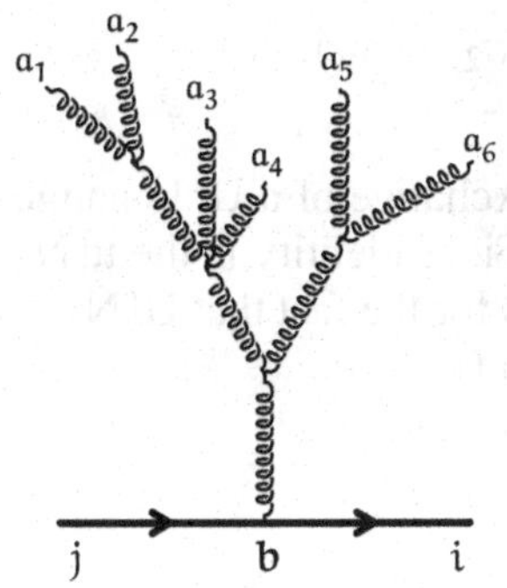

Figure 4.2 Color-ordered amplitude with a quark–antiquark line, to which a gluon tree is attached.

58.c Generalize this result to an arbitrary number of gluon trees attached to the quark line, in order to show that

$$\mathcal{M}_{n+2}(\overline{q}12\cdots nq) = \sum_{\sigma\in\mathfrak{S}_n} \left(t_f^{a_{\sigma_1}}\cdots t_f^{a_{\sigma_n}}\right)_{ij} \mathcal{A}_{n+2}(\overline{q}\sigma_1\sigma_2\cdots\sigma_n q),$$

where $\mathcal{A}_{n+2}(\overline{q}\sigma_1\sigma_2\cdots\sigma_n q)$ is a cyclic-ordered amplitude with a quark–antiquark pair and n gluons (in the sum, the permutation σ affects only the gluons).

58.a Consider first a graph such as the one displayed in Figure 4.2, where a single gluon tree is attached to a quark line: First, the color structure of the gluon tree can be decomposed as follows:

$$\mathcal{M}_{n+1}(b12\cdots n) = 2\sum_{\sigma\in\mathfrak{S}_n} \mathrm{tr}\left(t_f^b t_f^{a_{\sigma_1}}\cdots t_f^{a_{\sigma_n}}\right) \mathcal{A}_{n+1}(b\sigma_1\sigma_2\cdots\sigma_n).$$

In this expression, we denote by b the color of the (off-shell) gluon leg attached to the quark line. Since the purely gluonic color decomposition contains a sum over all the permutations of the color indices that are not equivalent up to a circular permutation (i.e., the elements of $\mathfrak{S}_{n+1}/\mathbb{Z}_{n+1}$), we have fixed the matrix t_f^b to be in the first position inside the trace and we sum over the permutations of the remaining adjoint indices. The color structure of the graph of Figure 4.2 is obtained simply by multiplying this by $(t_f^b)_{ij}$:

$$(t_f^b)_{ij}\,\mathcal{M}_{n+1}(b12\cdots n) = 2\sum_{\sigma\in\mathfrak{S}_n} (t_f^b)_{ij}\,\underbrace{\mathrm{tr}\left(t_f^b t_f^{a_{\sigma_1}}\cdots t_f^{a_{\sigma_n}}\right)}_{(t_f^b)_{kl}(t_f^{a_{\sigma_1}}\cdots t_f^{a_{\sigma_n}})_{lk}} \mathcal{A}_{n+1}(b\sigma_1\sigma_2\cdots\sigma_n).$$

Then, we use the Fierz identity for the generators of the fundamental representation of $\mathfrak{su}(N)$:

$$(t_f^b)_{ij}(t_f^b)_{kl} = \tfrac{1}{2}\,\delta_{il}\delta_{jk} - \tfrac{1}{2N}\,\delta_{ij}\delta_{kl},$$

and the above color structure becomes

$$\sum_{\sigma\in\mathfrak{S}_n}\Big(\big(t_f^{a_{\sigma_1}}\cdots t_f^{a_{\sigma_n}}\big)_{ij}-\tfrac{1}{N}\delta_{ij}\,\mathrm{tr}\,\big(t_f^{a_{\sigma_1}}\cdots t_f^{a_{\sigma_n}}\big)\Big)\,\mathcal{A}_{n+1}(b\sigma_1\sigma_2\cdots\sigma_n).$$

The second term, in N^{-1}, is in fact zero because it describes the exchange of a $U(1)$ gauge boson between the quark and the gluon tree (the first term of the Fierz identity is the $u(N)$ contribution, and the second term is the $u(1)$ correction that accounts for the fact that $U(N) = SU(N)\times U(1)$). Diagrammatically, the resulting color structure reads

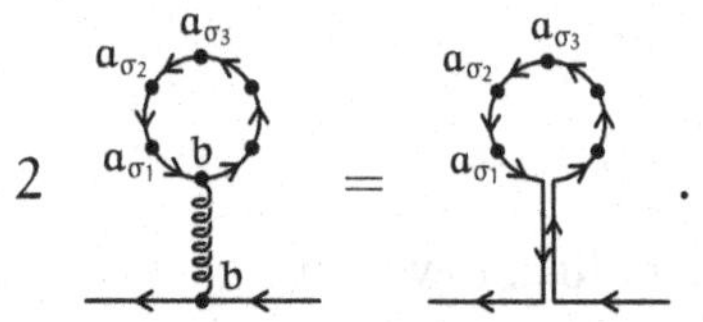

Therefore, after we introduce the quark and antiquark spinors to obtain a full-fledged amplitude, we have

$$\mathcal{M}_{n+2}(\overline{q}12\cdots nq)=\sum_{\sigma\in\mathfrak{S}_n}\big(t_f^{a_{\sigma_1}}\cdots t_f^{a_{\sigma_n}}\big)_{ij}\,\mathcal{A}_{n+2}(\overline{q}[\sigma_1\sigma_2\cdots\sigma_n]q),\tag{4.31}$$

where $\mathcal{A}_{n+2}(\overline{q}[\sigma_1\sigma_2\cdots\sigma_n]q)$ is a cyclic-ordered amplitude with n gluons between an antiquark and a quark. Note that here the n gluons are all attached to the quark via a single tree (this is indicated by enclosing their labels within square brackets). In the sum over the permutations, only the gluons are reshuffled, while the quark and antiquark stay in fixed positions.

In the general case, the n gluons can be distributed into several (up to n) trees attached at distinct points along the quark line. Let us denote by $I_1,\ldots,I_p$ (with $p\le n$) a partition of $[1,n]$ into p subsets, each set containing the indices of the gluons that belong to the corresponding tree. In this case, the Fierz identity still leads to a product of t_f^a's:

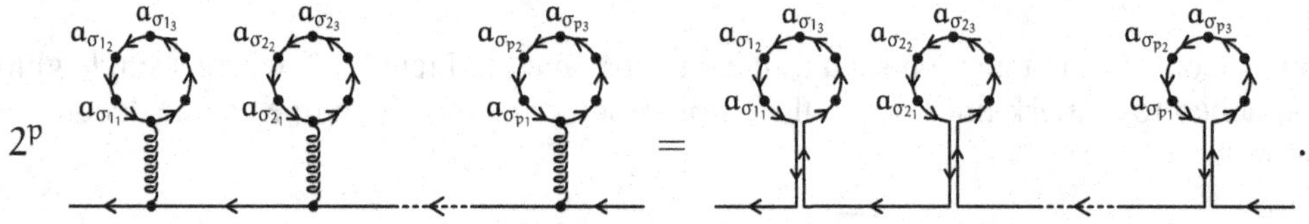

The p color-ordered trees attached to the quark line combine into a color-ordered amplitude with the n gluons between the antiquark and the quark. Thus, for this partition of the n gluons into p trees, the right-hand side of (4.31) becomes

$$\sum_{\substack{\sigma_1\in\mathfrak{S}_{I_1},\cdots\\ \cdots,\sigma_p\in\mathfrak{S}_{I_p}}}\Big(\prod_{i\in I_1}t_f^{\sigma_{1i}}\prod_{i\in I_2}t_f^{\sigma_{2i}}\cdots\prod_{i\in I_p}t_f^{\sigma_{pi}}\Big)_{ij}\mathcal{A}_{n+2}\Big(\overline{q}\,\Big[\prod\nolimits_{i\in I_1}\sigma_{1i}\Big]\cdots\Big[\prod\nolimits_{i\in I_p}\sigma_{pi}\Big]\,q\Big),$$

where we also indicate by square brackets the gluons that are attached via the same tree to the quark line. The main point is that the ordering of the fundamental generators in the product is the same as the ordering of the gluons in the partial amplitude. In this formula, we have independent permutations of the gluon indices inside each tree, but no permutations of the gluons among trees. When we sum over all the ways of assigning the n gluons into trees, we obtain the following decomposition:

$$\mathcal{M}_{n+2}(\overline{q}12\cdots nq)=\sum_{\sigma\in\mathfrak{S}_n}\big(t_f^{a_{\sigma_1}}\cdots t_f^{a_{\sigma_n}}\big)_{ij}\,\mathcal{A}_{n+2}(\overline{q}\sigma_1\sigma_2\cdots\sigma_nq),$$

where now the color-ordered amplitude on the right-hand side does not have any restriction on the number of trees attached to the quark line.

59. Quark-Antiquark-Gluon Amplitudes The goal of this problem is to extend the spinor-helicity formalism to amplitudes containing massless quarks and antiquarks.

59.a Show that massless helicity eigenstate spinors can be written as

$$u_+(p) = |p\rangle, \quad u_-(p) = |p], \quad \overline{u_+}(p) = -[p|, \quad \overline{u_-}(p) = \langle p|.$$

Hint: check that the proposed spinors obey the Dirac equation, are eigenstates of γ^5*, are normalized according to* $u_h^\dagger(p)u_{h'}(p) = 2\,E_p\,\delta_{hh'}$ *and satisfy* $\overline{u_h} = u_h^\dagger\gamma^0$*.*

59.b Express the combination $-ig\,\overline{u_h}(p)\gamma^\mu u_{h'}(p')$ in the spinor-helicity language.

59.c Show that the three-point amplitudes with a quark–antiquark pair and a gluon are given by

$$\mathcal{A}_3(1_{\bar q}^- 2_q^+ 3^+) = ig\sqrt{2}\,\frac{[23]^2}{[12]}, \quad \mathcal{A}_3(1_{\bar q}^- 2_q^+ 3^-) = ig\sqrt{2}\,\frac{\langle 31\rangle^2}{\langle 12\rangle}.$$

59.d Check that the momentum dependence of these three-point functions could have been guessed from little-group scaling.

59.e Work out the color structure of the $q\bar q \to Q\bar Q$ tree amplitude. Then, calculate the color-stripped amplitude $\mathcal{A}_4(1_{\bar q}^- 2_q^+ 3_{\bar Q}^- 4_Q^+)$ and show that

$$\mathcal{A}_4(1_{\bar q}^- 2_q^+ 3_{\bar Q}^- 4_Q^+) = ig^2\,\frac{\langle 13\rangle^2}{\langle 12\rangle\langle 34\rangle}.$$

59.a The following matrices are a valid representation of the Dirac matrices in four dimensions (see Problem 53):

$$\gamma^\mu = \begin{pmatrix} 0 & \sigma^\mu \\ \overline{\sigma}^\mu & 0 \end{pmatrix}.$$

Therefore, we have

$$\not p \equiv p_\mu\gamma^\mu = \begin{pmatrix} 0 & p_\mu\sigma^\mu \\ p_\mu\overline{\sigma}^\mu & 0 \end{pmatrix} = \begin{pmatrix} 0 & |p]\langle p| \\ -|p\rangle[p| & 0 \end{pmatrix}.$$

This implies the following identities:

$$\not p\begin{pmatrix} |p] \\ 0 \end{pmatrix} = \not p\begin{pmatrix} 0 \\ |p\rangle \end{pmatrix} = 0, \quad \begin{pmatrix} 0 & \langle p| \end{pmatrix}\not p = \begin{pmatrix} [p| & 0 \end{pmatrix}\not p = 0.$$

Therefore, the column vectors and line vectors in this equation are linear combinations of the free spinors $u_s(p)$ and $\overline{u}_s(p)$, respectively. Recall also that, in this representation, we have

$\gamma^5 = \text{diag}\,(-1, 1)$ (see Problem 53). Therefore,

$$\gamma^5 \begin{pmatrix} |p] \\ 0 \end{pmatrix} = -\begin{pmatrix} |p] \\ 0 \end{pmatrix}, \quad \gamma^5 \begin{pmatrix} 0 \\ |p\rangle \end{pmatrix} = +\begin{pmatrix} 0 \\ |p\rangle \end{pmatrix},$$
$$\begin{pmatrix} 0 & \langle p| \end{pmatrix}\gamma^5 = +\begin{pmatrix} 0 & \langle p| \end{pmatrix}, \quad \begin{pmatrix} [p| & 0 \end{pmatrix}\gamma^5 = -\begin{pmatrix} [p| & 0 \end{pmatrix}.$$

Since, for massless particles, the notions of chirality and helicity are equivalent, this means that we can perform the following identifications:

$$u_+(p) = \alpha \begin{pmatrix} 0 \\ |p\rangle \end{pmatrix}, \quad u_-(p) = \beta \begin{pmatrix} |p] \\ 0 \end{pmatrix},$$
$$\overline{u_+}(p) = \gamma \begin{pmatrix} [p| & 0 \end{pmatrix}, \quad \overline{u_-}(p) = \delta \begin{pmatrix} 0 & \langle p| \end{pmatrix},$$

with $\alpha, \beta, \gamma, \delta$ some constant prefactors. With a slight abuse of notation, this may also be written more simply as

$$u_+(p) = \alpha|p\rangle, \quad u_-(p) = \beta|p], \quad \overline{u_+}(p) = \gamma[p|, \quad \overline{u_-}(p) = \delta\langle p|.$$

The unknown prefactors should be determined so that the following relationships hold:

$$\overline{u_h}(p)u_{h'}(p) = 0,$$
$$u_h^\dagger(p)u_{h'}(p) = 2\,E_p\,\delta_{hh'},$$
$$\sum_{h=\pm} u_h(p)\overline{u_h}(p) = \not{p}.$$

The first of these identities is always satisfied and does not constrain the coefficients. Consider now the second identity. For instance, we have

$$u_+^\dagger(p)u_+(p) = \overline{u_+}(p)\gamma^0 u_+(p) = \alpha\gamma\,[p|\sigma^0|p\rangle$$
$$= \alpha\gamma\,\text{tr}\,(\sigma^0\overline{P}) = -\alpha\gamma\,p_\mu \underbrace{\text{tr}\,(\sigma^0\overline{\sigma}^\mu)}_{2\,g^{0\mu}} = -2\alpha\gamma E_p.$$

Therefore, we must have $\alpha\gamma = -1$. Working out the other three cases of this identity, we obtain also the constraint $\beta\delta = +1$ (two out of four cases are trivially satisfied and provide no constraint). Consider now the final identity. We have

$$u_+(p)\overline{u_+}(p) + u_-(p)\overline{u_-}(p) = \begin{pmatrix} 0 & \underbrace{\beta\delta}_{+1}\,|p]\langle p| \\ \underbrace{\alpha\gamma}_{-1}\,|p\rangle[p| & 0 \end{pmatrix} = \not{p}.$$

Therefore, all the identities are satisfied provided that we have $\alpha\gamma = -1$ and $\beta\delta = +1$.

h	h'	$-ig\,\overline{u_h}(p)\gamma^\mu u_{h'}(p')$
$+$	$+$	$ig\,[p\vert\sigma^\mu\vert p'\rangle \neq 0$
$+$	$-$	$ig\,[p\vert\gamma^\mu\vert p'] = 0$
$-$	$+$	$-ig\,\langle p\vert\gamma^\mu\vert p'\rangle = 0$
$-$	$-$	$-ig\,\langle p\vert\overline{\sigma}^\mu\vert p'] \neq 0$

Table 4.2 Spinor-helicity expressions for the coupling of a gluon to an on-shell quark–antiquark current.

Additional constraints are provided by requiring that

$$\overline{u_+}(p) = u_+^\dagger(p)\gamma^0, \quad \overline{u_-}(p) = u_-^\dagger(p)\gamma^0.$$

We have

$$u_+^\dagger(p)\gamma^0 = \alpha^* \begin{pmatrix} 0 \\ |p\rangle \end{pmatrix}^\dagger \gamma^0 = \alpha^* \Big(0 \;\; \underbrace{|p\rangle^\dagger}_{-[p|}\Big) \begin{pmatrix} 0 & 1 \\ 1 & 0 \end{pmatrix} = -\alpha^* \big([p| \;\; 0\big).$$

Likewise,

$$u_-^\dagger(p)\gamma^0 = \beta^* \begin{pmatrix} |p] \\ 0 \end{pmatrix}^\dagger \gamma^0 = \beta^* \Big(\underbrace{|p]^\dagger}_{\langle p|} \;\; 0\Big) \begin{pmatrix} 0 & 1 \\ 1 & 0 \end{pmatrix} = \beta^* \big(0 \;\; \langle p|\big).$$

Therefore, the coefficients must obey

$$\gamma = -\alpha^*, \quad \delta = \beta^*, \quad |\alpha| = |\beta| = 1,$$

which is sufficient to determine the spinors up to a phase. For instance, we may choose

$$u_+(p) = |p\rangle, \quad u_-(p) = |p], \quad \overline{u_+}(p) = -[p|, \quad \overline{u_-}(p) = \langle p|. \tag{4.32}$$

59.b Consider now a gluon of Lorentz index μ attached to an on-shell quark current, which brings a factor $-ig\,\overline{u_h}(p)\gamma^\mu u_{h'}(p')$. We list in Table 4.2 the four possible combinations of quark helicities. (When $p = p'$, the two non-zero couplings are both equal to $-2igp^\mu$.) Therefore, the absorption or emission of a gluon does not flip the helicity of a massless quark. Note that, with this notation, one of the fermions is incoming and the other one is outgoing, in contrast with the "all incoming" convention used in the rest of this chapter. In amplitudes, we will denote by i_q^h or $i_{\bar{q}}^h$ an incoming quark or antiquark of helicity h. To switch to this convention, we just need to flip the notation for the helicity of the quark line in the current:

$$i_q^+ \longleftrightarrow u_-(p_i), \quad i_q^- \longleftrightarrow u_+(p_i).$$

59.c Thus, the non-zero quark–antiquark–gluon tree amplitudes are

$$\mathcal{A}_3(1^+_{\bar{q}}2^-_q3^\pm), \quad \mathcal{A}_3(1^-_{\bar{q}}2^+_q3^\pm),$$

where the subscripts q and $\bar{q}$ indicate which lines are the quark and the antiquark. In order to calculate them, we use the following Fierz identity:

$$\left\langle 1\right|\overline{\sigma}^\mu\left|2\right]\left\langle 3\right|\overline{\sigma}_\mu\left|4\right] = 2\left\langle 13\right\rangle\left[24\right].$$

We first obtain

$$\begin{aligned}\mathcal{A}_3(1^-_{\bar{q}}2^+_q3^+) &= -ig\left(\overline{u_-}(p_1)\gamma_\mu u_-(p_2)\right)\epsilon^\mu_+(p_3;q) = -ig\,\frac{\left\langle 1\right|\overline{\sigma}_\mu\left|2\right]\left\langle q\right|\overline{\sigma}^\mu\left|3\right]}{\sqrt{2}\left\langle q3\right\rangle}\\ &= ig\sqrt{2}\,\frac{\left\langle q1\right\rangle\left[23\right]}{\left\langle q3\right\rangle} = ig\sqrt{2}\,\frac{\left[23\right]^2}{\left[12\right]}\end{aligned} \tag{4.33}$$

(momentum conservation implies $\left\langle q1\right\rangle\left[12\right] + \left\langle q3\right\rangle\left[32\right] = 0$),

$$\begin{aligned}\mathcal{A}_3(1^-_{\bar{q}}2^+_q3^-) &= -ig\left(\overline{u_-}(p_1)\gamma_\mu u_-(p_2)\right)\epsilon^\mu_-(p_3;q) = -ig\,\frac{\left\langle 1\right|\overline{\sigma}_\mu\left|2\right]\left\langle 3\right|\overline{\sigma}^\mu\left|q\right]}{\sqrt{2}\left[3q\right]}\\ &= -ig\sqrt{2}\,\frac{\left\langle 13\right\rangle\left[2q\right]}{\left[3q\right]} = ig\sqrt{2}\,\frac{\left\langle 31\right\rangle^2}{\left\langle 12\right\rangle},\end{aligned}$$

$$\begin{aligned}\mathcal{A}_3(1^+_{\bar{q}}2^-_q3^+) &= -ig\left(\overline{u_+}(p_1)\gamma_\mu u_+(p_2)\right)\epsilon^\mu_+(p_3;q) = ig\,\frac{\left[1\right|\sigma_\mu\left|2\right\rangle\left\langle q\right|\overline{\sigma}^\mu\left|3\right]}{\sqrt{2}\left\langle q3\right\rangle}\\ &= -ig\,\frac{\left\langle 2\right|\overline{\sigma}_\mu\left|1\right]\left\langle q\right|\overline{\sigma}^\mu\left|3\right]}{\sqrt{2}\left\langle q3\right\rangle} = ig\sqrt{2}\,\frac{\left\langle q2\right\rangle\left[13\right]}{\left\langle q3\right\rangle} = ig\sqrt{2}\,\frac{\left[31\right]^2}{\left[21\right]},\end{aligned}$$

and

$$\begin{aligned}\mathcal{A}_3(1^+_{\bar{q}}2^-_q3^-) &= -ig\left(\overline{u_+}(p_1)\gamma_\mu u_+(p_2)\right)\epsilon^\mu_-(p_3;q) = ig\,\frac{\left[1\right|\sigma_\mu\left|2\right\rangle\left\langle 3\right|\overline{\sigma}^\mu\left|q\right]}{\sqrt{2}\left[3q\right]}\\ &= -ig\,\frac{\left\langle 2\right|\overline{\sigma}_\mu\left|1\right]\left\langle 3\right|\overline{\sigma}^\mu\left|q\right]}{\sqrt{2}\left[3q\right]} = -ig\sqrt{2}\,\frac{\left\langle 23\right\rangle\left[1q\right]}{\left[3q\right]} = ig\sqrt{2}\,\frac{\left\langle 31\right\rangle^2}{\left\langle 21\right\rangle}.\end{aligned}$$

59.d The momentum dependence of these three-point functions is in fact completely constrained by little-group scaling and the expected dimension of the result. Let us illustrate this in the case of the function $\mathcal{A}_3(1^-_{\bar{q}}2^+_q3^+)$, for which the helicities are $h_1 = -\frac{1}{2}, h_2 = +\frac{1}{2}$ and $h_3 = +1$. For these helicities, little-group scaling implies one of the following two behaviors:

$$\mathcal{A}_3(1^-_{\bar{q}}2^+_q3^+) \propto \underbrace{\left\langle 12\right\rangle^1\left\langle 23\right\rangle^{-2}\left\langle 31\right\rangle^0}_{\text{mass dimension }-1} \quad\text{or}\quad \underbrace{\left[12\right]^{-1}\left[23\right]^2\left[31\right]^0}_{\text{mass dimension }+1}.$$

Since spinors have mass dimension $1/2$ and polarization vectors are dimensionless, as is the gauge coupling g, this three-point function should have mass dimension $+1$. Therefore, only the second form, expressed in terms of square brackets, gives the correct dimension. It has indeed the same momentum dependence as the result in Eq. (4.33).

59.e Consider the $\bar{q}q \to \bar{Q}Q$ process, in which q and Q denote two distinct flavors of quarks. At tree level, there is a single diagram, with a gluon connecting the two quark lines. Since the gluon propagator is diagonal in color, the color factors of this graph are given by the $\mathfrak{su}(N)$ Fierz identity:

$$(t_f^a)_{ij}(t_f^a)_{kl} = \tfrac{1}{2}\delta_{il}\delta_{jk} - \tfrac{1}{2N}\delta_{ij}\delta_{kl},$$

and the full amplitude factorizes as follows:

$$\mathcal{M}_4(1_{\bar{q}}^{i-}2_q^{j+}3_{\bar{Q}}^{k-}4_Q^{l+}) = \left(\tfrac{1}{2}\delta_{il}\delta_{jk} - \tfrac{1}{2N}\delta_{ij}\delta_{kl}\right)\mathcal{A}_4(1_{\bar{q}}^-2_q^+3_{\bar{Q}}^-4_Q^+).$$

The color-stripped amplitude $\mathcal{A}_4(1_{\bar{q}}^-2_q^+3_{\bar{Q}}^-4_Q^+)$ is given by

$$\begin{aligned}\mathcal{A}_4(1_{\bar{q}}^-2_q^+3_{\bar{Q}}^-4_Q^+) &= (-ig)^2\left(\overline{u_-}(p_1)\gamma_\mu u_-(p_2)\right)\left(\overline{u_-}(p_3)\gamma^\mu u_-(p_4)\right)\frac{i}{(p_1+p_2)^2} \\ &= -\frac{ig^2}{(p_1+p_2)^2}\langle 1|\overline{\sigma}_\mu|2]\langle 3|\overline{\sigma}^\mu|4] \\ &= -ig^2\frac{\langle 13\rangle[24]}{\langle 12\rangle[12]} = ig^2\frac{\langle 13\rangle^2}{\langle 12\rangle\langle 34\rangle},\end{aligned}$$

where we have used momentum conservation in the form of $\langle 31\rangle[12] + \langle 34\rangle[42] = 0$ in the last equality.

60. Quark-Antiquark-Gluon Amplitudes (Continued) This is a continuation of Problem 59, in which we explore applications of the BCFW recursion to amplitudes with a quark–antiquark pair.

60.a Use the BCFW recursion in order to obtain the following four-point tree amplitude:

$$\mathcal{A}_4(1_{\bar{q}}^-2_q^+3^-4^+) = 2g^2\frac{\langle 13\rangle^3\langle 23\rangle}{\langle 12\rangle\langle 23\rangle\langle 34\rangle\langle 41\rangle}.$$

60.b Show that the same amplitude with flipped gluon helicities reads

$$\mathcal{A}_4(1_{\bar{q}}^-2_q^+3^+4^-) = 2g^2\frac{\langle 14\rangle^3\langle 24\rangle}{\langle 12\rangle\langle 23\rangle\langle 34\rangle\langle 41\rangle}.$$

60.c Calculate the five-point amplitude $\mathcal{A}_5(1_{\bar{q}}^-2_q^+3^-4^+5^-)$ at tree level, and show that

$$\mathcal{A}_5(1_{\bar{q}}^-2_q^+3^-4^+5^-) = -2\sqrt{2}ig^3\frac{[24]^3[14]}{[12][23][34][45][51]}.$$

60.a Four-Point Amplitude $\mathcal{A}_4(1_{\bar{q}}^- 2_q^+ 3^- 4^+)$ The basic ideas of the BCFW method also apply to amplitudes that contain fermions. In the case of the color-ordered amplitude $\mathcal{A}_4(1_{\bar{q}}^- 2_q^+ 3^- 4^+)$, we can apply a shift to the gluon lines, 3 and 4:

$$\widehat{p}_3 \equiv p_3 + zk, \quad \widehat{p}_4 \equiv p_4 - zk, \quad |k\rangle = |3\rangle, \quad |k] = |4].$$

This shift ensures that the two polarization vectors $\epsilon_-(\widehat{3}; q_3)$ and $\epsilon_+(\widehat{4}; q_4)$ both decrease like z^{-1} at large z. Since the coupling of a gluon to a quark line is independent of z, this choice of shift guarantees that the shifted amplitude vanishes when $|z| \to \infty$. In the present case, there is a unique (fermionic) intermediate propagator that may vanish, and the amplitude therefore factorizes as follows:

$$\mathcal{A}_4(1_{\bar{q}}^- 2_q^+ 3^- 4^+) = \sum_{h=\pm} \mathcal{A}_3(1_{\bar{q}}^- -\widehat{K}_q^h \widehat{4}^+) \frac{i}{(p_2+p_3)^2} \mathcal{A}_3(\widehat{K}_{\bar{q}}^{-h} 2_q^+ \widehat{3}^-), \tag{4.34}$$

where $\widehat{K} \equiv p_1 + \widehat{p}_4 = -p_2 - \widehat{p}_3$ is the shifted momentum of the intermediate propagator. Let us clarify a subtlety here: the propagator of the shifted intermediate quark has a numerator $\widehat{\not{K}}$, which may be written as

$$\widehat{\not{K}} = |\widehat{K}]\langle\widehat{K}| - |\widehat{K}\rangle[\widehat{K}|.$$

The worrisome minus sign on the right-hand side of this formula is in fact absorbed into the minus sign of $\overline{u_+}(p) = -[p|$ (see Eq. (4.32) in Problem 59), so that the factor $\widehat{\not{K}}$ is simply the sum over the two possible helicity assignments for the intermediate fermion.

Note that only the choice $h = +$ is allowed in (4.34). Moreover, with the shift chosen above, we have

$$|\widehat{3}\rangle = |3\rangle, \quad |\widehat{3}] = |3] + z|4], \quad |\widehat{4}\rangle = |4\rangle - z|3\rangle, \quad |\widehat{4}] = |4]. \tag{4.35}$$

Using the results of Problem 59 for three-point amplitudes involving fermions, we first obtain

$$\mathcal{A}_4(1_{\bar{q}}^- 2_q^+ 3^- 4^+) = -2ig^2 \frac{[-\widehat{K}4]^2}{[1-\widehat{K}]} \frac{\langle\widehat{K}3\rangle^2}{\langle\widehat{K}2\rangle} \frac{1}{\langle 23\rangle[23]}.$$

Note that

$$-[4|\widehat{K}|3\rangle = -[4\widehat{K}]\langle\widehat{K}3\rangle = [42]\langle 23\rangle + [43]\underbrace{\langle 33\rangle}_{0} + z[44]\underbrace{\langle 33\rangle}_{0},$$

from which we get

$$[-\widehat{K}4]^2\langle\widehat{K}3\rangle^2 = -[42]^2\langle 23\rangle^2.$$

(We have used $|-\widehat{K}] = i|\widehat{K}]$.) Likewise, we have

$$-[1|\widehat{K}|2\rangle = -[1\widehat{K}]\langle\widehat{K}2\rangle = [12]\underbrace{\langle 22\rangle}_{0} + [13]\langle 32\rangle + z[14]\langle 32\rangle.$$

To simplify this expression, we need the value of z at the pole. It can be determined by

$$0 = \widehat{K}^2 = (p_1 + \widehat{p}_4) = \langle 1\widehat{4}\rangle[14] \quad \Leftrightarrow \quad 0 = \langle 1\widehat{4}\rangle = \langle 14\rangle - z\langle 13\rangle$$
$$\Leftrightarrow \quad z = \frac{\langle 14\rangle}{\langle 13\rangle}.$$

Therefore, we have

$$-[1\widehat{K}]\langle\widehat{K}2\rangle = [13]\langle 32\rangle + \frac{\langle 14\rangle}{\langle 13\rangle}[14]\langle 32\rangle = \frac{\langle 32\rangle(s_{13}+s_{14})}{\langle 13\rangle} = -\frac{\langle 32\rangle s_{12}}{\langle 13\rangle}.$$

(We have used $s_{12} + s_{13} + s_{14} = 0$.) Finally, this gives

$$[1-\widehat{K}]\langle\widehat{K}2\rangle = i\frac{\langle 32\rangle s_{12}}{\langle 13\rangle}.$$

Replacing all these intermediate results in the formula for the four-point amplitude, we obtain

$$\mathcal{A}_4(1^-_{\bar{q}}2^+_q3^-4^+) = -2g^2\frac{[24]^2\langle 13\rangle}{[12][23]\langle 12\rangle}. \tag{4.36}$$

This expression can be simplified a bit more by noting that momentum conservation implies

$$\frac{[24]}{[12]} = -\frac{\langle 13\rangle}{\langle 34\rangle}, \quad \frac{[24]}{[23]} = -\frac{\langle 13\rangle}{\langle 14\rangle}.$$

This leads to

$$\mathcal{A}_4(1^-_{\bar{q}}2^+_q3^-4^+) = 2g^2\frac{\langle 13\rangle^3\langle 23\rangle}{\langle 12\rangle\langle 23\rangle\langle 34\rangle\langle 41\rangle}.$$

An alternative route for simplifying (4.36) is to use

$$\frac{\langle 13\rangle}{\langle 12\rangle} = -\frac{[24]}{[34]},$$

which leads to the following expression:

$$\mathcal{A}_4(1^-_{\bar{q}}2^+_q3^-4^+) = 2g^2\frac{[24]^3}{[12][23][34]}.$$

60.b Four-Point Amplitude $\mathcal{A}_4(1^-_{\bar{q}}2^+_q3^+4^-)$ The method is the same as in the preceding case, and we shall skip some details. The main difference is that the above shift is not legitimate since it does not ensure that the shifted amplitude goes to zero when $|z| \to \infty$ (if we insist on using this shift anyway, the BCFW factorization appears to give zero – in fact, the entire result would come from the integral on the circle at infinity). Instead, we use the following shift:

$$|k\rangle = |4\rangle, \quad |k] = |3],$$
$$|\widehat{3}\rangle = |3\rangle + z|4\rangle, \quad |\widehat{3}] = |3], \quad |\widehat{4}\rangle = |4\rangle, \quad |\widehat{4}] = |4] - z|3].$$

The pole condition may be determined by

$$0 = \widehat{K}^2 = (p_1 + \widehat{p}_4)^2 = \langle 14\rangle[1\widehat{4}] \quad \Leftrightarrow \quad z = \frac{[14]}{[13]}.$$

We also have

$$\langle 1\text{–}\widehat{K}\rangle[\widehat{K}2] = i\, s_{12}\frac{[32]}{[13]}.$$

The BCFW recursion produces a single term:

$$\mathcal{A}_4(1^-_{\bar{q}}2^+_q3^+4^-) = -2ig^2\frac{\langle 41\rangle^2}{\langle 1\text{–}\widehat{K}\rangle}\frac{[23]^2}{[\widehat{K}2]}\frac{1}{\langle 23\rangle[23]} = 2g^2\frac{\langle 14\rangle^2}{\langle 12\rangle\langle 13\rangle}\frac{[13]}{[12]}$$
$$= 2g^2\frac{\langle 14\rangle^3\langle 24\rangle}{\langle 12\rangle\langle 23\rangle\langle 34\rangle\langle 41\rangle}.$$

(The last equality uses $[13]/[12] = -\langle 24\rangle/\langle 34\rangle$, which is obtained from momentum conservation.)

60.c Five-Point Amplitude $\mathcal{A}_5(1^-_{\bar{q}}2^+_q3^-4^+5^-)$ Let us now consider the tree-level five-point amplitude $\mathcal{A}_5(1^-_{\bar{q}}2^+_q3^-4^+5^-)$. We can apply the shift defined by Eqs. (4.35) to lines 3 and 4, which produces two terms in the BCFW recursion:

$$\mathcal{A}_5(1^-_{\bar{q}}2^+_q3^-4^+5^-) = \mathcal{A}_3(\widehat{4}^+5^-\text{–}\widehat{K}^-)\frac{i}{(p_4+p_5)^2}\mathcal{A}_4(\widehat{K}^+1^-_{\bar{q}}2^+_q\widehat{3}^-)$$
$$+ \mathcal{A}_4(\widehat{4}^+5^-1^-_{\bar{q}}\text{–}\widehat{K}^+_q)\frac{i}{(p_2+p_3)^2}\mathcal{A}_3(\widehat{K}^-_{\bar{q}}2^+_q\widehat{3}^-).$$

(In the first term, the intermediate propagator is a gluon, while it is a fermion in the second term.) In this equation, we have included only the helicity assignments of the intermediate states that produce non-zero results. Note that all the sub-amplitudes that appear on the right-hand side are known.

Let us start with the first term. In this term, we have $\widehat{K} = \widehat{p}_4 + p_5$, and the pole condition can be written as

$$0 = \widehat{K}^2 = \langle\widehat{4}5\rangle\underbrace{[\widehat{4}5]}_{[45]} \quad \Leftrightarrow \quad 0 = \langle\widehat{4}5\rangle.$$

Also using momentum conservation, we obtain $\langle\widehat{K}4\rangle = \langle\widehat{K}5\rangle = 0$, which implies that the three-gluon amplitude on the left of the first term is zero:

$$\mathcal{A}_3(\widehat{4}^+5^- -\widehat{K}^-) \propto \frac{\langle 5 -\widehat{K}\rangle^3}{\langle -\widehat{K}\widehat{4}\rangle\langle \widehat{4}5\rangle} \propto \frac{(z-z_*)^3}{(z-z_*)^2} \underset{z\to z_*}{\longrightarrow} 0.$$

Therefore, we need only consider the second term, and we have

$$\mathcal{A}_5(1_{\bar{q}}^- 2_q^+ 3^- 4^+ 5^-) = -2\sqrt{2}ig^3 \frac{\langle 15\rangle^3 \langle \widehat{K}5\rangle}{\langle 1\widehat{K}\rangle\langle \widehat{K}\widehat{4}\rangle\langle \widehat{4}5\rangle\langle 51\rangle} \frac{1}{\langle 23\rangle[23]} \frac{\langle \widehat{K}3\rangle^2}{\langle \widehat{K}2\rangle}.$$

(Note that $|-\widehat{K}\rangle = -i|\widehat{K}\rangle$ if $|-\widehat{K}] = i|\widehat{K}]$; the sign difference in the prefactor is necessary in order to have $|\widehat{K}]^\dagger = \langle \widehat{K}|$ and $|-\widehat{K}]^\dagger = \langle -\widehat{K}|$.) In this term, the intermediate momentum is $\widehat{K} = -p_2 - \widehat{p}_3$, and the pole condition reads

$$0 = \widehat{K}^2 = \underbrace{\langle 2\widehat{3}\rangle}_{\langle 23\rangle} [2\widehat{3}] \quad \Leftrightarrow \quad [2\widehat{3}] = 0 \quad \Leftrightarrow \quad z = -\frac{[23]}{[24]}.$$

Using momentum conservation in the form of $\widehat{K} + p_2 + \widehat{p}_3$ or $p_1 + \widehat{p}_4 + p_5 = \widehat{K}$, we also obtain

$$\frac{\langle \widehat{K}3\rangle}{\langle \widehat{K}2\rangle} = -\frac{[24]}{[\widehat{3}4]}, \quad \frac{\langle \widehat{K}3\rangle}{\langle 23\rangle} = -\frac{[24]}{[\widehat{K}4]}, \quad \frac{\langle \widehat{K}5\rangle}{\langle \widehat{K}\widehat{4}\rangle} = -\frac{[41]}{[51]},$$

$$\langle 1\widehat{K}\rangle[\widehat{K}4] = -\langle 12\rangle[24] - \langle 13\rangle \underbrace{[\widehat{3}4]}_{[34]} = \langle 15\rangle[54].$$

(The final equality in the second line uses momentum conservation in the form of $p_1 + p_2 + p_3 + p_4 + p_5 = 0$.) At this point, we have

$$\mathcal{A}_5(1_{\bar{q}}^- 2_q^+ 3^- 4^+ 5^-) = 2\sqrt{2}ig^3 \frac{\langle 15\rangle[24]^2[41]}{[23][34][45][51]\langle \widehat{4}5\rangle}.$$

We also have

$$\langle \widehat{4}5\rangle = \langle 45\rangle - z\langle 35\rangle = \frac{[24]\langle 45\rangle + [23]\langle 35\rangle}{[24]} = \frac{[12]\langle 15\rangle}{[24]},$$

which leads to

$$\mathcal{A}_5(1_{\bar{q}}^- 2_q^+ 3^- 4^+ 5^-) = -2\sqrt{2}ig^3 \frac{[24]^3[14]}{[12][23][34][45][51]}.$$

Note the similarity of this result with the anti-MHV amplitudes (gluonic amplitudes with two positive helicities, and any number of negative ones). In the present case, the amplitude is also expressible as a single term that depends solely on square brackets.

Additional Note: There is an interesting general correspondence between the tree-level amplitudes with a quark–antiquark pair and their counterpart with gluons only. Let us recall here some results for three-, four- and five-point gluon amplitudes:

$$\mathcal{A}_3(1^-2^+3^-) = g\sqrt{2}\,\frac{\langle 31\rangle^3}{\langle 12\rangle\langle 23\rangle}, \quad \mathcal{A}_3(1^-_{\bar{q}}2^+_q3^-) = ig\sqrt{2}\,\frac{\langle 31\rangle^2}{\langle 12\rangle},$$

$$\mathcal{A}_4(1^-2^+3^-4^+) = 2ig^2\,\frac{[24]^4}{[12]\,[23]\,[34]\,[41]}, \quad \mathcal{A}_4(1^-_{\bar{q}}2^+_q3^-4^+) = 2g^2\,\frac{[24]^3}{[12]\,[23]\,[34]},$$

$$\mathcal{A}_5(1^-2^+3^-4^+5^-) = -2\sqrt{2}g^3\,\frac{[24]^4}{[12]\,[23]\,[34]\,[45]\,[51]},$$

$$\mathcal{A}_5(1^-_{\bar{q}}2^+_q3^-4^+5^-) = -2\sqrt{2}ig^3\,\frac{[24]^3\,[14]}{[12]\,[23]\,[34]\,[45]\,[51]}.$$

It turns out that we have, in all these cases,

$$\mathcal{A}_n(1^-_{\bar{q}}2^+_q3^-[4^+5^-\cdots n^-]) = \mathcal{A}_n(1^-2^+3^-[4^+5^-\cdots n^-]) \times i\,\frac{[14]}{[24]},$$

where the dots denote any number of negative-helicity gluons. (In the case of the three-point functions, we used the fact that $\langle 23\rangle/\langle 31\rangle = [1q]/[2q]$ for any q, in particular for $q = p_4$.) Although we have proven this correspondence only for $n = 3, 4, 5$, it is in fact true for any n. As an alternative to a proof by explicit calculation of the relevant amplitudes, as we have done here, this identity also follows from considerations in supersymmetric extensions of Yang–Mills theory.

61. Gluon-Gluon Scattering in the Spinor-Helicity Framework The goal of this problem is to use the spinor-helicity framework and the Parke–Taylor formula in order to calculate the color/helicity summed squared amplitude for gluon–gluon elastic scattering at tree level,

$$\sum_{\substack{a,b,c,d\\ \text{helicities}}} \left|\mathcal{M}_4(1_a2_b3_c4_d)\right|^2 = 16g^4N^2(N^2-1)\Big(3 - \frac{tu}{s^2} - \frac{us}{t^2} - \frac{st}{u^2}\Big).$$

Although this is of course a very well-known result, the traditional approach for obtaining it leads to quite intricate intermediate expressions.

61.a Establish the *Kleiss–Kuijf relation*,

$$\mathcal{A}_4(1^-2^-3^+4^+) + \mathcal{A}_4(1^-3^+2^-4^+) + \mathcal{A}_4(1^-2^-4^+3^+) = 0.$$

61.b Using also the cyclic invariance (e.g., $\mathcal{A}_4(1^-2^-3^+4^+) = \mathcal{A}_4(2^-3^+4^+1^-)$) and mirror symmetry (e.g., $\mathcal{A}_4(1^-2^-3^+4^+) = \mathcal{A}_4(4^+3^+2^-1^-)$) of the color-ordered amplitudes, prove that the full-color four-point amplitude can be written in terms of only two color-ordered ones,

$$\mathcal{M}_4(1^{h_1}_a2^{h_2}_b3^{h_3}_c4^{h_4}_d) = f^{cae}f^{bde}\mathcal{A}_4(1^{h_1}3^{h_3}2^{h_2}4^{h_4}) - f^{abe}f^{cde}\mathcal{A}_4(1^{h_1}2^{h_2}3^{h_3}4^{h_4}).$$

61.c Use these results and the Parke–Taylor formula to obtain the announced expression for gluon–gluon scattering.

61.a The basic color decomposition of a four-point tree amplitude has $(4-1)! = 6$ terms, and its square would therefore have 36 terms. But, as we shall see, this can be drastically reduced thanks to various symmetries of color-ordered amplitudes. Consider first the following sum of partial amplitudes:

$$\begin{aligned} \mathcal{A}_4(1^-2^-3^+4^+) + \mathcal{A}_4(1^-3^+2^-4^+) &\propto \frac{\langle 12\rangle^4}{\langle 12\rangle\langle 23\rangle\langle 34\rangle\langle 41\rangle} + \frac{\langle 12\rangle^4}{\langle 13\rangle\langle 32\rangle\langle 24\rangle\langle 41\rangle} \\ &\propto \frac{\langle 12\rangle^4}{\langle 23\rangle\langle 41\rangle}\,\frac{\langle 31\rangle\langle 24\rangle + \langle 12\rangle\langle 34\rangle}{\langle 12\rangle\langle 34\rangle\langle 31\rangle\langle 24\rangle} \\ &\propto -\frac{\langle 12\rangle^4}{\langle 23\rangle\langle 41\rangle}\,\frac{\langle 23\rangle\langle 14\rangle}{\langle 12\rangle\langle 34\rangle\langle 31\rangle\langle 24\rangle} \\ &\propto -\frac{\langle 12\rangle^4}{\langle 12\rangle\langle 24\rangle\langle 43\rangle\langle 31\rangle} \\ &= -\mathcal{A}_4(1^-2^-4^+3^+). \end{aligned}$$

(The unwritten prefactor is always $i(g\sqrt{2})^2$.) This linear relationship among color-ordered amplitudes is an example of the *Kleiss–Kuijf relations*. Note how the factor $\langle 12\rangle^4$ in the numerator plays no role in these manipulations. Therefore, the same identity holds for any choice of the pair of legs carrying the negative helicities.

61.b From the Parke–Taylor formula, we see that color-ordered amplitudes are cyclically invariant, as long as the two legs carrying the negative helicities stay the same. For instance, one has

$$\mathcal{A}_4(1^-2^-3^+4^+) = \mathcal{A}_4(2^-3^+4^+1^-).$$

Consider then the color-ordered amplitude with the mirror-symmetric ordering of the external legs, $4^+3^+2^-1^-$. According to the Parke–Taylor formula, it reads

$$\mathcal{A}_4(4^+3^+2^-1^-) \propto \frac{\langle 21\rangle^4}{\langle 43\rangle\langle 32\rangle\langle 21\rangle\langle 14\rangle}.$$

Since $\langle ij\rangle = -\langle ji\rangle$ and there is an even number of angle products in the numerator and denominator, we have

$$\mathcal{A}_4(1^-2^-3^+4^+) = \mathcal{A}_4(4^+3^+2^-1^-).$$

Now, start from the original color decomposition of a four-point amplitude, and eliminate all the terms that are not independent according to above identities:

$$\begin{aligned}\tfrac{1}{2}\,\mathcal{M}_4(1^-_a 2^-_b 3^+_c 4^+_d) &= \mathrm{tr}\,(t_f^a t_f^b t_f^c t_f^d)\,A_4(1^-2^-3^+4^+) + \mathrm{tr}\,(t_f^a t_f^b t_f^d t_f^c)\,A_4(1^-2^-4^+3^+)\\ &\quad + \mathrm{tr}\,(t_f^a t_f^c t_f^b t_f^d)\,A_4(1^-3^+2^-4^+) + \mathrm{tr}\,(t_f^a t_f^c t_f^d t_f^b)\,A_4(1^-3^+4^+2^-)\\ &\quad + \mathrm{tr}\,(t_f^a t_f^d t_f^b t_f^c)\,A_4(1^-4^+2^-3^+) + \mathrm{tr}\,(t_f^a t_f^d t_f^c t_f^b)\,A_4(1^-4^+3^+2^-)\\ &= \mathrm{tr}\,(t_f^a t_f^b t_f^c t_f^d + t_f^a t_f^d t_f^c t_f^b)\,A_4(1^-2^-3^+4^+)\\ &\quad + \mathrm{tr}\,(t_f^a t_f^b t_f^d t_f^c + t_f^a t_f^c t_f^d t_f^b)\,A_4(1^-2^-4^+3^+)\\ &\quad + \mathrm{tr}\,(t_f^a t_f^c t_f^b t_f^d + t_f^a t_f^d t_f^b t_f^c)\,A_4(1^-3^+2^-4^+).\end{aligned}$$

Using the Kleiss–Kuijf relation, the second term in this equation can be rewritten as a linear combination of the first and third terms:

$$\begin{aligned}\tfrac{1}{2}\,\mathcal{M}_4(1^-_a 2^-_b 3^+_c 4^+_d) &= \mathrm{tr}\,(t_f^a t_f^b t_f^c t_f^d + t_f^a t_f^d t_f^c t_f^b - t_f^a t_f^b t_f^d t_f^c - t_f^a t_f^c t_f^d t_f^b)\,A_4(1^-2^-3^+4^+)\\ &\quad + \mathrm{tr}\,(t_f^a t_f^c t_f^b t_f^d + t_f^a t_f^d t_f^b t_f^c - t_f^a t_f^b t_f^d t_f^c - t_f^a t_f^c t_f^d t_f^b)\,A_4(1^-3^+2^-4^+).\end{aligned}$$

The color traces can be easily evaluated in terms of the structure constants. For instance, the first one can be rewritten as

$$\begin{aligned}&\mathrm{tr}\,(t_f^a t_f^b t_f^c t_f^d + t_f^a t_f^d t_f^c t_f^b - t_f^a t_f^b t_f^d t_f^c - t_f^a t_f^c t_f^d t_f^b)\\ &\quad = \mathrm{tr}\,(t_f^a t_f^b [t_f^c, t_f^d] - t_f^a [t_f^c, t_f^d] t_f^b) = i\, f^{cde}\,\mathrm{tr}\,(t_f^a t_f^b t_f^e - t_f^a t_f^e t_f^b)\\ &\quad = i\, f^{cde}\,\mathrm{tr}\,([t_f^a, t_f^b] t_f^e) = \tfrac{i^2}{2}\, f^{abe} f^{cde}.\end{aligned}$$

The second color trace is calculated in the same way, leading to the following expression with only two terms:

$$\mathcal{M}_4(1^-_a 2^-_b 3^+_c 4^+_d) = f^{cae} f^{bde}\,A_4(1^-3^+2^-4^+) - f^{abe} f^{cde}\,A_4(1^-2^-3^+4^+). \tag{4.37}$$

61.c At this point, let us note that this decomposition is valid for all the possible helicity assignments (cyclic and mirror symmetries are readily checked to be independent of the helicity assignment, and we have pointed out earlier that this is also true of the Kleiss–Kuijf relation). When we square this expression and sum it over all the gluon color indices a, b, c, d, we encounter various types of color contractions:

$$\begin{aligned}f^{cae} f^{bde} f^{caf} f^{bdf} &= N^2 \delta_{ef} \delta_{ef} = N^2(N^2-1),\\ f^{abe} f^{cde} f^{abf} f^{cdf} &= N^2 \delta_{ef} \delta_{ef} = N^2(N^2-1),\\ -f^{cae} f^{bde} f^{abf} f^{cdf} &= \underbrace{f^{ace} f^{bed} f^{fdc}}_{\frac{N}{2} f^{abf}}\, f^{abf} = \frac{1}{2} N^2(N^2-1).\end{aligned}$$

Therefore, the color-summed squared modulus of (4.37) reads

$$\sum_{a,b,c,d} \left|\mathcal{M}_4(1_a 2_b 3_c 4_d)\right|^2 = N^2(N^2-1)\Big(\left|\mathcal{A}_4(1234)\right|^2 + \left|\mathcal{A}_4(1324)\right|^2 + \tfrac{1}{2}\mathcal{A}_4(1234)\mathcal{A}_4^*(1324) + \tfrac{1}{2}\mathcal{A}_4^*(1234)\mathcal{A}_4(1324)\Big).$$

Now, we use the Parke–Taylor formula for the partial amplitudes:

$$\mathcal{A}_4^{i^-j^-}(1234) = 2g^2 \frac{\langle ij\rangle^4}{\langle 12\rangle\langle 23\rangle\langle 34\rangle\langle 41\rangle} = 2g^2 \frac{[ij]^4}{[12]\,[23]\,[34]\,[41]},$$

$$\mathcal{A}_4^{i^-j^-}(1324) = 2g^2 \frac{\langle ij\rangle^4}{\langle 13\rangle\langle 32\rangle\langle 24\rangle\langle 41\rangle} = 2g^2 \frac{[ij]^4}{[13]\,[32]\,[24]\,[41]},$$

where the notation on the left-hand side indicates which are the two lines carrying the two negative helicities (recall that for four-point amplitudes, the only possibility is to have two helicities of each sign). On the right-hand side, we have used the form of the formulas that have a common denominator. From these amplitudes, it is easy to obtain the squares,

$$\left|\mathcal{A}_4^{i^-j^-}(1234)\right|^2 = 4g^4 \frac{s_{ij}^4}{s_{12}s_{23}s_{34}s_{41}} = 4g^4 \frac{s_{ij}^4}{s^2u^2},$$

$$\left|\mathcal{A}_4^{i^-j^-}(1234)\right|^2 = 4g^4 \frac{s_{ij}^4}{s_{13}s_{32}s_{24}s_{41}} = 4g^4 \frac{s_{ij}^4}{t^2u^2},$$

where we have introduced $s_{ij} \equiv \langle ij\rangle\,[ij] = (p_i + p_j)^2$ and the Mandelstam variables, $s \equiv s_{12} = s_{34}$, $t \equiv s_{13} = s_{24}$ and $u \equiv s_{14} = s_{23}$. The mixed products are slightly more complicated:

$$\mathcal{A}_4^{i^-j^-}(1234)\mathcal{A}_4^{i^-j^-*}(1324) = -4g^4 \frac{s_{ij}^4}{u^2\ \underbrace{\langle 12\rangle\,[24]\,\langle 43\rangle\,[31]}_{\mathrm{tr}\,(P_1\overline{P}_2P_3\overline{P}_4)=\frac{s^2+t^2-u^2}{2}}} = 4g^4 \frac{s_{ij}^4}{u^2st},$$

where we have used $s + t + u = 0$. The other mixed product has the same value. Therefore, we have

$$\sum_{a,b,c,d} \left|\mathcal{M}_4(1_a 2_b 3_c 4_d)\right|^2_{i^-j^-} = 4g^4N^2(N^2-1)\frac{s_{ij}^4}{u^2}\Big(\frac{1}{s^2} + \frac{1}{t^2} + \frac{1}{st}\Big).$$

By appropriately setting the values of (i,j), this expression gives the squared amplitudes for all the polarized gluon–gluon scattering processes at tree level. Summing over the helicities amounts to summing over the pairs (i, j) of indices that carry the two negative helicities:

$$\sum_{\substack{a,b,c,d\\ \text{helicities}}} \left|\mathcal{M}_4(1_a 2_b 3_c 4_d)\right|^2$$
$$= 4g^4N^2(N^2-1)\frac{s_{12}^4+s_{13}^4+s_{14}^4+s_{23}^4+s_{24}^4+s_{34}^4}{u^2}\left(\frac{1}{s^2}+\frac{1}{t^2}+\frac{1}{st}\right)$$
$$= 8g^4N^2(N^2-1)\frac{s^4+t^4+u^4}{u^2}\left(\frac{1}{s^2}+\frac{1}{t^2}+\frac{1}{st}\right)$$
$$= 16g^4N^2(N^2-1)\left(3-\frac{tu}{s^2}-\frac{us}{t^2}-\frac{st}{u^2}\right).$$

(Since $s+t+u=0$, there are many equivalent ways of writing this result. The expression in the final line is one of the most frequently encountered forms.) Even though this tree-level scattering process receives contributions from only four Feynman diagrams in the traditional approach, it is a rather hairy calculation due to the proliferation of Lorentz and color indices. The present approach brings an appreciable order into this calculation, by considerably simplifying the bookkeeping, and by hiding from the view the complicated expressions for the three-gluon and four-gluon vertices.

62. Soft Limit of Gluon Amplitudes The goal of this problem is to translate the eikonal formula (see for instance Eq. (3.51) in Problem 43) for the emission of a soft gluon from a hard process in the language of the spinor-helicity formalism (at tree level).

62.a Show that the insertion of a soft gluon of momentum k between adjacent lines i and j of a tree-level color-ordered amplitude brings an extra factor:

$$S(i,k^+,j) \equiv ig\sqrt{2}\,\frac{\langle ij\rangle}{\langle ik\rangle\langle kj\rangle} \qquad \text{if the gluon has a positive helicity,}$$
$$S(i,k^-,j) \equiv ig\sqrt{2}\,\frac{[ij]}{[ik]\,[kj]} \qquad \text{if the gluon has a negative helicity.}$$

In the next questions, we check on various examples studied in this chapter that, in the limit where a gluon becomes soft, amplitudes indeed factorize into such a soft factor and an amplitude with this gluon removed.

62.b Check this for the gluonic MHV amplitudes.

62.c Check the limit $p_5 \to 0$ of $\mathcal{A}_5(1^-_{\bar{q}}2^+_q3^-4^+5^-)$ derived in Problem 60.

62.d Check the limit $p_5 \to 0$ of $\mathcal{A}_6(1^-2^-3^-4^+5^+6^+)$ obtained in Problem 52.

62.a Here, we need to reproduce the calculation done in Problem 43 with the color-ordered Feynman rules. Let us start from an amplitude $\epsilon^{\nu}_{h'}(\mathbf{p})\Gamma_\nu(p,\dots)$, stripped of all the color factors (we focus on a single gluon line, and ignore the other external lines). As usual, the convention is that the momentum p is incoming. When we attach another gluon of momentum k to this external leg, we must consider the combination

$$\epsilon^{\nu}_{h'}(\mathbf{p};q')\epsilon^{\mu}_{h}(\mathbf{k};q)V_{\mu\nu\rho}(k,p,-(p+k))\frac{-i\,g^{\rho\rho'}}{(p+k)^2}\Gamma_{\rho'}(p+k,\dots),$$

where $V_{\mu\nu\rho}$ is the color-stripped three-gluon vertex. We use the Feynman gauge for the intermediate propagator. Then, we insert the explicit expression for the three-gluon vertex, and assume that the additional gluon is soft, i.e., $k \ll p, p+k \approx p$. In this approximation, the above expression becomes

$$\begin{aligned} &g\,\epsilon^{\nu}_{h'}(\mathbf{p};q')\epsilon^{\mu}_{h}(\mathbf{k};q)\Big(\underline{g_{\mu\rho}(-p_\rho)}+g_{\nu\rho}(2p_\mu)-\underline{g_{\rho\mu}p_\nu}\Big)\frac{-i\,g^{\rho\rho'}}{2p\cdot k}\Gamma_{\rho'}(p,\dots)\\ &\quad=-ig\,\frac{p\cdot\epsilon_h(\mathbf{k};q)}{p\cdot k}\,\epsilon^{\nu}_{h'}(\mathbf{p})\Gamma_\nu(\mathbf{p},\dots).\end{aligned}$$

On the left-hand side, the first underlined term vanishes thanks to the Ward–Takahashi identity obeyed by $\Gamma_\rho(p,\dots)$, and the second underlined term vanishes because the polarization vector $\epsilon^{\nu}_{h'}(\mathbf{p};q')$ is physical (i.e., transverse). Thus, attaching a soft gluon to an incoming line of momentum p simply corrects the original amplitude by a factor

$$-ig\,\frac{p\cdot\epsilon_h(\mathbf{k};q)}{p\cdot k}.$$

Consider now a color-ordered amplitude $\mathcal{A}_n(\cdots ij\cdots)$ (we do not need to specify the helicities of its external lines), and insert a soft gluon of momentum k and helicity h between the *adjacent* lines i and j. The ordering means that the gluon can only be attached to the line i or to the line j, and this insertion therefore corrects the amplitude by a factor

$$S(i,k^h,j)\equiv -ig\left\{\frac{p_i\cdot\epsilon_h(\mathbf{k};q)}{p_i\cdot k}+\frac{p_j\cdot\epsilon(\mathbf{k};q)}{p_j\cdot k}\right\}.$$

Let us first treat the case of a positive helicity, $h=+$. We have

$$S(i,k^+,j)=ig\sqrt{2}\left\{\frac{\langle q|\overline{P}_i|k]}{\langle qk\rangle\langle ik\rangle[ik]}+\frac{\langle q|\overline{P}_j|k]}{\langle qk\rangle\langle jk\rangle[jk]}\right\}.$$

(Recall that $p_\mu\overline{\sigma}^\mu=-\overline{P}$.) Then, this factor may be calculated as follows:

$$
\begin{aligned}
S(i,k^+,j) &= -i\frac{g\sqrt{2}}{\langle qk\rangle}\frac{\langle q|\overline{P}_i|k]\langle kj\rangle[jk]+\langle q|\overline{P}_j|k]\langle ki\rangle[ik]}{\langle ik\rangle[ik]\langle jk\rangle[jk]}\\
&= -i\frac{g\sqrt{2}}{\langle qk\rangle}\frac{\mathrm{tr}\,(\overline{P}_iK\overline{P}_j\eta+\overline{P}_jK\overline{P}_i\eta)}{\langle ik\rangle[ik]\langle jk\rangle[jk]},
\end{aligned}
$$

where $\eta \equiv |k]\langle q|$. The trace can be calculated using the result of Problem 53:

$$
\begin{aligned}
\mathrm{tr}\,(\overline{P}_iK\overline{P}_j\eta+\overline{P}_jK\overline{P}_i\eta) &= 4\left((p_i\cdot k)(p_j\cdot\eta)+(p_j\cdot k)(p_i\cdot\eta)-(p_i\cdot p_j)\underbrace{(k\cdot\eta)}_{0}\right)\\
&= [ik]\,[jk]\,(\langle ik\rangle\langle jq\rangle+\langle jk\rangle\langle iq\rangle).
\end{aligned}
$$

(We have used $2\,p_j\cdot\eta = -\mathrm{tr}\,(\overline{P}_j\eta) = \langle jq\rangle[jk]$.) At this stage, we have

$$
S(i,k^+,j) = -i\frac{g\sqrt{2}}{\langle qk\rangle}\frac{\langle ik\rangle\langle jq\rangle+\langle jk\rangle\langle iq\rangle}{\langle ik\rangle\langle jk\rangle} = ig\sqrt{2}\,\frac{\langle ij\rangle}{\langle ik\rangle\langle kj\rangle},
$$

where we have made the choice $q = p_i$ of auxiliary vector in order to obtain the last equality without effort (but the result does not depend on this choice, since it just amounts to a gauge redefinition of the polarization vector). Now, let us repeat this calculation for the case of a negative helicity:

$$
\begin{aligned}
S(i,k^-,j) &= ig\sqrt{2}\left\{\frac{\langle k|\overline{P}_i|q]}{[kq]\langle ik\rangle[ik]}+\frac{\langle k|\overline{P}_j|q]}{[kq]\langle jk\rangle[jk]}\right\}\\
&= -i\frac{g\sqrt{2}}{[kq]}\frac{\mathrm{tr}\,(\overline{P}_iK\overline{P}_j\eta'+\overline{P}_jK\overline{P}_i\eta')}{\langle ik\rangle[ik]\langle jk\rangle[jk]} \qquad (\eta'\equiv|q]\langle k|)\\
&= -i\frac{g\sqrt{2}}{[kq]}\frac{[jk]\,[iq]+[ik]\,[jq]}{[ik]\,[jk]} = ig\sqrt{2}\,\frac{[ij]}{[ik]\,[kj]}.
\end{aligned}
$$

Note that the *soft factors* $S(i,k^\pm,j)$ are invariant under rescalings of the spinors $|i\rangle$ and $|j\rangle$. In other words, they depend on the orientations of the momenta p_i and p_j, but not on their magnitudes.

62.b A particularly simple occurrence of these soft factors is in MHV amplitudes. Consider, for instance, an n-point MHV color-ordered amplitude, expressed via the Parke–Taylor formula:

$$
\mathcal{A}_n(\cdots i^-\cdots j^-\cdots) = -i(ig\sqrt{2})^{n-2}\frac{\langle ij\rangle^4}{\langle 12\rangle\langle 23\rangle\cdots\langle n1\rangle}. \tag{4.38}
$$

(We have explicitly indicated only the two legs that carry negative helicities; all the other legs carry positive ones.) Consider now the limit where a certain leg k has a momentum that goes to zero (by this, we mean that all the components of k^μ go to zero — we may think of rescaling $k\to\xi k$ with $\xi\to 0$). In this limit, the angle spinor $|k\rangle$ goes to zero like the square root of the scaling factor ξ, while all the other angle spinors stay constant. The spinor $|k\rangle$ appears twice in the denominator of (4.38). This spinor is absent from the numerator if $k \notin \{i, j\}$ and appears to the power 4 if $k \in \{i, j\}$. Therefore, the limit is as follows:

$$\mathcal{A}_n(\cdots i^- \cdots j^- \cdots) \underset{k\notin\{i,j\}}{\longrightarrow} \underbrace{ig\sqrt{2}\frac{\langle k-1\,k+1\rangle}{\langle k-1\,k\rangle\langle k\,k+1\rangle}}_{S(k-1,k^+,k+1)} \times \underbrace{\left[-i(ig\sqrt{2})^{n-1}\right]\frac{\langle ij\rangle^4}{\langle 12\rangle\cdots\langle k-1\,k+1\rangle\cdots\langle n1\rangle}}_{\text{MHV amplitude with } k^+ \text{ removed}},$$

$$\mathcal{A}_n(\cdots i^- \cdots j^- \cdots) \underset{k\in\{i,j\}}{\longrightarrow} 0.$$

In the first case, the factorization is in fact valid even without taking the limit $k \to 0$. In the second case, the trivial null limit could also have been written as

$$\mathcal{A}_n(\cdots i^- \cdots j^- \cdots) \underset{k=i}{\longrightarrow} \underbrace{ig\sqrt{2}\frac{[k-1\,k+1]}{[k-1\,k]\,[k\,k+1]}}_{S(k-1,k^-,k+1)} \times \underbrace{\mathcal{A}_{n-1}(\cdots j^- \cdots)}_{=0}.$$

(The second factor is zero because it is an amplitude with only one negative helicity.) Therefore, in all cases, when an external leg becomes soft, the amplitude factorizes as follows:

$$\mathcal{A}_n(\cdots) \underset{k\to 0}{\longrightarrow} S(k-1, k^h, k+1) \times \mathcal{A}_n(\cdots(k-1)(k+1)\cdots),$$

i.e., as the product of a soft factor that depends on the helicity of the soft gluon and the amplitude obtained by removing the leg k from the original amplitude.

62.c Another example where it is easy to observe the same factorization is the five-point amplitude with a quark–antiquark pair, derived in Problem 60:

$$\mathcal{A}_5(1^-_{\bar{q}}2^+_q3^-4^+5^-) = -2\sqrt{2}ig^3\frac{[24]^3\,[14]}{[12]\,[23]\,[34]\,[45]\,[51]},$$

which can be written as

$$\mathcal{A}_5(1^-_{\bar{q}}2^+_q3^-4^+5^-) = \underbrace{ig\sqrt{2}\frac{[41]}{[45]\,[51]}}_{S(4,5^-,1)} \times \underbrace{[2g^2]\frac{[24]^3}{[12]\,[23]\,[34]}}_{\mathcal{A}_4(1^-_{\bar{q}}2^+_q3^-4^+)}.$$

Here also, the amplitude (not only in the soft limit, but for all values of p_5, as it turns out) factorizes into a soft factor for the emission of the gluon of momentum p_5, and an amplitude with this gluon removed. Remarkably, this factorization works regardless of the precise nature (gluon or fermion) of the neighboring lines. Moreover, the soft factor does not depend on this nature, which is reminiscent of how the eikonal factors for soft emissions do not depend on the spin of the emitter in the traditional approach.

62.d As a last – rather non-trivial – example, let us consider the six-point amplitude derived in Problem 52:

$$\mathcal{A}_6(1^-2^-3^-4^+5^+6^+) = -\frac{4ig^4}{[2|P_3+P_4|5\rangle}\left\{\frac{\langle 1|\overline{P}_2+\overline{P}_3|4]^3}{[23]\,[34]\,\langle 56\rangle\langle 61\rangle(p_2+p_3+p_4)^2} + \frac{\langle 3|\overline{P}_4+\overline{P}_5|6]^3}{[61]\,[12]\,\langle 34\rangle\langle 45\rangle(p_3+p_4+p_5)^2}\right\}. \tag{4.39}$$

Consider, for instance, the limit where $p_5 \to 0$. We have

$$\begin{aligned}
&\mathcal{A}_6(1^-2^-3^-4^+5^+6^+)\\
&\underset{p_5\to 0}{=} -\frac{4ig^4}{[2|P_3+P_4|5\rangle}\frac{1}{[34]\,[61]}\left\{\frac{\langle 1|\overline{P}_2+\overline{P}_3|4]^3}{[23]\langle 56\rangle\langle 61\rangle^2}+\frac{\langle 3|\overline{P}_4|6]^3}{[12]\langle 45\rangle\langle 34\rangle^2}\right\}\\
&= -\frac{4ig^4}{[2|P_3+P_4|5\rangle}\frac{1}{[34]\,[61]}\left\{-\frac{\langle 1|\overline{P}_6|4]^3}{[23]\langle 56\rangle\langle 61\rangle^2}+\frac{\langle 3|\overline{P}_4|6]^3}{[12]\langle 45\rangle\langle 34\rangle^2}\right\}\\
&= -\frac{4ig^4}{[2|P_3+P_4|5\rangle}\frac{[46]^3}{[34]\,[61]}\left\{-\frac{\langle 61\rangle}{[23]\langle 56\rangle}+\frac{\langle 34\rangle}{[12]\langle 45\rangle}\right\}\\
&= -4ig^4\frac{[46]^3}{[12]\,[23]\,[34]\,[61]}\frac{\langle 46\rangle}{\langle 45\rangle\langle 56\rangle}\underbrace{\frac{[23]\langle 34\rangle\langle 56\rangle-[21]\langle 16\rangle\langle 45\rangle}{\langle 46\rangle[2|P_3+P_4|5\rangle}}_{\equiv A}.
\end{aligned}$$

Let us now evaluate the last factor:

$$\begin{aligned}
A &= \frac{[21]\langle 16\rangle[64]\langle 45\rangle-[23]\langle 34\rangle[46]\langle 65\rangle}{\langle 46\rangle[46]\,[2|P_3+P_5|5\rangle}\\
&= \frac{\mathrm{tr}\,(P_1\overline{P}_6P_4\overline{\eta})-\mathrm{tr}\,(P_3\overline{P}_4P_6\overline{\eta})}{2(p_4\cdot p_6)\,\mathrm{tr}\,((P_3+P_4)\overline{\eta})} \qquad \text{with } \overline{\eta}\equiv|5\rangle[2|.
\end{aligned}$$

The denominator of A expands to

$$\mathrm{deno}\,(A) = -4(p_4\cdot p_6)\,((p_3+p_4)\cdot\eta).$$

The numerator is a bit more complicated:

$$\begin{aligned}
\mathrm{num}\,(A) = 2\Big\{&(p_1\cdot p_6)(p_4\cdot\eta)+(p_1\cdot\eta)(p_4\cdot p_6)-(p_1\cdot p_4)(p_6\cdot\eta)+(p_3\cdot p_6)\\
&-(p_3\cdot\eta)(p_4\cdot p_6)-(p_3\cdot p_4)(p_6\cdot\eta)-i\epsilon(164\eta)+i\epsilon(346\eta)\Big\},
\end{aligned}$$

where we use the notation $\epsilon(ijk\eta)\equiv p_{i\mu}p_{j\nu}p_{k\rho}\eta_\sigma\epsilon^{\mu\nu\rho\sigma}$. Using momentum conservation and $p_5\approx 0$, this may be rearranged into

$$\begin{aligned}\text{num}(A) = &-4(p_4\cdot p_6)((p_3+p_4)\cdot\eta)\\ &+\underbrace{2\Big\{(p_2\cdot p_4)(p_6\cdot\eta)+\underbrace{(p_2\cdot\eta)}_{0}(p_4\cdot p_6)-(p_2\cdot p_6)(p_4\cdot\eta)-i\epsilon(246\eta)\Big\}}_{=\text{tr}(P_2\overline{P}_4P_6\overline{\eta})=0\quad\text{since }\overline{\eta}P_2=|5\rangle[22]\langle 2|=0}.\end{aligned}$$

Therefore, we have shown that $A = 1$, and that

$$\mathcal{A}_6(1^-2^-3^-4^+5^+6^+) \underset{p_5\to 0}{=} \underbrace{ig\sqrt{2}\frac{\langle 46\rangle}{\langle 45\rangle\langle 56\rangle}}_{S(4,5^+,6)} \times \underbrace{\big[-2\sqrt{2}g^3\big]\frac{[46]^3}{[12][23][34][61]}}_{\mathcal{A}_5(1^-2^-3^-4^+6^+)}.$$

Of course, this factorization in the soft limit $p_5 \to 0$ is something we expected to happen on general grounds from the discussion at the beginning of this problem. Thus, one may view this verification as a consistency check of the expression (4.39) for the six-point amplitude. For the test to be more stringent, one should also check the soft limits for the other external momenta (however, keep in mind that these tests are useful for spotting mistakes but cannot be a proof that the formula is correct).

63. Collinear Limit of Gluon Amplitudes In this problem, we study the limit of tree-level color-ordered amplitudes when two external momenta become collinear. As we shall see, this limit is controlled by a set of universal *splitting amplitudes*. They can be used to recover, without much effort, the momentum fraction dependence of the Altarelli–Parisi splitting functions.

63.a Explain why color-ordered amplitudes become singular when the momenta of two adjacent lines become collinear, but not for non-adjacent lines.

63.b Show the following limit:

$$\mathcal{A}_5(1^-2^-3^+4^+5^+) \underset{2\|3}{\approx} \text{Sp}_+(2^-3^+) \times \mathcal{A}_4(1^-P^-4^+5^+),$$

where $\text{Sp}_+(2^-3^+) \equiv ig\,\dfrac{\sqrt{2}}{\langle 23\rangle}\dfrac{z_2^2}{\sqrt{z_2z_3}}$, with $P^\mu \equiv p_2^\mu + p_3^\mu$, $p_{2,3}^\mu \equiv z_{2,3}P^\mu$ in the collinear limit.

63.c By considering other examples, determine all the splitting amplitudes $\text{Sp}_\pm(1^\pm 2^\pm)$. Why are the splitting amplitudes $\text{Sp}_+(1^+2^+)$ and $\text{Sp}_-(1^-2^-)$ vanishing?

63.d Use these results to recover the z dependence of the gluonic Altarelli–Parisi splitting function,

$$P_{gg}(z) \propto \frac{z}{1-z} + \frac{1-z}{z} + z(1-z).$$

63.e Check the collinear limit $3 \parallel 4$ of the six-point amplitude obtained in Problem 52.

63.f By studying the collinear limits of amplitudes with a quark–antiquark pair (see Problem 60), determine the splitting amplitudes involving quarks, and the z dependence of the corresponding Altarelli–Parisi splitting functions.

63.a A color-ordered amplitude can have a pole when a sum of adjacent momenta goes on-shell, since this configuration leads to the vanishing of the denominator of an internal propagator. In the case of two adjacent momenta $p_{1,2}$, this denominator reads

$$s_{12} = (p_1 + p_2)^2 = E_1 E_2 (1 - \cos\theta_{12}),$$

where θ_{12} is the angle between $\mathbf{p}_1$ and $\mathbf{p}_2$ (recall that $p_{1,2}$ are both on-shell and massless). Therefore, this denominator can vanish when $\theta_{12} = 0$, i.e., when the two momenta are collinear. (This argument suggests that the amplitude has a singularity of the form $\langle 12\rangle^{-1}[12]^{-1}$. However, as we shall see, the actual singularity is weaker because of a partial cancellation from the numerator.) In cyclic-ordered amplitudes, nothing special happens when a pair of non-adjacent lines, for instance 1 and 3, become collinear. Indeed, the only denominators that appear in the amplitude are squares of sums of consecutive momenta, of the form $(\sum_{i=a}^{b} p_i)^2$.

63.b Consider now the five-point amplitude $\mathcal{A}_5(1^-2^-3^+4^+5^+)$. This is a MHV amplitude given by

$$\mathcal{A}_5(1^-2^-3^+4^+5^+) = -i(ig\sqrt{2})^3 \frac{\langle 12\rangle^3}{\langle 23\rangle\langle 34\rangle\langle 45\rangle\langle 51\rangle}.$$

Now, we let p_2 and p_3 become collinear. We see immediately that the amplitude is singular, because of the factor $\langle 23\rangle$ in the denominator, but not as singular as expected from the behavior of s_{23}. The strategy is to factor out this singular denominator and to evaluate the non-singular factors precisely at the collinear configuration. When p_2 and p_3 are collinear, we may write

$$P \equiv p_2 + p_3, \quad p_{2,3} \equiv z_{2,3} P, \quad z_2 + z_3 = 1.$$

Similar relationships exist among the angle spinors,

$$|2\rangle = \sqrt{z_2}|\mathbf{P}\rangle, \quad |3\rangle = \sqrt{z_3}|\mathbf{P}\rangle,$$

and it is immediate to obtain

$$\mathcal{A}_5(1^-2^-3^+4^+5^+) \underset{2\|3}{\approx} \underbrace{ig\frac{\sqrt{2}}{\langle 23\rangle}\frac{z_2^2}{\sqrt{z_2 z_3}}}_{\equiv \mathrm{Sp}_+(2^-3^+)} \times \underbrace{\left[-i(ig\sqrt{2})^2\right]\frac{\langle 1\mathbf{P}\rangle^3}{\langle \mathbf{P}4\rangle\langle 45\rangle\langle 51\rangle}}_{\mathcal{A}_4(1^-\mathbf{P}^-4^+5^+)}.$$

The prefactor $\mathrm{Sp}_+(2^-3^+)$, called a *splitting amplitude*, describes how an on-shell gluon of momentum P and positive helicity splits into the two collinear gluons 2^- and 3^+, with the respective momentum fractions $z_{2,3}$ (note that the splitting amplitude is also defined with the convention that all momenta are incoming).

63.c Starting from the same amplitude, we may also consider the collinear limit 3 || 4. We obtain

$$\mathcal{A}_5(1^-2^-3^+4^+5^+) \underset{3\|4}{\approx} \underbrace{ig\frac{\sqrt{2}}{\langle 34\rangle}\frac{1}{\sqrt{z_3z_4}}}_{\equiv\, \mathrm{Sp}_-(3^+4^+)} \times \underbrace{\left[-i(ig\sqrt{2})^2\right]\frac{\langle 12\rangle^3}{\langle 2P\rangle\langle P5\rangle\langle 51\rangle}}_{\mathcal{A}_4(1^-2^-P^+5^+)}.$$

In order to obtain the other splitting amplitudes, we can also start from an anti-MHV amplitude, such as

$$\mathcal{A}_5(1^+2^+3^-4^-5^-) = -i(ig\sqrt{2})^3\frac{[12]^3}{[23]\,[34]\,[45]\,[51]}.$$

The collinear limit 2 || 3 gives

$$\mathcal{A}_5(1^+2^+3^-4^-5^-) \underset{2\|3}{\approx} \underbrace{ig\frac{\sqrt{2}}{[23]}\frac{z_2^2}{\sqrt{z_2z_3}}}_{\mathrm{Sp}_-(2^+3^-)} \times \underbrace{\left[-i(ig\sqrt{2})^2\right]\frac{[1P]^3}{[P4]\,[45]\,[51]}}_{\mathcal{A}_4(1^+P^+4^-5^-)},$$

while the collinear limit 3 || 4 leads to

$$\mathcal{A}_5(1^+2^+3^-4^-5^-) \underset{3\|4}{\approx} \underbrace{ig\frac{\sqrt{2}}{[34]}\frac{1}{\sqrt{z_2z_3}}}_{\mathrm{Sp}_+(3^-4^-)} \times \underbrace{\left[-i(ig\sqrt{2})^2\right]\frac{[12]^3}{[2P]\,[P5]\,[51]}}_{\mathcal{A}_4(1^+2^+P^-5^-)}.$$

Note that the splitting amplitudes $\mathrm{Sp}_+(i^+j^+)$ and $\mathrm{Sp}_-(i^-j^-)$ did not appear in these examples. It turns out that they are in fact vanishing. Indeed, since the intermediate denominator s_{ij} vanishes in the collinear limit $i \parallel j$, the amplitude factorizes into two *on-shell* sub-amplitudes, one of them being the three-point splitting amplitude. Therefore, the selection rules of the on-shell three-point amplitudes also apply here, and the $+++$ and $---$ cases are forbidden.

Therefore, we can summarize here the set of gluonic splitting amplitudes:

$$\mathrm{Sp}_+(i^-j^+) = ig\frac{\sqrt{2}}{\langle ij\rangle}\frac{z_i^2}{\sqrt{z_iz_j}}, \quad \mathrm{Sp}_+(i^-j^-) = ig\frac{\sqrt{2}}{[ij]}\frac{1}{\sqrt{z_iz_j}}, \quad \mathrm{Sp}_+(i^+j^+) = 0,$$
$$\mathrm{Sp}_-(i^+j^-) = ig\frac{\sqrt{2}}{[ij]}\frac{z_i^2}{\sqrt{z_iz_j}}, \quad \mathrm{Sp}_-(i^+j^+) = ig\frac{\sqrt{2}}{\langle ij\rangle}\frac{1}{\sqrt{z_iz_j}}, \quad \mathrm{Sp}_-(i^-j^-) = 0.$$

63.d The Altarelli–Parisi splitting functions are *probabilities* of gluon splitting, summed over the helicities of the final state gluons i, j and averaged over the helicities of the initial gluon P:

$$P_{gg}(z_i, z_j) \propto \sum_{h,h_i,h_j} \left| \mathrm{Sp}_h(i^{h_i} j^{h_j}) \right|^2 \propto \frac{z_i^4 + z_j^4 + 1}{z_i z_j}.$$

(Note that $|\langle ij\rangle|^2 = |[ij]|^2$ for real momenta.) Since $z_i + z_j = 1$, a more standard notation is to denote $z_i = z, z_j = 1 - z$, so that

$$P_{gg}(z) \propto \frac{z^4 + (1-z)^4 + 1}{z(1-z)} \propto \frac{z}{1-z} + \frac{1-z}{z} + z(1-z).$$

Those familiar with the Altarelli–Parisi splitting kernels will recognize this z dependence (except for the virtual term proportional to $\delta(1-z)$, which cannot be obtained from this argument but can be guessed from unitarity).

63.e Consider now the six-point amplitude $\mathcal{A}_6(1^-2^-3^-4^+5^+6^+)$ derived in Problem 52:

$$\mathcal{A}_6(1^-2^-3^-4^+5^+6^+) = -\frac{4ig^4}{[2|\mathbf{P}_3 + \mathbf{P}_4|5\rangle} \left\{ \frac{\langle 1|\overline{\mathbf{P}}_2 + \overline{\mathbf{P}}_3|4]^3}{[23]\,[34]\,\langle 56\rangle\langle 61\rangle(p_2+p_3+p_4)^2} + \frac{\langle 3|\overline{\mathbf{P}}_4 + \overline{\mathbf{P}}_5|6]^3}{[61]\,[12]\,\langle 34\rangle\langle 45\rangle(p_3+p_4+p_5)^2} \right\}.$$

When $p_{3,4} \equiv z_{3,4}P$ (with $z_3 + z_4 = 1$), we have

$$\overline{\mathbf{P}}_3|4] = \langle 3|\overline{\mathbf{P}}_4 = 0, \quad (p_2+p_3+p_4)^2 = \langle 2P\rangle[2P], \quad (p_3+p_4+p_5)^2 = \langle P5\rangle[P5],$$

and this amplitude simplifies to

$$\mathcal{A}_6(1^-2^-3^-4^+5^+6^+) = \underbrace{ig\frac{\sqrt{2}}{[34]}\frac{z_4^2}{\sqrt{z_3 z_4}}}_{\mathrm{Sp}_-(3^-4^+)} \times \underbrace{[-i(ig\sqrt{2})^3]\frac{\langle 12\rangle^3}{\langle 2P\rangle\langle P5\rangle\langle 56\rangle\langle 61\rangle}}_{\mathcal{A}_5(1^-2^-P^+5^+6^+)}$$
$$+ \underbrace{ig\frac{\sqrt{2}}{\langle 34\rangle}\frac{z_3^2}{\sqrt{z_3 z_4}}}_{\mathrm{Sp}_+(3^-4^+)} \times \underbrace{[-i(ig\sqrt{2})^3]\frac{[56]^3}{[61]\,[12]\,[2P]\,[P5]}}_{\mathcal{A}_5(1^-2^-P^-5^+6^+)}.$$

This example shows that in general the collinear limit involves a sum over the two helicity assignments of the intermediate particle. As with the soft limits, the collinear limits can be used as an a posteriori check of the consistency of the expression for a given amplitude.

63.f In order to determine the splitting amplitudes that involve quarks, let us start from the following amplitudes:

$$\mathcal{A}_4(1_{\bar{q}}^-2_q^+3^-4^+) = 2g^2\frac{[24]^3}{[12][23][34]},$$
$$\mathcal{A}_4(1_{\bar{q}}^-2_q^+3^+4^-) = 2g^2\frac{\langle 14\rangle^3\langle 24\rangle}{\langle 12\rangle\langle 23\rangle\langle 34\rangle\langle 41\rangle}, \tag{4.40}$$

obtained in Problem 60. (Note that because of peculiarities of four-point kinematics – for instance $\langle 12\rangle[12] = \langle 34\rangle[34]$ – there are alternative ways of writing these expressions, which may obscure their singularities. Thus, we use the expressions that make the singularity in the channel under consideration more directly accessible.) Consider first the collinear limit $1 \parallel 2$:

$$\mathcal{A}_4(1_{\bar{q}}^-2_q^+3^-4^+) \underset{1\parallel 2}{\approx} \underbrace{-g\frac{\sqrt{2}}{[12]}z_2}_{\mathrm{Sp}_-(1_{\bar{q}}^-2_q^+)} \times \underbrace{[g\sqrt{2}]\frac{[4P]^3}{[P3][34]}}_{\mathcal{A}_3(P^+3^-4^+)},$$
$$\mathcal{A}_4(1_{\bar{q}}^-2_q^+3^+4^-) \underset{1\parallel 2}{\approx} \underbrace{g\frac{\sqrt{2}}{\langle 12\rangle}z_1}_{\mathrm{Sp}_+(1_{\bar{q}}^-2_q^+)} \times \underbrace{[g\sqrt{2}]\frac{\langle 4P\rangle^3}{\langle P3\rangle\langle 34\rangle}}_{\mathcal{A}_3(P^-3^+4^-)},$$

which provides the splitting amplitudes for a gluon splitting into a collinear quark–antiquark pair. Note that because of helicity conservation, a gluon (of helicity ± 1) cannot split into a quark–antiquark pair of opposing helicities (which, with the all-incoming convention, would correspond for instance to the splitting function $\mathrm{Sp}_+(1_{\bar{q}}^-2_q^-)$.) By summing the squares of the above two splitting amplitudes, we can determine the Altarelli–Parisi splitting function for a gluon becoming a quark:

$$\mathrm{P}_{qg}(z_i, z_j) \propto \sum_{h,h_i,h_j}\left|\mathrm{Sp}_h(1_{\bar{q}}^{h_i}2_q^{h_j})\right|^2 \propto z_i^2 + z_j^2$$
$$\propto z^2 + (1-z)^2 \quad (\text{with } z_i \equiv z \text{ and } z_j = 1-z).$$

Consider now the collinear limit $2 \parallel 3$ in the four-point amplitudes of Eqs. (4.40):

$$\mathcal{A}_4(1_{\bar{q}}^-2_q^+3^-4^+) \underset{2\parallel 3}{\approx} \underbrace{-ig\frac{\sqrt{2}}{[23]}\frac{z_2}{\sqrt{z_3}}}_{\mathrm{Sp}_{\bar{q}-}(2_q^+3^-)} \times \underbrace{[ig\sqrt{2}]\frac{[P4]^2}{[1P]}}_{\mathcal{A}_3(1_{\bar{q}}^-P_q^+4^+)},$$

$$\mathcal{A}_4(1^-_{\bar{q}}2^+_q3^+4^-) \underset{2\|3}{\approx} \underbrace{ig\frac{\sqrt{2}}{\langle 23\rangle}\frac{1}{\sqrt{z_3}}}_{\mathrm{Sp}_{\bar{q}^-}(2^+_q3^+)} \times \underbrace{[ig\sqrt{2}]\frac{\langle 14\rangle^2}{\langle 1P\rangle}}_{\mathcal{A}_3(1^-_{\bar{q}}P^+_q4^-)} .$$

Likewise, the collinear limit $1 \parallel 4$ in $\mathcal{A}_4(1^-_{\bar{q}}2^+_q3^-4^+)$ gives

$$\mathcal{A}_4(1^-_{\bar{q}}2^+_q3^-4^+) \underset{1\|4}{\approx} \underbrace{ig\frac{\sqrt{2}}{\langle 41\rangle}\frac{z_1}{\sqrt{z_4}}}_{\mathrm{Sp}_{q^+}(4^+1^-_{\bar{q}})} \times \underbrace{[ig\sqrt{2}]\frac{\langle P3\rangle^2}{\langle P2\rangle}}_{\mathcal{A}_3(P^-_{\bar{q}}2^+_q3^-)} .$$

(The amplitude $\mathcal{A}_4(1^-_{\bar{q}}2^+_q3^+4^-)$ has no collinear singularity when $1 \parallel 4$.) The missing splitting amplitude may be obtained from the five-point amplitude derived in Problem 60, and reads

$$\mathrm{Sp}_{q^+}(4^-1^-_{\bar{q}}) = -ig\frac{\sqrt{2}}{[41]}\frac{1}{\sqrt{z_4}}.$$

By summing the squares of these splitting amplitudes, we can recover the z dependence of the corresponding Altarelli–Parisi splitting functions. First, we have

$$\sum_{h,h_i,h_j} \left|\mathrm{Sp}_{\bar{q}^h}(i^{h_i}_q j^{h_j})\right|^2 \propto \frac{1+z_i^2}{z_j}.$$

This result contains both the quark-to-gluon and the quark-to-quark splitting functions, obtained respectively by setting $z_j = z, z_i = 1-z$ or $z_i = z, z_j = 1-z$:

$$P_{gq}(z) \propto \frac{1+(1-z)^2}{z}, \quad P_{qq}(z) \propto \frac{1+z^2}{1-z}.$$

---

64. Enumeration of One-Loop Graphs The goal of this problem is to count the one-loop graphs in various quantum field theories, by extending the method used in Problem 51.

64.a In order to count graphs with loops, it is sufficient to view a loop with n lines attached as a special n-point vertex with Feynman rule 1. Explain why the number of one-loop graphs can be obtained from the Taylor coefficients at order ℓ^1 of the solution of the equation

$$X = j + V'(X) + \ell\, V'_{\text{loop}}(X),$$

$$\text{with} \begin{cases} V(X) \underset{\phi^3}{=} \dfrac{X^3}{3!}, \quad V(X) \underset{\text{Yang–Mills}}{=} \dfrac{X^3}{3!} + \dfrac{X^4}{4!}, \\ V_{\text{loop}}(X) = \dfrac{V''(X)}{2} + \dfrac{(V''(X))^2}{4} - \dfrac{1}{2}\ln\left(1 - V''(X)\right). \end{cases}$$

64.b Let $X_0(j)$ be the solution of this equation at zeroth order in ℓ. Express the solution at order one in terms of X_0. Expand it in powers of j in order to extract the number of one-loop graphs in ϕ^3 scalar theory, and in Yang–Mills theory (with only a gluon loop).

64.c Adapt the same approach to counting the graphs with n external gluons and a ghost or quark loop. Determine the numbers of one-loop n-gluon graphs in QCD with one quark flavor.

64.d Extend the method of Problem 40 to count the one-loop color-ordered graphs in Yang–Mills theory. *Hint: write a stripped-down Berends–Giele equation for the generating function of the numbers of one-loop color-ordered graphs.*

64.a In order to generalize the approach of Problem 51 to one loop (for non-ordered graphs), let us first recall the general structure of the recursive equation we had to solve at tree level,

$$X = j + V'(X),$$

where $V(X)$ is the potential that encodes the vertices of the theory, with their respective symmetry factors,

$$V(X) \underset{\phi^3}{=} \frac{X^3}{3!}, \quad V(X) \underset{\text{Yang–Mills}}{=} \frac{X^3}{3!} + \frac{X^4}{4!}.$$

(Here, by Yang–Mills theory, we mean the pure gluonic sector. Graphs with a ghost loop may be counted separately.) To include the loops in this counting, we simply need to view the loops as providing a new set of vertices, encoded in a function $V_{\text{loop}}(X)$. The recursive equation becomes

$$X = j + V'(X) + \ell\, V'_{\text{loop}}(X),$$

where we have also introduced a parameter ℓ in order to keep track of the number of times these new vertices are inserted in the solution. The term of order X^1 in $V_{\text{loop}}(X)$ encodes the tadpoles, the term of order 2 encodes the self-energy insertions, the term of order 3 the loops with three legs attached, etc. Since there can be arbitrarily many legs attached around a loop, this function is an infinite series of powers of X. In a theory where the bare vertices are represented by the potential $V(X)$, the basic block which is attached around a loop is the second derivative $V''(X)$. Therefore, $V_{\text{loop}}(X)$ is in fact a series of powers of $V''(X)$:

$$V_{\text{loop}}(X) = \sum_{n\geq 1} s_n \left(V''(X)\right)^n.$$

The symmetry factors are

$$s_1 = 1, \quad s_2 = \frac{1}{2}, \quad s_{n\geq 3} = \frac{1}{2n}$$

(for $n \geq 3$, these terms have an order-n cyclic symmetry and a mirror symmetry; for $n = 2$, these two symmetries are the same; for $n = 1$, the symmetry group is limited to the identity), which leads to the following expression:

$$V_{\rm loop}(X) = V''(X) + \frac{(V''(X))^2}{2} + \sum_{n\geq 3} \frac{1}{2n}\,(V''(X))^n$$
$$= \frac{V''(X)}{2} + \frac{(V''(X))^2}{4} - \frac{1}{2}\ln\big(1 - V''(X)\big).$$

The recursive equation thus reads

$$X = j + V'(X) + \ell\,\frac{V'''(X)}{2}\left[1 + V''(X) + \frac{1}{1 - V''(X)}\right]. \tag{4.41}$$

64.b In order to obtain the number of one-loop graphs, we must solve this equation to the desired order in j (this depends on the number of external legs one wishes to reach), and to order one in ℓ. By writing $X \equiv X_0 + \ell X_1 + \cdots$, Eq. (4.41) leads to separate equations for the first two coefficients X_0 and X_1:

$$X_0 = j + V'(X_0),$$
$$X_1 = X_1 V''(X_0) + \frac{V'''(X_0)}{2}\left[1 + V''(X_0) + \frac{1}{1 - V''(X_0)}\right]. \tag{4.42}$$

(The second equation above is obtained by differentiating (4.41) with respect to ℓ and then setting $\ell = 0$.) The first equation above is simply the recursive equation derived in Problem 51, whose solution X_0 counts the tree graphs, and the second equation can be solved exactly in terms of X_0:

$$X_1 = \frac{V'''(X_0)}{2(1 - V''(X_0))}\left[1 + V''(X_0) + \frac{1}{1 - V''(X_0)}\right].$$

We obtain the following results:

$$X_1(j) \underset{\phi^3}{=} 1 + 2j + \frac{7j^2}{2} + \frac{13j^3}{2} + \frac{99j^4}{8} + \frac{191j^5}{8} + \frac{743j^6}{16}$$
$$+ \frac{1453j^7}{16} + \frac{22{,}819j^8}{128} + \frac{44{,}923j^9}{128} + \mathcal{O}(j^{10}),$$
$$X_1(j) \underset{\text{Yang–Mills}}{=} 1 + 3j + 7j^2 + \frac{33j^3}{2} + \frac{947j^4}{24} + \frac{191j^5}{2} + \frac{8383j^6}{36}$$
$$+ \frac{2283j^7}{4} + \frac{1{,}617{,}619j^8}{1152} + \frac{498{,}871j^9}{144} + \mathcal{O}(j^{10}).$$

From these series, we can read the number of one-loop Feynman graphs in these two theories. These numbers are listed in Table 4.3.

64.c In full-fledged QCD there are, in addition, graphs with a quark loop or a ghost loop. Their numbers (they are equal) can be obtained by adding an extra term to $V_{\rm loop}(X)$. The

Number of external legs	ϕ^3 theory	Yang–Mills (gluon loop)
2	2	3
3	7	14
4	39	99
5	297	947
6	2865	11,460
7	33,435	167,660
8	457,695	2,876,580
9	7,187,985	56,616,665
10	127,356,705	1,257,154,920

Table 4.3 Number of n-point one-loop graphs in ϕ^3 scalar field theory and in Yang–Mills theory (for the latter, only the graphs with a gluon loop are counted here).

easiest way to derive this term is to start from a "potential" $W(X, Y)$, where X tracks the gluons and Y tracks the field running in the loop (a ghost or a quark), defined as

$$W(X, Y) \equiv \tfrac{1}{2} XY^2.$$

(This expression follows from the fact that the gluon–ghost vertex has one gluon and a pair of ghosts, and likewise for the coupling to quarks.) To determine the new term in $V_{\text{loop}}(X)$, we should recall that the second derivative in the $V''(X)$ that appears in our previous expression for V_{loop} is in fact a derivative with respect to the variable that tracks the particle running in the loop. Therefore, the extra term due to a ghost or quark loop is obtained by replacing $V''(X)$ by $\partial_Y^2 W(X, Y) = X$, i.e.,

$$V_{\text{loop}}(X) \underset{\substack{\text{ghost loop}\\ \text{quark loop}}}{=} \frac{X}{2} + \frac{X^2}{4} - \frac{1}{2}\ln(1 - X).$$

With gluons, ghosts and n_f flavors of quarks running in the loop, the total V_{loop} reads

$$V_{\text{loop}}(X) = \underbrace{\frac{V''(X)}{2} + \frac{(V''(X))^2}{4} - \frac{1}{2}\ln\left(1 - V''(X)\right)}_{\text{gluon loop}} + \underbrace{\frac{1 + n_f}{2}\left[X + \frac{X^2}{2} - \ln(1 - X)\right]}_{\text{quark and ghost loops}},$$

which gives the following solution at order ℓ^1:

$$X_1 = \frac{V'''(X_0)}{2(1 - V''(X_0))}\left[1 + V''(X_0) + \frac{1}{1 - V''(X_0)}\right] + \frac{1 + n_f}{2}\left[\frac{1 + X_0 + \frac{1}{1 - X_0}}{1 - V''(X_0)}\right].$$

For $n_f = 1$, this leads to

$$X_1(j) \underset{\text{QCD}, n_f = 1}{=} 3 + 7j + 15j^2 + \frac{205\, j^3}{6} + \frac{637\, j^4}{8} + \frac{2263\, j^5}{12} + \frac{1805\, j^6}{4} + \frac{78,313\, j^7}{72} + \frac{1,012,205\, j^8}{384} + \frac{33,254,659\, j^9}{5184} + \mathcal{O}(j^{10}),$$

hence the numbers of one-loop graphs in QCD listed in Table 4.4.

Number of external gluons	Yang–Mills (ghost or quark loop)	QCD ($n_f = 1$)
2	2	7
3	8	30
4	53	205
5	482	1,911
6	5585	22,630
7	78,620	324,900
8	1,302,665	5,481,910
9	24,832,430	106,281,525
10	535,335,605	2,327,826,130

Table 4.4 Number of n-gluon graphs with a ghost or quark loop, and in QCD with one quark flavor (i.e., total number of graphs with a gluon loop, a ghost loop or a quark loop).

Additional Remarks:

- The above counting of one-loop graphs includes tadpole contributions that are usually discarded in the perturbative expansion. Another class of graphs which is included but not strictly necessary is the set of all the graphs with a self-energy correction on an external leg. Both types of contribution may be excluded from the counting by a small modification of the above results. For QCD, one should use

$$X_1 = \frac{V'''(X_0)}{2(1-V''(X_0))}\left[1+V''(X_0)+\frac{1}{1-V''(X_0)}\right]+\frac{1+n_f}{2}\left[\frac{1+X_0+\frac{1}{1-X_0}}{1-V''(X_0)}\right]$$
$$-\underbrace{\frac{2+n_f+X_0}{1-V''(X_0)}}_{\text{tadpoles}}-\underbrace{(2+n_f)\left(\frac{j}{1-V''(X_0)}+X_0\right)}_{\text{self-energies on external legs}},$$

which leads to

$$X_1(j) \underset{\text{QCD}, n_f=1}{=} \frac{6j^2}{2!}+\frac{63j^3}{3!}+\frac{711j^4}{4!}+\frac{9360j^5}{5!}+\frac{144,300j^6}{6!}$$
$$+\frac{2,565,360j^7}{7!}+\frac{51,802,485j^8}{8!}+\frac{1,172,660,580j^9}{9!}+\mathcal{O}(j^{10}).$$

(For instance, with these exclusions, the number of one-loop 10-gluon graphs in QCD with one quark flavor is reduced from 2,327,826,130 to 1,172,660,580.)

- In order to obtain the numbers of higher-order graphs (two loops and beyond), it is not sufficient to solve Eq. (4.41) to higher orders in ℓ. Doing so would only give the number of graphs that have independent loops, but the graphs with nested loops would not be counted.

64.d Most calculations of amplitudes start with the color decomposition of the amplitude, reducing the problem to the calculation of color-ordered amplitudes. Therefore, it is also useful to know the number of graphs when they are cyclically ordered. Let us recall that the generating function X_0 for color-ordered gluonic trees obeys

$$X_0 = j + X_0^2 + X_0^3. \tag{4.43}$$

(This equation, derived in Problem 51, is similar to the first of Eqs. (4.42) with all the symmetry factors equal to unity.) Ordered graphs with one gluon loop may be constructed in two steps. First, we build a gluon propagator G dressed by all the possible insertions of trees, which will become the loop when we connect its endpoints. This object satisfies a kind of Dyson equation:

$$G = \underbrace{X_0 + X_0^2}_{\text{first insertion}} + G\big(X_0 + X_0^2\big).$$

Diagrammatically, this expands to

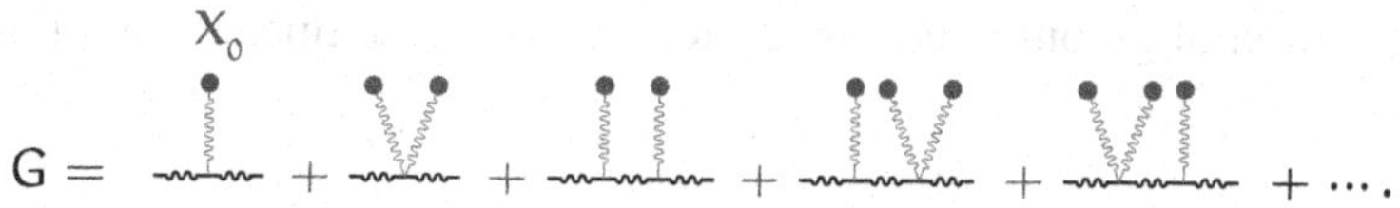

(Here, of course, we are not constructing the full propagator, but simply a generating function whose Taylor coefficients count how many graphs contribute at each order in j.) Then, the generating function $L(j)$ for graphs with one gluon loop may be obtained by an equation similar to the Berends–Giele recursion:

$$L = G + 2\,GX_0 + 2LX_0 + 3LX_0^2.$$

The diagrammatic content of this equation is the following:

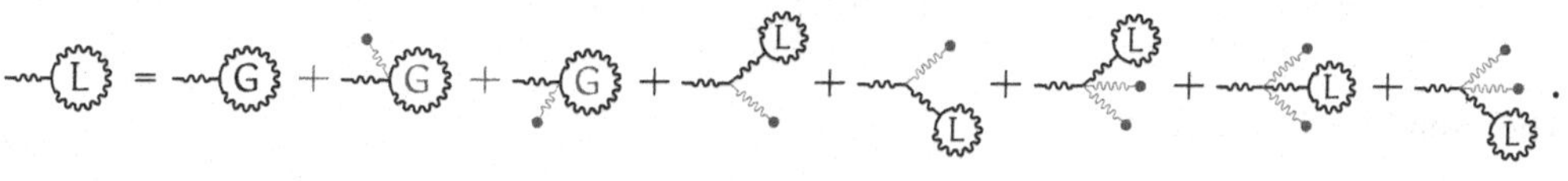

The first three terms are those where the final gluon propagator is directly attached to the loop, while the other terms are those where the loop is separated from the endpoint by at least one intermediate vertex. The numerical prefactors in the terms $2GX_0$, $2LX_0$ and $3LX_0^2$ are necessary, because the cyclic ordering implies that inserting the loop before or after X_0 (when viewing the insertions clockwise around the exposed vertex) leads to distinct color-ordered graphs. Note also that, since G contains at least one insertion of X_0, the first three terms do not

Number of external gluons	Tree graphs	Gluon loop	Ghost loop	One-loop total
2	1	1	1	2
3	1	7	4	11
4	3	37	19	56
5	10	191	91	282
6	38	980	447	1427
7	154	5034	2227	7261
8	654	25,929	11,219	37,148
9	2,871	133,969	56,992	190,961
10	12,925	694,265	291,433	985,698

Table 4.5 Numbers of tree and one-loop n-gluon color-ordered graphs in Yang–Mills theory.

produce tadpoles. These equations can be solved trivially in terms of X_0, to give

$$L(j) = \frac{(X_0 + X_0^2)(1 + 2X_0)}{(1 - X_0 - X_0^2)(1 - 2X_0 - 3X_0^2)}.$$

For a ghost loop, the insertion of gluonic trees on the loop can only be linear. Therefore, the loop propagator obeys

$$G_\chi = X_0 + G_\chi X_0, \qquad \text{i.e.,} \quad G_\chi =$$

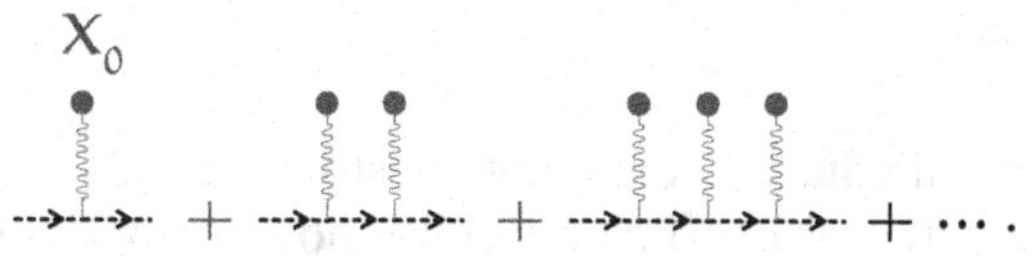

From this, the color-ordered graphs with a ghost loop are generated by a function $L_\chi(j)$ that satisfies

$$L_\chi = G_\chi + 2L_\chi X_0 + 3L_\chi X_0^2,$$

whose diagrammatic representation is

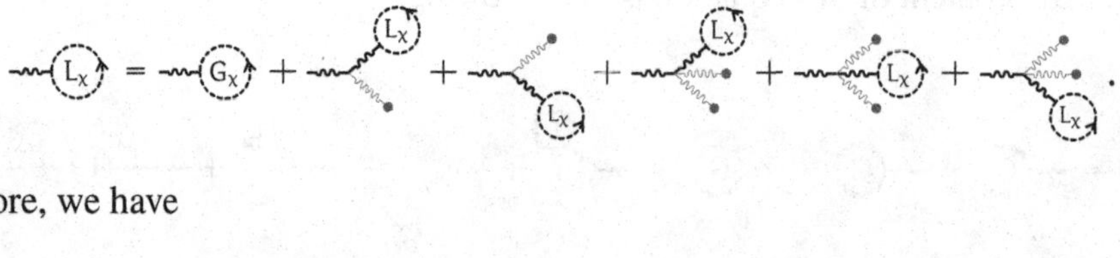

Therefore, we have

$$L_\chi(j) = \frac{X_0}{(1 - X_0)(1 - 2X_0 - 3X_0^2)}.$$

Solving Eq. (4.43) for $X_0(j)$ and injecting the solution into $L(j)$ and $L_\chi(j)$, we obtain the numbers listed in Table 4.5 for the one-loop color-ordered graphs in Yang–Mills theory.

65. Ultraviolet Finiteness of One-Loop All-Plus Amplitudes In Yang–Mills theory (or in QCD), explain why the all-plus amplitude $\mathcal{A}_n(1^+2^+\cdots n^+)$ cannot be ultraviolet divergent at one loop in four dimensions.

The Yang–Mills theory (and also quantum chromodynamics) is renormalizable in four dimensions. This implies that, at one loop, the ultraviolet divergent amplitudes can only be those that correspond to operators already present in the bare action. (Beyond one loop, some amplitudes may be ultraviolet divergent because of a divergent *sub-diagram*.)

An ultraviolet divergence in the amplitude $\mathcal{A}_n(1^+2^+\cdots n^+)$ at one loop would require a counterterm for some n-gluon operator that gives a non-zero result when all the legs are on-shell and contracted with positive helicity polarization tensors. Since the theory is renormalizable, this operator must also exist in the bare theory. Consequently, this operator would lead to a non-zero value of $\mathcal{A}_n(1^+2^+\cdots n^+)$ at tree level, which would contradict the known fact that all-plus amplitudes are zero in Yang–Mills theory at tree level. Therefore, $\mathcal{A}_n(1^+2^+\cdots n^+)$ must be ultraviolet finite at one loop in four dimensions. Of course, the same reasoning applies to the all-minus one-loop amplitudes.

66. Rationality of One-Loop All-Plus Amplitudes The goal of this problem is to use the method of generalized cuts in order to show that the all-plus amplitudes are purely rational at one loop in four dimensions.

66.a Explain why four-dimensional cuts cannot capture the rational part of an amplitude.

66.b Show that, in Yang–Mills theory, the one-loop amplitude $\mathcal{A}_4(1^+2^+3^+4^+)$ is purely rational in four dimensions.

66.c Extend this result to any n-point all-plus amplitude $\mathcal{A}_n(1^+2^+\cdots n^+)$ at one loop.

66.a As with the usual Cutkosky cuts, where only two propagators of the loop are cut, generalized cuts exist when the loop integral has branch cut singularities in some of the variables it depends upon. However, the rational term of an amplitude has only poles, and no branch cut singularities. Therefore, (generalized) cuts are totally blind to the rational term. Note, however, that a term which is rational in $d = 4$ dimensions usually develops branch cuts away from $d = 4$, because of factors such as $(-p^2)^{d-4} = 1 + (d-4)\ln(-p^2) + \cdots$.

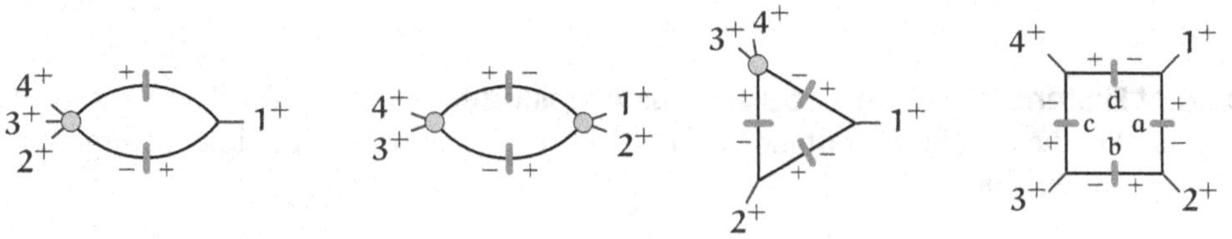

Figure 4.3 Two-, three- and four-propagator cuts of the one-loop four-point $+ + + +$ amplitude. The blobs denote sums of tree graphs.

66.b Consider now the possible cuts in the one-loop amplitude $\mathcal{A}_4(1^+2^+3^+4^+)$, listed in Figure 4.3. We have represented only one of each type of cut when there are several. On the cut propagators, we have chosen (when possible) helicities that allow us to have non-zero three-point functions (either $+ + -$ or $+ - -$). Our assignment of helicities is not the only choice, but one can easily check that all other choices also lead to vanishing factors. Let us discuss all these cuts one by one, from left to right:

- The first graph is a two-particle cut, which splits the amplitude into a three-point and a five-point sub-amplitude. In total, three negative helicities are needed for the two sub-amplitudes to be non-vanishing, but the cut propagators provide only two. Hence this cut vanishes.

- The second graph is another type of two-particle cut, with two four-point sub-amplitudes. To be non-zero, each four-point amplitude needs two negative helicities. Hence, we need four negative helicities in total, but the two cut propagators can provide only two. Thus, this cut is zero.

- Next, there is a three-particle cut, which divides the amplitude into a two three-point sub-amplitudes and one four-point sub-amplitude. In total, we need $1 + 1 + 2 = 4$ negative helicities for all the sub-amplitudes to be non-zero, but the three cut propagators provide only three. This cut is also zero.

- The last one is a four-particle cut, which splits the amplitude into four three-point sub-amplitudes. This time, counting of the necessary negative helicities is not sufficient to conclude that this cut is zero (the assignment of helicities shown in Figure 4.3 appears to work). But another constraint comes into play here. Recall that in four dimensions (this is necessary in order to use the spinor-helicity formalism), the $+ + -$ three-point amplitudes depend only on the square brackets, while all the angle brackets are zero. For instance, for the top right corner sub-amplitude, $\mathcal{A}_3(1^+\mathrm{a}^+\mathrm{d}^-)$, we have

$$\langle 1\mathrm{a}\rangle = \langle \mathrm{ad}\rangle = \langle \mathrm{d}1\rangle = 0.$$

For two-component spinors, this implies that the corresponding angle spinors are all proportional to each other:

$$|1\rangle \propto |a\rangle \propto |d\rangle.$$

Likewise, by considering the other three sub-amplitudes, we also have

$$|2\rangle \propto |b\rangle \propto |a\rangle, \quad |3\rangle \propto |c\rangle \propto |b\rangle, \quad |4\rangle \propto |d\rangle \propto |c\rangle.$$

Since proportionality is a transitive relationship, this implies that

$$|1\rangle \propto |2\rangle \propto |3\rangle \propto |4\rangle.$$

This is a constraint on the external momenta, which implies that the three Mandelstam invariants are zero, $s = t = u = 0$, which is not satisfied in general. Therefore, this cut is also zero.

Thus, we have shown that all the possible cuts of the one-loop amplitude $\mathcal{A}_4(1^+2^+3^+4^+)$ are zero in four dimensions, which implies that this amplitude is purely rational.

66.c Consider now the all-plus n-point amplitude at one loop, $\mathcal{A}_n(1^+2^+\cdots n^+)$, with $n \geq 5$. With a four-dimensional loop momentum, the maximal number of cut lines is four. The proof that all these cuts are zero follows the same reasoning as in the preceding case, and is even simpler in the case of the four-particle cuts.

- *Two-particle cuts* split the amplitude into a p-point $(3 \leq p \leq n)$ and an $(n-p+4)$-point sub-amplitude. In total, these two sub-amplitudes require at least three negative helicities, while the two cut propagators provide only two. These cuts are all zero.
- *Three-particle cuts* give a p-point, a q-point and an $(n-p-q+6)$-point sub-amplitude. At least four negative helicities are necessary, while the cut propagators carry only three. All these cuts are also zero.
- *Four-particle cuts* split the amplitude into a p-point, a q-point, an r-point and an $(n-p-q-r+8)$-point sub-amplitude. At least five negative helicities are necessary when $n \geq 5$, since at least one of the sub-amplitudes has four points or more. The cuts provide only four negative helicities, implying that these cuts are zero.

Therefore, $\mathcal{A}_n(1^+2^+\cdots n^+)$ is purely rational at one loop.

67. One-Loop Four-Point ++++ Amplitude from Generalized Unitarity The goal of this problem is to calculate analytically the all-plus four-gluon color-ordered amplitude at one loop, with a massless complex scalar running in the loop. Assume the loop momentum ℓ^μ to be in $d \equiv 4 - 2\epsilon$ dimensions, while the external gluons are kept strictly four-dimensional.

67.a Explain why the reduction of the corresponding loop integral is equivalent to that of the same loop with a massive scalar in four dimensions. What set of scalar integrals can appear in the result of this reduction?

67.b Consider now the four-point all-plus amplitude with a scalar loop, $\mathcal{A}_4^{\text{scalar}}(1^+2^+3^+4^+)$. Write its reduction to master integrals. Why are tadpoles and bubbles absent?

67.c Apply a four-particle cut to this amplitude to obtain the box coefficient,

$$D_{0123} = 8g^4\ell_\perp^4 \frac{[12]\,[34]}{\langle 12\rangle\langle 34\rangle}.$$

67.d By applying the three-particle cut (012), check that the triangle coefficient C_{012} is zero.

67.e Perform explicitly the loop momentum integration in the master integrals, to obtain

$$\mathcal{A}_4^{\text{scalar}}(1^+2^+3^+4^+) = -\frac{ig^4}{12\pi^2}\frac{[12]\,[34]}{\langle 12\rangle\langle 34\rangle}.$$

67.a Recall the propagator for a massless complex scalar and its trilinear coupling to a gauge field:

$$= \frac{i}{p^2 + i0^+},$$

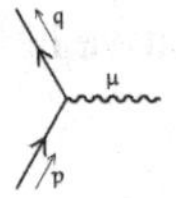

$$= -ig(p+q)^\mu.$$

(Since we are interested in color-ordered amplitudes, the vertex has been stripped of its color factor. There is in addition a $\phi^*\phi A_\mu A^\mu$ vertex that we shall not need in the discussion.) The loop momentum ℓ^μ can be decomposed into a component $\ell_\parallel^\mu$ that lives in the four physical dimensions, and a component $\ell_\perp^\mu$ in the $d-4$ extra dimensions:

$$\ell^\mu \equiv \underbrace{\ell_\parallel^\mu}_{4\text{ dim}} + \underbrace{\ell_\perp^\mu}_{d-4\text{ dim}}.$$

By construction, $\ell_\perp$ is orthogonal to $\ell_\parallel$, and also to all the external momenta and polarization vectors that are assumed to remain four-dimensional. Given this decomposition, the denominators of the propagators in the loop are all of the form

$$(\ell+q)^2 = (\ell_\parallel + q)^2 + \ell_\perp^2.$$

Interestingly, $\ell_\perp$ appears in the same way in all the denominators, via its square. Moreover, if we view the extra dimensions as space-like, this contribution is identical to a mass term $\mu^2 \equiv -\ell_\perp^2$.

Consider now the numerator of the loop integral. Its momentum dependence comes from the $\phi^*\phi A^\mu$ vertices, contracted with external trees made of gluons or with polarization vectors.

Since these are four-dimensional, these contractions depend on $\ell_{\|}$ but not on $\ell_{\perp}$. Therefore, a loop integral with m denominators has the following generic structure:

$$\underbrace{\int \frac{d^d\ell}{(2\pi)^d} \frac{\mathcal{N}(\ell)}{\prod_{i=0}^{m-1}(\ell+q_i)^2}}_{\text{d-dimensional massless}} = \int \frac{d^{d-4}\ell_{\perp}}{(2\pi)^{d-4}} \underbrace{\int \frac{d^4\ell_{\|}}{(2\pi)^4} \frac{\mathcal{N}(\ell_{\|})}{\prod_{i=0}^{m-1}((\ell_{\|}+q_i)^2+\ell_{\perp}^2)}}_{\text{four-dimensional massive}}.$$

Except for the additional integration over the extra dimensions of the loop momentum, the right-hand side contains the same loop integral, for a four-dimensional but massive scalar. The standard reduction techniques, applied to the latter integral, lead to a general decomposition in *boxes, triangles, bubbles* and *tadpoles.*

67.b This observation is handy for calculating some amplitudes that are purely rational, such as a one-loop amplitude with all external gluons having positive helicities. Since the result is rational, it can be picked by unitarity methods only if the loop momentum is allowed to be d-dimensional rather than four-dimensional. But this makes it difficult to use the spinor-helicity formalism to express the tree sub-amplitudes obtained after the cuts have been applied. With the above observation, we need to cut a four-dimensional massive loop instead. When the particle running in the loop is a scalar, simple expressions exist for the tree sub-amplitudes, some of which have been derived in Problem 57, even when the scalar is massive.

Let us now apply this to the calculation of the one-loop amplitude $\mathcal{A}_4^{\text{scalar}}(1^+2^+3^+4^+)$. From Problems 65 and 66, we know that this amplitude is ultraviolet finite and is a rational function of the external momenta. Splitting the loop momentum into its four-dimensional part $\ell_{\|}$ and the extra components $\ell_{\perp}$, we first have

$$\mathcal{A}_4^{\text{scalar}}(1^+2^+3^+4^+) = \lim_{d\to 4} \int \frac{d^{d-4}\ell_{\perp}}{(2\pi)^{d-4}} \underbrace{\mathcal{A}_4^{\text{scalar}}(1^+2^+3^+4^+;-\ell_{\perp}^2)}_{\text{massive 4d scalar loop}},$$

where the integrand on the right-hand side is the same amplitude evaluated with a four-dimensional loop momentum and a squared mass $-\ell_{\perp}^2$. Since we expect an ultraviolet finite result (see Problem 65), the reduction of the four-dimensional loop integral over $\ell_{\|}^{\mu}$ cannot contain bubbles or tadpoles, but only boxes and triangles. Therefore, the reduction applied to this integrand leads to

$$\begin{aligned}\mathcal{A}_4^{\text{scalar}}(1^+2^+3^+4^+;-\ell_{\perp}^2) &= D_{0123}L_{0123}^{(4)} \\ &+ C_{012}L_{012}^{(4)} + C_{023}L_{023}^{(4)} + C_{013}L_{013}^{(4)} + C_{123}L_{123}^{(4)},\end{aligned} \tag{4.44}$$

where we have defined

$$L_{i_1\cdots i_p}^{(4)} \equiv \int \frac{d^4\ell_{\|}}{(2\pi)^4} \frac{i^p}{((\ell_{\|}+q_{i_1})^2+\ell_{\perp}^2)\cdots((\ell_{\|}+q_{i_p})^2+\ell_{\perp}^2)},$$
$$q_0 \equiv 0, \quad q_1 \equiv p_1, \quad q_2 \equiv p_1+p_2, \quad q_3 \equiv p_1+p_2+p_3 = -p_4$$

($p_{1,2,3,4}$ are the momenta of the four external gluons). It is interesting to note the dimensions of the various objects in the decomposition of the amplitude (in the limit $\epsilon \to 0$):

$$\dim\left(\mathcal{A}_4^{\text{scalar}}\right) = 0,$$
$$\dim\left(L_{0123}^{(4)}\right) = -4, \quad \dim\left(D_{0123}\right) = 4,$$
$$\dim\left(L_{ijk}^{(4)}\right) = -2, \quad \dim\left(C_{ijk}\right) = 2.$$

Given the dimension of the coefficient D_{0123}, it may contain a factor $\ell_\perp^4$, which would produce an ultraviolet divergence in the integration over $\ell_\perp$, and therefore a pole in ϵ^{-1}. However, as we shall see, because the factor $\ell_\perp^4$ is restricted to the $d-4$ extra dimensions, it also brings an explicit factor ϵ into the numerator that compensates for the pole and thus leads to a finite result in the end.

67.c In order to determine the coefficients D_{0123} and C_{ijk}, we apply cuts simultaneously to the amplitude $\mathcal{A}_4^{\text{scalar}}(1^+2^+3^+4^+;-\ell_\perp^2)$ and to the various integrals $L_{i_1\cdots i_p}^{(4)}$. Let us denote by Δ_i the operator that cuts the propagator of denominator $(\ell_\parallel + q_i)^2 + \ell_\perp^2$:

$$\frac{i}{(\ell_\parallel + q_i)^2 + \ell_\perp^2} \xrightarrow[\Delta_i]{} \underbrace{\delta\left((\ell_\parallel + q_i) + \ell_\perp^2\right)}_{\equiv\, \delta_i}.$$

Consider first the four-line cut obtained by applying $\Delta_0\Delta_1\Delta_2\Delta_3$ to Eq. (4.44). All the triangles $L_{ijk}^{(4)}$ give zero since they have only three denominators. Therefore, we must have

$$\Delta_0\Delta_1\Delta_2\Delta_3\left[\mathcal{A}_4^{\text{scalar}}(1^+2^+3^+4^+;-\ell_\perp^2)\right] = D_{0123}\,\Delta_0\Delta_1\Delta_2\Delta_3\left[L_{0123}^{(4)}\right].$$

On the right-hand side, the application of the four cuts is immediate and gives

$$\Delta_0\Delta_1\Delta_2\Delta_3\left[L_{0123}^{(4)}\right] = \int \frac{d^4\ell_\parallel}{(2\pi)^4}\,\delta_0\delta_1\delta_2\delta_3. \tag{4.45}$$

Applying the four cuts to the full amplitude is equally straightforward, since the four-point amplitude has at most four denominators. We obtain

$$\Delta_0\Delta_1\Delta_2\Delta_3\left[\mathcal{A}_4^{\text{scalar}}(1^+2^+3^+4^+;-\ell_\perp^2)\right] =$$

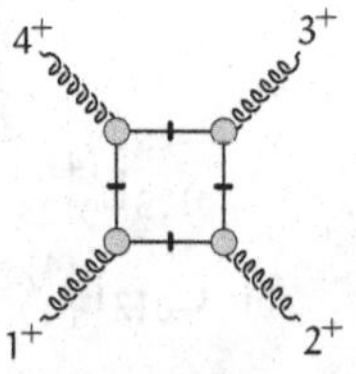

$$= 2\int \frac{d^4\ell_\parallel}{(2\pi)^4}\,\delta_0\delta_1\delta_2\delta_3\,\mathcal{A}_3(\ell_\parallel, 1^+, -(\ell_\parallel + q_1)^*)\,\mathcal{A}_3(\ell_\parallel + q_1, 2^+, -(\ell_\parallel + q_2)^*)$$
$$\times\,\mathcal{A}_3(\ell_\parallel + q_2, 3^+, -(\ell_\parallel + q_3)^*)\,\mathcal{A}_3(\ell_\parallel + q_3, 4^+, -\ell_\parallel^*), \tag{4.46}$$

where $\mathcal{A}_3(\ell_\parallel, i^+, -\ell_\parallel'^*)$ denotes the on-shell three-point amplitude (with a mass $-\ell_\perp^2$ for the scalar particles), with a scalar, an antiscalar and a gluon. The prefactor 2 accounts for the fact that the roles of the scalar and antiscalar may be interchanged. These three-point amplitudes have been calculated in Problem 57. In fact, we also evaluated there the product of two such amplitudes, leading to the following compact results:

$$\mathcal{A}_3(\ell_\parallel, 1^+, -(\ell_\parallel + q_1)^*)\,\mathcal{A}_3(\ell_\parallel + q_1, 2^+, -(\ell_\parallel + q_2)^*) = 2g^2\ell_\perp^2 \frac{[12]}{\langle 12\rangle},$$

$$\mathcal{A}_3(\ell_\parallel + q_2, 3^+, -(\ell_\parallel + q_3)^*)\,\mathcal{A}_3(\ell_\parallel + q_3, 4^+, -\ell_\parallel^*) = 2g^2\ell_\perp^2 \frac{[34]}{\langle 34\rangle}.$$

Comparing (4.45) and (4.46) therefore gives the following value for the box coefficient:

$$\mathrm{D}_{0123} = 8g^4\ell_\perp^4 \frac{[12]\,[34]}{\langle 12\rangle\langle 34\rangle}.$$

(Such a direct identification is permitted by the fact that the four delta functions freeze the integral over the four components of $\ell_\parallel$.)

67.d Let us now turn to the triple cuts. We can limit the discussion to the cut $\Delta_0\Delta_1\Delta_2$, and by symmetry we will be able to infer the other triple cuts without further calculation. When applied to the right-hand side of (4.44), this triple cut receives contributions from the box integral $\mathrm{L}^{(4)}_{0123}$ and from the triangle $\mathrm{L}^{(4)}_{012}$:

$$\Delta_0\Delta_1\Delta_2\left[\mathcal{A}_4^{\mathrm{scalar}}(1^+2^+3^+4^+;-\ell_\perp^2)\right] = \mathrm{D}_{0123}\,\Delta_0\Delta_1\Delta_2\left[\mathrm{L}^{(4)}_{0123}\right] + \mathrm{C}_{012}\,\Delta_0\Delta_1\Delta_2\left[\mathrm{L}^{(4)}_{012}\right].$$

The cuts on the right-hand side of this equation read

$$\Delta_0\Delta_1\Delta_2\left[\mathrm{L}^{(4)}_{0123}\right] = \int \frac{d^4\ell_\parallel}{(2\pi)^4}\,\frac{i}{(\ell_\parallel + q_3)^2 + \ell_\perp^2}\,\delta_0\delta_1\delta_2,$$

$$\Delta_0\Delta_1\Delta_2\left[\mathrm{L}^{(4)}_{012}\right] = \int \frac{d^4\ell_\parallel}{(2\pi)^4}\,\delta_0\delta_1\delta_2.$$

Applying this triple cut to the full amplitude gives

$$\Delta_0\Delta_1\Delta_2\left[\mathcal{A}_4^{\mathrm{scalar}}(1^+2^+3^+4^+;-\ell_\perp^2)\right] =$$

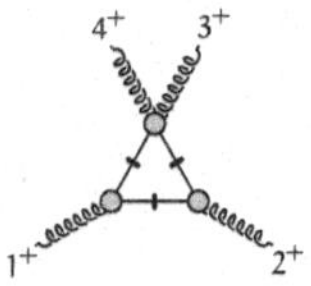

$$= 2\int \frac{d^4\ell_\parallel}{(2\pi)^4}\,\delta_0\delta_1\delta_2\,\mathcal{A}_3(\ell_\parallel, 1^+, -(\ell_\parallel + q_1)^*)\,\mathcal{A}_3(\ell_\parallel + q_1, 2^+, -(\ell_\parallel + q_2)^*)$$
$$\times\,\mathcal{A}_4(\ell_\parallel + q_2, 3^+4^+, -\ell_\parallel^*).$$

The four-point amplitude that appears in the last line has also been calculated in Problem 57:

$$\mathcal{A}_4(\ell_\parallel + q_2, 3^+4^+, -\ell_\parallel^*) = 2g^2\ell_\perp^2 \frac{[34]}{\langle 34\rangle} \frac{i}{(\ell_\parallel + q_3)^2 + \ell_\perp^2}.$$

Therefore, we have

$$\begin{aligned}
&\Delta_0\Delta_1\Delta_2 \left[\mathcal{A}_4^{\text{scalar}}(1^+2^+3^+4^+; -\ell_\perp^2)\right] \\
&= \underbrace{8g^4\ell_\perp^4 \frac{[12]\,[34]}{\langle 12\rangle\langle 34\rangle}}_{D_{0123}} \underbrace{\int \frac{d^4\ell_\parallel}{(2\pi)^4} \frac{i}{(\ell_\parallel + q_3)^2 + \ell_\perp^2} \delta_0\delta_1\delta_2}_{\Delta_0\Delta_1\Delta_2\left[L_{0123}^{(4)}\right]}.
\end{aligned}$$

This triple cut is in fact saturated by the triple cut of the box integral. Therefore, the coefficient of the triangle $L_{012}^{(4)}$ vanishes, $C_{012} = 0$.

67.e Likewise, all the other triangle coefficients are zero, and the amplitude is fully given by

$$\begin{aligned}
&\mathcal{A}_4^{\text{scalar}}(1^+2^+3^+4^+) \\
&\quad = 8g^4 \frac{[12]\,[34]}{\langle 12\rangle\langle 34\rangle} \lim_{d\to 4} \int \frac{d^{d-4}\ell_\perp}{(2\pi)^{d-4}} \int \frac{d^4\ell_\parallel}{(2\pi)^4} \frac{\ell_\perp^4}{((\ell_\parallel + q_0)^2 + \ell_\perp^2)\cdots((\ell_\parallel + q_3)^2 + \ell_\perp^2)} \\
&\quad = 8g^4 \frac{[12]\,[34]}{\langle 12\rangle\langle 34\rangle} \lim_{d\to 4} g_{\mu\nu}^{d-4} g_{\rho\sigma}^{d-4} \int \frac{d^d\ell}{(2\pi)^d} \frac{\ell^\mu\ell^\nu\ell^\rho\ell^\sigma}{(\ell + q_0)^2\cdots(\ell + q_3)^2}.
\end{aligned}$$

In the final line, it is convenient to combine the integrations over $\ell_\perp$ and $\ell_\parallel$ into a unique d-dimensional integral, by introducing the restriction $g_{\mu\nu}^{d-4}$ of the metric tensor to the extra dimensions. Its trace is equal to $d-4$, and it gives zero when contracted to a Lorentz index that lives in the four physical dimensions. Therefore, after the Lorentz contractions are performed, the factor $g_{\mu\nu}^{d-4} g_{\rho\sigma}^{d-4}$ will bring at least one power of ϵ. However, the integral is logarithmically divergent in four dimensions, and therefore has a pole ϵ^{-1}. Since the divergence is logarithmic, the coefficient of the pole is simply given by

$$\begin{aligned}
\int \frac{d^d\ell}{(2\pi)^d} \frac{\ell^\mu\ell^\nu\ell^\rho\ell^\sigma}{\ell^8} &= \tfrac{1}{d(d+2)}(g^{\mu\nu}g^{\rho\sigma} + g^{\mu\rho}g^{\nu\sigma} + g^{\mu\sigma}g^{\nu\rho}) \int \frac{d^d\ell}{(2\pi)^d} \frac{1}{\ell^4} \\
&= \tfrac{i}{24}(g^{\mu\nu}g^{\rho\sigma} + g^{\mu\rho}g^{\nu\sigma} + g^{\mu\sigma}g^{\nu\rho}) \frac{\Gamma(2-\frac{d}{2})}{(4\pi)^2\,\Gamma(2)} \\
&= \tfrac{i}{12(4\pi)^2(4-d)}(g^{\mu\nu}g^{\rho\sigma} + g^{\mu\rho}g^{\nu\sigma} + g^{\mu\sigma}g^{\nu\rho}).
\end{aligned}$$

(At every stage, we keep only the term that produces a pole in $d = 4$, and disregard the finite

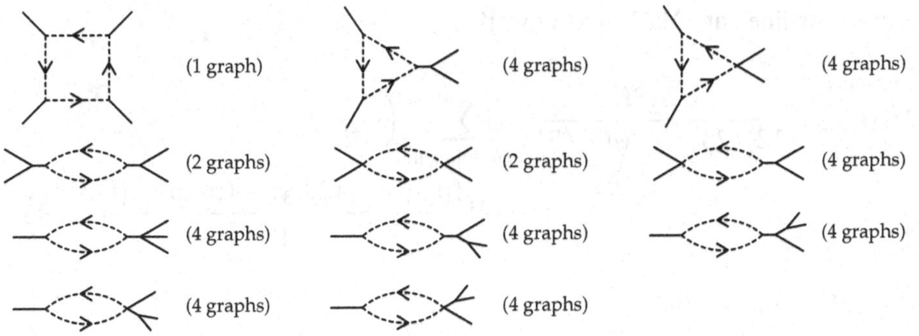

Figure 4.4 The 37 color-ordered graphs contributing to the amplitude $\mathcal{A}_4^{\text{scalar}}(1^+2^+3^+4^+)$ in the traditional approach (solid lines ≡ gluons, dashed lines ≡ scalars).

terms.) Contracting this with $g_{\mu\nu}^{d-4}g_{\rho\sigma}^{d-4}$ gives

$$g_{\mu\nu}^{d-4}g_{\rho\sigma}^{d-4}\int\frac{d^d\ell}{(2\pi)^d}\,\frac{\ell^\mu\ell^\nu\ell^\rho\ell^\sigma}{\ell^8}=\frac{i}{12(4\pi)^2(4-d)}g_{\mu\nu}^{d-4}g_{\rho\sigma}^{d-4}(g^{\mu\nu}g^{\rho\sigma}+g^{\mu\rho}g^{\nu\sigma}+g^{\mu\sigma}g^{\nu\rho})$$
$$=\frac{i}{12(4\pi)^2}(2-d),$$

so that we finally obtain the following remarkably simple result:

$$\mathcal{A}_4^{\text{scalar}}(1^+2^+3^+4^+)=-\frac{ig^4}{12\pi^2}\,\frac{[12]\,[34]}{\langle 12\rangle\langle 34\rangle}.$$

In the traditional approach based on Feynman diagrams, one would have to sum the contributions of 37 color-ordered graphs (see Figure 4.4) in order to obtain this one-loop amplitude. For reasons beyond the scope of this problem, this result also turns out to be the value of the $1^+2^+3^+4^+$ amplitude in Yang–Mills theory (i.e., with a gluon and a ghost loop instead of a scalar loop).
Source: Brandhuber, A., McNamara, S., Spence, B. and Travaglini, G. (2005), *J High Energy Phys* 10: 011.

68. One-Loop Five-Point $+++++$ Amplitude from Generalized Unitarity This problem continues along the lines of Problem 67. The goal is now to calculate the five-point all-plus one-loop amplitude $\mathcal{A}_5^{\text{scalar}}(1^+2^+3^+4^+5^+)$.

68.a Interpreting the contribution of the $d-4$ extra dimensions to the scalar propagator as a mass, write the reduction of this amplitude on a basis of master integrals (denote by ℓ_i the momenta carried by the loop propagators, with $\ell_i=\ell_{i-1}+p_i$, and by $D_i\equiv\ell_i^2+\ell_\perp^2$ the corresponding denominators).

68.b Consider the four-line cut (0123) and show that

$$D_{0123} = -\frac{2\sqrt{2}\,g^5 \ell_\perp^4}{\langle 12\rangle\langle 23\rangle\langle 34\rangle\langle 45\rangle\langle 51\rangle} \sum_{\text{cut solutions}} \Big(s_{12}s_{23} + 4i\frac{(p_3\cdot p_4)\epsilon(123\ell_3) - (p_1\cdot p_2)\epsilon(\ell_3 234)}{p_4\cdot \ell_3}\Big),$$

where $\epsilon(abcd) \equiv a_\mu b_\nu c_\rho d_\sigma \epsilon^{\mu\nu\rho\sigma}$.

68.c Find the two values of the loop momentum that solve these cut conditions. *Hint: it is convenient to express ℓ_3^μ on the basis made of $p_1^\mu, p_2^\mu, p_3^\mu$ and $P^\mu \equiv \epsilon^{\mu\nu\rho\sigma} p_{1\nu} p_{2\rho} p_{3\sigma}$.*

68.d Show that the coefficient D_{0123} of the box integral $L^{(4)}_{0123}$ is

$$D_{0123} = -\frac{4\sqrt{2}\,g^5 \ell_\perp^4}{\langle 12\rangle\langle 23\rangle\langle 34\rangle\langle 45\rangle\langle 51\rangle} \times \Big(s_{12}s_{23} + 4i\ell_\perp^2\,\epsilon(1234)\Big[\frac{1}{(\ell_3^+ + p_4)^2 + \ell_\perp^2} + \frac{1}{(\ell_3^- + p_4)^2 + \ell_\perp^2}\Big]\Big).$$

The other box coefficients are given by a relabeling of the external lines.

68.e Ignoring the triangles, perform explicitly the loop integration in order to obtain

$$\mathcal{A}_5^{\text{scalar}}(1^+\cdots 5^+) = \frac{ig^5\sqrt{2}}{24\pi^2}\,\frac{\big(s_{12}s_{23} + s_{23}s_{34} + s_{34}s_{45} + s_{45}s_{51} + s_{51}s_{12} + 4i\epsilon(1234)\big)}{\langle 12\rangle\langle 23\rangle\langle 34\rangle\langle 45\rangle\langle 51\rangle}.$$

68.f Extra: show that the triple cuts are satisfied without having to add triangle integrals (i.e., the triangle coefficients are zero, $C_{ijk} = 0$).

68.a The strategy is the same as in Problem 67. We first split the loop momentum into its four-dimensional part $\ell_\parallel$ and the extra components $\ell_\perp$, in order to write

$$\mathcal{A}_5^{\text{scalar}}(1^+2^+3^+4^+5^+) = \lim_{d\to 4}\int \frac{d^{d-4}\ell_\perp}{(2\pi)^{d-4}}\,\underbrace{\mathcal{A}_5^{\text{scalar}}(1^+2^+3^+4^+5^+;-\ell_\perp^2)}_{\text{massive 4d scalar loop}}.$$

Then, we decompose the amplitude in the integrand as a sum of boxes and triangles (we exclude pentagons since the loop momentum is strictly four-dimensional in this amplitude, and we exclude bubbles and tadpoles because we know the result should be ultraviolet finite):

$$\begin{aligned}
&\mathcal{A}_5^{\text{scalar}}(1^+2^+3^+4^+5^+;-\ell_\perp^2) \\
&\quad = D_{0123}L_{0123}^{(4)} + D_{0124}L_{0124}^{(4)} + D_{0134}L_{0134}^{(4)} + D_{0234}L_{0234}^{(4)} + D_{1234}L_{1234}^{(4)} \\
&\quad + C_{012}L_{012}^{(4)} + C_{013}L_{013}^{(4)} + C_{014}L_{014}^{(4)} + C_{023}L_{023}^{(4)} + C_{024}L_{024}^{(4)} \\
&\quad + C_{034}L_{034}^{(4)} + C_{123}L_{123}^{(4)} + C_{124}L_{124}^{(4)} + C_{134}L_{134}^{(4)} + C_{234}L_{234}^{(4)},
\end{aligned}$$

with the following notation:

$$\begin{aligned}
&L_{i_1\cdots i_p}^{(4)} \equiv \int \frac{d^4\ell_\parallel}{(2\pi)^4}\frac{i}{D_{i_1}}\cdots\frac{i}{D_{i_p}}, \\
&D_i \equiv (\ell_\parallel + q_i)^2 + \ell_\perp^2, \qquad q_0 \equiv 0, \quad q_1 \equiv p_1, \\
&q_2 \equiv p_1 + p_2, \quad q_3 \equiv p_1 + p_2 + p_3, \quad q_4 \equiv p_1 + p_2 + p_3 + p_4.
\end{aligned}$$

Although this decomposition contains many terms, all the box terms are related by cyclic permutations and it is sufficient to determine one of them. Moreover, we shall see later that the coefficients of the triangles are all zero.

68.b Let us focus on the box $L_{0123}^{(4)}$. It can be isolated by applying the four-line cut $\Delta_0\Delta_1\Delta_2\Delta_3$, since

$$\begin{aligned}
\Delta_0\Delta_1\Delta_2\Delta_3\left[D_{0123}L_{0123}^{(4)} + D_{0124}L_{0124}^{(4)} + \cdots\right] &= D_{0123}\,\Delta_0\Delta_1\Delta_2\Delta_3\left[L_{0123}^{(4)}\right] \\
&= D_{0123}\int\frac{d^4\ell_\parallel}{(2\pi)^4}\,\delta_0\delta_1\delta_2\delta_3, \qquad (4.47)
\end{aligned}$$

where we denote $\delta_i \equiv \delta(D_i)$. Let us now apply this cut directly to $\mathcal{A}_5^{\text{scalar}}(1^+2^+3^+4^+5^+;-\ell_\perp^2)$. This quadruple cut splits the amplitude into three three-point sub-amplitudes and a four-point one:

$$\Delta_0\Delta_1\Delta_2\Delta_3\left[\mathcal{A}_5^{\text{scalar}}(1^+2^+3^+4^+5^+;-\ell_\perp^2)\right] =$$

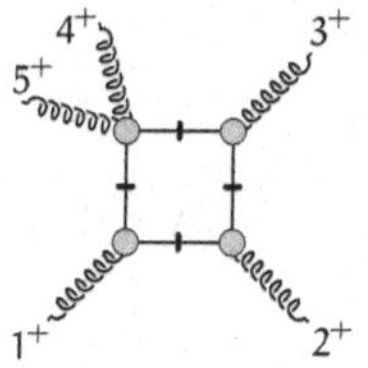

(4.48)

$$\begin{aligned}
&= 2\int\frac{d^4\ell_\parallel}{(2\pi)^4}\,\delta_0\delta_1\delta_2\delta_3\,\mathcal{A}_3(\ell_\parallel,1^+,-(\ell_\parallel+q_1)^*)\,\mathcal{A}_3(\ell_\parallel+q_1,2^+,-(\ell_\parallel+q_2)^*) \\
&\qquad \times \mathcal{A}_3(\ell_\parallel+q_2,3^+,-(\ell_\parallel+q_3)^*)\,\mathcal{A}_4(\ell_\parallel+q_3,4^+5^+,-\ell_\parallel^*).
\end{aligned}$$

(The prefactor 2 accounts for the possibility of exchanging the roles of the scalar and antiscalar.) The product of three-point amplitudes in the first line was evaluated in Problem 57:

$$\mathcal{A}_3(\ell_\parallel,1^+,-(\ell_\parallel+q_1)^*)\,\mathcal{A}_3(\ell_\parallel+q_1,2^+,-(\ell_\parallel+q_2)^*) = 2g^2\ell_\perp^2\frac{[12]}{\langle 12\rangle}.$$

In this problem, one can also find the expressions for the two amplitudes that appear in the second line:

$$\mathcal{A}_3(\ell_\parallel + q_2, 3^+, -(\ell_\parallel + q_3)^*) = ig\sqrt{2}\,\frac{\langle q|\overline{\ell}_2|3]}{\langle q3\rangle},$$

$$\mathcal{A}_4(\ell_\parallel + q_3, 4^+5^+, -\ell_\parallel^*) = 2g^2\ell_\perp^2\,\frac{[45]}{\langle 45\rangle}\frac{i}{D_4},$$

where we denote $\ell_i \equiv \ell_\parallel + q_i$ and $\overline{\ell}_i \equiv -\ell_{i\mu}\overline{\sigma}^\mu$. Choosing $q = p_2$ and noting that $\overline{\ell}_3 = \overline{\ell}_2 + |3\rangle[3|$, we have

$$\mathcal{A}_3(\ell_\parallel + q_2, 3^+, -(\ell_\parallel + q_3)^*) = ig\sqrt{2}\,\frac{\langle 2|\overline{\ell}_3|3]}{\langle 23\rangle}.$$

Therefore, the product of tree amplitudes that appears in (4.48) can be rewritten as

$$\begin{aligned}\mathcal{A}_3(\cdot)\mathcal{A}_3(\cdot)\mathcal{A}_3(\cdot)\mathcal{A}_4(\cdot) &= -4\sqrt{2}\,g^5\ell_\perp^4\,\frac{[12]\langle 2|\overline{\ell}_3|3]\langle 34\rangle[45]\langle 51\rangle}{D_4\,\langle 12\rangle\langle 23\rangle\langle 34\rangle\langle 45\rangle\langle 51\rangle}\\ &= -4\sqrt{2}\,g^5\ell_\perp^4\,\frac{\mathrm{tr}\,(\overline{\ell}_3 P_3 \overline{P}_4 P_5 \overline{P}_1 P_2)}{D_4\,\langle 12\rangle\langle 23\rangle\langle 34\rangle\langle 45\rangle\langle 51\rangle}\\ &= 4\sqrt{2}\,g^5\ell_\perp^4\,\frac{\mathrm{tr}\,(\overline{P}_1 P_2 \overline{P}_3 \ell_3 \overline{P}_4 P_5)}{D_4\,\langle 12\rangle\langle 23\rangle\langle 34\rangle\langle 45\rangle\langle 51\rangle},\end{aligned}$$

where we have used $\overline{\sigma}^\mu\sigma^\nu + \overline{\sigma}^\nu\sigma^\mu = 2g^{\mu\nu}$ in order to obtain $\overline{\ell}_3 P_3 = -\overline{P}_3\ell_3 + 2(\ell_3\cdot p_3) = -\overline{P}_3\ell_3$ (note that $0 = \ell_2^2 + \ell_\perp^2 = (\ell_3 - p_3)^2 + \ell_\perp^2 = -2\ell_3\cdot p_3$ thanks to the delta functions δ_2 and δ_3). Then, we eliminate P_5 by using momentum conservation:

$$\overline{P}_4 P_5 \overline{P}_1 = -\overline{P}_4(P_1 + P_2 + P_3 + P_4)\overline{P}_1 = -\overline{P}_4(P_2 + P_3)\overline{P}_1,$$

so that

$$\mathcal{A}_3(\cdot)\mathcal{A}_3(\cdot)\mathcal{A}_3(\cdot)\mathcal{A}_4(\cdot) = -2\sqrt{2}\,g^5\ell_\perp^4\,\frac{\mathrm{tr}\,(\overline{P}_1 P_2 \overline{P}_3 \ell_3 \overline{P}_4 (P_2 + P_3))}{(p_4\cdot\ell_3)\,\langle 12\rangle\langle 23\rangle\langle 34\rangle\langle 45\rangle\langle 51\rangle}.$$

(We have used $D_4 = (\ell_3 + p_4)^2 + \ell_\perp^2 = 2p_4\cdot\ell_3$.) The calculation of traces of products of six Dirac matrices can be avoided by using

$$P_2\overline{P}_1 P_2 = s_{12}\,P_2, \quad \overline{P}_3\ell_3\overline{P}_4 P_3 = -s_{34}\,\overline{\ell}_3 P_3$$

(valid since $p_2^2 = p_3^2 = p_3\cdot\ell_3 = 0$), which gives

$$\begin{aligned}\mathrm{tr}\,(\overline{P}_1 P_2 \overline{P}_3 \ell_3 \overline{P}_4 (P_2 + P_3)) &= s_{12}s_{23}(p_4\cdot\ell_3) + 2i(s_{34}\epsilon(123\ell_3) - s_{12}\epsilon(\ell_3 234))\\ &\quad + s_{34}(s_{12} + s_{13})(p_2\cdot\ell_3) - s_{23}s_{34}(p_1\cdot\ell_3),\end{aligned}$$

where we have introduced the notation $\epsilon(ij\ell_3 k) \equiv p_{i\mu}p_{j\nu}\ell_{3\rho}p_{k\sigma}\epsilon^{\mu\nu\rho\sigma}$. The cut conditions lead to

$$p_2 \cdot \ell_3 = \tfrac{1}{2}\, s_{23}, \quad p_1 \cdot \ell_3 = \tfrac{1}{2}\,(s_{12} + s_{13}),$$

implying that the two terms on the second line cancel, so that

$$\begin{aligned} D_{0123} &= \sum_{\text{cut solutions}} \mathcal{A}_3(\cdot)\mathcal{A}_3(\cdot)\mathcal{A}_3(\cdot)\mathcal{A}_4(\cdot) \\ &= -\frac{2\sqrt{2}\, g^5 \ell_\perp^4}{\langle 12\rangle\langle 23\rangle\langle 34\rangle\langle 45\rangle\langle 51\rangle} \sum_{\substack{\text{cut}\\ \text{solutions}}} \left(s_{12}s_{23} + 2i\frac{s_{34}\epsilon(123\ell_3) - s_{12}\epsilon(\ell_3 234)}{p_4 \cdot \ell_3} \right). \end{aligned} \tag{4.49}$$

(Note that the existence of two solutions to the four-cut equations implies that the right-hand side of (4.47) gives $2\,D_{0123}$.) Contrary to what happened in Problem 67, the result of applying the four cuts still contains an explicit dependence on the loop momentum (here, via the momentum ℓ_3).

68.c Therefore, we must solve the cut conditions in order to determine the allowed values of ℓ_3^μ on the cut, which we will then insert in the above expression. To that end, we shall express ℓ_3 on a basis made of $p_{1,2,3}$ and of a fourth vector orthogonal to them, for instance $P^\mu \equiv \epsilon^{\mu\nu\rho\sigma} p_{1\nu} p_{2\rho} p_{3\sigma}$. This vector satisfies

$$P \cdot p_1 = P \cdot p_2 = P \cdot p_3 = 0, \quad P^2 = \frac{stu}{4},$$

with $s \equiv (p_1 + p_2)^2$, $t \equiv (p_2 + p_3)^2$ and $u \equiv (p_1 + p_3)^2$. Writing

$$\ell_3 \equiv \alpha\, p_1 + \beta\, p_2 + \gamma\, p_3 + \delta\, P,$$

the four cut conditions

$$\ell_3^2 = (\ell_3 - p_3)^2 = (\ell_3 - p_3 - p_2)^2 = (\ell_3 - p_3 - p_2 - p_1)^2 = -\ell_\perp^2$$

are equivalent to

$$\begin{cases} \alpha\beta\, s + \alpha\gamma\, u + \beta\gamma\, t + \delta^2 \frac{stu}{4} = -\ell_\perp^2, \\ \alpha\beta\, s + \alpha(\gamma - 1)\, u + \beta(\gamma - 1)\, t + \delta^2 \frac{stu}{4} = -\ell_\perp^2, \\ \alpha(\beta - 1)\, s + \alpha(\gamma - 1)\, u + (\beta - 1)(\gamma - 1)\, t + \delta^2 \frac{stu}{4} = -\ell_\perp^2, \\ (\alpha - 1)(\beta - 1)\, s + (\alpha - 1)(\gamma - 1)\, u + (\beta - 1)(\gamma - 1)\, t + \delta^2 \frac{stu}{4} = -\ell_\perp^2. \end{cases}$$

This system is considerably simplified by taking differences of successive equations. This leads to a linear system for α, β, γ whose solution reads

$$\alpha = -\frac{t}{2u}, \quad \beta = \frac{1}{2}, \quad \gamma = 1 + \frac{s}{2u}.$$

Then, injecting these values in the first equation provides δ:

$$\delta = \pm\sqrt{\frac{st - 4\ell_\perp^2 u}{stu^2}}.$$

Therefore, there are two values of ℓ_3^μ that satisfy the cut conditions, $\ell_3^\pm \equiv \kappa_3 \pm \zeta_3$, with $\kappa_3 \equiv \alpha\, p_1 + \beta\, p_2 + \gamma\, p_3$ and $\zeta_3 \equiv |\delta|\, P$.

68.d In order to replace these two cut solutions in (4.49), we need

$$\begin{aligned}
&\epsilon(123\kappa_3) = 0, \quad \epsilon(\kappa_3 234) = \alpha\,\epsilon(1234),\\
&\epsilon(123\zeta_3) = -\frac{|\delta|}{4}\,stu, \quad \epsilon(\zeta_3 234) = \frac{|\delta|}{4}\big(st(p_3{\cdot}p_4) + tu(p_2{\cdot}p_4) - t^2(p_1{\cdot}p_4)\big),\\
&p_4\cdot\kappa_3 = \alpha(p_1{\cdot}p_4) + \beta(p_2{\cdot}p_4) + \gamma(p_3{\cdot}p_4),\\
&p_4\cdot\zeta_3 = -|\delta|\,\epsilon(1234).
\end{aligned}$$

After some straightforward but lengthy algebra, one gets

$$\begin{aligned}
D_{0123} &= -\frac{4\sqrt{2}\,g^5\ell_\perp^4}{\langle 12\rangle\langle 23\rangle\langle 34\rangle\langle 45\rangle\langle 51\rangle}\left(s_{12}s_{23} + 2i\ell_\perp^2\,\epsilon(1234)\frac{2(p_4\cdot\kappa_3)}{(p_4\cdot\kappa_3)^2 - (p_4\cdot\zeta_3)^2}\right)\\
&= -\frac{4\sqrt{2}\,g^5\ell_\perp^4}{\langle 12\rangle\langle 23\rangle\langle 34\rangle\langle 45\rangle\langle 51\rangle}\\
&\quad\times\left(s_{12}s_{23} + 4i\ell_\perp^2\,\epsilon(1234)\left[\frac{1}{(\ell_3^+ + p_4)^2 + \ell_\perp^2} + \frac{1}{(\ell_3^- + p_4)^2 + \ell_\perp^2}\right]\right).
\end{aligned}$$

68.e In order to evaluate the corresponding contribution to $\mathcal{A}_5^{\text{scalar}}(1^+2^+3^+4^+5^+)$, the main difficulty is the fact that $\ell_\parallel$ and $\ell_\perp$ do not appear on the same footing. Instead of performing this integral directly, an easier route is to use the above result in order to guess the d-dimensional integrand that would lead to the same four-line cut in the channel [0123],

$$\begin{aligned}
\mathcal{A}_5^{\text{scalar}}(1^+2^+3^+4^+5^+) &\overset{[0123]\text{-cut}}{\Longleftrightarrow} -\frac{4\sqrt{2}\,g^5}{\langle 12\rangle\langle 23\rangle\langle 34\rangle\langle 45\rangle\langle 51\rangle}\\
&\times \lim_{d\to 4}\left\{ s_{12}s_{23}\underbrace{\int\frac{d^d\ell}{(2\pi)^d}\frac{\big(\sum_\epsilon(\ell\cdot n_\epsilon)^2\big)^2}{D_0D_1D_2D_3}}_{\text{box}} + 8i\epsilon(1234)\underbrace{\int\frac{d^d\ell}{(2\pi)^d}\frac{\big(\sum_\epsilon(\ell\cdot n_\epsilon)^2\big)^3}{D_0D_1D_2D_3D_4}}_{\text{pentagon}}\right\}.
\end{aligned}$$

The notation $\overset{[0123]\text{-cut}}{\Longleftrightarrow}$ means that the cut [0123] gives the same result when applied to the left- and right-hand sides (but it does not imply that they are equal, since they may not agree in other cut channels). Interestingly, the second term on the right-hand side is a pentagon integral. That a pentagon master integral would arise at some point was to be expected, since we are in fact doing a calculation in $d = 4 - 2\epsilon$ dimensions. Therefore, we need to evaluate the following two integrals:

$$\begin{aligned}
&\lim_{d\to 4} g_{\mu\nu}^{d-4} g_{\rho\sigma}^{d-4}\int\frac{d^d\ell}{(2\pi)^d}\frac{\ell^\mu\ell^\nu\ell^\rho\ell^\sigma}{\ell^8} = -\frac{i}{6(4\pi)^2},\\
&\lim_{d\to 4} g_{\mu\nu}^{d-4} g_{\rho\sigma}^{d-4} g_{\tau\zeta}^{d-4}\int\frac{d^d\ell}{(2\pi)^d}\frac{\ell^\mu\ell^\nu\ell^\rho\ell^\sigma\ell^\tau\ell^\zeta}{\ell^{10}} = -\frac{i}{12(4\pi)^2}.
\end{aligned}$$

(A simple power counting indicates that we can replace all the denominators D_i by ℓ^2 in order to extract the value of the integrals at $\epsilon = 0$, since both integrals have a logarithmic divergence in $d = 4$, but the contractions with the tensor $g_{\mu\nu}^{d-4}$ that spans the extra dimensions brings a prefactor ϵ.) The first integral was calculated in Problem 67, and the second one is obtained by the same method. Therefore, the four-line cut [0123] tells us that the amplitude $\mathcal{A}_5^{\rm scalar}(1^+2^+3^+4^+5^+)$ must contain

$$\mathcal{A}_5^{\rm scalar}(1^+2^+3^+4^+5^+) \overset{[0123]\text{-cut}}{\Longleftrightarrow} \frac{ig^5\sqrt{2}}{24\pi^2\langle 12\rangle\langle 23\rangle\langle 34\rangle\langle 45\rangle\langle 51\rangle}\Big(s_{12}s_{23}+4i\epsilon(1234)\Big).$$

Consider now the cut [0124]. Its calculation is not necessary, since the result is obtained from the cut [0123] by a simple relabeling of the external lines:

$$\mathcal{A}_5^{\rm scalar}(1^+2^+3^+4^+5^+) \overset{[0124]\text{-cut}}{\Longleftrightarrow} -\frac{4\sqrt{2}\,g^5}{\langle 12\rangle\langle 23\rangle\langle 34\rangle\langle 45\rangle\langle 51\rangle}$$
$$\times \lim_{d\to 4}\left\{ s_{51}s_{12}\int \frac{d^d\ell}{(2\pi)^d}\,\frac{\big(\sum_\epsilon(\ell\cdot n_\epsilon)^2\big)^2}{D_0D_1D_2D_4} + 8i\,\underbrace{\epsilon(5123)}_{\epsilon(1234)}\int \frac{d^d\ell}{(2\pi)^d}\,\frac{\big(\sum_\epsilon(\ell\cdot n_\epsilon)^2\big)^3}{D_0D_1D_2D_3D_4}\right\}.$$

Note that the pentagon integral necessary to correctly reproduce the [0124] cut is the same as the one we encountered earlier in the cut [0123], while the box integral is a new contribution. The same pattern continues with all the four-line cuts: they are reproduced with a box and a pentagon, the pentagon being always the same while the boxes are all distinct. Obviously, we should not include the pentagon five times, since this would be overcounting its contribution. Therefore, summing over all the contributions encountered so far, we infer the following result for the amplitude:

$$\mathcal{A}_5^{\rm scalar}(1^+2^+3^+4^+5^+) \overset{\text{four-line cuts}}{\Longleftrightarrow} \frac{ig^5\sqrt{2}}{24\pi^2\langle 12\rangle\langle 23\rangle\langle 34\rangle\langle 45\rangle\langle 51\rangle}$$
$$\times\Big(s_{12}s_{23}+s_{23}s_{34}+s_{34}s_{45}+s_{45}s_{51}+s_{51}s_{12}+4i\epsilon(1234)\Big).$$

The derivation we have followed here is arguably intricate, but nevertheless orders of magnitude simpler than the calculation of the 191 color-ordered Feynman diagrams that contribute to this amplitude in the traditional approach.

68.f In order to confirm that the above result is complete, it is necessary to show that the contributions we have determined from the four-line cuts saturate the three-line cuts, so that we do not need to add triangle integrals. Let us consider the triple cut in the channel [012]. First, we need to evaluate this triple cut on the terms guessed from the quadruple cuts. The only terms that also contribute to this triple cut are those that contribute to the [0123] or [0124] quadruple cuts. Therefore, we have

$$\Delta_0\Delta_1\Delta_2\begin{bmatrix}\text{guessed}\\ \text{amplitude}\end{bmatrix} = -\frac{4i\sqrt{2}\,g^5}{\langle 12\rangle\langle 23\rangle\langle 34\rangle\langle 45\rangle\langle 51\rangle}\int\frac{d^d\ell}{(2\pi)^d}\,\frac{\delta_0\delta_1\delta_2}{D_3D_4}\,\ell_\perp^4$$
$$\times\Big\{s_{12}s_{23}D_4+s_{51}s_{12}D_3+8i\epsilon(1234)\ell_\perp^2\Big\}. \tag{4.50}$$

Then, we must apply the same cut to the full amplitude. It splits it into a pair of three-point tree amplitudes and a five-point tree amplitude:

$$\Delta_0\Delta_1\Delta_2\left[\mathcal{A}_5^{\text{scalar}}(1^+2^+3^+4^+5^+)\right] =$$

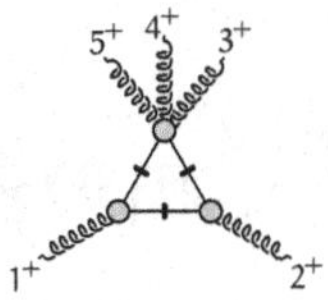

$$= 2\int \frac{d^d\ell}{(2\pi)^d}\,\delta_0\delta_1\delta_2\,\mathcal{A}_3(\ell_\parallel, 1^+, -(\ell_\parallel + q_1)^*)$$
$$\times\,\mathcal{A}_3(\ell_\parallel + q_1, 2^+, -(\ell_\parallel + q_2)^*)\mathcal{A}_5(\ell_\parallel + q_2, 3^+4^+5^+, -\ell_\parallel^*).$$

The results of Problem 57 lead to

$$\mathcal{A}_3(\ell_\parallel, 1^+, -(\ell_\parallel + q_1)^*)\,\mathcal{A}_3(\ell_\parallel + q_1, 2^+, -(\ell_\parallel + q_2)^*) = 2g^2\ell_\perp^2\frac{[12]}{\langle 12\rangle},$$

$$\mathcal{A}_5(\ell_\parallel + q_2, 3^+4^+5^+, -\ell_\parallel^*) = -\frac{2i\sqrt{2}g^3\ell_\perp^2}{D_3D_4}\,\frac{[5|(P_3+P_4)(\overline{\ell}_\parallel + \overline{Q}_2)|3]}{\langle 34\rangle\langle 45\rangle}.$$

Note that $(\overline{\ell}_\parallel + \overline{Q}_2)|3] = (\overline{\ell}_3 - |3\rangle[3|)|3] = \overline{\ell}_3|3]$. Therefore,

$$\Delta_0\Delta_1\Delta_2\left[\mathcal{A}_5^{\text{scalar}}(1^+2^+3^+4^+5^+)\right] = -\frac{4i\sqrt{2}g^5}{\langle 12\rangle\langle 23\rangle\langle 34\rangle\langle 45\rangle\langle 51\rangle}\int \frac{d^d\ell}{(2\pi)^d}\,\frac{\delta_0\delta_1\delta_2}{D_3D_4}\,\ell_\perp^4$$
$$\times\left\{2\langle 51\rangle[12]\langle 23\rangle[5|(P_3+P_4)\overline{\ell}_3|3]\right\}.$$

Using

$$2p_3\cdot\ell_3 = D_3,\quad 2p_2\cdot\ell_3 = s_{23},\quad 2p_1\cdot\ell_3 = s_{12}+s_{13},\quad 2p_4\cdot\ell_3 = D_4 - D_3,$$
$$s_{23}+s_{34}+s_{24} = (p_2+p_3+p_4)^2 = (p_1+p_5)^2 = s_{51},$$

the quantity in the second line of the preceding equation can be reduced to

$$2\langle 51\rangle[12]\langle 23\rangle[5|(P_3+P_4)\overline{\ell}_3|3] = s_{12}s_{23}D_4 + s_{51}s_{12}D_3$$
$$+\,4i\left(s_{34}\epsilon(123\ell_3) - s_{12}\epsilon(\ell_3 234)\right).$$

The first two terms are identical to the first two terms in the second line of (4.50) and need no further consideration. For the triple cut of the full amplitude and of the guessed one to agree, we thus need to show that

$$\int \frac{d^4\ell_3}{(2\pi)^4}\,\frac{\delta_0\delta_1\delta_2}{D_3D_4}\Big(s_{34}\epsilon(123\ell_3) - s_{12}\epsilon(\ell_3 234)\Big) = 2\ell_\perp^2\,\epsilon(1234)\int \frac{d^4\ell_3}{(2\pi)^4}\,\frac{\delta_0\delta_1\delta_2}{D_3D_4}.$$

The three delta functions reduce the integral to a one-dimensional one. Because of this, it is unclear at this point whether the equality can be made manifest at the level of the integrands or is only valid for the integrated results. As we did earlier, we parameterize ℓ_3^μ as

$$\ell_3^\mu \equiv \alpha p_1^\mu + \beta p_2^\mu + \gamma p_3^\mu + \delta P^\mu,$$

and the three cut equations read

$$\begin{aligned}
D_2 = 0: &\quad \alpha\beta\, s + \alpha(\gamma-1)\,u + \beta(\gamma-1)\,t + \delta^2 \tfrac{stu}{4} = -\ell_\perp^2,\\
D_1 = 0: &\quad \alpha(\beta-1)\,s + \alpha(\gamma-1)\,u + (\beta-1)(\gamma-1)\,t + \delta^2 \tfrac{stu}{4} = -\ell_\perp^2,\\
D_0 = 0: &\quad (\alpha-1)(\beta-1)\,s + (\alpha-1)(\gamma-1)\,u + (\beta-1)(\gamma-1)\,t + \delta^2 \tfrac{stu}{4} = -\ell_\perp^2.
\end{aligned}$$

Obviously, three equations do not uniquely determine four unknowns. Here, we choose to express β, γ and δ in terms of α:

$$\beta(\alpha) = 1 + \tfrac{u}{t}\alpha, \quad \gamma(\alpha) = 1 - \tfrac{s}{t}\alpha, \quad \delta(\alpha) = \pm 2\sqrt{\frac{su\alpha^2 - t\ell_\perp^2}{st^2u}}.$$

(There are two one-dimensional branches of solutions). The various factors appearing in the integrands, expressed in terms of $(\alpha, \beta, \gamma, \delta)$, read

$$\begin{aligned}
D_3 &= s_{13}\alpha + s_{23}\beta = 2s_{13}\alpha + s_{23},\\
D_4 &= \underbrace{(s_{13}+s_{14})\alpha + (s_{23}+s_{24})\beta + s_{34}\gamma}_{\equiv A} - 2\epsilon(1234)\delta,
\end{aligned}$$

$$s_{34}\epsilon(123\ell_3) - s_{12}\epsilon(\ell_3 234) = \underbrace{\left(s_{34}\epsilon(123P) - s_{12}\epsilon(P234)\right)}_{\equiv B}\delta - \underbrace{s_{12}\alpha}_{\equiv C}\epsilon(1234).$$

By summing over the two possible signs for δ, we obtain

$$\begin{aligned}
\sum_{\delta=\pm|\delta|} \frac{s_{34}\epsilon(123\ell_3) - s_{12}\epsilon(\ell_3 234)}{D_3 D_4} &= \frac{1}{D_3}\sum_{\epsilon=\pm} \frac{\epsilon B|\delta| - C\epsilon(1234)}{A - 2\epsilon|\delta|\epsilon(1234)}\\
&= \frac{\epsilon(1234)}{D_3}\frac{2(2B|\delta|^2 - AC)}{A^2 - 4|\delta|^2\epsilon^2(1234)}\\
&= \frac{\epsilon(1234)}{D_3}\frac{2B|\delta|^2 - AC}{A}\left\{\frac{1}{A - 2|\delta|\epsilon(1234)} + \frac{1}{A + 2|\delta|\epsilon(1234)}\right\}\\
&= \epsilon(1234)\left(\frac{2B|\delta|^2 - AC}{A}\right)\sum_{\delta=\pm|\delta|}\frac{1}{D_3 D_4}.
\end{aligned}$$

One may then check that the quantity between the curly brackets is of the form

$$\frac{2B|\delta|^2 - AC}{A} = 2\ell_\perp^2 + D_3\, f(D_3).$$

The first term gives the expected result, while the second one has a vanishing integral when integrated over the two branches of three-cut solutions.

Source: Brandhuber, A., McNamara, S., Spence, B. and Travaglini, G. (2005), *J High Energy Phys* 10: 011.

CHAPTER 5

Lattice, Finite T, Strong Fields

Introduction

In this chapter, we explore aspects of quantum field theory related to the topics of lattice field theory, field theory at finite temperature, and strong fields. The connection between the last two subjects is that they depart from quantum field theory in the vacuum (e.g., scattering amplitudes for high-energy physics) by considering situations with a large density of particles, in which the quantum field theory formalism needs to be supplemented with aspects of many-body physics. The reason for the inclusion of lattice field theory in this chapter is that, being a Euclidean formulation of QFT, it has a natural finite-temperature interpretation if one views the inverse of the extent of the time direction as a temperature.

Lattice Field Theory

Replacing continuous spacetime by a discrete grid of points provides a regularization of ultraviolet divergences (the inverse of the grid spacing acts as a momentum cutoff) and also turns a path integral into a finite set of integrals if the volume is also kept finite. Moreover, for Yang–Mills theories, there exist lattice discretizations of the action that preserve gauge invariance. Such formulations are in terms of *link variables* $U_\mu(x)$, i.e., elements of the gauge group that one may view as Wilson lines along the elementary edges of the lattice. The simplest of these discretizations is provided by the *Wilson action:*

$$\mathcal{S}_W[U] \equiv \frac{1}{g^2} \sum_{x\in \text{ lattice}} \sum_{(\mu,\nu)} \Big(\text{tr}\left(\text{Re}\,[\square]_{x;\mu\nu}\right) - N\Big), \tag{5.1}$$

where $[\square]_{x;\mu\nu}$ (known as a *plaquette*) is the product of the four link variables around an elementary square of a cubic lattice of base point x, in the $(\mu\nu)$ plane. This action is positive definite in Euclidean spacetime, allowing an evaluation of the path integral by Monte Carlo sampling (moreover, for a compact gauge group, the domain to be sampled is compact).

The Dirac action for fermions can also be discretized in a gauge invariant way:

$$\mathcal{S}_D[U,\overline{\psi},\psi] \equiv a^3 \sum_{x\in\,\text{lattice}} \overline{\psi}(x) \sum_{\mu} \gamma^\mu \frac{U_\mu^\dagger(x)\psi(x+\widehat{\mu}) - U_\mu(x-\widehat{\mu})\psi(x-\widehat{\mu})}{2}, \tag{5.2}$$

where a is the lattice grid spacing and $\widehat{\mu}$ denotes the displacement by one lattice spacing in the direction μ. Note that the Grassmann variables $\overline{\psi}, \psi$ can be integrated out analytically since this action is quadratic, giving the determinant of the Dirac operator multiplied by fermion propagators (i.e., inverses of the Dirac operator) that connect the $\overline{\psi}$'s and the ψ's of the observable under consideration.

A major complication with lattice fermions is that the above action leads to 2^d (in d dimensions) propagating modes for each fermion flavor – the extra ones are called *fermion doublers*. The existence of these doublers, which give unphysical contributions to fermion loops, is closely related to the *Nielsen–Ninomiya theorem*, one form of which states that a chirally invariant lattice fermion action necessarily has doublers. One of several possibilities for mitigating the effect of the doublers is to add a *Wilson term* to the action, of the form $a\,\overline{\psi}D^\mu D_\mu\psi$, whose effect is to give masses of order a^{-1} to the doublers (such a term evades the above theorem by explicitly breaking chiral symmetry).

Quantum Field Theory at Finite Temperature

The goal of QFT at finite temperature (or density) is to calculate expectation values of operators in the presence of a thermal bath,

$$\langle \mathcal{O} \rangle \equiv \frac{\mathrm{tr}\,(e^{-\beta\mathcal{H}}\mathcal{O})}{\mathrm{tr}\,(e^{-\beta\mathcal{H}})}, \tag{5.3}$$

where β is the inverse temperature (the natural system of units is extended by setting the Boltzmann constant to unity, which amounts to measuring temperatures in units of energy). When the ground state is unique, such an expectation value becomes a vacuum expectation value in the zero-temperature limit. The perturbative expansion is based on relating the Heisenberg field $\phi(x)$ to the free field $\phi_{\text{in}}(x)$ of the interaction picture, via Eqs. (1.10). Moreover, interactions are also present in the density operator, for which we have

$$e^{-\beta\mathcal{H}} = e^{-\beta\mathcal{H}_0}\, U(-\infty - i\beta, -\infty), \tag{5.4}$$

where $\mathcal{H}_0$ is the free Hamiltonian and U is the evolution operator that also relates ϕ and ϕ_{in}, extended to an imaginary time evolution. The perturbative expansion can be expressed seamlessly by defining the temporal contour $\mathcal{C} \equiv (-\infty, +\infty) \cup (+\infty, -\infty) \cup (-\infty, -\infty - i\beta)$. One may define a generating functional for expectation values of path-ordered products of fields defined on this contour:

$$Z[j] \equiv \mathrm{tr}\left(e^{-\beta\mathcal{H}}\, P\exp i\int_{\mathcal{C}} d^4x\, j(x)\phi(x)\right). \tag{5.5}$$

$Z[j]$ takes the following more explicit form where interactions have been factored out:

$$Z[j] = \exp\left(ig\int_{\mathcal{C}} d^4x\,\mathcal{L}_I\Big(\frac{\delta}{i\delta j(x)}\Big)\right)\exp\left(-\frac{1}{2}\int_{\mathcal{C}} d^4x d^4y\, j(x)j(y)\,G_0(x,y)\right), \tag{5.6}$$

with a free propagator G_0 defined as

$$G_0(x,y) = \int \frac{d^3p}{(2\pi)^3 2E_p}\Big\{e^{-ip\cdot(x-y)}\big(\theta_c(x^0,y^0)+n_B(E_p)\big) + e^{+ip\cdot(x-y)}\big(\theta_c(y^0,x^0)+n_B(E_p)\big)\Big\}. \tag{5.7}$$

$\theta_c(x^0,y^0)$ is a step function defined on the contour $\mathcal{C}$ (equal to 1 if x^0 is posterior to y^0 on the oriented contour), and $n_B(\omega) = (e^{\beta\omega}-1)^{-1}$ is the Bose–Einstein distribution. Eqs. (5.6) and (5.7) directly give the Feynman rules in coordinate space at finite temperature:

- Draw all the relevant graphs at a given order in the coupling constant λ and compute their symmetry factors.
- To each edge of a graph, associate a free propagator G_0.
- At each vertex of coordinate x, one has an integral $-i\lambda\int_{\mathcal{C}} d^4x$.

Kubo–Martin–Schwinger Symmetry

The special form of the canonical density operator $e^{-\beta\mathcal{H}}$ can be formally interpreted as a time evolution operator for an imaginary time shift $-i\beta$. This implies that path-ordered bosonic correlation functions take identical values at the two endpoints of the contour $\mathcal{C}$:

$$G(\dots,-\infty,\dots) = G(\dots,-\infty-i\beta,\dots), \tag{5.8}$$

a property known as *Kubo–Martin–Schwinger (KMS) symmetry*. One of its consequences is that one may replace the initial time $-\infty$ by any finite initial time t_I without altering physical quantities (this property means that, in a system in thermal equilibrium, no measurement can tell when the system was prepared in equilibrium).

Matsubara Formalism

Thermodynamical quantities can be accessed from the *partition function*, $Z \equiv \mathrm{tr}\,(e^{-\beta\mathcal{H}})$, which is obtained as the sum of all the vacuum graphs at finite temperature. Because vacuum graphs do not depend on any coordinate, they can be calculated by deforming the contour $\mathcal{C}$ and replacing it by the line segment $[0,-i\beta]$. In terms of the imaginary time $\tau \equiv ix^0$, KMS symmetry is a periodicity of period β, and we can decompose the τ dependence of all correlation functions as discrete Fourier series. The discrete conjugate frequencies, $\omega_n \equiv 2\pi n T$ $(n \in \mathbb{Z})$, are known as *Matsubara frequencies*. This allows the following formulation of the Feynman rules in momentum space:

- Draw all the relevant graphs at a given order in the coupling constant λ and compute their symmetry factors.

- Each edge carries a discrete frequency ω_n and a momentum $\mathbf{p}$, and is assigned the following free propagator:

$$G_0(\omega_n, \mathbf{p}) \equiv \frac{1}{\omega_n^2 + \mathbf{p}^2 + m^2}.$$

- Each vertex brings a factor λ and a combination of Kronecker symbol and delta functions, $\delta_{\sum_i \omega_{n_i}} \delta\big(\sum_i \mathbf{p}_i\big)$, to ensure that the sum of Matsubara frequencies and 3-momenta entering in the vertex is zero.
- The frequency ω_n and the momentum $\mathbf{p}$ running in a loop must be summed and integrated with $T \sum_{n \in \mathbb{Z}} \int \frac{d^3p}{(2\pi)^2}$.

Fermions, Conserved Charges

For fermions, KMS symmetry states that the corresponding correlation functions must be β-anti-periodic in imaginary time:

$$S(\ldots, 0, \ldots) = -S(\ldots, \beta, \ldots). \tag{5.9}$$

Therefore, the corresponding Matsubara frequencies are $\omega_n \equiv 2\pi(n+\frac{1}{2})T$, and the propagator in the Matsubara formalism reads

$$S_0(\omega_n, \mathbf{p}) = \frac{\omega_n \gamma^0 + p_i \gamma^i + m}{\omega_n^2 + \mathbf{p}^2 + m^2}. \tag{5.10}$$

When a QFT possesses a conserved quantity Q (e.g., electrical charge in QED), the canonical density operator can be extended to that of the *grand canonical ensemble*:

$$e^{-\beta \mathcal{H}} \quad \to \quad e^{-\beta(\mathcal{H} - \mu Q)}, \tag{5.11}$$

where μ is the *chemical potential* associated with the conservation law. Each line of a graph carries a certain amount q of this conserved charge, and KMS symmetry and the Matsubara frequencies are modified as follows:

$$\begin{aligned} &\text{Bosons:} && G(\ldots, 0, \ldots) = e^{\beta\mu q}\, G(\ldots, \beta, \ldots) && \omega_n = 2\pi n T - i\mu q, \\ &\text{Fermions:} && S(\ldots, 0, \ldots) = -e^{\beta\mu q}\, S(\ldots, \beta, \ldots) && \omega_n = 2\pi(n + \tfrac{1}{2})T - i\mu q. \end{aligned} \tag{5.12}$$

Schwinger--Keldysh Formalism

The Schwinger–Keldysh formalism provides an alternative way of formulating perturbation theory at finite temperature, in Minkowski spacetime. For bosons, its rules are the following:

- Draw all the relevant graphs at a given order in the coupling constant λ and compute their symmetry factors.
- Each external point and vertex can be one of two types: + or −. A vertex of type + brings a factor $-i\lambda$, and a vertex of type − brings a factor $+i\lambda$. Momentum conservation is enforced at each vertex by a factor $(2\pi)^4 \delta(\sum_i p_i)$.

- A point of type ϵ and a point of type ϵ' are connected by a free propagator $G^0_{\epsilon\epsilon'}(p)$. These four propagators are given by

$$G^0_{++}(p) = \frac{i}{p^2-m^2+i0^+} + 2\pi n_B(|p^0|)\delta(p^2-m^2), \quad G^0_{--}(p) = (G^0_{++}(p))^*,$$
$$G^0_{-+}(p) = 2\pi(\theta(p^0) + n_B(|p^0|))\delta(p^2-m^2), \quad G^0_{+-}(p) = G^0_{-+}(-p).$$

(When $T = 0$, note that these propagators are the same as those that appear in Cutkosky's cutting rules.) These propagators satisfy the following identities:

$$G^0_{++}(p) + G^0_{--}(p) = G^0_{-+}(p) + G^0_{+-}(p), \quad G^0_{-+}(p) = e^{\beta p^0}\, G^0_{+-}(p).$$

- The momentum running in a loop is integrated with $\int \frac{d^4k}{(2\pi)^4}$.

Hard Thermal Loops

There exists a class of one-loop corrections, known as *hard thermal loops* (HTL), that have the same order of magnitude as their tree-level counterparts when their external momenta are of order gT. HTLs are dominated by thermal excitations running in the loop and are therefore essentially classical. In the $SU(N)$ Yang–Mills theory, the HTLs can be encapsulated in a compact effective Lagrangian

$$\mathcal{L}_{HTL} = g^2N \int \frac{d^3p}{(2\pi)^3 p}\, n_B(|p|)\, F^a_{\nu\beta} \left[\frac{p^\beta p_\rho}{(p\cdot D)^2}\right]_{ab} F^{\rho\nu}_b, \tag{5.13}$$

which also highlights the fact that these thermal contributions are gauge invariant. After re-summation of the HTL polarization tensor,

$$\begin{aligned}
\Pi^{\mu\nu}_{HTL}(k) &= P^{\mu\nu}_T\,\Pi_T(k) + P^{\mu\nu}_L\,\Pi_L(k),\\
\Pi_T(k) &= \frac{g^2NT^2}{6}\frac{k^0}{k}\left[\frac{k^0}{k} + \frac{1}{2}\left(1-\frac{k_0^2}{k^2}\right)\ln\left(\frac{k^0+k}{k^0-k}\right)\right],\\
\Pi_L(k) &= \frac{g^2NT^2}{3}\left(1-\frac{k_0^2}{k^2}\right)\left[1-\frac{k^0}{2k}\ln\left(\frac{k^0+k}{k^0-k}\right)\right],
\end{aligned} \tag{5.14}$$

with $P^{\mu\nu}_T \equiv g^{\mu\nu} - \frac{k^\mu k^\nu}{k^2} - \frac{V^\mu V^\nu}{V^2}$, $P^{\mu\nu}_L \equiv \frac{V^\mu V^\nu}{V^2}$, $V^\mu \equiv k^2 U^\mu - (k\cdot U)\, k^\mu$

(where U^μ is the velocity of the plasma rest frame), the gluon propagator exhibits a number of thermally induced collective phenomena:

- A minimal energy, known as the *plasmon mass*, is necessary to create a gauge boson at rest. In a purely gluonic plasma, $m_p^2 = g^2NT^2/9$.
- The gluon dispersion curves are modified by the thermal medium, and are different for the transverse and longitudinal modes (the latter decouple when $T \to 0$).

- The self-energy of a zero-frequency longitudinal gluon does not go to zero when $k \to 0$, leading to *Debye screening* (i.e., an exponential attenuation) of the Coulomb field of a static color charge. For a gluon, this damping is proportional to $e^{-m_D|x|}$ with $m_D^2 = g^2NT^2/3$ (m_D is called the *Debye mass*).

- The HTL self-energy of space-like gluons has a non-zero imaginary part, which is responsible for *Landau damping* (an exponential attenuation of plasma oscillations).

Out-of-Equilibrium Systems

When formulated in coordinate space, the Schwinger–Keldysh formalism can also cope with systems that are not in thermal equilibrium. For instance, when the initial state is the vacuum (the system may be brought off-equilibrium by an external source), the propagators have the following representation:

$$
\begin{aligned}
&G^0_{-+}(x,y) = \int \frac{d^3k}{(2\pi)^3 2E_k}\, e^{-ik\cdot(x-y)}, \quad G^0_{+-}(x,y) = \int \frac{d^3k}{(2\pi)^3 2E_k}\, e^{+ik\cdot(x-y)},\\
&G^0_{++}(x,y) = \theta(x^0-y^0)G^0_{-+}(x,y) + \theta(y^0-x^0)G^0_{+-}(x,y),\\
&G^0_{--}(x,y) = \big(G^0_{++}(x,y)\big)^*. \qquad (5.15)
\end{aligned}
$$

The generating functional $Z^{SK}[j_+, j_-]$ of correlation functions in this formalism is closely related to the generating functional $Z[j]$ of the ordinary time-ordered ones, by

$$
Z^{SK}[j_+,j_-] = \exp\left[\int d^4x d^4y\, G^0_{+-}(x,y)\,(\square_x + m^2)(\square_y + m^2)\frac{\delta^2}{\delta j_+(x)\delta j_-(y)}\right] Z[j_+]\, Z^*[j_-]. \qquad (5.16)
$$

This formula shows that any graph in the Schwinger–Keldysh formalism may be obtained by stitching an ordinary Feynman graph with the complex conjugate of another graph, by means of off-diagonal propagators G^0_{+-}, in agreement with our earlier observation that the Schwinger–Keldysh formalism at $T = 0$ is identical to Cutkosky's cutting rules.

In the presence of sources that create a strong field Φ, the propagators of Eqs. (5.15) dressed with insertions of the field Φ read

$$
\begin{aligned}
&G_{-+}(x,y) = \int \frac{d^3k}{(2\pi)^3 2E_k}\, a_{-k}(x) a_{+k}(y), \quad G^0_{+-}(x,y) = \int \frac{d^3k}{(2\pi)^3 2E_k}\, a_{+k}(x) a_{-k}(y),\\
&\big(\square_x + m^2 - \mathcal{L}_I''(\Phi(x))\big)\, a_{\pm k}(x) = 0, \qquad \lim_{x^0 \to -\infty} a_{\pm k}(x) = e^{\mp ik\cdot x}. \qquad (5.17)
\end{aligned}
$$

(The propagators G_{++} and G_{--} are then given by the same linear combinations of G_{-+} and G_{+-} as in vacuum.) The functions $a_{\pm k}(x)$, known as *mode functions*, form a basis of the space of small field perturbations propagating on top of the background field Φ. Up to one loop, all inclusive observables may be expressed in terms of the background field Φ and the mode functions $a_{\pm k}$.

About the Problems of this Chapter

- **Problem 69** derives the *Banks–Casher relation*, which relates the expectation value of the chiral condensate to the spectral density at null energy of the Dirac operator.
- In **Problem 70**, we discuss the *Nielsen–Ninomiya theorem*, which states that massless fermions on a lattice must have equal numbers of right-handed and left-handed degrees of freedom. This obstruction prevents the formulation of lattice chiral gauge theories.
- In **Problem 71**, we determine the expression of the Wigner transform of the convolution product of two two-point functions. This result plays a role in the derivation of kinetic equations from the Kadanoff–Baym equations, and is also essential in the phase-space formulation of quantum mechanics.
- The interplay between *center symmetry* and the deconfinement transition in Yang–Mills theory is studied in **Problem 72**.
- The hard thermal loop contribution to the photon polarization tensor is derived in **Problem 73** using classical charged particles and a kinetic equation that describes the evolution of their distribution under the effect of an external electromagnetic field. This approach is extended in **Problem 74** to the gluon hard thermal loop.
- We study photon radiation by a hot plasma of charged particles in **Problem 75**. For soft photons, the spectrum is modified by the *Landau–Pomeranchuk–Migdal effect.*
- In **Problem 76**, we discuss how a microcanonical equilibrium distribution may emerge semi-classically from the non-linearity of the field equation of motion. The relationship between the equation of state and a microcanonical distribution is also studied.
- The combinatorial aspects of multi-particle production are studied in **Problem 77**. These are very generic properties, true no matter what the underlying quantum field theory is. We also derive a diagrammatic interpretation in terms of cutting rules for some of the quantities that appear in this discussion.
- In **Problem 78**, these questions are made more quantitative, by studying a functional that generates the complete distribution of produced particles. From this, we determine the first moment of the distribution in terms of classical fields.
- The next two problems are devoted to the study of the *Schwinger mechanism* in scalar QED. In **Problem 79**, we follow the approach of **Problem 78** to calculate the functional that generates the full distribution of produced particles. These results are rederived in **Problem 80** using *Bogoliubov transformations.*
- In **Problem 81**, we use Bogoliubov transformations to relate the vacuum states of two observers whose relative motion is uniformly accelerated. This leads to the *Unruh effect.* By analogy, the same line of reasoning leads to *Hawking radiation*, which is a thermal spectrum seen by an observer at rest in the gravitational field of a black hole.
- **Problem 82** proposes a field theory approach to *Anderson localization*, namely the fact that electrons do not propagate in a disordered medium. In **Problem 83**, we present simple scaling arguments suggesting that strong localization always happens in large one- and two-dimensional systems, and depends on the disorder strength in three dimensions.

69. Banks–Casher Relation Quantum chromodynamics, if restricted to the up and down quarks, has a nearly exact chiral symmetry thanks to the very small masses of these quarks. The three pions ($\pi^0, \pi^\pm$) are also much lighter than the other hadrons, a property which is understood as a consequence of them being the Nambu–Goldstone bosons that result from the spontaneous breaking of chiral symmetry (from $\mathrm{SU}(2) \times \mathrm{SU}(2)$ to $\mathrm{SU}(2)$, hence three pions corresponding to the three broken symmetries – note that their masses are not exactly zero, because the symmetry was not exact to begin with). The goal of this problem is to establish a relationship, known as the *Banks–Casher relation*, between the value of the chiral condensate $\langle\overline{\psi}\psi\rangle$ (which should be zero if chiral symmetry holds, and which may be viewed as the order parameter for its spontaneous breaking) and the spectral density at zero energy of the Dirac operator. Consider Euclidean QCD in the continuum.

69.a Show that the expectation value $\langle\overline{\psi}(x)\psi(x)\rangle$ can be written as

$$\langle\overline{\psi}(x)\psi(x)\rangle = \lim_{V\to\infty} \frac{1}{ZV} \int [DA^\mu]\, e^{-\mathcal{S}_{YM}[A]} \det(i\not{D} - m)\, \mathrm{tr}_V \left[\frac{1}{i\not{D} - m}\right],$$

where V is a four-dimensional Euclidean volume to which the trace is restricted, and Z is the same integral without the trace and the prefactor $1/V$.

69.b Recalling the properties of the Euclidean Dirac operator, write the trace as follows:

$$\mathrm{tr}_V \left[\frac{1}{i\not{D} - m}\right] = \sum_n \frac{1}{i\lambda_n - m} = -\sum_n \frac{m}{\lambda_n^2 + m^2},$$

where the λ_n are real numbers.

69.c Show that

$$\langle\overline{\psi}(x)\psi(x)\rangle = -\lim_{V\to\infty} \int d\lambda\, \frac{m\, \rho_V(\lambda)}{\lambda^2 + m^2},$$

$$\text{with } \rho_V(\lambda) \equiv \frac{1}{ZV} \int [DA^\mu]\, e^{-\mathcal{S}_{YM}[A]} \det(i\not{D} - m) \sum_n \delta(\lambda - \lambda_n).$$

69.d Show that, in the limits of infinite volume and zero quark mass, we have $\langle\overline{\psi}(x)\psi(x)\rangle = -\pi \lim_{V\to\infty} \rho_V(0)$.

69.a Note first that the equal-point correlation function $\langle\overline{\psi}(x)\psi(x)\rangle$ is independent of the point x thanks to translation invariance. Therefore, we may write

$$\langle\overline{\psi}(x)\psi(x)\rangle = \lim_{V\to\infty}\frac{1}{V}\int_V d^4x\,\langle\overline{\psi}(x)\psi(x)\rangle,$$

where V is a finite volume in Euclidean $\mathbb{R}^4$. Then, it admits the following path integral representation:

$$\begin{aligned}\langle\overline{\psi}(x)\psi(x)\rangle &= \frac{1}{ZV}\int[DA^\mu D\psi D\overline{\psi}]\,e^{-\mathcal{S}_{YM}-\mathcal{S}_D}\int_V d^4x\,\overline{\psi}(x)\psi(x)\\ &= \lim_{V\to\infty}\frac{1}{ZV}\int[DA^\mu]\,e^{-\mathcal{S}_{YM}}\det(i\not{D}-m)\int_V d^4x\,\langle x|\frac{1}{i\not{D}-m}|x\rangle\\ &= \lim_{V\to\infty}\frac{1}{ZV}\int[DA^\mu]\,e^{-\mathcal{S}_{YM}}\det(i\not{D}-m)\,\mathrm{tr}_V\left[\frac{1}{i\not{D}-m}\right],\end{aligned} \tag{5.18}$$

$$\text{where } Z\equiv\int[DA^\mu D\psi D\overline{\psi}]\,e^{-\mathcal{S}_{YM}-\mathcal{S}_D}.$$

In this formula, it is implicitly understood that the integrand is calculated for a finite volume theory, in which the spectrum of the Dirac operator is discrete. The limit $V\to\infty$ ensures that we recover the original theory in the end.

69.b The Euclidean Dirac matrices are anti-Hermitian, $\gamma^{\mu\dagger}=-\gamma^\mu$, as is the derivative ∂_μ. Therefore, the Euclidean Dirac operator is Hermitian, and it can therefore be diagonalized in an orthonormal basis with real eigenvalues:

$$\not{D}\,\psi_n(x)=\lambda_n\psi_n(x),\quad \lambda_n\in\mathbb{R},\quad \int_V d^4x\,\psi_m^*(x)\psi_n(x)=\delta_{mn}.$$

(Note that the eigenvalues depend implicitly on the background gauge field A^μ.) The trace of the inverse of the Dirac operator can be expressed in terms of these eigenvalues as follows:

$$\mathrm{tr}_V\left[\frac{1}{i\not{D}-m}\right]=\sum_n\frac{1}{i\lambda_n-m}.$$

On the other hand, since $\gamma^5\gamma^\mu\gamma^5=\gamma^{\mu\dagger}=-\gamma^\mu$, we can write

$$\begin{aligned}\mathrm{tr}_V\left[\frac{1}{i\not{D}-m}\right]&=\mathrm{tr}_V\left[\gamma^5\gamma^5\frac{1}{i\not{D}-m}\right]=\mathrm{tr}_V\left[\gamma^5\frac{1}{i\not{D}-m}\gamma^5\right]\\ &=\mathrm{tr}_V\left[\frac{1}{-i\not{D}-m}\right]=\sum_n\frac{1}{-i\lambda_n-m}.\end{aligned}$$

By taking the half-sum of the two representations of the trace, we arrive at

$$\mathrm{tr}_V\left[\frac{1}{i\not{D}-m}\right]=-\sum_n\frac{m}{\lambda_n^2+m^2}=-\int d\lambda\left[\sum_n\delta(\lambda-\lambda_n)\right]\frac{m}{\lambda^2+m^2}.$$

69.c Inserting this expression in (5.18), we immediately get

$$\big\langle\overline{\psi}(x)\psi(x)\big\rangle = -\lim_{V\to\infty}\int d\lambda\,\frac{m\,\rho_V(\lambda)}{\lambda^2+m^2},$$

$$\text{with}\ \ \rho_V(\lambda)\equiv\frac{1}{ZV}\int[DA^\mu]\,e^{-\mathcal{S}_{YM}[A]}\,\det\big(i\not{D}-m\big)\,\sum_n\delta(\lambda-\lambda_n).$$

The integrand in the definition of $\rho_V(\lambda)$ is a distribution that has a delta peak at each eigenvalue of the finite volume Dirac operator, and its integral over some range $[\lambda_1,\lambda_2]$ counts the number of eigenvalues in that range, per unit of volume V. The average over all configurations of the gauge field A^μ washes out the delta peaks and turns $\rho_V(\lambda)$ into a smooth distribution that provides the average density of eigenvalues of the Dirac operator:

$$\int_{\lambda_1}^{\lambda_2} d\lambda\ \rho_V(\lambda) = \left\langle\frac{\#\text{ of eigenvalues in }[\lambda_1,\lambda_2]}{V}\right\rangle_A .$$

Note that this distribution has a gap at the lower end of the spectrum because of the finite volume.

69.d As the volume increases, the gap at the lower end decreases, and $\rho_V(\lambda)$ has support in the entire range $[0,+\infty)$ in the infinite-volume limit. Moreover, when the quark mass goes to zero, we have

$$\lim_{m\to 0^+}\frac{m}{\lambda^2+m^2} = \pi\,\delta(\lambda).$$

Therefore, we have

$$\lim_{m\to 0^+}\big\langle\overline{\psi}(x)\psi(x)\big\rangle = -\pi\lim_{V\to\infty}\rho_V(0).$$

This relation is known as the *Banks–Casher relation*. The left hand side is a quantity known as the *chiral condensate*. In the massless limit, the Dirac Lagrangian is invariant under chiral transformations, since it does not contain an operator $\overline{\psi}\psi$ (which breaks chiral symmetry). A non-zero expectation value of this operator therefore signals that the ground state of the theory breaks chiral symmetry. The Banks–Casher formula relates the chiral condensate, which quantifies the spontaneous breaking of chiral symmetry, to the spectral density at the lower end of the spectrum of the Dirac operator (the latter is controlled by how quickly eigenvalues accumulate near $\lambda = 0$ when the volume becomes infinite).
Source: Banks, T. and Casher, A. (1980), *Nucl Phys B* 169: 103.

70. Nielsen–Ninomiya Theorem The goal of this problem is to prove the *Nielsen–Ninomiya theorem*, an obstruction to the construction of chiral gauge theories on the lattice. Consider an infinite square lattice of spacing a. The conjugate momenta p^μ are continuous variables with components in the range $\big[-\frac{\pi}{a},+\frac{\pi}{a}\big]$ and any function of p^μ is periodic with period $2\pi/a$. In other words, we may view momentum space as a four-dimensional torus T_4. The inverse propagator of a massless *left-handed* fermion can be written as $S^{-1}(p)=\sum_\mu\gamma^\mu P_L F_\mu(p)$, where $F^\mu(p)$ is a real continuous vector field on T_4 and $P_L\equiv(1-\gamma^5)/2$.

70.a Explain why such a fermion exists if the vector field $F_\mu(p)$ has a zero p_*^μ in the vicinity of which it behaves as $F_\mu(p) = (p - p_*)_\mu + \mathcal{O}((p - p_*)^2)$. Note that the field lines of F_μ are all pointing away from p_*.

70.b Consider now a zero of $F_\mu(p)$ at which the field lines are incoming in one direction and outgoing in the other three directions, such as $F_\mu(p) = (-1)^{\delta_{\mu 1}}(p-p_*)_\mu + \mathcal{O}((p-p_*)^2)$. Show that this zero in fact describes a right-handed particle. *Hint: apply a similarity transformation* $\gamma^1\gamma^5$. Use this result to explain why, in one dimension, there must be equal numbers of left-handed and right-handed fermions in the spectrum of the lattice Dirac operator.

70.c More generally, explain the relationship between the handedness of a fermion and the number of "ingoing directions" of the vector field F_μ at a zero p_*.

70.d The *index* of a zero of a vector field is defined as $(-1)^n$, where n is the number of dimensions along which the field lines are flowing away from the zero. The *Poincaré–Hopf theorem* states that the sum of the indices of the zeroes of a vector field is equal to the Euler characteristic of the manifold on which it is defined, which is zero for a torus. Use this result to conclude that this lattice setup in fact has equal numbers of left-handed and right-handed fermions.

70.a The form $S^{-1}(p) = \sum_\mu \gamma^\mu P_L F_\mu(p)$ for the inverse propagator assumes (discrete) translation invariance on the lattice, and that the lattice is infinite, so that the conjugate momentum space is continuous rather than discrete. Because of the finite lattice spacing, the momentum p^μ is defined on a four-dimensional torus T_4 (each component of the momentum is bounded, and the bounds in the positive and negative directions are identified) rather than $\mathbb{R}^4$. The function $F_\mu(p)$ is real if the Lagrangian is Hermitian, and is continuous provided the theory is quasi-local (i.e., when the action is written as $\mathcal{S} \equiv \sum_{x,y} \overline{\psi}(x)h(x,y)\psi(y)$, the function $h(x,y)$ should decrease fast enough). Since $F_\mu(p)$ has as many components ($\mu = 1, 2, 3, 4$) as its argument p, it is a vector field. Moreover, $F_\mu(p)$ may be viewed locally as the gradient of a scalar function $\mathcal{F}(p)$, $F^\mu(p) = \partial_{p_\mu}\mathcal{F}(p)$.

A massless fermion corresponds to a momentum p_* where the inverse Dirac propagator vanishes *linearly* (think of the propagator $i/\not{p}$ in the continuous theory). This means that all the components of $F_\mu(p)$ vanish simultaneously at p_*,

$$F_\mu(p) \sim f_{\mu\nu}(p - p_*)^\nu.$$

By an appropriate change of coordinates, the matrix of coefficients $f_{\mu\nu}$ can be made diagonal, with only positive coefficients. Up to a rescaling (by positive factors) of the components of p_μ, this implies the announced behavior near p_*. These zeroes of the vector field $F_\mu(p)$ correspond to minima of $\mathcal{F}(p)$. About such a zero, the field lines of F_μ are all oriented outwards from p_*. Note that, because we have included an explicit projector P_L in the definition of the inverse propagator, these zeroes describe left-handed fermions.

70.b Let us now flip the orientation of the field lines along one direction, i.e., consider a zero for which the field lines are flowing away along three directions and into the zero along the remaining direction (for instance the direction $\mu = 1$):

$$F_\mu(p) = (-1)^{\delta_{\mu 1}}(p - p_*)_\mu + \mathcal{O}((p - p_*)^2).$$

As suggested in the statement of the problem, we should study the following combination:

$$\begin{aligned}(\gamma^1\gamma^5)^{-1}(-1)^{\delta_{\mu 1}}\gamma^\mu P_L(\gamma^1\gamma^5) &= \gamma^5\gamma^1(-1)^{\delta_{\mu 1}}\gamma^\mu P_L\gamma^1\gamma^5 \\ &= \begin{Bmatrix} \mu = 1: & -\gamma^5\gamma^1\gamma^1 P_L\gamma^1\gamma^5 = \gamma^1 P_R \\ \mu \neq 1: & +\gamma^5\gamma^1\gamma^\mu P_L\gamma^1\gamma^5 = \gamma^\mu P_R \end{Bmatrix} = \gamma^\mu P_R.\end{aligned}$$

(Note that we also have $(\gamma^1\gamma^5)^{-1}(-1)^{\delta_{\mu 1}}\gamma^\mu P_R(\gamma^1\gamma^5) = \gamma^\mu P_L$.) Therefore, this type of zero describes (in disguise, i.e., up to a similarity) a massless fermion of right-handed chirality. This observation is sufficient to prove the Nielsen–Ninomiya theorem in one dimension: indeed, a smooth vector field on a circle must have an equal number of zeroes where the field lines go towards the zero (corresponding to left-handed fermions) and zeroes where the field lines go away from the zero (corresponding to right-handed fermions).

70.c Consider now a zero such that the vector field is flowing into the zero along two directions and away from it along two directions. If the two dimensions along which the field is flowing inwards are $\mu = 1, 2$, we need now to consider $F_\mu(p) = (-1)^{\delta_{\mu 1}}(-1)^{\delta_{\mu 2}}(p - p_*)_\mu$. We have

$$\begin{aligned}&(\gamma^2\gamma^5)^{-1}(-1)^{\delta_{\mu 2}}\Big[(\gamma^1\gamma^5)^{-1}(-1)^{\delta_{\mu 1}}\gamma^\mu P_L(\gamma^1\gamma^5)\Big](\gamma^2\gamma^5) \\ &\quad = (\gamma^2\gamma^5)^{-1}(-1)^{\delta_{\mu 2}}\gamma^\mu P_R(\gamma^2\gamma^5) = \gamma^\mu P_L.\end{aligned}$$

Thus, this zero corresponds to a left-handed fermion. The pattern should be clear by now: the chirality of the fermion flips every time we add a direction along which the vector field is flowing into the corresponding zero. In more mathematical terms, if $a_1, \ldots, a_p$ are distinct integers in $[1, 4]$, we have

$$\begin{aligned}&(\gamma^{a_p}\gamma^5)^{-1}\cdots(\gamma^{a_1}\gamma^5)^{-1}\Big[(-1)^{\mu a_1}\cdots(-1)^{\mu a_p}\gamma^\mu P_L\Big](\gamma^{a_1}\gamma^5)\cdots(\gamma^{a_p}\gamma^5) \\ &\qquad = \begin{cases} \gamma^\mu P_L & \text{if } p \text{ is even,} \\ \gamma^\mu P_R & \text{if } p \text{ is odd.}\end{cases}\end{aligned}$$

In words, the zero describes a left-handed fermion if the field lines are flowing into the zero along an even number of directions, and a right-handed fermion for an odd number of directions.

70.d The *index* of a zero of a vector field is defined as $(-1)^n$, where n is the number of dimensions along which the field lines are flowing away from the zero (see Figure 5.1). Therefore, in d dimensions, we have the following correspondence between the index of a zero

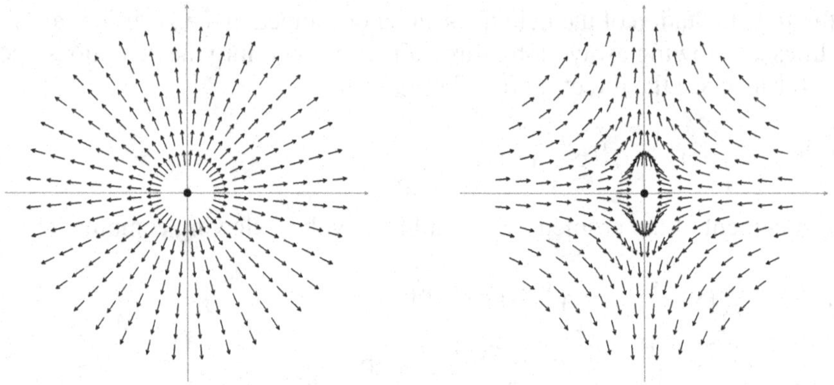

Figure 5.1 Two-dimensional vector fields with zeroes of index +1 (Left: $F_\mu(p) = p_\mu$) and −1 (Right: $F_\mu(p) = (-1)^{\delta_{\mu 1}} p_\mu$).

of F_μ and the chirality of the fermion it describes:

$$\begin{aligned} \text{index} = (-1)^d &\quad\longleftrightarrow\quad \text{left-handed,} \\ \text{index} = (-1)^{d+1} &\quad\longleftrightarrow\quad \text{right-handed.} \end{aligned}$$

The usefulness of this concept in the present discussion stems from the *Poincaré–Hopf theorem*, which relates the sum of the indices of the zeroes of a smooth vector field defined on a manifold $\mathcal{M}$ *without boundary* to a topological invariant of this manifold, namely its *Euler characteristic*, denoted by $\chi(\mathcal{M})$:

$$\sum_{\text{zeroes } z_*} \text{index}\,(z_*) = \chi(\mathcal{M}).$$

For two-dimensional manifolds, the Euler characteristic can be easily visualized and calculated from any triangulation of the surface, since it is given by

$$\chi(M) = (\text{number of vertices}) - (\text{number of edges}) + (\text{number of faces}).$$

(See Problem 42 for a discussion of the main properties of this quantity, in particular the fact that it does not depend on the chosen triangulation.) From this definition, one may see that the Euler characteristic of a 2-sphere is $\chi = 2$ and that it is $\chi = 0$ for a two-dimensional torus with one handle. This implies that any smooth vector field on such a torus must have equal numbers of zeroes of index +1 and −1, as illustrated in Figure 5.2. In the discussion of chiral fermions on the lattice, this result implies that any such propagator would in fact describe equal numbers of left-handed and right-handed fermions in two dimensions, despite the fact that we have put an explicit projector P_L in it. This obstruction to the possibility of defining chiral fermions on the lattice is known as the *Nielsen–Ninomiya theorem.*

The Euler characteristic is in fact zero for one-holed tori in any dimension (but this is harder to visualize in dimensions higher than two), in particular for the torus T_4 on which the vector $F_\mu(p)$ that enters in the inverse fermion propagator is defined. Thus, the Nielsen–Ninomiya theorem is also valid in four dimensions.

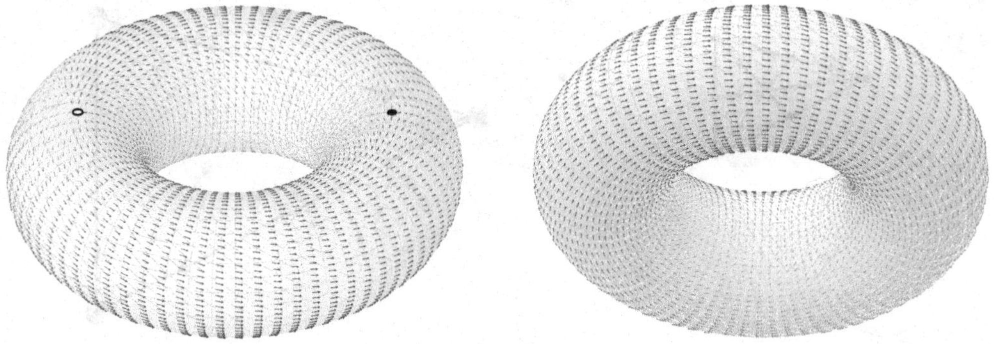

Figure 5.2 Illustration of the Poincaré–Hopf theorem: smooth vector field on a torus in two dimensions with a zero of index $+1$ (open circle) and a zero of index -1 (filled circle).

Sketch of a Proof of the Poincaré–Hopf Theorem: The first step in proving the Poincaré–Hopf theorem is to show that the sum of the indices of the zeroes is the same for all the smooth vector fields. In fact, one can show (see, for instance, Milnor, J. W. (1997), *Topology from the Differentiable Viewpoint* (Princeton University Press)) that the sum of the indices of any smooth vector field on a d-dimensional manifold $\mathcal{M}$ is equal to twice the degree of the Gauss mapping from $\mathcal{M}$ to the sphere S_d (loosely speaking, this degree counts how many times the Gauss mapping wraps around S_d), and is therefore a property of $\mathcal{M}$ itself. Then one could invoke the *Gauss–Bonnet theorem* (see Problem 22) to relate this quantity to the Euler characteristic.

A more elementary approach is to construct a special vector field on $\mathcal{M}$ for which it is easy to relate the sum of the indices to the Euler characteristic $\chi(\mathcal{M})$. For this, we start from a triangulation of the manifold $\mathcal{M}$. For a two-dimensional manifold, this is a decomposition of $\mathcal{M}$ into planar triangles, while in three dimensions it is a decomposition into tetrahedrons, and more generally a decomposition in d-simplices in d dimensions. Note that, in a triangulation, d-dimensional simplices meet at a $(d-1)$-dimensional simplex, etc. On such a triangulation, consider the vector field of Figure 5.3 (illustrated here in two dimensions). In this figure, the triangle in boldface shows the configuration of the field lines in one cell of the triangulation, and all the other triangles simply replicate the same pattern. This field has zeroes at all the vertices of the triangulation, at the midpoints of all the edges, and at the centers of all the triangles. Moreover, if we count the number of dimensions along which the field lines are oriented away from these zeroes, we have

$$\begin{aligned} \text{index}\,(\text{vertex}) &= +1, \\ \text{index}\,(\text{edge midpoint}) &= -1, \\ \text{index}\,(\text{face center}) &= +1, \end{aligned}$$

and the sum of the indices at all these zeroes reads

$$\sum_{\text{zeroes } z_*} \text{index}\,(z_*) = (\text{number of vertices}) - (\text{number of edges}) + (\text{number of faces}),$$

which is nothing but the definition of the Euler characteristic for a two-dimensional manifold.

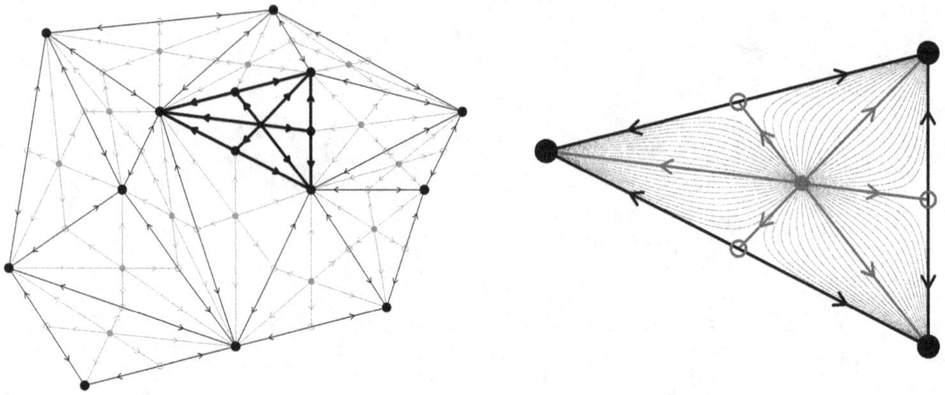

Figure 5.3 Special vector field used in the proof of the Poincaré–Hopf theorem (here, in two dimensions). Left: layout of the main field lines on top of a triangulation of the surface. Right: detail of the field lines in one elementary triangle.

In higher dimensions, one may construct a similar vector field, with zeroes at the centers of all the k-simplices ($k \le d$) that make up the triangulation, and the above pattern generalizes to

$$\text{index (center of k-dim simplex)} = (-1)^k.$$

Therefore, for such a field configuration in d dimensions, we have

$$\sum_{\text{zeroes } z_*} \text{index}\,(z_*) = \sum_{k=0}^{d} (-1)^k \times (\text{number of k-dim simplices}),$$

which is again the definition of the Euler characteristic of a d-dimensional manifold. Although explicit, the right-hand side of this equation is less practical for calculating the Euler characteristic in dimensions greater than two because it is hard to visualize triangulations of higher-dimensional manifolds.
Source: Karsten, L. H. (1981), *Phys Lett B* 104: 315.

71. Wigner Transform of a Convolution Product The Wigner transform of a two-point function $F(x, y)$ is a function $F(X, p)$ obtained by Fourier transforming $F(x, y)$ with respect to the difference $x - y$ while keeping the dependence on the midpoint coordinate $X \equiv (x + y)/2$:

$$F(X, p) \equiv \int d^d s \; e^{ip\cdot s}\, F(X + \tfrac{s}{2}, X - \tfrac{s}{2}).$$

Consider a pair of two-point functions $F(x, y)$ and $G(x, y)$ and their convolution $H(x, y) \equiv \int d^d z\, F(x, z) G(z, y)$. Show that their respective Wigner transforms are related by

$$H(X, p) = F(X, p) \exp\left\{\frac{i}{2}\big[\overleftarrow{\partial}_X \overrightarrow{\partial}_p - \overrightarrow{\partial}_X \overleftarrow{\partial}_p\big]\right\} G(X, p). \tag{5.19}$$

This result plays a role in the derivation of the Boltzmann equation from the Kadanoff–Baym equations. It is also an important ingredient in the formulation of quantum mechanics in phase-space, proposed by Moyal and Groenewold (Groenewold, H. J. (1946), *Physica* 12: 405; Moyal, J. E. and Bartlett, M. S. (1949), *Math Proc Cambridge Philos Soc* 45: 99). *Hint: one may instead prove an equivalent relationship between the operators* $\mathbf{F}, \mathbf{G}, \mathbf{H}$ *obtained from* F, G, H *via Weyl mapping (see Problem 18).*

The formula announced in the statement of the problem is intimately related to Weyl quantization prescription and to the fact that the Wigner transform is the inverse of Weyl mapping (this result is shown in Problem 18). Note first that a two-point function like $F(x, y)$ may always be viewed as the coordinate representation of an operator $\mathbf{F}$ (in this problem, we use boldface letters for quantum operators). In particular, we have

$$F(x,z) = \langle x|\mathbf{F}|z\rangle, \quad G(z,y) = \langle z|\mathbf{G}|y\rangle,$$
$$H(x,y) = \int d^dz\, \langle x|\mathbf{F}|z\rangle\langle z|\mathbf{G}|y\rangle = \langle x|\mathbf{FG}|y\rangle.$$

The last equality uses the completeness of the position eigenstates, and shows that the function $H(x, y)$, defined as the convolution product of F and G, is the coordinate space representation of the product of operators $\mathbf{FG}$.

In order to streamline the notation, let us collectively denote the phase-space coordinates by $u \equiv (X, p)$, and do the same with the corresponding quantum operators, denoting them by $\mathbf{u} \equiv (\mathbf{X}, \mathbf{p})$. The commutators among these operators have the following generic form:

$$[\mathbf{u}_m, \mathbf{u}_n] = i\,\theta_{mn},$$

where the θ_{mn} are commuting numbers. Let us also denote $\partial_m \equiv \partial/\partial u_m$. Equipped with this notation, the "product" on the right-hand side of (5.19) can be rewritten compactly as

$$F(X,p)\,\exp\Big\{\frac{i}{2}\big[\,\overleftarrow{\partial}_X\overrightarrow{\partial}_p - \overrightarrow{\partial}_X\overleftarrow{\partial}_p\,\big]\Big\}\,G(X,p) = \underbrace{F(u)e^{\frac{i}{2}\theta_{mn}\overleftarrow{\partial}_m\overrightarrow{\partial}_n}G(u)}_{\equiv\,(F\star G)(u)}.$$

The $\star$ operation is known as the *Moyal product.* The identity we wish to prove is that the Wigner transform of a product of operators is equal to the Moyal product of their respective Wigner transforms:

$$(5.19) \quad \Longleftrightarrow \quad \text{Wigner}\,(\mathbf{FG}) = \text{Wigner}\,(\mathbf{F}) \star \text{Wigner}\,(\mathbf{G}).$$

Since the Wigner transform is the inverse of Weyl mapping (see Problem 18), this identity is equivalent to

$$\text{Weyl}\,(F \star G) = \underbrace{\text{Weyl}\,(F)}_{\mathbf{F}}\;\underbrace{\text{Weyl}\,(G)}_{\mathbf{G}},$$

where Weyl $(\cdot)$ denotes the Weyl mapping of a function defined on the classical phase-space. In terms of the compact notation we have introduced, the Weyl mapping of a function A can be written in two steps,

$$\widetilde{A}(\tau) \equiv \int \frac{d^{2d}u}{(2\pi)^{2d}}\, e^{-iu\cdot\tau}\, A(u), \quad \text{Weyl}\,(A)(\mathbf{u}) \equiv \int d^{2d}\tau\; e^{i\mathbf{u}\cdot\tau}\,\widetilde{A}(\tau).$$

The first step is a Fourier transform whose result is an ordinary function of the variables τ (Fourier conjugates of the classical phase-space variables). The second one looks like the inverse Fourier transform, but this time one uses the quantum operators $\mathbf{u}$ instead of u, and the result is an operator.

Then, we can rearrange the expression for the Moyal product as follows:

$$
\begin{aligned}
(F \star G)(u) &= F(u)\, e^{\frac{i}{2}\theta_{mn}\overleftarrow{\partial}_m \overrightarrow{\partial}_n}\, G(u) \\
&= \int d^{2d}\tau d^{2d}\sigma\, \widetilde{F}(\tau)\, \widetilde{G}(\sigma)\, e^{iu\cdot\tau}\, e^{\frac{i}{2}\theta_{mn}\overleftarrow{\partial}_m \overrightarrow{\partial}_n}\, e^{iu\cdot\sigma} \\
&= \int d^{2d}\tau d^{2d}\sigma\, \widetilde{F}(\tau)\, \widetilde{G}(\sigma)\, e^{iu\cdot\tau}\, e^{\frac{i}{2}\theta_{mn}(i\tau_m)(i\sigma_n)}\, e^{iu\cdot\sigma} \\
&= \int d^{2d}\tau d^{2d}\sigma\, \widetilde{F}(\tau)\, \widetilde{G}(\sigma)\, e^{iu\cdot(\tau+\sigma)}\, e^{-\frac{i}{2}\theta_{mn}\tau_m\sigma_n}.
\end{aligned}
$$

Let us now calculate the Weyl mapping of this function. In fact, since it is already written as a Fourier integral, this just amounts to replacing the phase-space coordinates u by the quantum operators $\mathbf{u}$ in the integrand:

$$
\mathrm{Weyl}(F \star G)(\mathbf{u}) = \int d^{2d}\tau d^{2d}\sigma\, \widetilde{F}(\tau)\, \widetilde{G}(\sigma)\, e^{i\mathbf{u}\cdot(\tau+\sigma)}\, e^{-\frac{i}{2}\theta_{mn}\tau_m\sigma_n}. \tag{5.20}
$$

To simplify this expression, we use the Baker–Campbell–Hausdorff formula to write

$$
e^{i\mathbf{u}\cdot(\tau+\sigma)} = e^{i\mathbf{u}\cdot\tau}\, e^{i\mathbf{u}\cdot\sigma}\, e^{\frac{1}{2}[\mathbf{u}_m,\mathbf{u}_n]\tau_m\sigma_n} = e^{i\mathbf{u}\cdot\tau}\, e^{i\mathbf{u}\cdot\sigma}\, e^{\frac{i}{2}\theta_{mn}\tau_m\sigma_n}.
$$

We see that the last factor of this formula cancels the last factor of (5.20), so that the result in fact factorizes:

$$
\mathrm{Weyl}(F \star G)(\mathbf{u}) = \underbrace{\left(\int d^{2d}\tau\, e^{i\mathbf{u}\cdot\tau}\, \widetilde{F}(\tau)\right)}_{\mathrm{Weyl}\,(F)(\mathbf{u})} \times \underbrace{\left(\int d^{2d}\sigma\, e^{i\mathbf{u}\cdot\sigma}\, \widetilde{G}(\sigma)\right)}_{\mathrm{Weyl}\,(G)(\mathbf{u})},
$$

which proves the announced result.

Quantum Mechanics in Phase-Space: The Moyal product in fact allows a complete reformulation of quantum mechanics in terms of functions that depend on the classical phase-space variables:

Phase-space function:		Operator:
$F(X, p)$	$\underset{\text{Wigner transform}}{\overset{\text{Weyl mapping}}{\rightleftarrows}}$	$\mathbf{F}$
$F(X, p) \star G(X, p)$	$\longleftrightarrow$	$\mathbf{F}\mathbf{G}$

In this framework, known as the *Moyal–Groenewold quantization*, quantum operators are replaced by classical phase-space functions obtained via the Wigner transform, the product of operators is replaced by the Moyal product of phase-space functions, and their commutator is replaced by a Moyal commutator, $[F, G]_\star \equiv F \star G - G \star F$. Although this formalism is technically challenging to work with, it offers a more intuitive picture of what "quantization" means, since the Moyal commutator is a deformation of the classical Poisson bracket $\{F, G\}$

controlled by the continuously varying parameter $\hbar$. Indeed, from the expression for the Moyal product, the Moyal commutator reads

$$\begin{aligned} F \star G - G \star F &= \frac{2i}{\hbar} F \sin\left\{\frac{i\hbar}{2}\left[\overleftarrow{\partial}_x \overrightarrow{\partial}_p - \overrightarrow{\partial}_x \overleftarrow{\partial}_p\right]\right\} G \\ &= \{F, G\} + \frac{\hbar^2}{24} F\left[\overleftarrow{\partial}_x \overrightarrow{\partial}_p - \overrightarrow{\partial}_x \overleftarrow{\partial}_p\right]^3 G + \cdots . \end{aligned}$$

(We have reinstated Planck's constant in this formula, using the fact that it has the dimensions of an action.) From this expression, we also see that

$$F \star G - G \star F = \{F, G\}$$

if at least one of the functions F, G is a polynomial of degree two in X, p. Thus, the Moyal–Groenewold quantization of a theory with a quadratic Hamiltonian (e.g., a harmonic oscillator) is particularly simple: its phase-space density obeys the classical Liouville evolution equation even when $\hbar \neq 0$.

---

72. Center Symmetry and Deconfinement The goal of this problem is to study an extension of gauge symmetry called *center symmetry*, in a $SU(N)$ Yang–Mills theory at finite temperature in the imaginary time formalism. In particular, we show that the deconfinement transition may be viewed as a breaking of center symmetry.

72.a The *center* of a gauge group is the subgroup made of the elements that commute with the entire group. In the case of $SU(N)$, show that the center is $\mathbb{Z}_N \equiv \{e^{2i\pi k/N}\}_{0\le k<N}$, i.e., the set of the N-th roots of unity (multiplied by the identity matrix). *Hint: first, show that the center of* $U(N)$ *is the complex unit circle.*

72.b At finite temperature, legitimate gauge transformations should be singled-valued in imaginary time, i.e., be β-periodic, $\Omega(\beta, x) = \Omega(0, x)$. Consider now a more general transformation, called a *center transformation*, defined as a gauge transformation with a non-periodic Ω. Instead, $\Omega(\tau, x)$ is assumed to obey $\Omega(\beta, x) = \xi\Omega(0, x)$ with $\xi \in \mathbb{Z}_N$. Show that center transformations also preserve the periodicity of the gauge field.

72.c A *Polyakov loop* is the trace of a Wilson loop that wraps around the imaginary time direction, at fixed position x:

$$L(x) \equiv N^{-1}\,\mathrm{tr}\left\langle P \exp \int_0^\beta d\tau\, A^0(\tau, x)\right\rangle.$$

Show that $L(x)$ is invariant under the usual (periodic) gauge transformations, but not invariant under center transformations. What does this imply for the center transformations?

72.d Explain why $L(x)$ is related to the free energy F_q of an infinitely massive quark added to the system at the point x by $|L(x)| = e^{-\beta F_q}$ (i.e., the energy necessary for bringing the quark from infinity, arbitrarily slowly). From this interpretation, conclude that $L(x)$ should be zero in the confining phase of Yang–Mills theory, and that one may view deconfinement as a breaking of the center symmetry.

72.e Why is the center symmetry explicitly broken if dynamical quarks in the fundamental representation are added to the theory? What about adding matter fields in the adjoint representation?

72.a Consider first the case of the group $\mathrm{U}(N)$ of $N \times N$ unitary matrices. A matrix A is in the center of $\mathrm{U}(N)$ if $A = PAP^{-1}$ for all unitary matrices P. Since A is unitary, it can be diagonalized by a unitary transformation, which implies that we can reduce the discussion to diagonal matrices. Therefore, let us now assume that A is diagonal, but not proportional to the identity. This means that there exists a pair of indices $i \neq j$ such that $A_{ii} \neq A_{jj}$. Now, denote by Q the matrix obtained by swapping the columns i, j of the identity, and define $B \equiv QAQ^{-1}$. Note that $Q = Q^\dagger = Q^t$ and $QQ = 1$, hence Q is unitary. A simple calculation shows that $B_{ii} = A_{jj}$ and $B_{jj} = A_{ii}$, which implies that $B \neq A$, i.e.,

$$QAQ^{-1} \neq A,$$

in contradiction to the fact that A was assumed to be in the center of the group. Thus, our hypothesis that A is not proportional to the identity was inconsistent, and we conclude that

$$\text{center}\left(\mathrm{U}(N)\right) = \left\{c \cdot 1_{N\times N}, |c| = 1\right\}.$$

(The constraint on the modulus of c is necessary to have unitary matrices.)

Consider now the case of $\mathrm{SU}(N)$. It is useful to recall that any matrix of $\mathrm{U}(N)$ can be written as the product of a phase by a matrix of $\mathrm{SU}(N)$:

$$U = e^{i\theta} V, \quad U \in \mathrm{U}(N), \quad V \in \mathrm{SU}(N), \quad \theta \in \mathbb{R}.$$

If A is in the center of $\mathrm{SU}(N)$ and U is in $\mathrm{U}(N)$, we have

$$UAU^{-1} = \underbrace{VAV^{-1}}_{=A} = A.$$

Therefore A is also in the center of $\mathrm{U}(N)$ and must be of the form $c \cdot 1_{N\times N}$. Since the determinant of such a matrix is c^N, c must be one of the N-th roots of unity in order to have a matrix of $\mathrm{SU}(N)$. This group is denoted $\mathbb{Z}_N$.

72.b Consider a transformation $\Omega(\tau, x)$ that has the following boundary condition:

$$\Omega(\beta, x) = \xi\, \Omega(0, x), \quad \text{with } \xi \in \mathbb{Z}_N.$$

Calculate the transformed gauge field at $\tau = \beta$ (assuming that the gauge field was β-periodic before applying the transformation):

$$\begin{aligned}
&\Omega^\dagger(\beta,x)A^\mu(\beta,x)\Omega(\beta,x)+\frac{i}{g}\Omega^\dagger(\beta,x)\partial^\mu\Omega(\beta,x)\\
&\quad=\Omega^\dagger(0,x)\xi^*A^\mu(0,x)\xi\Omega(0,x)+\frac{i}{g}\Omega^\dagger(0,x)\xi^*\partial^\mu\xi\Omega(0,x)\\
&\quad=\Omega^\dagger(0,x)A^\mu(0,x)\Omega(0,x)+\frac{i}{g}\Omega^\dagger(0,x)\partial^\mu\Omega(0,x).
\end{aligned}$$

In the final step, the ξ and ξ^* cancel because we are allowed to move them through A^μ since they belong to the center of $SU(N)$. Despite the fact that this transformation is not β-periodic, it nevertheless preserves the periodicity of the gauge fields. Therefore, these more general transformations may be viewed as an extended symmetry of Yang–Mills theory (called *center symmetry*) at finite temperature.

72.c Recall that a Wilson line $W_\gamma(y,x)$ along any path γ between the points x and y transforms into $\Omega(y)W_\gamma(y,x)\Omega^\dagger(x)$ under a gauge transformation. Therefore, the trace of a Wilson loop is gauge invariant (and the Polyakov loop is just a special kind of Wilson loop). But the cancellation of $\Omega(\beta,\mathbf{x})$ with $\Omega^\dagger(0,\mathbf{x})$ inside the trace works only when Ω is β-periodic, and therefore it fails with the generalized transformations considered above. In this case, the Polyakov loop is not invariant, but is multiplied by an element $\xi\in\mathbb{Z}_N$ of the center:

$$L(\mathbf{x})\to\xi\,L(\mathbf{x}).$$

Since the Polyakov loop is gauge invariant, it should be considered as an observable of the theory. Thus, the center transformations can alter the physical content of the system (i.e., they are not just a reparameterization of redundant degrees of freedom). The only exception to this statement occurs when $L(\mathbf{x})=0$, since this value of the Polyakov loop is left invariant by center transformations.

72.d The energy gain F_q of the system due to bringing (infinitely slowly, to avoid disturbing the system) a heavy quark of mass M at the point $\mathbf{x}$ from infinity may be written as follows:

$$e^{-\beta(F_q+M)}=\frac{\sum_\alpha\big\langle\alpha+q(\mathbf{x})\big|e^{-\beta H}\big|\alpha+q(\mathbf{x})\big\rangle}{\sum_\alpha\big\langle\alpha\big|e^{-\beta H}\big|\alpha\big\rangle}$$

(the factor $e^{-\beta M}$, that contains the rest energy of the quark, has been explicitly added so that F_q measures only the energy due to the interactions of the quark with its environment), where $\big|\alpha+q(\mathbf{x})\big\rangle$ denotes the pure gluon state $\big|\alpha\big\rangle$ to which a quark (of spin s and color i) is added at the location $\mathbf{x}$:

$$\big|\alpha+q(\mathbf{x})\big\rangle\equiv\underbrace{\int\frac{d^3\mathbf{p}}{(2\pi)^3 2E_\mathbf{p}}\,e^{-i\mathbf{p}\cdot\mathbf{x}}\,f(\mathbf{p})\,v_s(\mathbf{p})b^\dagger_{si\mathbf{p}}}_{\equiv\,\psi_{si}(\mathbf{x})}\big|\alpha\big\rangle.$$

The function $f(\mathbf{p})$ is a wave-packet normalized according to

$$\int \frac{d^3p}{(2\pi)^3}\,|f(\mathbf{p})|^2 = 1,$$

which ensures that $\langle\alpha + q(\mathbf{x})|\alpha + q(\mathbf{x})\rangle = \langle\alpha|\alpha\rangle$. Interpreting the canonical density operator as an imaginary time evolution, we can write

$$\psi^\dagger_{si}(\tau,\mathbf{x})\,e^{-\beta H} = e^{-\beta H}\,\psi^\dagger_{si}(\tau+\beta,\mathbf{x}),$$

and the quark free energy becomes

$$e^{-\beta(F_q+M)} = \frac{\mathrm{tr}\,(e^{-\beta H}\,\psi^\dagger_{si}(\beta,\mathbf{x})\psi_{si}(0,\mathbf{x}))}{\mathrm{tr}\,(e^{-\beta H})}.$$

(Note that there is no sum on the spin s and the color i in this expression.) Loosely speaking, the right-hand side is the thermal expectation value of the propagator of the heavy quark between the times $\tau = 0$ and $\tau = \beta$ (the problem is translation invariant, and we may thus choose the first time to be 0). To promote $\psi_{si}(\mathbf{x})$ to a time-dependent object, we use Dirac's equation. In the limit of large mass, $M \to \infty$, where $p^0 \approx M$, $|\mathbf{p}| \ll M$, the Dirac equation simplifies to

$$((\partial_\tau - gA^0)\gamma^0 - M)\psi = 0.$$

Moreover, in this kinematical limit, we also have $\gamma^0 v_s(\mathbf{p}) = -v_s(\mathbf{p})$. Therefore, for a spinor proportional to $v_s(\mathbf{p})$, the Dirac equation becomes

$$(\partial_\tau - gA^0 + M)\psi = 0,$$

whose solution reads

$$\psi(\tau,\mathbf{x}) = e^{-M\tau}\,\underbrace{\mathrm{T}\exp\left(g\int_0^\tau d\tau'\,A^0(\tau',\mathbf{x})\right)}_{\equiv\, U(\tau,\mathbf{x})}\,\psi(0,\mathbf{x}).$$

(The time ordering is necessary because A^0 is a matrix in $\mathfrak{su}(N)$.) Therefore, we can write

$$\psi_{si}(\tau,\mathbf{x}) \equiv e^{-M\tau}\sum_j U_{ij}(\tau,\mathbf{x})\int\frac{d^3p}{(2\pi)^3 2E_p}\,e^{-ip\cdot x}\,f(\mathbf{p})\,v_s(\mathbf{p})b^\dagger_{sjp},$$

which leads to

$$\begin{aligned}
e^{-\beta F_q} &= e^{\beta M}\sum_j \frac{\mathrm{tr}\,(e^{-\beta H}\,\psi^\dagger_{sj}(0,\mathbf{x})U_{ji}(\beta,\mathbf{x})e^{-\beta M}\psi_{si}(0,\mathbf{x}))}{\mathrm{tr}\,(e^{-\beta H})}\\
&= \frac{1}{N}\sum_{i,j}\frac{\mathrm{tr}\,(e^{-\beta H}\,\psi^\dagger_{sj}(0,\mathbf{x})U_{ji}(\beta,\mathbf{x})\psi_{si}(0,\mathbf{x}))}{\mathrm{tr}\,(e^{-\beta H})}.
\end{aligned}$$

In the last equality, we have used the fact that the free energy is independent of the quark color i in order to introduce an average over i. In the evaluation of this expectation value, the product of the two free spinors simply yields

$$\psi_{sj}^\dagger(0,\mathbf{x})\psi_{si}(0,\mathbf{x}) = \int \frac{d^3p\,d^3q}{(2\pi)^6 4E_p E_q}\, e^{i(\mathbf{p}-\mathbf{q})\cdot\mathbf{x}}\, \underbrace{b_{sjp}b_{siq}^\dagger}_{-b_{siq}^\dagger b_{sjp}+(2\pi)^3 2E_p\delta_{ij}\delta(\mathbf{p}-\mathbf{q})}\, f^*(\mathbf{p})f(\mathbf{q})\, v_s^\dagger(\mathbf{p})v_s(\mathbf{q}).$$

The term in $-b_{siq}^\dagger b_{sjp}$ gives zero when inserted in the trace, because it has a fermion annihilation operator acting to the right on a pure gluonic state. Thus, only the term in $(2\pi)^3 2E_p\delta_{ij}\delta(\mathbf{p}-\mathbf{q})$ contributes:

$$\psi_{sj}^\dagger(0,\mathbf{x})\psi_{si}(0,\mathbf{x}) \to \delta_{ij}\underbrace{\int \frac{d^3p}{(2\pi)^3 2E_p}\,|f(\mathbf{p})|^2\,\underbrace{v_s^\dagger(\mathbf{p})v_s(\mathbf{p})}_{=2E_p}}_{=1} = \delta_{ij}.$$

The factor δ_{ij} then leads precisely to the trace of the Polyakov loop over its color indices.

In the confining phase, isolated quarks cannot exist; in other words, it would cost an infinite amount of energy to bring one from infinity (i.e., $F_q = +\infty$). Therefore, the expectation value of the Polyakov loop should be zero in this phase, and center symmetry is preserved (since zero times any element of $\mathbb{Z}_N$ is still zero). On the other hand, isolated quarks cost only a finite energy in the deconfined phase, and the center symmetry is broken since the expectation value of Polyakov loops is not zero. Therefore, one may view the expectation value of the Polyakov loop as an order parameter for the deconfinement transition in pure Yang–Mills theory.

72.e Under the transformation $\Omega(\tau,\mathbf{x})$, a fermion in the fundamental representation transforms as

$$\psi(\tau,\mathbf{x}) \to \Omega^\dagger(\tau,\mathbf{x})\psi(\tau,\mathbf{x}).$$

Therefore, the center transformations do not preserve the β-anti-periodicity of the spinors, implying that the center symmetry is explicitly broken by the presence of dynamical quarks in the fundamental representation. Therefore, in QCD with quarks, the expectation value of the Polyakov loop is no longer a good order parameter for the deconfinement transition.

If Φ_a is a field in the adjoint representation, it transforms as

$$\Phi_a \to \Omega_{ab}\Phi_b, \quad \text{i.e.,} \quad \Phi_a t_r^a \to \Omega_r^\dagger \Phi_a t_r^a \Omega_r.$$

In the second form, it is clear that the $\xi \in \mathbb{Z}_N$ of the center transformation cancels out, leaving a transformed field which is still β-periodic (or anti-periodic if it is a fermion). Therefore, a theory with matter fields in the adjoint representation would be invariant under the center symmetry.

Source: McLerran, L. D. and Svetitsky, B. (1981), *Phys Rev D* 24: 450.

73. Photon Hard Thermal Loop from Classical Particles In this problem, we use the concept of a classical particle developed in Problem 31 in order to derive the effective Lagrangian that generates hard thermal loops in QED. This derivation illustrates the classical nature of hard thermal loops, and is also a good workout in preparation for the QCD case (see Problem 74).

73.a Find a suitable generalization of the equations of motion discussed in Problem 37 to describe the evolution of the position and momentum of a massless charged particle in an external electromagnetic field.

73.b Consider a collection of classical electrons and positrons, with a phase-space density $f(x, p)$. Why do we have $df/d\tau = 0$? Using the equations of motion of the classical particles, turn this equation into a kinetic equation.

73.c Consider now an expansion of f, $f = f_0 + f_1 + \cdots$, where f_0 is a distribution independent of the external field. Formally solve the kinetic equation in order to obtain the first correction in the external field, f_1.

73.d Determine the electromagnetic current J^μ induced by f_1 (assume that f_0 produces no current). Then, determine an effective Lagrangian $\mathcal{L}_{\rm eff}$ such that $J^\mu = \partial\mathcal{L}_{\rm eff}/\partial A_\mu$.

73.e Show that its second derivative, $-\delta\mathcal{L}_{\rm eff}/\delta A_\mu \delta A_\nu$, is the photon polarization tensor $\Pi^{\mu\nu}_{\rm HTL}$ in the hard thermal loop approximation.

73.a In the classical particle approximation of Problem 37, the charged particles feel the effect of the background electromagnetic field, but their own radiation is not considered. Therefore, the particles do not feed the electromagnetic field, and there are no interactions between the charged particles. Since we are going to consider the ultra-relativistic limit, we need to adapt the equations of motion to the case of massless particles. Formally, we can start from the equations obtained in Problem 37, and define $\tau \equiv t/m$ before completely eliminating m. This leads to

$$p^\mu = \frac{dx^\mu}{d\tau}, \quad \frac{dp^\mu}{d\tau} = e\, F^{\mu\nu} p_\nu.$$

Note that since the particles are massless, τ is not really a time but simply some parameter that maps the trajectory of a particle (in fact, this parameter has the dimensions of a squared time).

73.b Let us denote by $f(x, p)$ the phase-space density of these charged particles. By the *Liouville theorem*, f is constant along the trajectories in phase-space (equivalently, the volume in phase-space is preserved by a Hamiltonian flow). Then, we can rewrite the time derivative of f more explicitly:

$$0 = \frac{df}{d\tau} = \frac{dx^\mu}{d\tau}\frac{\partial f}{\partial x^\mu} + \frac{dp^\mu}{d\tau}\frac{\partial f}{\partial p^\mu} = p^\mu \frac{\partial f}{\partial x^\mu} + e\, F^{\mu\nu} p_\nu \frac{\partial f}{\partial p^\mu}$$
$$= p^\mu \Big(\frac{\partial}{\partial x^\mu} - e\, F_\mu{}^\nu \frac{\partial}{\partial p^\nu}\Big) f.$$

This type of kinetic equation is known as a *Vlasov equation*. The first term, $p \cdot \partial_x f$, describes the free streaming of particles, i.e., the motion at constant momentum of free particles. The second term describes the deflection of their trajectories due to the force the particles feel from the external field in which they propagate.

73.c The Vlasov equation can be solved iteratively by writing

$$f = f_0 + f_1 + \cdots,$$

where f_0 is some distribution of free particles (it must obey $p \cdot \partial_x f_0 = 0$), f_1 is the first correction due to the external field, etc. In the first iteration, we have

$$p \cdot \partial_x f_1 = e\, p_\mu F^{\mu\nu} \frac{\partial f_0}{\partial p^\nu}.$$

Indeed, if f_1 is the correction of first order in the external field, we do not need to keep it on the right-hand side, so that the left- and right-hand sides have the same order in e. This equation can be solved formally to give

$$f_1 = e\, p_\mu \frac{1}{p \cdot \partial_x} F^{\mu\nu} \frac{\partial f_0}{\partial p^\nu}.$$

73.d The next step is to define the electromagnetic current associated with the distribution f:

$$J^\mu(x) \equiv e \int \frac{d^3\mathbf{p}}{(2\pi)^3} \frac{p^\mu}{p} f(x, p).$$

One may check that this current is conserved if f obeys the Vlasov equation:

$$\partial_\mu J^\mu = e \int \frac{d^3\mathbf{p}}{(2\pi)^3 p} (p \cdot \partial) f = e^2 \int \frac{d^3\mathbf{p}}{(2\pi)^3 p} p_\mu F^{\mu\nu} \frac{\partial f}{\partial p^\nu} = -e^2 \int \frac{d^3\mathbf{p}}{(2\pi)^3 p} \underbrace{\underbrace{\frac{\partial p_\mu}{\partial p^\nu}}_{g_{\mu\nu}} F^{\mu\nu}}_{=0} f = 0.$$

With the iterative solution obtained above, the current reads

$$J^\mu = e \int \frac{d^3\mathbf{p}}{(2\pi)^3 p} p^\mu \left(f_0 + e\, p_\rho \frac{1}{p \cdot \partial_x} F^{\rho\nu} \frac{\partial f_0}{\partial p^\nu} + \cdots \right).$$

If we assume that the free distribution f_0 does not produce any current, the first term is zero (physically, this is realized when the free distribution f_0 has equal distributions of electrons and positrons). Therefore, at order one in the external field, the current reads

$$\begin{aligned} J^\mu &= e^2 \int \frac{d^3\mathbf{p}}{(2\pi)^3 p} \frac{p^\mu p_\rho}{p \cdot \partial_x} F^{\rho\nu} \frac{\partial f_0}{\partial p^\nu} = -e^2 \int \frac{d^3\mathbf{p}}{(2\pi)^3 p} \frac{\partial}{\partial p^\nu} \left(\frac{p^\mu p_\rho}{p \cdot \partial_x} \right) F^{\rho\nu} f_0 \\ &= -e^2 \int \frac{d^3\mathbf{p}}{(2\pi)^3 p} \left(\frac{\delta_\nu{}^\mu p_\rho}{p \cdot \partial_x} + \underbrace{\frac{p^\mu g_{\nu\rho}}{p \cdot \partial_x}}_{\to 0} - \frac{p^\mu p_\rho \partial_\nu}{(p \cdot \partial_x)^2} \right) F^{\rho\nu} f_0. \end{aligned}$$

Note that the middle term gives zero after contraction with the antisymmetric tensor $F^{\rho\nu}$. Now, we would like to determine an effective Lagrangian $\mathcal{L}_{\text{eff}}$ such that

$$J^\mu = \frac{\partial \mathcal{L}_{\text{eff}}}{\partial A_\mu}.$$

Since J^μ is linear in the external field, this effective Lagrangian must have one more power of A. Moreover, in order to be gauge invariant, this additional A dependence should arise via a power of the field strength. In order to find this Lagrangian, recall first that

$$\frac{\partial F_{\alpha\beta}}{\partial A_\mu} = \frac{\partial(\partial_\alpha A_\beta - \partial_\beta A_\alpha)}{\partial A_\mu} = \delta_\beta{}^\mu\,\partial_\alpha - \delta_\alpha{}^\mu\,\partial_\beta.$$

Then, consider

$$\begin{aligned}\frac{\partial}{\partial A_\mu}\Big(F_{\nu\beta}\frac{p^\beta p_\rho}{(p\cdot\partial)^2}F^{\rho\nu}\Big) &= (\delta_\beta{}^\mu\partial_\nu - \delta_\nu{}^\mu\partial_\beta)\frac{p^\beta p_\rho}{(p\cdot\partial)^2}F^{\rho\nu} + F_{\nu\beta}\frac{p^\beta p_\rho}{(p\cdot\partial)^2}(g^{\mu\nu}\partial^\rho - g^{\rho\mu}\partial^\nu)\\ &= \frac{p^\mu p_\rho}{(p\cdot\partial)^2}(\partial_\nu F^{\rho\nu} + F^{\rho\nu}\partial_\nu) - 2\frac{p_\rho}{p\cdot\partial}F^{\rho\mu}\\ &= -\frac{p^\mu p_\rho}{(p\cdot\partial)^2}\big(\underbrace{\partial_\nu F^{\rho\nu}}_{0}\big) - 2\Big(\frac{\delta_\nu{}^\mu p_\rho}{p\cdot\partial} - \frac{p^\mu p_\rho\partial_\nu}{(p\cdot\partial)^2}\Big)F^{\rho\nu}.\end{aligned}$$

(Note that the derivatives act on everything located to their right, including the zeroth-order phase-space distribution f_0, not yet included in this formula. To obtain the final line, we used $F^{\rho\nu}\partial_\nu\cdots = \partial_\nu F^{\rho\nu}\cdots - (\partial_\nu F^{\rho\nu})\cdots$. Finally, we have $\partial_\nu F^{\rho\nu} = 0$ because the charged particles are not feeding the external field in this approximation (and we assume that the sources producing this field are far away). Up to a factor 2, we recognize the structure of the integrand in the current J^μ. Therefore, the effective Lagrangian whose derivative gives J^μ reads

$$\mathcal{L}_{\text{eff}} = \frac{e^2}{2}\int\frac{d^3p}{(2\pi)^3 p}\,F_{\nu\beta}\frac{p^\beta p_\rho}{(p\cdot\partial)^2}F^{\rho\nu}\,f_0.$$

73.e Let us now consider

$$\Pi^{\mu\nu} = -\frac{\partial J^\mu}{\partial A_\nu} = e^2\int\frac{d^3p}{(2\pi)^3 p}\Big(g^{\mu\nu} - \frac{p^\mu\partial^\nu + p^\nu\partial^\mu}{p\cdot\partial} + \frac{p^\mu p^\nu\partial^2}{(p\cdot\partial)^2}\Big)\,f_0.$$

To recover the HTL result, we replace f_0 by the equilibrium fermion distribution,

$$f_0(x,p) = 2\times 2\times\frac{1}{e^{\beta|p^0|}+1}$$

(the two prefactors 2 account for electrons and anti-electrons, and for the fact they have two spin states), and we Fourier transform the external coordinates (which simply amounts to the replacement $\partial \to ik$), so that

$$\Pi^{\mu\nu}(k) = 4e^2\int\frac{d^3p}{(2\pi)^3 p}\Big(g^{\mu\nu} - \frac{p^\mu k^\nu + p^\nu k^\mu}{p\cdot k} + \frac{p^\mu p^\nu k^2}{(p\cdot k)^2}\Big)\frac{1}{e^{\beta p}+1}$$

($p_0 = p$ in this equation). Note that the factor between the parentheses is homogeneous of degree zero in p, and therefore depends only on the orientation $\widehat{p}$ of p. By writing

$d^3p = p^2 dp d^2\widehat{p}$, we can easily factorize the radial and angular integrals:

$$\Pi^{\mu\nu}(k) = 2e^2 \int_0^{+\infty} \frac{dp\, p}{\pi^2} \frac{1}{e^{\beta p}+1} \times \int \frac{d^2\widehat{p}}{4\pi} \Big(g^{\mu\nu} - \frac{\widehat{P}^\mu k^\nu + \widehat{P}^\nu k^\mu}{\widehat{P}\cdot k} + \frac{\widehat{P}^\mu \widehat{P}^\nu k^2}{(\widehat{P}\cdot k)^2} \Big),$$

where $\widehat{P} \equiv (1, \widehat{p})$. The radial integral reads

$$\begin{aligned} 2e^2 \int_0^{+\infty} \frac{dp\, p}{\pi^2} \frac{1}{e^{\beta p}+1} &= \frac{2e^2T^2}{\pi^2} \int_0^{+\infty} dx\, x \Big(e^{-x} - e^{-2x} + e^{-3x} - e^{-4x} + \cdots \Big) \\ &= \frac{2e^2T^2}{\pi^2} \Big(\frac{1}{1^2} - \frac{1}{2^2} + \frac{1}{3^2} - \cdots \Big) = \frac{2e^2T^2}{\pi^2} \frac{\zeta(2)}{2} = \frac{e^2T^2}{6}. \end{aligned}$$

In order to compare with the known results for the photon hard thermal loop, it is enough to check the component Π^{00} and the trace $\Pi_\mu{}^\mu$. The angular integration reduces to three basic integrals,

$$\int \frac{d^2\widehat{p}}{4\pi} = 1, \qquad \int \frac{d^2\widehat{p}}{4\pi} \frac{1}{\widehat{P}\cdot k} = \frac{1}{2k} \ln \left| \frac{k^0 + k}{k^0 - k} \right|,$$

$$\int \frac{d^2\widehat{p}}{4\pi} \underbrace{\frac{1}{(\widehat{P}\cdot k)^2}}_{-\frac{\partial}{\partial k^0}\frac{1}{\widehat{P}\cdot k}} = \frac{1}{(k^0)^2 - \boldsymbol{k}^2},$$

and we obtain

$$\begin{aligned} \Pi_\mu{}^\mu(k) &= \frac{e^2T^2}{6} \int \frac{d^2\widehat{p}}{4\pi} \Big(4 - 2\frac{\widehat{P}\cdot k}{\widehat{P}\cdot k} + \frac{\overbrace{\widehat{P}^2}^{0}\, k^2}{(\widehat{P}\cdot k)^2} \Big) = \frac{e^2T^2}{3}, \\ \Pi^{00}(k) &= \frac{e^2T^2}{6} \int \frac{d^2\widehat{p}}{4\pi} \Big(1 - 2\frac{k^0}{\widehat{P}\cdot k} + \frac{k^2}{(\widehat{P}\cdot k)^2} \Big) = \frac{e^2T^2}{3} \Big(1 - \frac{k^0}{2k} \ln \left| \frac{k^0 + k}{k^0 - k} \right| \Big). \end{aligned}$$

These results provide the two independent terms that enter into the photon hard thermal loop. Although the gain provided by this approach to calculating the HTL contribution to the photon polarization tensor is marginal, it becomes appreciable when applied to the case of QCD, and does not require much more effort (as opposed to an approach in terms of Feynman diagrams). Moreover, the present discussion highlights the fact that, in the HTL approximation, the particle running in the loop is in fact classical.

74. Gluon Hard Thermal Loop from Colored Classical Particles This problem is a continuation of Problem 73. Extend the method of that problem (with the help of the results of Problem 37) in order to compute the gluon polarization tensor in the hard thermal loop approximation. *Hint: note that the current must now be defined by*

$$J^\mu \equiv \int \frac{d^3p}{(2\pi)^3} \int dQ\, Q\, \frac{p^\mu}{p}\, f(x,p,Q),$$

where Q is the "non-Abelian classical charge" introduced in Problem 37, and show that the phase-space distribution obeys the following Vlasov equation:

$$p^\mu \left(\frac{\partial}{\partial x^\mu} - Q^a F^a_{\mu\nu} \frac{\partial}{\partial p_\nu} + ig \left(A^{adj}_\mu\right)_{ab} Q^b \frac{\partial}{\partial Q^a}\right) f = 0.$$

Since the method is the same as in Problem 73, our derivation will just go through the main steps and will skip some details. The starting point is the equation of motion for classical colored particles:

$$p^\mu = \frac{dx^\mu}{d\tau}, \qquad \frac{dp^\mu}{d\tau} = Q^a\, F^{\mu\nu}_a p_\nu.$$

The classical charges Q^a obey the Wong equations:

$$\frac{dQ^a}{d\tau} = ig\, p^\mu \left(A^{adj}_\mu\right)_{ab} Q^b.$$

An important difference compared to QED is that, since the charge Q is not constant, the phase-space distribution may depend on Q as if it were a standard coordinate, $f(x,p,Q)$. Again, the Liouville theorem implies that the distribution is constant along phase-space trajectories, which can be rewritten as

$$\begin{aligned} 0 = \frac{df}{d\tau} &= \left(\frac{dx^\mu}{d\tau}\frac{\partial}{\partial x^\mu} + \frac{dp^\mu}{d\tau}\frac{\partial}{\partial p^\mu} + \frac{dQ^a}{d\tau}\frac{\partial}{\partial Q^a}\right) f \\ &= p^\mu \left(\frac{\partial}{\partial x^\mu} - Q^a F^a_{\mu\nu}\frac{\partial}{\partial p_\nu} + ig\left(A^{adj}_\mu\right)_{ab} Q^b \frac{\partial}{\partial Q^a}\right) f. \end{aligned}$$

Compared to the QED case, the Vlasov equation contains an additional term because the classical color charge is not constant in an external field.

Then, we expand $f \equiv f_0 + f_1 + \cdots$, assuming that f_0 obeys the free Vlasov equation, $(p\cdot\partial)\, f_0 = 0$. The solution at order one in the external field is then given by

$$f_1 = \frac{p^\mu}{p\cdot\partial}\left(Q^a F^a_{\mu\nu}\frac{\partial}{\partial p_\nu} - ig\left(A^{adj}_\mu\right)_{ab} Q^b \frac{\partial}{\partial Q^a}\right) f_0.$$

Anticipating the fact that at a later stage f_0 will be replaced by an equilibrium distribution, we can assume that f_0 does not depend on Q^a and therefore drop the last term, to obtain

$$f_1 = \frac{p^\mu}{p\cdot\partial}\, Q^a \left[F^a_{\mu\nu}\right]_{\text{abel.}} \frac{\partial}{\partial p_\nu} f_0.$$

Note that, to be consistent with the expansion in powers of the external field, we must keep only the Abelian part of the field strength, since the commutator is of higher order in the gauge field. From f_1, we get the following current:

$$
\begin{aligned}
J^{\mu b} &= \int \frac{d^3p\, dQ}{(2\pi)^3 p}\, Q^b\, \frac{p^\mu p^\rho}{p\cdot\partial}\, Q^a \left[F^a_{\rho\nu}\right]_{\text{abel.}} \frac{\partial}{\partial p_\nu} f_0 \\
&= -\int \frac{d^3p\, dQ}{(2\pi)^3 p}\, Q^b\, \frac{\partial}{\partial p_\nu}\left(\frac{p^\mu p^\rho}{p\cdot\partial}\right) Q^a \left[F^a_{\rho\nu}\right]_{\text{abel.}} f_0 \\
&= \int \frac{d^3p\, dQ}{(2\pi)^3 p}\, Q^b \left(\frac{p^\mu p^\rho \partial^\nu}{(p\cdot\partial)^2} - \frac{g^{\nu\mu} p^\rho}{p\cdot\partial}\right) Q^a \left[F^a_{\rho\nu}\right]_{\text{abel.}} f_0.
\end{aligned}
$$

At this point, we need only one property of the integration over the color charge Q, namely that the integration of Q^aQ^b is diagonal in color, and proportional to the trace of a pair of generators in the representation of the particles under consideration,

$$
\int dQ\; Q^a Q^b = g^2\, \mathcal{N}_r\, \delta_{ab} \qquad \left(\mathcal{N}_f = \tfrac{1}{2}, \mathcal{N}_{\text{adj}} = N\right),
$$

which gives

$$
J^{\mu a} = g^2 \int \frac{d^3p}{(2\pi)^3 p} \left(\frac{p^\mu p^\rho \partial^\nu}{(p\cdot\partial)^2} - \frac{g^{\nu\mu} p^\rho}{p\cdot\partial}\right) \left[F^a_{\rho\nu}\right]_{\text{abel.}} \sum_r \mathcal{N}_r f_{0r}, \tag{5.21}
$$

where we have included a sum over the representations, in order to include both gluons and quarks (we denote by f_{0r} the corresponding equilibrium distributions).

Then, the gluon polarization tensor reads

$$
\Pi^{\mu a,\nu b} = -\frac{\partial J^{\mu a}}{\partial A^b_\nu} = g^2 \int \frac{d^3p}{(2\pi)^3 p} \left(\frac{g^{\mu\sigma} p^\rho}{p\cdot\partial} - \frac{p^\mu p^\rho \partial^\sigma}{(p\cdot\partial)^2}\right) \frac{\partial \left[F^a_{\rho\sigma}\right]_{\text{abel.}}}{\partial A^b_\nu} \sum_r \mathcal{N}_r f_{0r}.
$$

Therefore, we have

$$
\begin{aligned}
\Pi^{\mu a,\nu b}(k) &= g^2 \delta_{ab} \int \frac{d^3p}{(2\pi)^3 p} \left(\frac{g^{\mu\sigma} p^\rho}{p\cdot k} - \frac{p^\mu p^\rho k^\sigma}{(p\cdot k)^2}\right) \left(\delta_\sigma{}^\nu k_\rho - \delta_\rho{}^\nu k_\sigma\right) \sum_r \mathcal{N}_r f_{0r} \\
&= g^2 \delta_{ab} \int \frac{d^3p}{(2\pi)^3 p} \left(g^{\mu\nu} - \frac{p^\mu k^\nu + p^\nu k^\mu}{p\cdot k} + \frac{p^\mu p^\nu k^2}{(p\cdot k)^2}\right) \sum_r \mathcal{N}_r f_{0r}.
\end{aligned}
$$

The structure of this integral is very similar to the Abelian case, the only modification being the phase-space distribution. With N_f flavors of quarks in the fundamental representation and with gluons, we have

$$
\sum_r \mathcal{N}_r f_{0r} = \frac{2N}{e^{\beta|p^0|} - 1} + \frac{2N_f}{e^{\beta|p^0|} + 1}.
$$

(For gluons, there are two polarization states and $\mathcal{N}_{\text{adj}} = N$. For quarks, there is a factor $2N_f$ due to N_f flavors of quarks and antiquarks, a factor 2 for their two spin states, and $\mathcal{N}_f = \frac{1}{2}$.)

After factorizing the radial and angular integrals, we obtain

$$\Pi^{\mu a,\nu b}(k) = 2g^2\delta_{ab}\int_0^{+\infty}\frac{dp\,p}{\pi^2}\left(\frac{N_f/2}{e^{\beta p}+1}+\frac{N/2}{e^{\beta p}-1}\right)$$
$$\times\int\frac{d^2\widehat{p}}{4\pi}\left(g^{\mu\nu}-\frac{\widehat{P}^\mu k^\nu+\widehat{P}^\nu k^\mu}{\widehat{P}\cdot k}+\frac{\widehat{P}^\mu\widehat{P}^\nu k^2}{(\widehat{P}\cdot k)^2}\right).$$

The angular integral is the same as in the Abelian case. In the radial integral, the term with the Fermi–Dirac distribution is also the same, except for the change $e^2 \rightarrow g^2\delta_{ab}\frac{N_f}{2}$,

$$g^2\delta_{ab}N_f\int_0^{+\infty}\frac{dp\,p}{\pi^2}\,\frac{1}{e^{\beta p}+1} = \frac{g^2T^2N_f}{12}\delta_{ab},$$

and we just need to calculate the second term in the radial integral:

$$\begin{aligned}
g^2\delta_{ab}N\int_0^{+\infty}\frac{dp\,p}{\pi^2}\,\frac{1}{e^{\beta p}-1} &= \frac{g^2T^2N}{\pi^2}\delta_{ab}\int_0^{+\infty}dx\,x\,\frac{1}{e^x-1}\\
&= \frac{g^2T^2N}{\pi^2}\delta_{ab}\underbrace{\int_0^{+\infty}dx\,x\left(e^{-x}+e^{-2x}+e^{-3x}+\cdots\right)}_{\frac{1}{1^2}+\frac{1}{2^2}+\frac{1}{3^3}+\cdots=\zeta(2)=\frac{\pi^2}{6}}\\
&= \frac{g^2T^2N}{6}\delta_{ab}.
\end{aligned}$$

Therefore, the HTL gluon polarization tensor is identical to that of the photon, with only the following change of prefactor:

$$\frac{e^2T^2}{6} \quad\rightarrow\quad \frac{g^2T^2}{6}\left(N+\frac{N_f}{2}\right)\delta_{ab}.$$

In the non-Abelian case, the present approach based on classical transport has a very clear advantage over the calculation of Feynman graphs (there are four of them). Moreover, from the current in Eq. (5.21), we can guess an effective Lagrangian whose derivative with respect to the field gives the current:

$$\mathcal{L}_{\text{eff}} = \frac{g^2}{2}\int\frac{d^3p}{(2\pi)^3p}\left[F^a_{\nu\beta}\right]_{\text{abel.}}\frac{p^\beta p_\rho}{(p\cdot\partial)^2}\left[F^{\rho\nu}_a\right]_{\text{abel.}}\sum_r \mathcal{N}_r f_{0r}.$$

This effective Lagrangian is sufficient to recover the polarization tensor by taking two derivatives. But we can get a lot more out of it by noticing that it is not gauge invariant, but can be promoted into a gauge invariant one by very simple substitutions:

$$\left[F^a_{\nu\beta}\right]_{\text{abel.}} \rightarrow F^a_{\nu\beta}, \qquad p\cdot\partial \rightarrow p\cdot D,$$

which leads to

$$\mathcal{L}_{\text{eff}} = \frac{g^2}{2}\int\frac{d^3p}{(2\pi)^3p}F^a_{\nu\beta}\left[\frac{p^\beta p_\rho}{(p\cdot D)^2}\right]_{ab}F^{\rho\nu}_b\sum_r \mathcal{N}_r f_{0r}.$$

Note that this effective Lagrangian now contains terms of arbitrarily high order in the gauge field. In fact, by differentiating it n times with respect to the field, it provides the n-gluon HTL

vertex. The main reason why it is possible to capture the HTL results in such a simple setting is that when the momentum circulating in the loop is much larger than the external momenta, then all the couplings of the external lines to the loop become eikonal, and therefore universal (in particular, independent of the spin). In this approximation, the only thing that distinguishes the various fields that can circulate in the loop is their phase-space distribution, their color representation and the corresponding number of degrees of freedom, while all the kinematical dependence is universal.
Sources: Kelly, P. F., Liu, Q., Lucchesi, C. and Manuel, C. (1994), *Phys Rev Lett* 72: 3461; Pisarski, R. D. (1998), Kinetic theory of hot gauge theories: Overview, details & extensions, in Sánchez, N. and Zichichi, A. (eds.), *Current Topics in Astrofundamental Physics: Primordial Cosmology*, NATO ASI Series, vol. 511 (Springer), p. 195.

75. Landau–Pomeranchuk–Migdal Effect The goal of this problem is to study the emission of soft photons by a hot plasma of electrons and anti-electrons, and in particular to discuss the *Landau–Pomeranchuk–Migdal effect*, an interference effect that reduces the emission of soft photons in a hot medium.

75.a Recall the relationship between the photon emission rate and the imaginary part of the retarded photon self-energy.

75.b Explain why this self-energy can be obtained as

$$-i\Pi_R^{ij}(x,y) = -\frac{\delta J^i_{\rm ind}(x)}{\delta A_j(y)},$$

where $J^i_{\rm ind}$ is the electromagnetic current induced by the application of an external potential A^j. Write this current in terms of the deviations from equilibrium $\delta f_e, \delta f_{\bar{e}}$ of the electron and anti-electron distributions.

75.c Show that δf_e obeys the following *Boltzmann–Vlasov equation*:

$$(v\cdot\partial_x)\delta f_e - \tfrac{e}{T}\,(v\cdot E)\,n_F(p)(1-n_F(p)) = \delta\mathcal{C}_{e,p}[\delta f],$$

where $\delta\mathcal{C}_{e,p}$ is a collision term and n_F is the Fermi–Dirac distribution.

75.d By explicitly writing the collision term with two-body collisions, show that

$$\delta\mathcal{C}_{e,p} = -e^3 n_F(p)(1-n_F(p))\int\frac{d^2\ell_\perp}{(2\pi)^2}\,\frac{m_D^2}{\ell_\perp^2(\ell_\perp^2+m_D^2)}\big[F(x,p)-F(x,p-\ell_\perp)\big],$$

where m_D is the photon Debye mass and $\delta f_e \equiv \frac{e}{T}n_F(1-n_F)F$.

75.e Apply a Fourier transform to the x dependence, and discuss qualitatively the effect of the integral on $\ell_\perp$ in the collision term.

75.f Extra: in the regime where $m \gg m_D$, transform the Boltzmann–Vlasov equation into a diffusion equation that one may solve analytically.

75.a We consider a hot plasma of electrically charged electrons and anti-electrons of mass m (this could be their in-vacuum mass or, at very high temperature, the mass they acquire in the plasma due to their interactions with their environment), interacting via photon exchanges (this point is not crucial and one could easily generalize this discussion to interactions via gluon exchanges). The photon emission rate from a sample of this plasma can be obtained from the in-medium photon self-energy, as follows:

$$2k_0(2\pi)^3\frac{dN_\gamma}{d^4x d^3k} = \sum_{\text{pol. }\lambda} \epsilon_\mu^{(\lambda)}(k)\epsilon_\nu^{(\lambda)}(k)\frac{2\,\text{Im}\left(-i\Pi_R^{\mu\nu}(K)\right)}{e^{\beta k_0}-1}.$$

There are various variants of this formula, differing in which form of the self-energy is used. Here, we choose the formula that involves the retarded self-energy, because we will want later to obtain it from linear response theory. To be definite, let us choose the photon momentum k to be aligned with the x^3 direction. For on-shell photons, the sum over the polarizations simply produces a projector on the plane x^1, x^2 transverse to the photon momentum:

$$\sum_{\text{pol. }\lambda} \epsilon_\mu^{(\lambda)}(k)\epsilon_\nu^{(\lambda)}(k)\Pi^{\mu\nu}(K) = \sum_{i=1,2} \Pi^{ii}(K).$$

75.b The retarded photon self-energy can be viewed as a retarded correlation function between a pair of electromagnetic currents (this is correct in linear gauges, where there is no spurious triple or quartic photon coupling). Then, by the same reasoning as for the derivation of Kubo's formula (see, for instance, Mahan, G. D. (2000), *Many-Particle Physics* (Kluwer Academic)), the retarded current–current correlator can be obtained by looking at the induced current (the response) that appears in the system when we apply an external field A^j (the source):

$$-i\Pi_R^{ij}(x,y) = -\frac{\delta J^i_{\text{ind}}(x)}{\delta A_j(y)}, \quad \text{with } J^i_{\text{ind}} \equiv e\langle\overline{\psi}\gamma^i\psi\rangle.$$

Since we want the photon self-energy in equilibrium, we need only consider arbitrarily small external fields (in other words, we calculate the induced current only at linear order in the applied external field).

In thermal equilibrium, there are no net currents in the system (since the fermions and anti-fermions have opposite charges and the same momentum distribution). The induced current can be obtained as

$$J^i_{\text{ind}}(x) = 2e\int\frac{d^3p}{(2\pi)^3}\, v^i\left(\delta f_e(x,p) - \delta f_{\bar{e}}(x,p)\right),$$

where $v^i \equiv p^i/E_p$ and δf_e (resp. $\delta f_{\bar{e}}$) is the deviation from Fermi–Dirac of the electron (resp. anti-electron) distribution induced by the external field (think of an electric field, which pulls electrons and anti-electrons in opposite directions). The prefactor 2 is due to the two spin states of an electron.

75.c The external field acts on the fermions of the system via the Lorentz force. The other effect that affects the fermion distribution is their collisions, which are responsible for the relaxation to equilibrium. Therefore, the equation that controls f and $\bar{f}$ is a Boltzmann–Vlasov equation:

$$\underbrace{(v \cdot \partial_x) f_e}_{\text{free streaming}} + \underbrace{e\left(\mathbf{E} + \mathbf{v} \times \mathbf{B}\right) \cdot \nabla_p f_e}_{\text{Vlasov term}} = \underbrace{\mathcal{C}_{e,p}[f_e, f_{\bar{e}}]}_{\text{collisions}},$$

where $v^\mu \equiv (1, v^i)$. The Vlasov term, proportional to the Lorentz force, is responsible for the change of the momenta of the particles as they move in the external field. The anti-electron distribution obeys the same equation with the change $e \to -e$ to account for the fact that they feel the opposite Lorentz force. In this problem, we need only consider small deviations of the distribution about the equilibrium Fermi–Dirac distribution, $f_e \equiv n_F + \delta f_e$, $f_{\bar{e}} \equiv n_F + \delta f_{\bar{e}}$, and we can linearize the above Boltzmann–Vlasov equation in $\delta f_e, \delta f_{\bar{e}}$. Recalling that the deviations $\delta f_e, \delta f_{\bar{e}}$ are of the same order as the external electromagnetic field $\mathbf{E}, \mathbf{B}$, and that the collision term vanishes when evaluated with the equilibrium Fermi–Dirac distribution, we obtain

$$(v \cdot \partial_x)\delta f_e + e\left(\mathbf{E} + \mathbf{v} \times \mathbf{B}\right) \cdot \underbrace{\nabla_p n_F}_{-\frac{1}{T} n_F(1-n_F)\mathbf{v}} = \delta\mathcal{C}_{e,p}[\delta f],$$

where we denote by $\delta\mathcal{C}$ the linearized collision term. Since the Fermi–Dirac distribution is isotropic, its derivative with respect to momentum is proportional to the velocity $\mathbf{v}$, implying that only the electric field $\mathbf{E}$ contributes. In the following, it will prove convenient to parameterize the deviations $\delta f_2, \delta f_{\bar{e}}$ as follows:

$$\delta f_e(x, \mathbf{p}) \equiv \tfrac{e}{T} n_F(\mathbf{p})(1 - n_F(\mathbf{p}))\, F(x, \mathbf{p}),$$
$$\delta f_{\bar{e}}(x, \mathbf{p}) \equiv -\tfrac{e}{T} n_F(\mathbf{p})(1 - n_F(\mathbf{p}))\, F(x, \mathbf{p}),$$

where the function $F(x, \mathbf{p})$ is the same in both cases.

75.d Before linearization, the collision term with two-body collisions $\mathbf{pq} \to \mathbf{p'q'}$ reads

$$\mathcal{C}_{e,p}[f, \bar{f}] = \sum_{\alpha=e,\bar{e}} \int_{p',q,q'} (2\pi)^4 \delta(p + q - p' - q') \frac{1}{16E_pE_{p'}E_qE_{q'}} \left|\mathcal{M}_{pq\to p'q'}\right|^2$$
$$\times \Big\{(1-f_e(x,\mathbf{p}))f_e(x,\mathbf{p}')(1-f_\alpha(x,\mathbf{q}))f_\alpha(x,\mathbf{q}')$$
$$- f_e(x,\mathbf{p})(1-f_e(x,\mathbf{p}'))f_\alpha(x,\mathbf{q})(1-f_\alpha(x,\mathbf{q}'))\Big\},$$

where we have used the shorthand $\int_q \equiv \int \frac{d^3q}{(2\pi)^3}$. Note that we sum over the two kinds of scattering partners, electrons and anti-electrons. Next, we expand the distribution functions about n_F, keeping only terms linear in the deviations. Since δf_e and $\delta f_{\bar{e}}$ are opposite, the deviations of the distribution f_α of the scattering partner cancel when we sum over electrons and anti-electrons. Therefore, we get

$$\delta\mathcal{C}_{e,p} = -\frac{2e}{T}\int_{p',q,q'} (2\pi)^4 \delta(p + q - p' - q') \frac{1}{16E_pE_{p'}E_qE_{q'}} \left|\mathcal{M}_{pq\to p'q'}\right|^2$$
$$\times\, n_F(\mathbf{p})(1-n_F(\mathbf{p}'))n_F(\mathbf{q})(1-n_F(\mathbf{q}'))\,\Big[F(x,\mathbf{p})-F(x,\mathbf{p}')\Big].$$

Note that the momentum transfer $\ell \equiv p - p' = q' - q$ is soft (of the order of the Debye mass, i.e., $\sim eT$) compared to the typical momentum of the electrons and anti-electrons (of the order of the temperature T). This implies that the scattering amplitude is dominated by the t-channel diagram. Consequently, there is no difference between scattering off an electron and off an anti-electron, hence the prefactor 2. Then, in order to rewrite the integrals in the collision terms in terms of ℓ, we can use

$$\begin{aligned}&\int \frac{d^3p'}{(2\pi)^2 2E_{p'}}\frac{d^3q}{(2\pi)^2 2E_q}\frac{d^3q'}{(2\pi)^2 2E_{q'}}(2\pi)^4\delta(p+q-p'-q')\\ &\approx \int \frac{d^4\ell}{(2\pi)^4}\frac{d^3q}{(2\pi)^3 2E_q}\frac{1}{4E_pE_q}(2\pi)^2\delta(\ell_0-\boldsymbol{\ell}\cdot\boldsymbol{v})\delta(\ell_0-\boldsymbol{\ell}\cdot\boldsymbol{v}_q),\end{aligned}$$

where we have neglected terms quadratic in the soft momentum ℓ and where we denote $\boldsymbol{v}_q \equiv \boldsymbol{q}/E_q$. We can also use the approximation

$$\begin{aligned}n_F(\boldsymbol{p})(1-n_F(\boldsymbol{p}')) &\approx n_F(\boldsymbol{p})(1-n_F(\boldsymbol{p})),\\ n_F(\boldsymbol{q})(1-n_F(\boldsymbol{q}')) &\approx n_F(\boldsymbol{q})(1-n_F(\boldsymbol{q}),\end{aligned}$$

and note that the integrated squared matrix element (at fixed transfer ℓ) can be expressed in terms of the hard thermal loop spectral function (i.e., the imaginary part of the HTL photon propagator):

$$2\int \frac{d^3q}{(2\pi)^3}2\pi\delta(\ell_0-\boldsymbol{\ell}\cdot\boldsymbol{v}_q)\,n_F(\boldsymbol{q})(1-n_F(\boldsymbol{q}))\frac{|\mathcal{M}_{pq\to p'q'}|^2}{16E_p^2E_q^2} \approx e^2\frac{T}{\ell_0}v_\mu v_\nu\,\rho^{\mu\nu}_{\text{HTL}}(\ell),$$

with

$$\rho^{\mu\nu}_{\text{HTL}}(\ell) \equiv \sum_{\alpha=T,L} P^{\mu\nu}_\alpha(\ell)\,\frac{2\,\text{Im}\,\Pi_\alpha(\ell)}{(\ell^2-\text{Re}\,\Pi_\alpha(\ell))^2+(\text{Im}\,\Pi_\alpha(\ell))^2}$$

($v^\mu \equiv (1, \boldsymbol{v}^i)$, and the projectors $P^{\mu\nu}_{T,L}$ were defined following Eq. (5.14)). Therefore, the linearized collision term becomes

$$\begin{aligned}\delta\mathcal{C}_{e,p} &= -\frac{e^3}{T}n_F(\boldsymbol{p})(1-n_F(\boldsymbol{p}))\\ &\times\int \frac{d^4\ell}{(2\pi)^4}2\pi\delta(\ell_0-\boldsymbol{\ell}\cdot\boldsymbol{v})\frac{T}{\ell_0}v_\mu v_\nu\,\rho^{\mu\nu}_{\text{HTL}}(\ell)\left[F(x,\boldsymbol{p})-F(x,\boldsymbol{p}-\boldsymbol{\ell})\right].\end{aligned}$$

Here, it is important to recognize that $F(x, \boldsymbol{p})$ is strongly peaked when $\boldsymbol{p}$ is aligned with the photon momentum, i.e., with the z axis. As a consequence, the variations of F are much faster when $\mathbf{p}$ varies in the transverse direction than in the longitudinal direction, implying that

$$F(x,\mathbf{p}) - F(x,\mathbf{p}-\boldsymbol{\ell}) \approx F(x,\mathbf{p}) - F(x,\mathbf{p}-\boldsymbol{\ell}_\perp).$$

For the same reason, we can approximate $\delta(\ell_0 - \boldsymbol{\ell}\cdot\mathbf{v}) \approx \delta(\ell_0 - \ell_z)$. Using the explicit form of the projectors $P^{\mu\nu}_{T,L}(\ell)$, we have

$$v_\mu v_\nu P^{\mu\nu}_L(\ell) = -v_\mu v_\nu P^{\mu\nu}_T(\ell) \underset{\ell_0=\ell_z}{=} \frac{\ell_\perp^2}{\ell_0^2 + \ell_\perp^2}.$$

It is convenient to introduce the variable $x \equiv \ell_0/|\boldsymbol{\ell}|$, since the HTL photon self-energies depend only on x. The Jacobian of this change of variable is given by

$$\frac{d\ell_0 d^2\boldsymbol{\ell}_\perp}{\ell_0} \frac{\ell_\perp^2}{\ell_0^2 + \ell_\perp^2} = \frac{dx d^2\boldsymbol{\ell}_\perp}{x}.$$

Therefore, the collision term reads

$$\delta\mathcal{C}_{e,\mathbf{p}} = -\,e^3 n_F(\mathbf{p})(1-n_F(\mathbf{p}))\int \frac{d^2\boldsymbol{\ell}_\perp}{(2\pi)^2}\left[F(x,\mathbf{p}) - F(x,\mathbf{p}-\boldsymbol{\ell}_\perp)\right]$$
$$\times \underbrace{\frac{2}{\pi}\int_0^1 \frac{dx}{x}\left\{\frac{\mathrm{Im}\,\Pi_L(x)}{(\ell_\perp^2+\mathrm{Re}\,\Pi_L(x))^2+(\mathrm{Im}\,\Pi_L(x))^2} - \frac{\mathrm{Im}\,\Pi_T(x)}{(\ell_\perp^2+\mathrm{Re}\,\Pi_T(x))^2+(\mathrm{Im}\,\Pi_T(x))^2}\right\}}_{=\frac{m_D^2}{\ell_\perp^2(\ell_\perp^2+m_D^2)}}.$$

The identity used in the second line, where m_D is the Debye screening mass, is an exact sum rule satisfied by the photon HTL self-energies.

75.e Combining the above results, the Boltzmann–Vlasov equation can be written as follows in terms of F:

$$(v\cdot\partial_x)F(x,\mathbf{p}) + e^2T\int \frac{d^2\boldsymbol{\ell}_\perp}{(2\pi)^2}\frac{m_D^2}{\ell_\perp^2(\ell_\perp^2+m_D^2)}\left[F(x,\mathbf{p}) - F(x,\mathbf{p}-\boldsymbol{\ell}_\perp)\right] = \mathbf{v}\cdot\mathbf{E}.$$

Using $\mathbf{E} \equiv -\partial_t\mathbf{A}$ and applying a Fourier transform to turn x into the photon momentum K, this equation leads to

$$\frac{\delta F(K,\mathbf{p})}{\delta A_j(K)} = \frac{k^0 v^j}{K\cdot v} + i\frac{e^2T}{K\cdot v}\int \frac{d^2\boldsymbol{\ell}_\perp}{(2\pi)^2}\frac{m_D^2}{\ell_\perp^2(\ell_\perp^2+m_D^2)}\left[\frac{\delta F(K,\mathbf{p})}{\delta A_j(K)} - \frac{\delta F(K,\mathbf{p}-\boldsymbol{\ell}_\perp)}{\delta A_j(K)}\right] \tag{5.22}$$

(recall that $\mathbf{v} \equiv \mathbf{p}/E_\mathbf{p}$), whose solution must be inserted into

$$\mathrm{Im}\left(-i\Pi_R^{ij}(K)\right) = -\frac{4e^2}{T}\int \frac{d^3\mathbf{p}}{(2\pi)^3} n_F(\mathbf{p})(1-n_F(\mathbf{p}))\,\mathbf{v}^i\,\mathrm{Im}\left(\frac{\delta F(K,\mathbf{p})}{\delta A_j(K)}\right).$$

The integral term in (5.22) takes into account the collisions of the emitter with the other particles in the plasma, while the first term on the right-hand side of this equation would give the contribution of the process $e \to e\gamma$. Note that this collisionless contribution gives a purely real $\delta F/\delta A_j$ and therefore does not contribute to the photon yield, as expected from trivial kinematics.

The equation (5.22) could in principle be solved iteratively in powers of e^2, by inserting inside the integral the solution obtained at the previous order. Assuming that $\delta F/\delta A_j$ is of order one as in the collisionless limit, then the integral in (5.22) is also of order one (the typical $\ell_\perp$ is of the order of the Debye mass m_D). We therefore see that multiple collisions all contribute to the same order if $v \cdot K \lesssim e^2 T$. On the other hand, simple kinematics leads to

$$v \cdot K \approx k_0 \frac{m^2 + p_\perp^2}{2 p_z^2}. \tag{5.23}$$

Note that $v \cdot K$ can be interpreted as the inverse of the photon *formation time*, i.e., the time necessary for the photon wave-packet to become separated from that of its emitter. In more physical terms, multiple scatterings are important when the photon formation time is comparable to or larger than the mean free path, i.e., the distance traveled by the emitter between two successive collisions (in this process, the relevant collisions are small-angle collisions, separated by $(e^2T)^{-1}$). This effect is known as the *Landau–Pomeranchuk–Migdal effect.* In this regime, there is no point in solving (5.22) perturbatively as a series in e^2 because all orders are in fact comparable. From (5.23), we see that $v \cdot K$ may be small when the electron momentum is much larger than its mass, $p_z \gg m$, and its transverse momentum $p_\perp$ is small (quasi-collinear emission of the photon). This is precisely what happens in a plasma at very high temperature, where the relevant mass is a thermal mass of order $m \sim eT$, the typical p_z is of order T, and photon emission happens predominantly at small angles $\theta \sim e$ (i.e., $p_\perp \sim eT$).

The effect of these multiple scatterings on the photon spectrum can be estimated roughly as follows. On the one hand, the photon formation length is $\ell \sim 1/v \cdot K$. On the other hand, the transverse momentum accumulated by the emitter over this distance is $p_\perp^2 \sim m_D^2\,\ell/\lambda$, where λ is the electron mean free path. This leads to

$$\frac{\ell}{\lambda} \sim \frac{p_z}{m_D\sqrt{k_0\lambda}} \underset{k_0\to 0}{\gg} 1.$$

This implies that it takes a large number (growing like $1/\sqrt{k_0}$) of collisions to produce a single soft photon. In other words, the soft photon spectrum is reduced by a factor $\sqrt{k_0}$ compared to what one would obtain by multiplying the one-collision spectrum by the number of collisions (this is known as the *Bethe–Heitler limit*). This suppression is a quantum effect, due to the destructive interference between photon emissions occurring at various points along the trajectory of the emitting electron (the photon formation length controls how far apart two emissions that still interfere can be); see Figure 5.4 for an illustration.

75.f This discussion can be made more quantitative analytically when $m \gg m_D$ (for this to be true, the bare mass of the electron must dominate over its thermal mass). In this case, we may write

$$\int \frac{d^2\ell_\perp}{(2\pi)^2} \frac{m_D^2}{\ell_\perp^2(\ell_\perp^2+m_D^2)} \left[\frac{\delta F(K,p)}{\delta A_j(K)} - \frac{\delta F(K,p-\ell_\perp)}{\delta A_j(K)}\right]$$

$$\approx \int \frac{d^2\ell_\perp}{(2\pi)^2} \frac{m_D^2}{\ell_\perp^2(\ell_\perp^2+m_D^2)} \frac{1}{2}(\ell_\perp \cdot \nabla_p)^2 \frac{\delta F(K,p)}{\delta A_j(K)} \approx \frac{m_D^2}{8\pi} \ln\left(\frac{m}{m_D}\right) \nabla_p^2 \frac{\delta F(K,p)}{\delta A_j(K)}.$$

Here, we have used the separation of scale between the typical $p_\perp \sim m$ and the Debye mass m_D in order to make a leading logarithmic approximation of the integral over $\ell_\perp$ (the logarithm

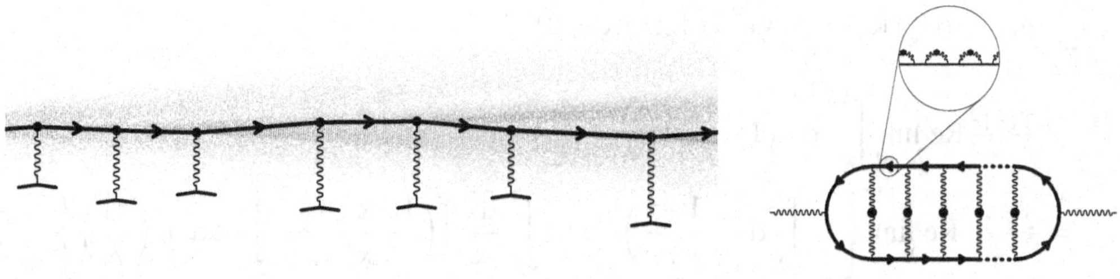

Figure 5.4 Left: illustration of the Landau–Pomeranchuk–Migdal effect. Right: the Feynman graphs contributing to the photon self-energy in the LPM regime (the black dots on the internal photons indicate the hard thermal loop re-summation).

must be large for this approximation to be justified). In this approximation, the equation (5.22) that sums the multiple collisions of the emitter becomes a diffusion equation in the transverse velocity space ($\mathbf{v}_\perp \equiv \mathbf{p}_\perp/p$):

$$ik_0\Big(\tfrac{1}{2}\big(v_\perp^2 + \tfrac{m^2}{p^2}\big)\frac{\delta F}{\delta A_j} - v^j\Big) + D\boldsymbol{\nabla}^2_{v_\perp}\frac{\delta F}{\delta A_j} = 0,$$

where the diffusion constant is defined as $D \equiv \frac{e^2 T m_D^2}{8\pi p^2}\ln(m/m_D)$. In order to solve this diffusion equation, let us define $\delta F/\delta A_j \equiv v^j(\varphi(s)+1)/s$ with $s \equiv \frac{1}{2}v_\perp^2$ (note that we must have $\varphi(0) = -1$ for this to be defined at $\mathbf{v}_\perp = 0$). In terms of φ and s, we have

$$s\varphi'' + i(\alpha + \beta s)\varphi + i\alpha = 0, \quad \text{with } \alpha \equiv \tfrac{k_0 m^2}{4Dp^2},\ \beta \equiv \tfrac{k_0}{2D}.$$

This is a differential equation with polynomial coefficients, which may be solved by taking a Laplace transform,

$$\nu(\lambda) \equiv \int_0^\infty ds\, e^{-\lambda s}\varphi(s).$$

First, we obtain a differential equation for $\nu(\lambda)$:

$$\nu' + \frac{2\lambda - i\alpha}{\lambda^2 + i\beta}\nu + \frac{1 - i\frac{\alpha}{\lambda}}{\lambda^2 + i\beta} = 0.$$

Since this is a first-order equation, it can be solved analytically:

$$\nu(\lambda) = -\frac{1}{\lambda_1^2 - \lambda^2}\Big(\frac{\lambda_1 + \lambda}{\lambda_1 - \lambda}\Big)^{\mu}\int_\lambda^{\lambda_1} d\xi\,\big(1 - i\tfrac{\alpha}{\xi}\big)\Big(\frac{\lambda_1 - \xi}{\lambda_1 + \xi}\Big)^{\mu},$$

with $\lambda_1 \equiv \sqrt{\beta}e^{-i\pi/4}$ and $\mu \equiv (\alpha/2\sqrt{\beta})e^{-i\pi/4}$. In terms of this function, we have

$$\begin{aligned}
\sum_{i=1,2} \operatorname{Im}\big(-i\Pi_R^{ii}(K)\big) &= -\tfrac{2}{3}e^2T^2 \operatorname{Im}\nu(\lambda=0)\\
&= \tfrac{2}{3}\tfrac{e^2T^2}{\beta}\operatorname{Re}\lim_{\lambda\to 0}\int_\lambda^{\lambda_1} d\xi\,\big(1-i\tfrac{\alpha}{\xi}\big)\Big(\frac{\lambda_1-\xi}{\lambda_1+\xi}\Big)^\mu\\
&= \tfrac{2}{3}\tfrac{e^2T^2}{\beta}\operatorname{Re}\lim_{\lambda\to 0}\underbrace{\Big\{\lambda_1\int_0^1 dx\,\Big(\frac{1-x}{1+x}\Big)^\mu - i\alpha\int_0^1\frac{dx}{x}\Big[\Big(\frac{1-x}{1+x}\Big)^\mu - 1\Big] - i\alpha\ln\Big(\frac{\lambda_1}{\lambda}\Big)\Big\}}_{\sim\sqrt{\beta}\quad \text{when } k_0\to 0}\\
&\underset{k_0\to 0}{\propto} \sqrt{k_0}.
\end{aligned}$$

In the penultimate line, we have directly set $\lambda = 0$ in the terms that are regular in this limit, and isolated a term that has a logarithmic singularity. But note that the singularity only contributes to the imaginary part, so that the real part has a finite limit (λ should approach zero along the positive real axis in order to obtain the correct result). When we combine this result with the soft limit of the Bose–Einstein factor that multiplies $\operatorname{Im}(-i\Pi_R^{ii}(K))$ in the photon emission spectrum, we obtain a spectrum in $1/\sqrt{k_0}$, instead of $1/k_0$ for bremsstrahlung induced by a single scattering. Thus, the effect of multiple scatterings is to reduce the emission of soft photons by a factor $\sqrt{k_0}$, as we had anticipated from our earlier qualitative discussion.

Let us finish with a general remark about the approach employed here to derive this effect. Naively, it may seem paradoxical to be able to derive a quantum interference effect by means of kinetic theory, which assumes that electrons behave completely classically. But in the LPM effect, the only object whose coherence matters is the photon wavefunction, while the electron can be totally classical.

Sources: Baym, G., Blaizot, J.-P., Gelis, F. and Matsui, T. (2007), *Phys Lett B* 644: 48; Migdal, A. B. (1956), *Phys Rev* 103: 1811.

76. Microcanonical Equilibration in ϕ^4 Scalar Theory Consider a massless scalar field theory with ϕ^4 interaction in four spacetime dimensions. The goal of this problem is to study some aspects of classical dynamics in this field theory, in relation to thermal equilibration.

76.a What is the canonical energy–momentum tensor in this theory? What is its trace?

76.b What is the equation of state relating pressure to energy density for a scale invariant system in statistical equilibrium?

76.c Consider a spatially homogeneous classical field in this theory. Show that this field is a periodic function of time, and find the relationship between the period and the amplitude of the field oscillations. Show that the trace of the energy–momentum tensor is non-zero for this field, but vanishes when averaged over one period.

76.d Consider now a statistical ensemble of such homogeneous classical fields, obtained by choosing an ensemble of initial conditions and letting the fields evolve according to the classical equation of motion. How does the expectation value of the trace of the energy–momentum tensor evolve with time?

76.a The Lagrangian for this theory reads

$$\mathcal{L} = \frac{1}{2}(\partial_\mu\phi)(\partial^\mu\phi) - \frac{g^2}{4!}\phi^4.$$

Note that, in four spacetime dimensions, the coupling constant g is dimensionless. Therefore, in the absence of a mass term, this theory is scale invariant. Its classical equation of motion is a non-linear Klein–Gordon equation:

$$\Box\phi + \frac{g^2}{6}\phi^3 = 0,$$

and its canonical energy–momentum tensor is given by

$$T^{\mu\nu} = (\partial^\mu\phi)(\partial^\nu\phi) - g^{\mu\nu}\mathcal{L}.$$

The trace of $T^{\mu\nu}$ reads

$$\begin{aligned} T_\mu{}^\mu &= (\partial_\mu\phi)(\partial^\mu\phi) - 4\mathcal{L} = -(\partial_\mu\phi)(\partial^\mu\phi) + \frac{g^2}{6}\phi^4 \\ &= \phi\underbrace{\left(\Box\phi + \frac{g^2}{6}\phi^3\right)}_{=0} - \partial_\mu(\phi\partial^\mu\phi). \end{aligned} \tag{5.24}$$

In the final form, we have chosen to exhibit a term that vanishes thanks to the field equation of motion, plus a remainder which is not zero in general.

76.b In a scale invariant system (this is the case of massless scalar theory with a quartic coupling in four dimensions, at the classical level), the equation of state that relates the energy density ϵ and the pressure p should be $\epsilon = 3p$ (in four spacetime dimensions). This is equivalent to a vanishing trace of the energy–momentum tensor, *after averaging over the ensemble of fields.* In other words, a single field configuration in general has a non-zero $T_\mu{}^\mu$, but for an ensemble of fields in thermal equilibrium, the expectation value of the trace of the energy–momentum tensor must vanish, $\langle T_\mu{}^\mu\rangle = 0$.

76.c Consider now a spatially homogeneous field ϕ, that obeys the classical equation of motion,

$$\ddot\phi + \frac{g^2}{6}\phi^3 = 0. \tag{5.25}$$

(We denote the time derivatives by a dot.) This is a non-linear ordinary differential equation, that may be solved in terms of elliptic cosines. However, a qualitative analysis of the behavior of the solutions may be performed without writing the explicit solution. Indeed, this equation is analogous to the equation of motion of a point particle in a one-dimensional convex potential $\sim \phi^4$, and it is well known that the solutions are periodic functions of time. The period of the oscillations may be determined by noting that this equation of motion leads to a constant energy. Multiplying (5.25) by $\dot{\phi}$, we obtain

$$\epsilon \equiv \frac{1}{2}\dot{\phi}^2 + \frac{g^2}{4!}\phi^4 = \text{const} = \frac{g^2}{4!}\phi_{max}^4,$$

where ϕ_{max} is the amplitude of the oscillations of the field. Therefore, we have

$$\dot{\phi} = \pm\sqrt{\frac{g^2}{12}\left(\phi_{max}^4 - \phi^4\right)}.$$

A quarter period corresponds to ϕ varying from 0 to ϕ_{max}. Therefore, the period of the oscillations is given by

$$T = \frac{8\sqrt{3}}{g}\int_0^{\phi_{max}} \frac{d\phi}{\sqrt{\phi_{max}^4 - \phi^4}} = \frac{8\sqrt{3}}{g\,\phi_{max}}\int_0^1 \frac{dx}{\sqrt{1-x^4}}. \tag{5.26}$$

(The remaining integral is an elliptic integral whose precise value is not important here.) The most notable property of this result is that the period is not constant, but is inversely proportional to the amplitude of the oscillations. This could have been guessed without any calculation. Indeed, Eq. (5.25) describes an anharmonic oscillator. Since the coupling constant is dimensionless and the field has mass dimension one, the period must be proportional to ϕ_{max}^{-1} on purely dimensional grounds (this would not be true with a ϕ^6 potential, for which the coupling constant would have a mass dimension -2). For such a classical field configuration, the trace of the energy–momentum tensor is

$$T_\mu{}^\mu = -\frac{d(\phi\dot{\phi})}{dt} \neq 0.$$

Although it is not vanishing, it has the remarkable property of being a total time derivative. Therefore, if we average it over one oscillation period, we have

$$\overline{T_\mu{}^\mu} \equiv -\frac{1}{T}\int_0^T dt\,\frac{d(\phi\dot{\phi})}{dt} = -\left[\phi\dot{\phi}\right]_0^T = 0,$$

thanks to the periodicity of ϕ and $\dot{\phi}$. This time averaging usually does not correspond to a typical (instantaneous) measurement. However, a collection of such fields, all having the same energy but with random phase shifts uniformly distributed around the orbit, would have a vanishing *ensemble average* of $T_\mu{}^\mu$. Indeed, the ensemble average for this system would just be an average over the phase shifts of the individual fields, which would be equivalent to the average over one period if the phases were uniformly distributed. Note that an ensemble having a single energy and a uniform distribution of phases is in *microcanonical equilibrium* (its phase-space density uniformly covers the manifold allowed by energy conservation).

76.d Consider now an ensemble of such classical fields, specified by choosing some initial configuration for every field in the ensemble (but a priori not in a microcanonical equilibrium state initially). A crucial question is whether this ensemble evolves towards a state where the trace of the energy–momentum tensor is zero, so that we recover the expected equation of state. There are two cases:

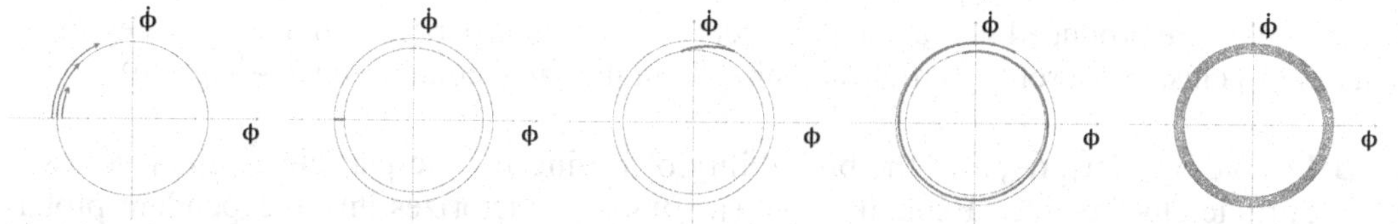

Figure 5.5 Leftmost panel: illustration of the fact that the phase velocity depends on the field amplitude. Four panels on the right: successive stages of the evolution in phase-space of an ensemble of classical fields in ϕ^4 scalar field theory.

- If this ensemble is mono-energetic (i.e., all fields in the ensemble have exactly the same energy), then this microcanonical equilibration will never happen. Indeed, in this case, all the fields are periodic functions with exactly the same period, and therefore their relative phase shifts remain constant with time. If the distribution of fields does not uniformly cover the constant energy manifold initially, this will never happen at later times.

- If the ensemble has some (no matter how small) energy spread ΔE, then microcanonical equilibration happens. Indeed, a spread in E is equivalent to a spread in ϕ_{max}, or a spread in the periods T. This implies that fields that have nearby (but not exactly equal) energies rotate at different frequencies on their respective orbits. Therefore, a cloud of points in phase-space will spread along the orbits and eventually distribute uniformly in the domain allowed by energy conservation. When this has happened, ensemble averages become equivalent to time averages, and we have $\langle T_\mu{}^\nu \rangle = 0$, i.e., the equation of state $\epsilon = 3p$ (for the ensemble, not for individual fields). An illustration of this decoherence phenomenon is shown in Figure 5.5. In the leftmost panel, we show how trajectories with different energies have different rotation speeds in the $(\phi, \dot{\phi})$ plane. The next four panels in Figure 5.5 show that an initial ensemble of configurations with a small dispersion in energy progressively spreads around the average orbit to form a microcanonical ensemble.

In this toy example, reaching the microcanonical equilibrium appears to be essentially a decoherence phenomenon. However, one should keep in mind that we have considered only spatially homogeneous configurations (the study is much more complicated with fields that depend also on space). In more general situations, the relationship between microcanonical equilibration and decoherence has been argued to follow from the *eigenstate thermalization hypothesis*, which conjectures that measurements in an energy eigenstate are equal to those made in the microcanonical ensemble at the same energy. In particular, this is expected to happen in systems that are classically chaotic: for such systems, *Berry's conjecture* states that the eigenfunctions of high-lying energy eigenstates are random superpositions of plane waves, which leads to predictions in agreement with microcanonical equilibration.

77. Combinatorics of Multi-particle Production The goal of this problem is to study the combinatorial aspects of multi-particle production, which become relevant in the strong-field regime since many particles may be produced. Except for the final question, this discussion is very general and independent of the details of a particular quantum field theory. Let us assume that particles are produced in *clusters*, i.e., sets of correlated particles (in terms of Feynman diagrams, a cluster is a set of particles coming from the same connected sub-diagram).

77.a Denote by $P(n_1, n_2, \ldots)$ the probability of having n_1 one-particle clusters, n_2 two-particle clusters, etc. Explain why $P(n_1, n_2, \ldots)$ factorizes into independent probabilities for n_1, n_2, etc. Explain why these independent probabilities are Poissonian, i.e.,

$$P(n_1, n_2, \ldots) = \prod_{r=1}^{\infty} \frac{\overline{n_r}^{n_r} e^{-\overline{n_r}}}{n_r!}.$$

77.b Let $N \equiv n_1 + n_2 + \cdots$ be the total number of clusters. What is the probability of N?

77.c Now, let $\mathcal{P}_n$ be the probability of producing n particles. Determine $\mathcal{P}_n$ from the $\overline{n_r}$.

77.d Define $\mathcal{F}(z) \equiv \sum_{n \geq 0} \mathcal{P}_n z^n$, the generating function of this distribution of probabilities. Calculate $\mathcal{F}(z)$ in terms of the $\overline{n_r}$. If $\mathcal{F}(z)$ is given, how can $\mathcal{P}_n$ and $\overline{n_r}$ be determined?

77.e In a theory of scalar fields coupled to an external source, what is the diagrammatic interpretation of the $\overline{n_r}$?

77.a By the definition of a cluster, there cannot be any correlation between clusters since they are disconnected Feynman graphs. It is this absence of correlation that implies the factorization of $P(n_1, n_2, \ldots)$ into separate functions of n_1, n_2, etc. For the same reason, if several clusters of the same size are produced, they are also uncorrelated, which implies that the distribution of the clusters of a given size is Poissonian. Therefore, the distribution $P(n_1, n_2, \ldots)$ depends only on the mean values $\overline{n_1}, \overline{n_2}, \ldots$ and is a product of Poisson distributions:

$$P(n_1, n_2, \ldots) = \prod_{r=1}^{\infty} \frac{\overline{n_r}^{n_r} e^{-\overline{n_r}}}{n_r!}.$$

77.b If $N \equiv n_1 + n_2 + \cdots$ and $P(N)$ is the probability of having N clusters in total, we can write

$$P(N) = \sum_{n_1, n_2, \cdots \geq 0} \delta_{N, n_1 + n_2 + \cdots} \; P(n_1, n_2, \ldots).$$

Although the distribution $P(n_1, n_2, \ldots)$ is factorized, the Kronecker symbol entangles the various n_r. They can be disentangled by the following representation of the delta:

$$\delta_{N, n_1 + n_2 + \cdots} = \oint_{|z|=1} \frac{dz}{2i\pi z} \, z^{n_1 + n_2 + \cdots - N},$$

which leads to

$$
\begin{aligned}
P(N) &= \sum_{n_1,n_2,\ldots\geq 0} \oint_{|z|=1} \frac{dz}{2i\pi z}\, z^{n_1+n_2+\cdots-N}\, P(n_1,n_2,\ldots) \\
&= \oint_{|z|=1} \frac{dz}{2i\pi z}\, z^{-N} \prod_{r=1}^{\infty} \Big(\sum_{n_r\geq 0} z^{n_r} \frac{\overline{n_r}^{\,n_r} e^{-\overline{n_r}}}{n_r!} \Big) \\
&= e^{-\overline{N}} \oint_{|z|=1} \frac{dz}{2i\pi z}\, z^{-N} e^{z\overline{N}} \qquad \Big(\text{with } \overline{N} \equiv \sum_{r\geq 1} \overline{n_r}\Big) \\
&= \frac{\overline{N}^{N} e^{-\overline{N}}}{N!}.
\end{aligned}
$$

In other words, N is Poisson distributed, with a mean value $\overline{N}$ equal to the sum of the mean values of the n_r. (This is a standard result: the sum of Poisson distributed random variables is another Poissonian random variable, with a mean equal to the sum of the means.)

77.c Consider now the probability $\mathcal{P}_n$ of producing n particles. In terms of the distribution of clusters, this can be obtained as

$$
\begin{aligned}
\mathcal{P}_n &= \sum_{\substack{n_1,n_2,\cdots\geq 0\\ n_1+2n_2+\cdots=n}} P(n_1,n_2,\ldots) = \oint_{|z|=1} \frac{dz}{2i\pi z}\, z^{-n} \prod_{r=1}^{\infty} \sum_{n_r\geq 0} z^{r\,n_r} \frac{\overline{n_r}^{\,n_r} e^{-\overline{n_r}}}{n_r!} \\
&= e^{-\overline{N}} \oint_{|z|=1} \frac{dz}{2i\pi z}\, z^{-n} \exp\Big(\sum_{r\geq 1} \overline{n_r} z^r\Big) = e^{-\overline{N}} \sum_{p=0}^{n} \frac{1}{p!} \sum_{r_1+\cdots+r_p=n} \overline{n_{r_1}} \cdots \overline{n_{r_p}}. \qquad (5.27)
\end{aligned}
$$

In this formula, p is the number of clusters (there can be at most n clusters; $p = 0$ is only allowed for $n = 0$). The second sum is over all the ways of dividing n particles into p clusters of sizes $r_1, \ldots, r_p$. Note that the r_i do not have to be distinct.

Let us make a comment on $\mathcal{P}_0$, the probability of not producing any particle. In quantum field theory, this would be the probability of vacuum-to-vacuum transition, $\left|\left\langle 0_{\rm out} \middle| 0_{\rm in}\right\rangle\right|^2$. From (5.27), we see that $\mathcal{P}_0 = \exp(-\overline{N})$, where $\overline{N}$ is the mean number of clusters. In the special case of a Poisson distribution, where particles are all produced uncorrelated, the mean number of clusters is also the mean number of particles. But in the general case, the vacuum survival probability is less suppressed than what one would naively infer from the mean particle multiplicity.

77.d The generating function of this distribution is

$$
\begin{aligned}
\mathcal{F}(z) &\equiv \sum_{n=0}^{\infty} \mathcal{P}_n z^n = \sum_{n=0}^{\infty} z^n e^{-\overline{N}} \sum_{p=0}^{n} \frac{1}{p!} \sum_{r_1+\cdots+r_p=n} \overline{n_{r_1}} \cdots \overline{n_{r_p}} \\
&= e^{-\overline{N}} \sum_{p\geq 0} \frac{1}{p!} \Big(\sum_r \overline{n_r} z^r \Big)^p = e^{-\overline{N}} \exp\Big(\sum_r \overline{n_r} z^r \Big).
\end{aligned}
$$

An alternative derivation of this generating function (which may be viewed as a justification for the last step in obtaining (5.27)) consists in starting from the penultimate expression for $\mathcal{P}_n$

in (5.27):

$$\mathcal{F}(z) = \sum_{n\geq 0} z^n e^{-\overline{N}} \oint_{|y|=1} \frac{dy}{2i\pi y}\, y^{-n} \exp\Big(\sum_{r\geq 1} \overline{n_r} y^r\Big)$$
$$= e^{-\overline{N}} \oint_{|y|=1} \frac{dy}{2i\pi}\, \frac{1}{y-z} \exp\Big(\sum_{r\geq 1} \overline{n_r} y^r\Big) = e^{-\overline{N}} \exp\Big(\sum_{r\geq 1} \overline{n_r} z^r\Big).$$

An interesting question is the inverse problem: given a function $\mathcal{F}(z)$ (obtained via some approximate calculation, or modeling), how can we obtain the probability distribution $\mathcal{P}_n$? A naive approach is to note that the $\mathcal{P}_n$ are the Taylor coefficients of the generating function, i.e.,

$$\mathcal{P}_n = \frac{1}{n!} \frac{d^n \mathcal{F}}{dz^n}\bigg|_{z=0}.$$

Although this formula is certainly correct, it is not very practical beyond very small values of n unless $\mathcal{F}(z)$ is a function whose derivatives are easily evaluated analytically (numerically evaluating successive derivatives by nested finite differences is very unstable). If $\mathcal{F}(z)$ is given analytically but its derivatives must be evaluated numerically, a much better alternative, stable even for very large values of n, is to extract $\mathcal{P}_n$ as follows:

$$\mathcal{P}_n = \oint_{|z|=1} \frac{dz}{2i\pi z}\, z^{-n}\, \mathcal{F}(z) = \int_0^{2\pi} \frac{d\varphi}{2\pi}\, e^{-in\varphi}\, \mathcal{F}(e^{i\varphi}),$$

i.e., via the Fourier coefficients of $\mathcal{F}(z)$ evaluated on the unit circle (these can be evaluated very efficiently via discrete Fourier transforms). The average number of clusters of size r can be obtained as

$$\overline{n_r} = \oint_{|z|=1} \frac{dz}{2i\pi z}\, z^{-r} \ln \mathcal{F}(z) = \int_0^{2\pi} \frac{d\varphi}{2\pi}\, e^{-ir\varphi} \ln \mathcal{F}(e^{i\varphi}).$$

(We see here that taking the logarithm of the generating function gives access to information about the connected structures in the multi-particle distribution – very much like the logarithm of $Z[j]$ in QFT generates the connected graphs.) This may be used to uncover some aspects of the underlying dynamics from observation of the multiplicity probability distribution $\mathcal{P}_n$:

$$\mathcal{P}_n \text{ (measured)} \quad \longrightarrow \quad \mathcal{F}(z) \quad \longrightarrow \quad \ln \mathcal{F}(z) \quad \longrightarrow \quad \overline{n_r}.$$

77.e So far, the discussion in this problem has been completely independent of the details of any particular model. The relations we have derived express in a very general way the combinatorics of multi-particle production, and remain valid even where perturbation theory would not apply. In order to make contact with quantum field theory, let us consider the example of a scalar field theory (in the diagrams below, we assume cubic couplings, but this is not important) coupled to an external source j. First, note that the probability of not producing any particles is $\mathcal{P}_0 = e^{-\overline{N}}$. In quantum field theory, this probability is the square of the vacuum-to-vacuum transition amplitude,

$$\mathcal{P}_0 = \left|\langle 0_{\text{out}} | 0_{\text{in}} \rangle_j\right|^2.$$

(The subscript j is a reminder of the fact this amplitude should be evaluated in the presence of the external source j.) This transition amplitude is the sum of the vacuum graphs of the theory. This sum can be conveniently written as the exponential of the sum of *connected vacuum graphs*:

$$\langle 0_{\text{out}} | 0_{\text{in}} \rangle_j = e^{iV[j]},$$

with

$iV[j] =$ 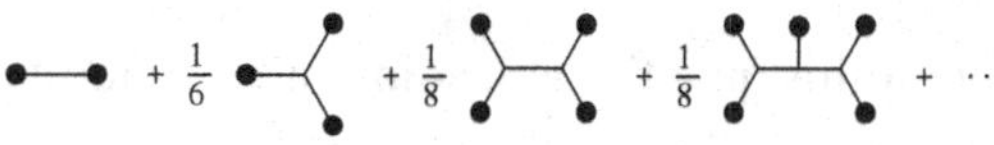 $+ \cdots.$

(The black dots denote the external sources j. We have represented only a few tree-level graphs, but of course this is an infinite series involving also loop corrections to all orders.) The squared modulus of this transition amplitude is then given by

$$\left|\langle 0_{\text{out}} | 0_{\text{in}} \rangle_j\right|^2 = e^{-2\,\text{Im}\,V[j]}.$$

Using Cutkosky's cutting rules, we can write the imaginary part inside the exponential as a sum over the cut connected vacuum graphs:

$2\,\text{Im}\,V[j] =$ 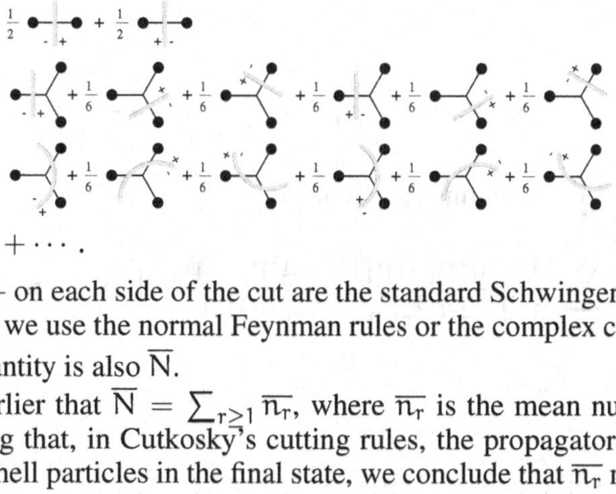

$+ \cdots.$

The labels + and − on each side of the cut are the standard Schwinger–Keldysh notation for indicating whether we use the normal Feynman rules or the complex conjugated ones. Since $\mathcal{P}_0 = e^{-\overline{N}}$, this quantity is also $\overline{N}$.

We learned earlier that $\overline{N} = \sum_{r\geq 1} \overline{n_r}$, where $\overline{n_r}$ is the mean number of clusters of r particles. Recalling that, in Cutkosky's cutting rules, the propagators traversed by the cut correspond to on-shell particles in the final state, we conclude that $\overline{n_r}$ must be the sum of cut connected vacuum graphs where exactly r propagators are cut. For instance:

$\overline{n_1} =$ 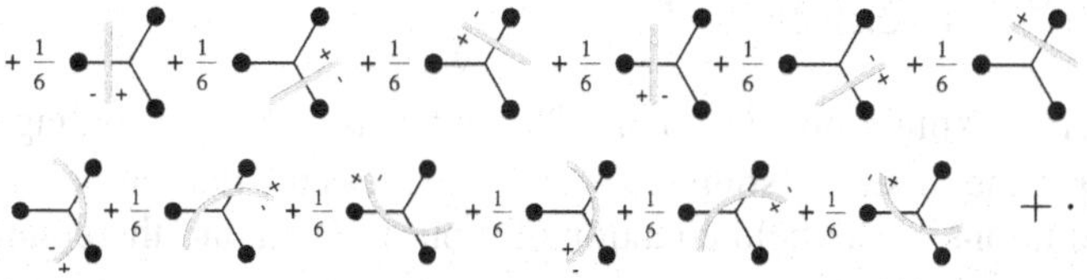 $+ \cdots,$

$\overline{n_2} =$ 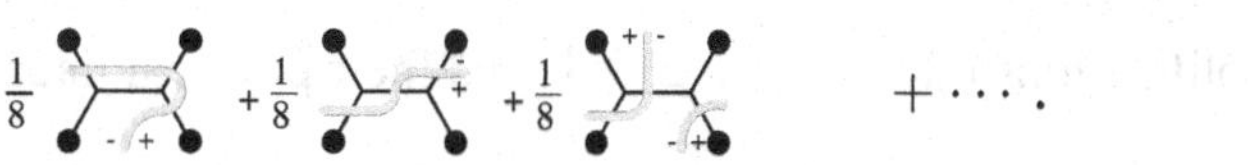 $+ \cdots,$

$\overline{n_3} =$ 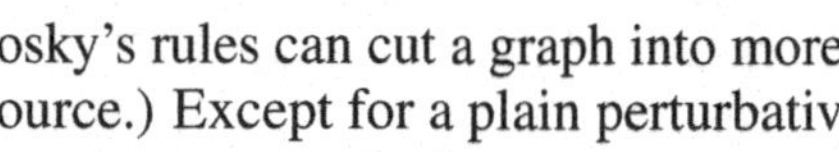 $+ \cdots.$

(Note that Cutkosky's rules can cut a graph into more than two parts when the field is coupled to an external source.) Except for a plain perturbative expansion (justified when the external source is weak, but not in the strong source regime), these quantities cannot be accessed easily in the underlying quantum field theory, since even the tree-level approximation requires that we sum an infinite set of diagrams. Physically, the main source of difficulty is that these quantities are *exclusive*, i.e., they are defined by selecting a subset of final states that fulfill a certain condition (here, having a definite number r of particles) and vetoing all the others. It is much easier to evaluate *inclusive quantities*, defined as weighted sums over all the final states. The simplest of them is the first moment $\sum_{r\geq 1} r\,\overline{n}_r$, which is also the mean number of produced particles:

$$\sum_{r\geq 1} r\,\overline{n}_r = \sum_{n\geq 0} n\,\mathcal{P}_n.$$

(See Problem 78 for a derivation of this quantity at leading order in terms of solutions of the classical field equations of motion.)

---

78. Generating Functional for Particle Production In this problem, a continuation of Problem 77, we consider the theory of a scalar field coupled to an external source j. Denote by $\left\langle 0_{\mathrm{out}} \middle| 0_{\mathrm{in}} \right\rangle_j \equiv \exp(iV[j])$ the vacuum-to-vacuum transition amplitude in the presence of the source j.

78.a Explain why $\left\langle 0_{\mathrm{out}} \middle| T\,\phi(x_1)\cdots\phi(x_n) \middle| 0_{\mathrm{in}} \right\rangle = \dfrac{\delta}{i\delta\eta(x_1)}\cdots\dfrac{\delta}{i\delta\eta(x_n)}\, e^{iV[j+\eta]}\Big|_{\eta=0}$.

78.b Define the following generating functional:

$$\mathcal{F}[z(\mathbf{p})] \equiv \sum_{n\geq 0} \frac{1}{n!} \int \frac{d^3\mathbf{p}_1\, z(\mathbf{p}_1)}{(2\pi)^3 2E_{\mathbf{p}_1}} \cdots \frac{d^3\mathbf{p}_n\, z(\mathbf{p}_n)}{(2\pi)^3 2E_{\mathbf{p}_n}} \left| \left\langle \mathbf{p}_1 \cdots \mathbf{p}_{n\,\mathrm{out}} \middle| 0_{\mathrm{in}} \right\rangle_j \right|^2.$$

Show that

$$\mathcal{F}[z(\mathbf{p})] = \exp\left(\int_{x,y} \overline{G}^0_{+-}(x,y)\,\square_x \square_y \frac{\delta^2}{\delta\eta_+(x)\,\delta\eta_-(y)} \right) e^{iV[j+\eta_+]}\, e^{-iV^*[j+\eta_-]}\Big|_{\eta_\pm=0},$$

with $\overline{G}^0_{+-}(x,y) \equiv \displaystyle\int \frac{d^3\mathbf{p}}{(2\pi)^3 2E_{\mathbf{p}}}\, z(\mathbf{p})\, e^{ip\cdot(x-y)}$.

78.c Using Eq. (5.16), explain how to obtain $\mathcal{F}[z]$ in terms of Feynman diagrams. At leading order (in the regime of strong source, $j \sim g^{-1}$), express the derivative $\delta\mathcal{F}/\delta z(\mathbf{p})$ in terms of classical solutions of the field equation of motion. What are the boundary conditions?

78.d Discuss the simplifications that occur at $z \equiv 1$, and their physical interpretation.

78.a Recall that, in a theory with an interaction Lagrangian $\mathcal{L}_I$, the vacuum-to-vacuum transition amplitude can be expressed as

$$\langle 0_{\mathrm{ou}}|0_{\mathrm{in}}\rangle = \langle 0_{\mathrm{in}}|U(+\infty,-\infty)|0_{\mathrm{in}}\rangle = \langle 0_{\mathrm{in}}|\mathrm{T}\,\exp i\int_{-\infty}^{+\infty} d^4x\;\mathcal{L}_I(\phi_{\mathrm{in}}(x))|0_{\mathrm{in}}\rangle. \quad (5.28)$$

On the other hand, the generating functional of time-ordered correlation functions of the field is given by

$$Z[\eta] = \langle 0_{\mathrm{in}}|\mathrm{T}\,\exp i\int_{-\infty}^{+\infty} d^4x\;\big((\mathcal{L}_I(\phi_{\mathrm{in}}(x)) + \eta(x)\phi_{\mathrm{in}}(x)\big)|0_{\mathrm{in}}\rangle. \quad (5.29)$$

In the special case where the interaction Lagrangian is the sum of a self-interaction potential $U(\phi)$ and a coupling to an external source $j\phi$,

$$\mathcal{L}_I(\phi) = -U(\phi) + j\phi,$$

we see that (5.29) is in fact a vacuum-to-vacuum transition amplitude (5.28) in the presence of a source $j+\eta$, i.e.,

$$Z[\eta] = \langle 0_{\mathrm{ou}}|0_{\mathrm{in}}\rangle_{j+\eta} \equiv e^{iV[j+\eta]}.$$

Therefore, we indeed have

$$\langle 0_{\mathrm{out}}|\mathrm{T}\,\phi(x_1)\cdots\phi(x_n)|0_{\mathrm{in}}\rangle = \frac{\delta}{i\delta\eta(x_1)}\cdots\frac{\delta}{i\delta\eta(x_n)}\,e^{iV[j+\eta]}\Big|_{\eta=0}.$$

78.b Then, the LSZ reduction formula gives

$$\langle p_1\cdots p_{n\,\mathrm{out}}|0_{\mathrm{in}}\rangle = i^n\int\prod_{i=1}^{n}\left(d^4x_i\;e^{p_i\cdot x_i}\square_{x_i}\right)\langle 0_{\mathrm{out}}|\mathrm{T}\,\phi(x_1)\cdots\phi(x_n)|0_{\mathrm{in}}\rangle.$$

(For notational simplicity, we assume a massless field, but there is no conceptual difficulty in adding a mass.) Next, we square this expression (be careful to use distinct dummy integration variables in the complex conjugate), and we use the above formula for the time-ordered correlator. This leads to

$$\left|\langle p_1\cdots p_{n\,\mathrm{out}}|0_{\mathrm{in}}\rangle\right|^2 = C_{p_1}\cdots C_{p_n}\,e^{iV[j+\eta_+]}e^{-iV^*[j+\eta_-]}\Big|_{\eta_\pm=0},$$

where we have defined

$$C_p \equiv \int d^4x d^4y\;e^{ip\cdot(x-y)}\,\square_x\square_y\,\frac{\delta^2}{\delta\eta_+(x)\delta\eta_-(y)}.$$

The only notable aspect of this formula is the necessity of having distinct sources η_+ and η_- in the amplitude and its complex conjugate, so that their actions do not get mixed up. From this expression, we easily obtain the generating functional $\mathcal{F}[z(\mathbf{p})]$:

$$\mathcal{F}[z(\mathbf{p})] = \exp\Big(\int \frac{d^3\mathbf{p}\, z(\mathbf{p})}{(2\pi)^3 2E_\mathbf{p}}\, \mathbf{C}_\mathbf{p}\Big)\, e^{iV[j+\eta_+]} e^{-iV^*[j+\eta_-]}\Big|_{\eta_\pm=0},$$

which is equivalent to the proposed formula,

$$\mathcal{F}[z(\mathbf{p})] = \exp\Big(\int d^4x d^4y\, \overline{G}^0_{+-}(x,y)\square_x\square_y \frac{\delta^2}{\delta\eta_+(x)\delta\eta_-(y)}\Big)\, e^{iV[j+\eta_+]} e^{-iV^*[j+\eta_-]}\Big|_{\eta_\pm=0},$$

with

$$\overline{G}^0_{+-}(x,y) \equiv \int \frac{d^3\mathbf{p}}{(2\pi)^3 2E_\mathbf{p}}\, z(\mathbf{p})\, e^{ip\cdot(x-y)}.$$

78.c By comparing this expression for $\mathcal{F}[z]$ with Eq. (5.16), we see that – before the sources $\eta_\pm$ are set to zero – the right-hand side is the generating functional for path-ordered correlation functions in the Schwinger–Keldysh formalism, with a modified off-diagonal propagator $G^0_{+-} \to \overline{G}^0_{+-}$. This modification amounts to multiplying all the off-diagonal ($+-$ and $-+$) propagators by $z(\mathbf{p})$ in momentum space. Setting $\eta_\pm = 0$ leaves only the vacuum diagrams constructed with these modified Schwinger–Keldysh rules (and the logarithm of $\mathcal{F}[z]$ is the sum of all these vacuum graphs that are connected). This modification of the Schwinger–Keldysh rules for obtaining the generating functional $\mathcal{F}[z]$ is quite natural: indeed, the off-diagonal propagators $G^0_{-+}(p)$ and $G^0_{+-}(p)$ describe the on-shell particles in the final state. Weighting these propagators by $z(\mathbf{p})$ is precisely what we need in order to keep track of the produced particles and their momenta.

Let us mention a subtlety here: the sum of the vacuum graphs is equal to unity in the Schwinger–Keldysh formalism, as a consequence of unitarity. However, this property applies only to the unmodified rules, for which the free propagators obey

$$G^0_{++} + G^0_{--} = G^0_{+-} + G^0_{-+}.$$

This identity is no longer true after the modification $G^0_{+-} \to \overline{G}^0_{+-}$, implying that $\mathcal{F}[z]$ is non-trivial. Note however that this property implies that $\mathcal{F}[z=1] = 1$, which is consistent with unitarity since $\mathcal{F}[1]$ is the sum of all the probabilities $\mathcal{P}_n$ of producing n particles:

$$\mathcal{F}[1] = \sum_{n\geq 0} \mathcal{P}_n = 1.$$

For the subsequent discussion, it is convenient to denote

$$e^{W_j[\eta_+,\eta_-]} \equiv \exp\Big(\int d^4x d^4y\, \overline{G}^0_{+-}(x,y)\square_x\square_y \frac{\delta^2}{\delta\eta_+(x)\delta\eta_-(y)}\Big)\, e^{iV[j+\eta_+]} e^{-iV^*[j+\eta_-]}.$$

As we have discussed above, this is the generating functional for the modified Schwinger–Keldysh formalism (in the presence of an external source j), and $W_j[\eta_+,\eta_-]$ is the generating functional for the connected graphs. Since $z(\mathbf{p})$ appears only in the propagator $\overline{G}^0_{+-}$, it is straightforward to differentiate this expression with respect to $z(\mathbf{p})$:

$$\frac{\delta}{\delta z(\mathbf{p})} e^{W_j[\eta_+,\eta_-]} = \frac{1}{(2\pi)^3 2E_\mathbf{p}} \int d^4x d^4y \, e^{i\mathbf{p}\cdot(x-y)} \square_x \square_y \frac{\delta^2}{\delta\eta_+(x)\delta\eta_-(y)} e^{W_j[\eta_+,\eta_-]}$$
$$= \frac{e^{W_j[\eta_+,\eta_-]}}{(2\pi)^3 2E_\mathbf{p}} \int d^4x d^4y \, e^{i\mathbf{p}\cdot(x-y)} \square_x \square_y \Big[\Phi_+(x)\Phi_-(y) + \mathcal{G}_{+-}(x,y)\Big],$$

where we have defined

$$\Phi_+(x) \equiv \frac{\delta W_j[\eta_+,\eta_-]}{\delta\eta_+(x)}, \quad \Phi_-(y) \equiv \frac{\delta W_j[\eta_+,\eta_-]}{\delta\eta_-(y)},$$
$$\mathcal{G}_{+-}(x,y) \equiv \frac{\delta^2 W_j[\eta_+,\eta_-]}{\delta\eta_+(x)\delta\eta_-(y)}.$$

In other words, $\Phi_\pm$ are one-point connected functions and $\mathcal{G}_{+-}$ is a two-point connected function in the modified Schwinger–Keldysh formalism. Note that this expression is valid to all orders.

To restrict this result to the lowest order in the regime of strong sources (i.e., $gj \sim 1$), note that

$$\Phi_\pm \sim \mathcal{O}(g^{-1}), \quad \mathcal{G}_{+-} \sim \mathcal{O}(g^0).$$

Therefore, the leading order is obtained solely from the term in $\Phi_+(x)\Phi_-(y)$, with both factors evaluated at tree level. At tree level, the fields $\Phi_\pm$ obey the classical field equation of motion,

$$\square\Phi_\pm + U'(\Phi_\pm) = j,$$

with intricate z-dependent boundary conditions at $x^0 = \pm\infty$ for the coefficients of the Fourier decomposition of $\Phi_\pm$:

$$\Phi_\epsilon(x) \equiv \int \frac{d^3\mathbf{p}}{(2\pi)^3 2E_\mathbf{p}} \Big[f^{(+)}_\epsilon(x^0,\mathbf{p})\, e^{-i\mathbf{p}\cdot x} + f^{(-)}_\epsilon(x^0,\mathbf{p})\, e^{+i\mathbf{p}\cdot x}\Big],$$
$$f^{(+)}_+(-\infty,\mathbf{p}) = f^{(-)}_-(-\infty,\mathbf{p}) = 0,$$
$$f^{(-)}_+(+\infty,\mathbf{p}) = z(\mathbf{p})\, f^{(-)}_-(+\infty,\mathbf{p}),$$
$$f^{(+)}_-(+\infty,\mathbf{p}) = z(\mathbf{p})\, f^{(+)}_+(+\infty,\mathbf{p}).$$

Because they impose constraints at both ends of the time evolution and the field evolution in between is non-linear, these boundary conditions lead to a very hard problem in general.

78.d The above results simplify considerably in the very special case of $z \equiv 1$. In this case, the two classical fields become degenerate,

$$\Phi_+(x) = \Phi_-(x),$$

and the boundary conditions become very simple retarded boundary conditions,

$$\lim_{x^0\to-\infty} \Phi_\pm(x) = 0.$$

Note that $z = 1$ is a very special argument for the generating functional, since $\mathcal{F}[1] = 1$ and

$$\left.\frac{\delta\mathcal{F}[z]}{\delta z(\mathbf{p})}\right|_{z=1} = \underbrace{\frac{1}{(2\pi)^3 2E_{\mathbf{p}}}\sum_{n\geq 0}\frac{1}{n!}\int\frac{d^3\mathbf{p}_1}{(2\pi)^3 2E_{\mathbf{p}_1}}\cdots\frac{d^3\mathbf{p}_n}{(2\pi)^3 2E_{\mathbf{p}_n}}\left|\left\langle\mathbf{p}\mathbf{p}_1\cdots\mathbf{p}_{n\,\mathrm{out}}\middle|0_{\mathrm{in}}\right\rangle\right|^2}_{\equiv\frac{dN_1}{d^3\mathbf{p}}}.$$

This quantity, defined as a sum over all the final states, weighted by the number of particles of the state, and where all but one of the momenta are integrated out, is called the *inclusive single particle spectrum* (experimentally, it would be obtained by making a histogram of the distribution of produced particles according to their momenta). The integral over momentum of this quantity is nothing but the mean number of produced particles,

$$\int d^3\mathbf{p}\,\frac{dN_1}{d^3\mathbf{p}} = \sum_{n\geq 1} n\,\mathcal{P}_n.$$

Setting $z = 0$ instead of $z = 1$ in the derivative of $\mathcal{F}[z]$ would instead produce the probability of producing *exactly one* particle of momentum $\mathbf{p}$. Although they are both expressible in terms of classical fields at leading order, the inclusive spectrum is much easier to calculate, thanks to the retarded boundary conditions. The connection between the inclusive nature of the observable and the fact that the boundary conditions are retarded is in fact quite intuitive. Since inclusive observables do not put any restriction on the final state, there is also no constraint on the underlying fields at $x^0 = +\infty$. In contrast, in the calculation of exclusive observables (such as the probability of producing exactly one particle, which vetoes all other final states), the underlying fields must be constrained somehow at $x^0 = +\infty$.

79. Correlations in the Schwinger Mechanism The Schwinger mechanism is a vacuum instability in the presence of a homogeneous electric field, by which the field decays into pairs of charged particles. The goal of this problem is to study the distribution of produced particles and their correlations, in the simpler setting of scalar QED. Consider the Lagrangian

$$\mathcal{L} \equiv (D_\mu\phi)(D^\mu\phi)^* - m^2\phi\phi^*,$$

without self-interactions for the scalar field (this simplifying assumption avoids interactions among the produced particles). In this problem, we treat the photon field as a fixed background $A_\mu(x)$, i.e., we disregard radiation by the charged scalars.

79.a Consider the functional

$$\mathcal{F}[z,\bar{z}] \equiv \sum_{m,n=0}^{\infty}\frac{1}{m!n!}\prod_{\substack{1\leq i\leq m\\1\leq j\leq n}}\int_{\mathbf{p}_i,\mathbf{q}_j} z(\mathbf{p}_i)\bar{z}(\mathbf{q}_j)\,\Big|\big\langle\underbrace{\mathbf{p}_1\cdots\mathbf{p}_m}_{\text{scalars}}\underbrace{\mathbf{q}_1\cdots\mathbf{q}_n}_{\text{antiscalars}}{}_{\mathrm{out}}\big|0_{\mathrm{in}}\big\rangle\Big|^2.$$

(We use the shorthand $\int_{\mathbf{p}} \equiv \int\frac{d^3\mathbf{p}}{(2\pi)^3 2E_{\mathbf{p}}}$ for the invariant on-shell measure.) Check that its derivatives at $z = \bar{z} = 1$ give the (multi-)particle spectra. What are the Feynman graphs contributing to $\mathcal{F}[z,\bar{z}]$?

79.b Show that

$$\ln \mathcal{F}[z,\bar{z}] = \text{const} - \text{tr}\,\ln\Big(1 - \mathcal{T}_+\,(zG^0_{+-})\,\mathcal{T}_-\,(\bar{z}G^0_{-+})\Big),$$

where $\mathcal{T}_+$ is the (amputated) Feynman propagator in the external field, and $\mathcal{T}_-$ its complex conjugate.

79.c Show that

$$\ln \mathcal{F}[z,\bar{z}] = \text{const} - \text{tr}\,\ln\Big(1 - \mathcal{T}^*_R\, zG^0_{+-}\,\mathcal{T}_R G^0_{-+}\big(1+\mathcal{T}_R G^0_{-+}\big)^{-1}\bar{z}\big(1+\mathcal{T}^*_R G^0_{-+}\big)^{-1}\Big),$$

where $\mathcal{T}_R$ is the (amputated) retarded propagator in the external field.

79.d Show that

$$\big(1+\mathcal{T}_R G^0_{-+}\big)^{-1}\big(1+\mathcal{T}^*_R G^0_{-+}\big)^{-1} = \big(1+\mathcal{T}^*_R G^0_{+-}\mathcal{T}_R G^0_{-+}\big)^{-1},$$

and use this identity to obtain

$$\ln \mathcal{F}[z,1] = \text{const} - \text{tr}\,\ln\Big(1 + (1-z)G^0_{+-}\mathcal{T}_R G^0_{-+}\mathcal{T}^*_R\Big).$$

79.e Let us specialize now to the case of a spatially homogeneous background field. Simplify the generating functional to obtain

$$\ln \mathcal{F}[z,\bar{z}] = \text{const} - V\int \frac{d^3p}{(2\pi)^3}\,\ln\Big(1 + \big(1 - z(\mathbf{p})\bar{z}(-\mathbf{p})\big)\,f_{\mathbf{p}}\Big),$$

and use it to study some properties of the distribution of produced particles.

79.f Discuss the relationship between the vacuum survival probability and the mean number of produced scalar particles.

79.a In this setup, even if the initial state is the empty state, scalars and antiscalars may be produced by the external electromagnetic field that acts as a source. In the limit of a static external field, this is the scalar QED analogue of the Schwinger mechanism (by which a homogeneous electric field spontaneously decays into electron–positron pairs). In order to keep track of the produced particles, we define a generating functional that consists of a weighted sum over all the possible final states, with the vacuum as initial state:

$$\mathcal{F}[z,\bar{z}] \equiv \sum_{m,n=0}^{\infty} \frac{1}{m!n!} \prod_{\substack{1\le i\le m\\ 1\le j\le n}} \int_{\mathbf{p}_i,\mathbf{q}_j} z(\mathbf{p}_i)\bar{z}(\mathbf{q}_j)\,\Big|\big\langle \underbrace{\mathbf{p}_1\cdots\mathbf{p}_m}_{\text{scalars}}\,\underbrace{\mathbf{q}_1\cdots\mathbf{q}_n}_{\text{antiscalars}}{}_{\text{out}}\big|0_{\text{in}}\big\rangle\Big|^2,$$

where $z(\mathbf{p})$ and $\bar{z}(\mathbf{q})$ are two independent functions that separately count the scalars and the antiscalars. (We have used the shorthand notation $\int_{\mathbf{p}} \equiv \int \frac{d^3p}{(2\pi)^3 2E_{\mathbf{p}}}$ for the invariant on-shell phase-space integral.) Note that in this definition we do not include final states with photons, since we are not interested in photon production in this problem (processes with an extra photon integrated out would be of higher order). Unitarity trivially implies $\mathcal{F}[1,1] = 1$.

By differentiating with respect to z or $\bar{z}$, we obtain the single particle and anti-particle spectra,

$$\frac{dN_1^+}{d^3p} = \left.\frac{\delta\mathcal{F}[z,\bar{z}]}{\delta z(p)}\right|_{z=\bar{z}=1}, \qquad \frac{dN_1^-}{d^3q} = \left.\frac{\delta\mathcal{F}[z,\bar{z}]}{\delta \bar{z}(q)}\right|_{z=\bar{z}=1},$$

and second derivatives give the two particles and two anti-particles spectra,

$$\frac{dN_2^{++}}{d^3p_1 d^3p_2} = \left.\frac{\delta^2\mathcal{F}[z,\bar{z}]}{\delta z(p_1)\delta z(p_2))}\right|_{z=\bar{z}=1}, \qquad \frac{dN_2^{--}}{d^3q_1 d^3q_2} = \left.\frac{\delta^2\mathcal{F}[z,\bar{z}]}{\delta \bar{z}(q_1)\delta \bar{z}(q_2))}\right|_{z=\bar{z}=1},$$

as well as a mixed spectrum

$$\frac{dN_2^{+-}}{d^3p d^3q} = \left.\frac{\delta^2\mathcal{F}[z,\bar{z}]}{\delta z(p)\delta \bar{z}(q))}\right|_{z=\bar{z}=1}.$$

Note that when taking successive derivatives with respect to the same variable (z or $\bar{z}$), one obtains multi-particle spectra in which all the particles are necessarily distinct. This implies that, when integrating these spectra, one obtains the so-called *factorial moments*. For instance,

$$\int d^3p_1 d^3p_2 \, \frac{dN_2^{++}}{d^3p_1 d^3p_2} = \overline{N^+(N^+-1)},$$

and more generally

$$\int d^3p_1 \cdots d^3p_n \, \frac{dN_n^{+\cdots+}}{d^3p_1 \cdots d^3p_n} = \overline{N^+(N^+-1)\cdots(N^+-n+1)}.$$

79.b As we have discussed in Problem 78, the generating functional $\mathcal{F}[z,\bar{z}]$ is the sum of the vacuum graphs in a modified Schwinger–Keldysh formalism (in which the off-diagonal propagators are multiplied by z or $\bar{z}$ in momentum space). The logarithm of $\mathcal{F}[z,\bar{z}]$ is therefore the sum of the connected vacuum diagrams. The lowest order for these graphs is to only have one scalar loop embedded in the background electromagnetic field. At this order, we need only perform the infinite sum over the photon field attachments around a scalar loop, with all the possible assignments for the + and − Schwinger–Keldysh vertices. This sum can be organized by having alternating blocks with only + or − vertices:

$\ln \mathcal{F}[z,\bar{z}] = \text{const} +$ 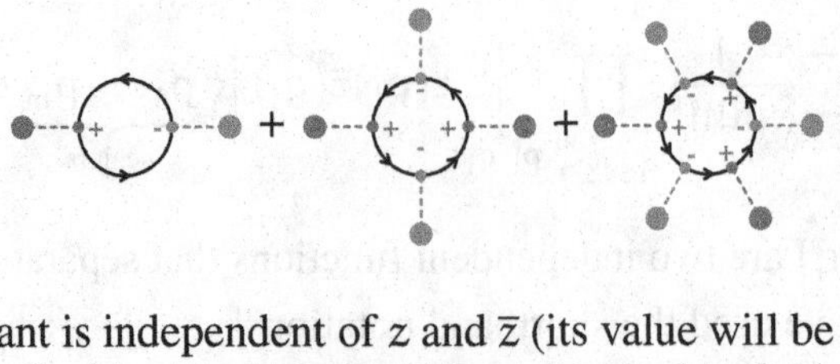 $+\cdots,$ (5.30)

where the unspecified constant is independent of z and $\bar{z}$ (its value will be adjusted to satisfy unitarity, i.e., $\ln\mathcal{F}[1,1] = 0$). In this equation, the objects attached to the loop already re-sum infinite sequences of $+$ or $-$ vertices:

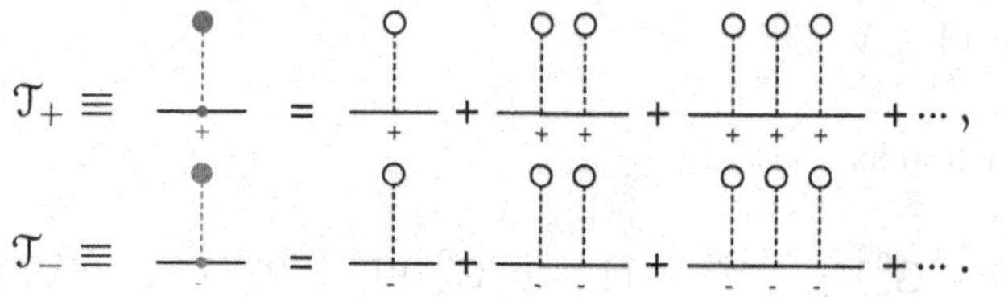

(5.31)

Therefore, these objects do not contain any G^0_{+-} or G^0_{-+} propagators, and are independent of z or $\bar{z}$. The z and $\bar{z}$ dependence is carried by the propagators that appear explicitly in the loops in Eq. (5.30). In the diagrammatic representation used in Eqs. (5.31), the dotted lines are a shorthand for the sum of the two possible interactions that exist in scalar QED:

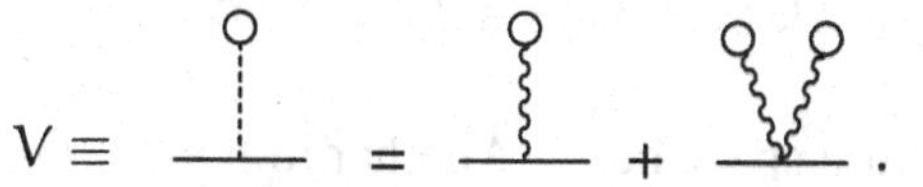

(The wavy lines terminated by a circle denote the external electromagnetic potential.) The sum in Eq. (5.30) can be written in the following compact form:

$$\ln \mathcal{F}[z,\bar{z}] = \text{const} + \sum_{n=1}^{\infty} \frac{1}{n} \,\text{tr}\, \Big(\mathcal{T}_+ (zG^0_{+-})\, \mathcal{T}_- (\bar{z}G^0_{-+})\Big)^n$$
$$= \text{const} - \text{tr} \ln \Big(1 - \mathcal{T}_+ (zG^0_{+-})\, \mathcal{T}_- (\bar{z}G^0_{-+})\Big),$$

where we have made explicit that the factors z and $\bar{z}$ come along with the off-diagonal propagators G^0_{+-} and G^0_{-+} (the functions z and $\bar{z}$ carry the same momentum as the propagator they are attached to). The trace denotes an integration over all the spacetime coordinates of the vertices around the loop. The factor $1/n$ is a symmetry factor, absorbed in the second line in the Taylor expansion of the logarithm.

At this point, we have an expression for $\ln \mathcal{F}[z,\bar{z}]$ that involves infinite series of time-ordered (G^0_{++}) or anti-time-ordered (G^0_{--}) propagators hidden in the objects $\mathcal{T}_\pm$. To be a bit more explicit, $\mathcal{T}_+$ satisfies a Lipmann–Schwinger equation of the form

$$\mathcal{T}_+ = V + VG^0_{++}\mathcal{T}_+ = V + \mathcal{T}_+ G^0_{++} V \Rightarrow \mathcal{T}_+ = \big(1 - VG^0_{++}\big)^{-1} V = V\big(1 - G^0_{++}V\big)^{-1},$$

and $\mathcal{T}_-$ is its complex conjugate,

$$\mathcal{T}_- = \big(1 - V^* G^0_{--}\big)^{-1} V^* = V^* \big(1 - G^0_{--} V^*\big)^{-1}.$$

In other words, $\mathcal{T}_+$ is the Feynman propagator dressed by the external field, and amputated of its external legs. The evaluation of this propagator may be recast into solving the classical equation of motion for ϕ in the presence of an external field, but with non-trivial boundary conditions at $x^0 = \pm\infty$, which makes the problem practically intractable.

79.c In contrast, the same equation of motion would be straightforward to solve (at least numerically) with retarded boundary conditions. It turns out that, at this order of approximation, there is a way to express $\mathcal{T}_\pm$ in terms of an analogous retarded object. Let us define $\mathcal{T}_R$ by the following Lipmann–Schwinger equation:

$$\mathcal{T}_R = V + \mathcal{T}_R G^0_R V = V + VG^0_R \mathcal{T}_R, \quad \mathcal{T}^*_R = V^* + \mathcal{T}^*_R G^{0*}_R V^* = V^* + V^* G^{0*}_R \mathcal{T}^*_R,$$

where G^0_R is the free retarded scalar propagator. These definitions imply

$$\mathcal{T}_R = V(1 - G_R^0 V)^{-1} = (1 - VG_R^0)^{-1}V,$$
$$\mathcal{T}_R^* = V^*(1 - G_R^{0*}V^*)^{-1} = (1 - V^*G_R^{0*})^{-1}V^*,$$

and can also be conveniently rewritten as

$$V = (1 - VG_R^0)\mathcal{T}_R = \mathcal{T}_R(1 - G_R^0 V), \quad V^* = (1 - V^*G_R^{0*})\mathcal{T}_R^* = \mathcal{T}_R^*(1 - G_R^{0*}V^*).$$

Moreover, the retarded propagator and its complex conjugate are related to the Schwinger–Keldysh propagators by

$$G_R^0 = G_{++}^0 - G_{+-}^0 = G_{-+}^0 - G_{--}^0, \quad G_R^{0*} = G_{--}^0 - G_{+-}^0 = G_{-+}^0 - G_{++}^0.$$

(We are considering the propagators in momentum space here.) Another important property, valid in any unitary theory, is $V^* = -V$ (the background field insertions contain a factor $-i$ and the external field is real). Combining these identities, we have

$$\begin{aligned}
\mathcal{T}_+ &= -(1 + V^*G_{++}^0)^{-1}V^* = -(1 + V^*G_{++}^0)^{-1}(1 - V^*G_R^{0*})\mathcal{T}_R^* \\
&= -(1 + (1 - V^*G_R^*)^{-1}V^*G_{-+}^0)^{-1}\mathcal{T}_R^* \\
&= -(1 + \mathcal{T}_R^*G_{-+}^0)^{-1}\mathcal{T}_R^*.
\end{aligned}$$

Likewise, for $\mathcal{T}_-$, we have

$$\mathcal{T}_- = -\mathcal{T}_R(1 + G_{-+}^0\mathcal{T}_R)^{-1}.$$

Therefore, the generating functional reads

$$\ln \mathcal{F}[z, \bar{z}] = \text{const} - \text{tr}\,\ln\Big(1 - \mathcal{T}_R^*\, zG_{+-}^0\, \mathcal{T}_R G_{-+}^0 \underline{(1 + \mathcal{T}_R G_{-+}^0)^{-1}\bar{z}(1 + \mathcal{T}_R^* G_{-+}^0)^{-1}}\Big), \tag{5.32}$$

where we have used the cyclic invariance of the trace to rearrange the expression.

79.d This expression can be simplified further when we set $\bar{z} = 1$, i.e., when we are only interested in the scalars and not in the antiscalars (of course, there exists another equally simple expression for the reverse situation). Using the Lipmann–Schwinger equations for $\mathcal{T}_R$ and $\mathcal{T}_R^*$ and the fact that $V^* = -V$, we have

$$\begin{aligned}
\mathcal{T}_R + \mathcal{T}_R^* &= VG_R^0\mathcal{T}_R - \mathcal{T}_R^*G_R^{0*}V \\
&= (-\mathcal{T}_R^* - \mathcal{T}_R^*G_R^{0*}V)G_R^0\mathcal{T}_R - \mathcal{T}_R^*G_R^{0*}(\mathcal{T}_R - VG_R^0\mathcal{T}_R) \\
&= -\mathcal{T}_R^*(G_R^0 + G_R^{0*})\mathcal{T}_R = -\mathcal{T}_R^*(G_{-+}^0 - G_{+-}^0)\mathcal{T}_R.
\end{aligned}$$

Thanks to this identity, we can write

$$\begin{aligned}(1+\mathcal{T}_R G^0_{-+})^{-1}(1+\mathcal{T}^*_R G^0_{-+})^{-1} &= \Big((1+\mathcal{T}^*_R G^0_{-+})(1+\mathcal{T}_R G^0_{-+})\Big)^{-1} \\ &= \Big(1+(\mathcal{T}_R+\mathcal{T}^*_R)G^0_{-+}+\mathcal{T}^*_R G^0_{-+}\mathcal{T}_R G^0_{-+}\Big)^{-1} \\ &= (1+\mathcal{T}^*_R G^0_{+-}\mathcal{T}_R G^0_{-+})^{-1},\end{aligned}$$

and

$$G^0_{-+}(1+\mathcal{T}_R G^0_{-+})^{-1}(1+\mathcal{T}^*_R G^0_{-+})^{-1} = (1+G^0_{-+}\mathcal{T}^*_R G^0_{+-}\mathcal{T}_R)^{-1} G^0_{-+},$$

so that

$$\begin{aligned}\ln \mathcal{F}[z,1] &= \text{const} - \text{tr}\ln\Big(1 - G^0_{-+}\mathcal{T}^*_R\, z G^0_{+-}\,\mathcal{T}_R(1+G^0_{-+}\mathcal{T}^*_R G^0_{+-}\mathcal{T}_R)^{-1}\Big) \\ &= \text{const}' - \text{tr}\ln\Big(1+(1-z)\underbrace{G^0_{+-}\mathcal{T}_R G^0_{-+}\mathcal{T}^*_R}_{\equiv \mathcal{M}}\Big).\end{aligned}$$

Therefore, all the successive derivatives of this generating functional can be expressed in terms of the object $\mathcal{M}$. The trace is over the momentum circulating in the loop. With all the momentum dependence explicitly written, we have

$$\Big[(1-z)\mathcal{M}\Big]_{pp'} = 2\pi\,\theta(p^0)\delta(p^2-m^2)(1-z(\mathbf{p}))\int\frac{d^3k}{(2\pi)^3 2E_k}\,\mathcal{T}_R(p,-k)\mathcal{T}^*_R(-k,p').$$

(The propagators G^0_{-+} and G^0_{+-} form a cut through the loop. The choice of labeling $-k$ the intermediate momentum is such that both p and k are flowing in the same direction through this cut.) Moreover, the retarded scattering matrix $\mathcal{T}_R$ can be computed by solving the equation of motion in the background field, with retarded boundary conditions. More precisely, it can be expressed in terms of mode functions as follows:

$$\begin{aligned}&\mathcal{T}_R(p,-k) = \lim_{x_0\to+\infty}\int d^3x\; e^{ip\cdot x}\,(\partial_{x_0} - iE_p)\, a_k(x) \\ &(D^2+m^2)\, a_k(x) = 0, \qquad \lim_{x_0\to-\infty} a_k(x) = e^{ik\cdot x}.\end{aligned}$$

In other words, in order to obtain $\mathcal{T}_R(p,-k)$, one should start in the remote past with a negative energy plane wave of momentum $\mathbf{k}$, evolve it over the background field until late times, and project it on a positive energy plane wave of momentum $\mathbf{p}$.

79.e Let us now consider the special case of a spatially homogeneous external field. In this case, there is no 3-momentum flowing from the external field, and therefore the 3-momentum is constant around the scalar loop. This means that the factors z and $\bar{z}$ in Eq. (5.32) carry the same momentum, and more importantly that we can modify to our liking their position in the cyclic ordering of the expression. For instance, it is convenient to write

$$\ln \mathcal{F}[z,\bar{z}] = \text{const} - \text{tr}\,\ln\Big(1 - z\bar{z}\,\mathcal{T}^*_R\,G^0_{+-}\,\mathcal{T}_R\,\underbrace{G^0_{-+}(1+\mathcal{T}_R G^0_{-+})^{-1}(1+\mathcal{T}^*_R G^0_{-+})^{-1}}_{(1+G^0_{-+}\mathcal{T}^*_R G^0_{+-}\mathcal{T}_R)^{-1}G^0_{-+}}\Big)$$

$$= \text{const}' - \text{tr}\,\ln\Big(1 + (1 - z\bar{z})\,\mathcal{M}\Big),$$

with $\mathcal{M}$ already defined. Another simplification is that the scattering matrix becomes diagonal in momentum:

$$\mathcal{T}_R(\mathbf{p}, -\mathbf{k}) \equiv 2iE_\mathbf{p}\,(2\pi)^3\delta(\mathbf{p}+\mathbf{k})\,\beta_\mathbf{p},$$

and all the information about the background field is contained in the coefficients $\beta_\mathbf{p}$. This leads to

$$\Big[(1 - z\bar{z})\,\mathcal{M}\Big]_{pp'} = 2\pi\theta(p^0)\delta(p^2 - m^2)(2\pi)^3 2E_\mathbf{p}\delta(\mathbf{p} - \mathbf{p}')(1 - z(\mathbf{p})\bar{z}(-\mathbf{p}))\big|\beta_\mathbf{p}\big|^2.$$

Since a unique 3-momentum runs through the loop in the spatially homogeneous case, this formula implies that particles are correlated only if they have identical momenta, and are correlated with anti-particles of the opposite momentum (because of the respective arguments of the functions z and $\bar{z}$). Therefore, we obtain

$$\ln \mathcal{F}[z,\bar{z}] = \text{const} - V\int \frac{d^3\mathbf{p}}{(2\pi)^3}\,\ln\Big[1 + (1 - z(\mathbf{p})\bar{z}(-\mathbf{p}))\,f_\mathbf{p}\Big],$$

where we denote $f_\mathbf{p} \equiv |\beta_\mathbf{p}|^2$. The prefactor V is the overall volume of the system, which results from a $(2\pi)^3\delta(0)$ in momentum space.

By differentiating this formula with respect to z and/or $\bar{z}$, we obtain the following results for the one- and two-particle spectra:

$$\frac{dN_1^+}{d^3\mathbf{p}} = \underbrace{\frac{V}{(2\pi)^3}\,f_\mathbf{p}}_{\equiv n_\mathbf{p}},$$

$$\frac{dN_2^{++}}{d^3\mathbf{p}_1 d^3\mathbf{p}_2} - \frac{dN_1^+}{d^3\mathbf{p}_1}\frac{dN_1^+}{d^3\mathbf{p}_2} = \delta(\mathbf{p}_1 - \mathbf{p}_2)\,n_{\mathbf{p}_1}\,f_{\mathbf{p}_1},$$

$$\frac{dN_2^{+-}}{d^3\mathbf{p}d^3\mathbf{q}} - \frac{dN_1^+}{d^3\mathbf{p}}\frac{dN_1^-}{d^3\mathbf{q}} = \delta(\mathbf{p} + \mathbf{q})\,n_\mathbf{p}\,(1 + f_\mathbf{p})\,.$$

In the limit where $f_\mathbf{p} \ll 1$, the right-hand side of the two-particle correlation would simplify to a form consistent with a Poisson distribution. In contrast, when the occupation number $f_\mathbf{p}$ is not small, deviations from a Poisson distribution arise because of Bose–Einstein correlations.

Since there is no correlation except among particles with the same momentum or anti-particles with the opposite momentum, one can also derive the probability distribution $P_\mathbf{p}(m,n)$ of producing m particles of momentum $\mathbf{p}$ and n anti-particles of momentum $-\mathbf{p}$,

$$P_\mathbf{p}(m,n) = \delta_{m,n}\,\frac{1}{1+f_\mathbf{p}}\left(\frac{f_\mathbf{p}}{1+f_\mathbf{p}}\right)^n,$$

a distribution known as the *discrete Bose–Einstein distribution*. The main difference between such a distribution and a Poisson distribution is the existence of large multiplicity tails, which are due to stimulated emission.

79.f From the results of Problem 77, we know that $P_0 = \exp(-\overline{N}_{\text{clusters}})$. It is only when particles are produced without any correlations that the mean number of clusters is equal to the mean number of produced particles $\overline{N^+}$ (since each particle is its own cluster in that case). Therefore, we expect in general that $P_0 \neq \exp(-\overline{N^+})$ due to correlations. This discrepancy can be made more explicit by using the formulas obtained above. By evaluating the generating functional at $z = \overline{z} = 0$, one obtains the following expression for the vacuum-to-vacuum transition probability (i.e., the vacuum survival probability):

$$P_0 = \exp\big(-\overline{N}_{\text{clusters}}\big) \quad \text{with } \overline{N}_{\text{clusters}} = V \int \frac{d^3p}{(2\pi)^3} \ln(1 + f_p),$$

while the total particle multiplicity is

$$\overline{N^+} = V \int \frac{d^3p}{(2\pi)^3} f_p.$$

Only when the occupation number is small in all modes can we expand the logarithm in P_0 and obtain $P_0 \approx \exp(-\overline{N^+})$, but this relationship is not exact in very strong fields. This is a limitation of the methods that give only P_0 (for instance by providing a way to calculate the imaginary part of the effective action in a background field): in general, knowledge of P_0 is not sufficient to obtain the momentum dependence of the spectrum of produced particles (proportional to f_p).
Source: Fukushima, K., Gelis, F. and Lappi, T. (2009), *Nucl Phys A* 831: 184.

80. Schwinger Mechanism from Bogoliubov Transformations The goal of this problem is to rederive some of the results of Problem 79 in the canonical formalism, and to introduce the concept of *Bogoliubov transformation* (quite generally, this is a mapping between the creation and annihilation operators on the Fock space of an observer and those of another observer).

80.a Write the Hamiltonian of the theory of a complex scalar field coupled to an external electromagnetic potential. Specialize it to the gauge $A^0 = 0$. Derive the Hamilton–Jacobi equations of motion.

80.b Decompose the field operator ϕ in the *in* and *out* bases. Show that there is a linear relationship between the *in* and *out* creation and annihilation operators,

$$a_{k,\text{out}} = \alpha_k \, a_{k,\text{in}} + \beta_k \, b^\dagger_{-k,\text{in}}, \quad b^\dagger_{k,\text{out}} = \alpha^*_{-k} \, b^\dagger_{k,\text{in}} + \beta^*_{-k} \, a_{-k,\text{in}},$$
with $|\alpha_k|^2 - |\beta_k|^2 = 1$.

80.c Show that the *in* vacuum state can be expressed in terms of *out* objects as follows:

$$\big|0_{\text{in}}\big\rangle = e^{-\frac{V}{2}\int \frac{d^3p}{(2\pi)^3} \ln(1+|\beta_p|^2)} \, e^{\int \frac{d^3p}{(2\pi)^3} \frac{\beta_p}{\alpha^*_p} a^\dagger_{p,\text{out}} b^\dagger_{-p,\text{out}}} \big|0_{\text{out}}\big\rangle,$$

where V is the volume of the system.

80.d Use this formula to calculate the inclusive one-particle and two-particle spectra, and check that they agree with the results of Problem 79.

80.a Let us start from the Lagrangian density,

$$\mathcal{L} = (D_\mu\phi)^*(D^\mu\phi) - m^2\phi^*\phi.$$

The conjugate momentum of the fields ϕ, ϕ^* are respectively

$$\Pi \equiv \frac{\partial\mathcal{L}}{\partial\partial_0\phi} = (D_0\phi)^*, \quad \Pi^* \equiv \frac{\partial\mathcal{L}}{\partial\partial_0\phi^*} = D_0\phi, \tag{5.33}$$

from which we obtain the Hamiltonian

$$\begin{aligned} H &= \int d^3x \left(\Pi^*\partial_0\phi^* + \Pi\partial_0\phi - \mathcal{L}\right) \\ &= \int d^3x \Big(\Pi^*\Pi + ieA_0\Pi\phi - ieA_0\Pi^*\phi^* + m^2\phi^*\phi + (\mathbf{D}\phi)^*(\mathbf{D}\phi)\Big). \end{aligned}$$

Here, ϕ and Π are operators in the Heisenberg picture satisfying the equal-time commutation relation,

$$\big[\phi(x), \Pi(y)\big] \underset{x^0=y^0}{=} i\,\delta(\mathbf{x} - \mathbf{y}).$$

It is convenient for the following discussion to choose a gauge where $A_0 = 0$, because in this gauge there is a non-ambiguous correspondence between the time dependence of a wavefunction and the physical energy of the corresponding particle (in particular, the concepts of positive and negative energies are invariant under the residual gauge freedom). In the temporal gauge, we have the following Hamiltonian:

$$H = \int d^3x \left(\Pi^*\Pi + m^2\phi^*\phi + (\mathbf{D}\phi)^*(\mathbf{D}\phi)\right).$$

The dynamics of the scalar fields are determined by the Hamilton–Jacobi equations of motion,

$$\begin{aligned} \partial_0\phi &= i\,[H, \phi] = \Pi^*, \\ \partial_0\Pi &= i\,[H, \Pi] = (\mathbf{D}^2 - m^2)\phi^*. \end{aligned}$$

Note that, since we are not considering self-interactions of the scalar fields, these equations of motion are linear. Moreover, since the background electromagnetic field is treated as a classical field given once and for all, there are no back-reaction effects (e.g., no photon loop corrections on the propagator of the scalar). Thanks to the absence of loop corrections, solving these equations of motion as if they were classical equations of motion is sufficient in order to obtain the relationship between the field operators at $x^0 = -\infty$ and $x^0 = +\infty$. In other words, solving exactly the simplified QFT problem considered here can be reduced to solving a set of linear partial differential equations.

80.b Let us now decompose the field operator on the basis of creation or annihilation operators. For this, we may choose either the *in* or the *out* basis. Let us mention here an important subtlety. Even if we assume that the external electromagnetic field vanishes when $x^0 = \pm\infty$, there may be a non-zero gauge potential A_i at asymptotic times, $A_i^{\text{in,out}}$. (These gauge potentials are constants, since there is no physical electromagnetic field at asymptotic times.) These asymptotic potentials modify the dispersion relation that relates the on-shell energy and the wavenumber $\mathbf{k}$, since the equation of motion of the field contains a covariant derivative $\mathbf{D}$ rather than the ordinary derivative ∂. For this reason, we must replace the usual $E_{\mathbf{k}}$ by $E_{\mathbf{k}}^{\text{in,out},+} = \sqrt{m^2 + (\mathbf{k} + e\mathbf{A}^{\text{in,out}})^2}$ for particles and by $E_{\mathbf{k}}^{\text{in,out},-} = \sqrt{m^2 + (\mathbf{k} - e\mathbf{A}^{\text{in,out}})^2}$ for anti-particles. Since we have $E_{\mathbf{k}}^{\text{in,out},-} = E_{-\mathbf{k}}^{\text{in,out},+}$, we can simplify the subsequent notation by systematically using only the energy of the anti-particles, which we denote more simply by $E_{\mathbf{k}}^{\text{in,out}} \equiv E_{\mathbf{k}}^{\text{in,out},+}$ (with $\mathbf{k}$ changed into $-\mathbf{k}$ when necessary). (In a detector, one measures the kinetic momentum of a particle, $\mathbf{p} \equiv \mathbf{k} \pm e\mathbf{A}$. If particles are measured in the final state, it is convenient to choose a residual gauge in which $\mathbf{A}^{\text{out}} = 0$, so that the wavenumber and the kinetic momentum are identical at $x^0 = +\infty$.)

With this in mind, the field operator may be decomposed in two different ways:

$$\phi(x) = \int \frac{d^3\mathbf{k}}{(2\pi)^3} \left[\frac{a_{\mathbf{k},\text{in}}}{\sqrt{2E_{\mathbf{k}}^{\text{in}}}} \phi_{\mathbf{k},\text{in}}^{+}(x) + \frac{b_{\mathbf{k},\text{in}}^{\dagger}}{\sqrt{2E_{-\mathbf{k}}^{\text{in}}}} \phi_{-\mathbf{k},\text{in}}^{-}(x) \right]$$
$$= \int \frac{d^3\mathbf{k}}{(2\pi)^3} \left[\frac{a_{\mathbf{k},\text{out}}}{\sqrt{2E_{\mathbf{k}}^{\text{out}}}} \phi_{\mathbf{k},\text{out}}^{+}(x) + \frac{b_{\mathbf{k},\text{out}}^{\dagger}}{\sqrt{2E_{-\mathbf{k}}^{\text{out}}}} \phi_{-\mathbf{k},\text{out}}^{-}(x) \right], \tag{5.34}$$

where the functions $\phi_{\mathbf{k},\text{in}}^{\pm}$ and $\phi_{\mathbf{k},\text{out}}^{\pm}$ should be determined from the equations of motion (they cannot be simple plane waves because of the external field). Note that we have adopted here a slightly different normalization of the creation and annihilation operators than in the rest of this book. Indeed, because we now have $E_{\mathbf{k}}^{\text{in}} \neq E_{\mathbf{k}}^{\text{out}}$, this choice of normalization turns out to be a bit simpler.

As usual, we want the Heisenberg representation field operator $\phi(x)$ to go to the *in* interaction picture field operator when $x^0 \to -\infty$ and to the *out* interaction picture field operator at $x^0 \to +\infty$. These boundary conditions are realized if the functions $\phi_{\mathbf{k},\text{in}}^{\pm}$ and $\phi_{\mathbf{k},\text{out}}^{\pm}$ have the following asymptotic behavior:

$$\begin{aligned}
\phi_{\mathbf{k},\text{in}}^{+}(x) &= e^{-iE_{\mathbf{k}}^{\text{in}}x^0 + i\mathbf{k}\cdot\mathbf{x}} && \text{for } x^0 \to -\infty,\\
\phi_{\mathbf{k},\text{in}}^{-}(x) &= e^{iE_{\mathbf{k}}^{\text{in}}x^0 + i\mathbf{k}\cdot\mathbf{x}} && \text{for } x^0 \to -\infty,\\
\phi_{\mathbf{k},\text{out}}^{+}(x) &= e^{-iE_{\mathbf{k}}^{\text{out}}x^0 + i\mathbf{k}\cdot\mathbf{x}} && \text{for } x^0 \to +\infty,\\
\phi_{\mathbf{k},\text{out}}^{-}(x) &= e^{iE_{\mathbf{k}}^{\text{out}}x^0 + i\mathbf{k}\cdot\mathbf{x}} && \text{for } x^0 \to +\infty.
\end{aligned}$$

The corresponding annihilation operators $a_{\text{in,out}}, b_{\text{in,out}}$ remove an in-state or out-state particle and anti-particle, respectively. Note that, for further convenience, we have chosen the spatial dependence in both $\phi_{\mathbf{k}}^{+}(x)$ and $\phi_{\mathbf{k}}^{-}(x)$ to be $e^{+i\mathbf{k}\cdot\mathbf{x}}$, and thus the usual negative energy plane wave $e^{-i\mathbf{k}\cdot\mathbf{x}}$ corresponds to $\phi_{-\mathbf{k}}^{-}(x)$. The canonical commutation relation for ϕ and Π is satisfied provided that we have

$$\left[a_{\mathbf{k},\text{in}}, a_{\mathbf{p},\text{in}}^{\dagger}\right] = \left[b_{\mathbf{k},\text{in}}, b_{\mathbf{p},\text{in}}^{\dagger}\right] = \left[a_{\mathbf{k},\text{out}}, a_{\mathbf{p},\text{out}}^{\dagger}\right] = \left[b_{\mathbf{k},\text{out}}, b_{\mathbf{p},\text{out}}^{\dagger}\right] = (2\pi)^3\delta(\mathbf{k} - \mathbf{p}).$$

(Note the absence of factors $2E_{\mathbf{k}}$ here, because of our choice of normalization in (5.34).)

Because the equation of motion satisfied by the field is linear, there is a linear transformation that relates the coefficient functions in the *in* and *out* decompositions. Moreover, if the background field depends only on time, then this linear transformation does not mix values of

$\mathbf{k}$. Let us introduce coefficient $\alpha_{\mathbf{k}}$ and $\beta_{\mathbf{k}}$ (known as *Bogoliubov coefficients*) such that

$$\lim_{x^0\to+\infty} \phi^+_{\mathbf{k},\text{in}}(x) \equiv \sqrt{\frac{E^{\text{in}}_{\mathbf{k}}}{E^{\text{out}}_{\mathbf{k}}}}\left(\alpha_{\mathbf{k}}\, e^{-iE^{\text{out}}_{\mathbf{k}}x^0+i\mathbf{k}\cdot\mathbf{x}} + \beta^*_{\mathbf{k}}\, e^{iE^{\text{out}}_{\mathbf{k}}x^0+i\mathbf{k}\cdot\mathbf{x}}\right).$$

In other words, $\alpha_{\mathbf{k}}$ and $\beta_{\mathbf{k}}$ are defined by decomposing the time dependence of $\phi^+_{\mathbf{k},\text{in}}$ on the basis of the $e^{\pm iE^{\text{out}}_{\mathbf{k}}x^0}$ at large x^0. In the absence of an external field, one would simply have $\alpha_{\mathbf{k}} = 1$ and $\beta_{\mathbf{k}} = 0$. Since $(\phi^+_{\mathbf{k},\text{in}}(x^0, -\mathbf{x}))^*$ satisfies both the same initial condition as $\phi^-_{\mathbf{k},\text{in}}(x)$ and the same equation of motion, they are therefore equal everywhere, and we have also

$$\lim_{x^0\to+\infty} \phi^-_{\mathbf{k},\text{in}}(x) = \sqrt{\frac{E^{\text{in}}_{\mathbf{k}}}{E^{\text{out}}_{\mathbf{k}}}}\left(\alpha^*_{\mathbf{k}}\, e^{iE^{\text{out}}_{\mathbf{k}}x^0+i\mathbf{k}\cdot\mathbf{x}} + \beta_{\mathbf{k}}\, e^{-iE^{\text{out}}_{\mathbf{k}}x^0+i\mathbf{k}\cdot\mathbf{x}}\right).$$

Note now that the plane waves that appear in the integrands of the preceding two equations are nothing but the asymptotic behavior of the waves $\phi^\pm_{\mathbf{k},\text{out}}$. Since the waves $\phi^\pm_{\mathbf{k},\text{in}}$ and the $\phi^\pm_{\mathbf{k},\text{out}}$ obey the same linear equation of motion (the Klein–Gordon equation), these linear relationships are valid at all times (and not just at $x^0 = +\infty$):

$$\phi^+_{\mathbf{k},\text{in}}(x) = \sqrt{\frac{E^{\text{in}}_{\mathbf{k}}}{E^{\text{out}}_{\mathbf{k}}}}\left(\alpha_{\mathbf{k}}\, \phi^+_{\mathbf{k},\text{out}}(x) + \beta^*_{\mathbf{k}}\, \phi^-_{\mathbf{k},\text{out}}(x)\right),$$

$$\phi^-_{\mathbf{k},\text{in}}(x) = \sqrt{\frac{E^{\text{in}}_{\mathbf{k}}}{E^{\text{out}}_{\mathbf{k}}}}\left(\alpha^*_{\mathbf{k}}\, \phi^-_{\mathbf{k},\text{out}}(x) + \beta_{\mathbf{k}}\, \phi^+_{\mathbf{k},\text{out}}(x)\right).$$

By inserting these formulas in (5.34), we obtain similar relationships between the creation and annihilation operators in the two bases:

$$a_{\mathbf{k},\text{out}} = \alpha_{\mathbf{k}}\, a_{\mathbf{k},\text{in}} + \beta_{\mathbf{k}}\, b^\dagger_{-\mathbf{k},\text{in}}, \quad b^\dagger_{\mathbf{k},\text{out}} = \alpha^*_{-\mathbf{k}}\, b^\dagger_{\mathbf{k},\text{in}} + \beta^*_{-\mathbf{k}}\, a_{-\mathbf{k},\text{in}}.$$

This linear relationship is known as a *Bogoliubov transformation.* Consistency with the commutation relations requires that

$$|\alpha_{\mathbf{k}}|^2 - |\beta_{\mathbf{k}}|^2 = 1.$$

(One can also check that this condition implies that the charge operator is the same in the two bases: $Q_{\text{in}} = Q_{\text{out}}$.) The above relations can be inverted as follows:

$$a_{\mathbf{k},\text{in}} = \alpha^*_{\mathbf{k}}\, a_{\mathbf{k},\text{out}} - \beta_{\mathbf{k}}\, b^\dagger_{-\mathbf{k},\text{out}}, \quad b_{\mathbf{k},\text{in}} = \alpha^*_{-\mathbf{k}}\, b_{\mathbf{k},\text{out}} - \beta_{-\mathbf{k}}\, a^\dagger_{-\mathbf{k},\text{out}}.$$

80.c The vacuum is defined as the state annihilated by all the annihilation operators:

$$a_{\mathbf{k}}|0\rangle = b_{\mathbf{k}}|0\rangle = 0.$$

(We ignore the *in* and *out* labels for this generic discussion.) A properly normalized state with one particle of momentum $\mathbf{k}$ can be constructed as

$$|\mathbf{k}\rangle = \frac{a^\dagger_{\mathbf{k}}}{\sqrt{V}}|0\rangle,$$

where V is the spatial volume. (Note that here we are departing momentarily from the convention used in the rest of the book, namely to have one-particle states normalized to $2E_{\mathbf{k}}$

particles per unit volume. The state defined here contains exactly one particle in the volume V.) Particles of momentum $\mathbf{k}$ can then be counted with the following particle number operator:

$$\frac{d\widehat{N}}{d^3\mathbf{k}} = \frac{a_\mathbf{k}^\dagger a_\mathbf{k}}{(2\pi)^3}.$$

To check the consistency of these definitions, note that

$$\langle \mathbf{p}|\frac{d\widehat{N}}{d^3\mathbf{k}}|\mathbf{p}\rangle = \langle 0|\frac{a_\mathbf{p}}{\sqrt{V}}\frac{a_\mathbf{k}^\dagger a_\mathbf{k}}{(2\pi)^3}\frac{a_\mathbf{p}^\dagger}{\sqrt{V}}|0\rangle = \delta(\mathbf{p}-\mathbf{k}).$$

(We use the commutation relations in order to obtain the last equality.) The integral over $d^3\mathbf{k}$ indeed gives 1, i.e., the correct total number of particles in the state.

After having set these conventions, let us return to the problem of particle production. We consider the situation where there are no particles at $x^0 \to -\infty$, i.e., the system is prepared in the *in* vacuum state defined by

$$a_{\mathbf{p},\mathrm{in}}\big|0_{\mathrm{in}}\big\rangle = b_{\mathbf{p},\mathrm{in}}\big|0_{\mathrm{in}}\big\rangle = 0.$$

In the Heisenberg picture, the states do not evolve (the operators do), and therefore the system stays in this state forever. What changes with time is the number operator used to count particles. For instance, at $x^0 = +\infty$, the operator $a_{\mathbf{k},\mathrm{out}}$ in general no longer annihilates $\big|0_{\mathrm{in}}\big\rangle$ and an observer would see particles. In order to count the number of outgoing particles, we must thus count the number of *out* particles contained in the state $\big|0_{\mathrm{in}}\big\rangle$. For this, it is useful to derive an expression for $\big|0_{\mathrm{in}}\big\rangle$ in terms of the *out* quantities $a_{\mathbf{p},\mathrm{out}}$, $b_{\mathbf{p},\mathrm{out}}$ and $\big|0_{\mathrm{out}}\big\rangle$. That is, we should solve

$$\begin{aligned} a_{\mathbf{p},\mathrm{in}}\big|0_{\mathrm{in}}\big\rangle &= \big(\alpha_\mathbf{p}^*\, a_{\mathbf{p},\mathrm{out}} - \beta_\mathbf{p} b^\dagger_{-\mathbf{p},\mathrm{out}}\big)\big|0_{\mathrm{in}}\big\rangle = 0,\\ b_{\mathbf{p},\mathrm{in}}\big|0_{\mathrm{in}}\big\rangle &= \big(\alpha_{-\mathbf{p}}^*\, b_{\mathbf{p},\mathrm{out}} - \beta_{-\mathbf{p}} a^\dagger_{-\mathbf{p},\mathrm{out}}\big)\big|0_{\mathrm{in}}\big\rangle = 0. \end{aligned}$$

Let us consider the following ansatz:

$$\big|0_{\mathrm{in}}\big\rangle \equiv C\,\exp\Big(\int \frac{d^3\mathbf{k}}{(2\pi)^3}\,\chi_\mathbf{k}\, a^\dagger_{\mathbf{k},\mathrm{out}}\, b^\dagger_{-\mathbf{k},\mathrm{out}}\Big)\,\big|0_{\mathrm{out}}\big\rangle,$$

where C is a normalization constant. (This state is a coherent state in which each particle is paired with an anti-particle of opposite momentum – a necessary condition for any representation of the vacuum state because of charge and momentum conservation.) Requesting that this state be annihilated by $a_{\mathbf{k},\mathrm{in}}$ and $b_{-\mathbf{k},\mathrm{in}}$ puts some constraints on the function $\chi_\mathbf{k}$. In order to see this, let us denote

$$A^\dagger \equiv \int \frac{d^3k}{(2\pi^3)}\, \chi_k\, a^\dagger_{k,\text{out}}\, b^\dagger_{-k,\text{out}},$$

$$B_p \equiv \alpha^*_p a_{p,\text{out}} - \beta_p b^\dagger_{-p,\text{out}},$$

$$C_p \equiv \left[B_p, A^\dagger\right] = \chi_p \alpha^*_p b^\dagger_{-p,\text{out}}.$$

Note that C_p commutes with $A^\dagger$. Therefore, we have

$$\begin{aligned} B_p A^\dagger &= A^\dagger B_p + C, \\ B_p (A^\dagger)^2 &= (A^\dagger)^2 B_p + 2C_p A^\dagger, \\ &\cdots \\ B_p (A^\dagger)^n &= (A^\dagger)^n B_p + nC_p (A^\dagger)^{n-1}, \\ B_p e^{A^\dagger} &= e^{A^\dagger}(B_p + C_p). \end{aligned}$$

Then, we have

$$\left(\alpha^*_p a_{p,\text{out}} - \beta_p b^\dagger_{-p,\text{out}}\right)\big|0_{\text{in}}\big\rangle = e^{A^\dagger}(B_p + C_p)\big|0_{\text{out}}\big\rangle = e^{A^\dagger}(-\beta_p + \chi_p \alpha^*_p) b^\dagger_{-p,\text{out}}\big|0_{\text{out}}\big\rangle.$$

For this to be zero, we must have $\chi_p = \beta_p/\alpha^*_p$. (One may check that the second condition leads to the same constraint.)

The next step is to determine the normalization constant so that $\big\langle 0_{\text{in}}\big|0_{\text{in}}\big\rangle = 1$, i.e.,

$$|C|^2 = \left\langle 0_{\text{out}}\big|e^{A} e^{A^\dagger}\big|0_{\text{out}}\right\rangle^{-1}.$$

An easier way to handle this calculation is to discretize the integrals over the momentum in A and $A^\dagger$. The consistent way of doing this is to write

$$A^\dagger = \frac{1}{V}\sum_k \chi_k A^\dagger_k, \quad A = \frac{1}{V}\sum_k \chi^*_k A_k,$$

$$\left[A_p, A^\dagger_q\right] = V\delta_{pq}\underbrace{\left(a^\dagger_{p,\text{out}} a_{p,\text{out}} + b_{-p,\text{out}} b^\dagger_{-p,\text{out}}\right)}_{\equiv N_p},$$

$$\left[N_p, A^\dagger_q\right] = 2V\delta_{pq}A^\dagger_p, \quad \left[N_p, A_q\right] = -2V\delta_{pq}A_p,$$

where δ_{pq} is now a discrete Kronecker symbol. We have

$$\begin{aligned} \left\langle 0_{\text{out}}\big|e^{A} e^{A^\dagger}\big|0_{\text{out}}\right\rangle &= \left\langle 0_{\text{out}}\Big|\prod_p e^{\chi^*_p \frac{A_p}{V}} \prod_q e^{\chi_q \frac{A^\dagger_q}{V}}\Big|0_{\text{out}}\right\rangle \\ &= \left\langle 0_{\text{out}}\Big|\prod_p e^{\chi^*_p \frac{A_p}{V}} e^{\chi_p \frac{A^\dagger_p}{V}}\Big|0_{\text{out}}\right\rangle. \end{aligned}$$

Noting that

$$\left\langle 0_{\text{out}}\Big|\left(\tfrac{A_p}{V}\right)^m \left(\tfrac{A^\dagger_p}{V}\right)^n\Big|0_{\text{out}}\right\rangle = \delta_{mn}(n!)^2$$

(the factorial factors come from the fact that $\left(\frac{A^\dagger_p}{V}\right)^n\big|0_{\text{out}}\big\rangle$ is a state with n pairs of particles and anti-particles with identical momenta, but this state is not normalized to account for the fact that the n pairs are identical), we obtain

$$\langle 0_{out}|e^{A}e^{A^\dagger}|0_{out}\rangle = \prod_{p}(1-|\chi_p|^2)^{-1} = \prod_{p}(1+|\beta_p|^2)$$
$$= e^{\sum_k \ln(1+|\beta_p|^2)} = \exp\left(V\int\frac{d^3p}{(2\pi)^3}\ln\left(1+|\beta_p|^2\right)\right).$$

(We have restored the continuous notation in the last expression.) Therefore, the normalized *in* vacuum state expressed in terms of *out* objects reads

$$|0_{in}\rangle = e^{-\frac{V}{2}\int\frac{d^3p}{(2\pi)^3}\ln(1+|\beta_p|^2)}\, e^{\int\frac{d^3p}{(2\pi)^3}\frac{\beta_p}{\alpha_p^*}a^\dagger_{p,out}b^\dagger_{-p,out}}|0_{out}\rangle. \tag{5.35}$$

The physical meaning of this formula is that, for an observer measuring particles in the final state, the initially empty state has evolved into a coherent state filled with scalar–antiscalar pairs of opposite momenta.

80.d Using (5.35), it is now straightforward to calculate the single and double inclusive spectra by taking expectation values of the number operator. The spectrum of particles is

$$\frac{dN_1^+}{d^3p} = \langle 0_{in}|\frac{a^\dagger_{p,out}a_{p,out}}{(2\pi)^3}|0_{in}\rangle.$$

The easiest way to evaluate this expression is to note that the expression (5.35) for the initial vacuum state behaves as a coherent state. In particular, it is a kind of "generalized eigenstate" of the *out* annihilation operators:

$$a_{p,out}|0_{in}\rangle = \chi_p b^\dagger_{-p,out}|0_{in}\rangle, \quad \langle 0_{in}|a^\dagger_{p,out} = \chi_p^*\langle 0_{in}|b_{-p,out}.$$

(The proof is the same as in the case of the usual coherent states, since $b^\dagger_{-p,out}$ and $b_{-p,out}$ are transparent to the action of $a_{p,out}$ and $a^\dagger_{p,out}$.) Using these identities, we can relate the particle spectrum at $\mathbf{p}$ to the anti-particle spectrum at $-\mathbf{p}$:

$$\frac{dN_1^+}{d^3p} = |\chi_p|^2\frac{dN_1^-}{d^3(-p)} + \frac{V}{(2\pi)^3}|\chi_p|^2.$$

However, since the external field is spatially homogeneous and the initial state is the empty state, these two spectra must be equal. Therefore, we get

$$\frac{dN_1^+}{d^3p} = \frac{V}{(2\pi)^3}\frac{|\chi_p|^2}{1-|\chi_p|^2} = \frac{V}{(2\pi)^3}|\beta_p|^2.$$

This expression exactly coincides with the one obtained in Problem 79. The two-particle spectrum (equal sign) is given by

$$\frac{dN_2^{++}}{d^3p\,d^3q} = \langle 0_{in}| \left(\frac{a^\dagger_{p,out}a_{p,out}}{(2\pi)^3}\frac{a^\dagger_{q,out}a_{p,out}}{(2\pi)^3} - \delta(p-q)\frac{a^\dagger_{p,out}a_{p,out}}{(2\pi)^3}\right)|0_{in}\rangle$$
$$= \frac{1}{(2\pi)^6}\langle 0_{in}|a^\dagger_{p,out}a^\dagger_{q,out}a_{p,out}a_{q,out}|0_{in}\rangle$$
$$= \frac{dN_1^+}{d^3p}\frac{dN_1^+}{d^3q} + \delta(p-p)\frac{V}{(2\pi)^3}|\beta_p|^4\,.$$

Note that in the first line, we are explicitly subtracting the delta function contribution in order to agree with the definition used in Problem 79, where the bi-spectrum was defined so that its integral is the mean value of $N^+(N^+-1)$.

Additional Note: In order to uncover the whole probability distribution, we can first factorize the Fock space into a direct product of the Fock spaces for different momenta k. An arbitrary state $|\Psi\rangle$ can be written as a tensor product,

$$|\Psi\rangle \equiv \bigotimes_k |\Psi\rangle_k.$$

Since particles and anti-particles are produced in pairs of opposite momenta, it is natural to group together particles of momentum k and anti-particles of momentum $-k$ under the same label k. Thus we define the vacuum states $|0,0\rangle_k$ of the subspaces k as

$$a_{k,out}|0,0\rangle_k = b_{-k,out}|0,0\rangle_k = 0,$$

and $|0_{out}\rangle$ is a tensor product of the vacuum states of the different modes,

$$|0_{out}\rangle = \bigotimes_k |0,0\rangle_k\,.$$

An outgoing state with m particles of momentum k and n anti-particles of momentum $-k$ at $x^0 \to \infty$ is then denoted by

$$|m,n\rangle_k = \left(\frac{a^\dagger_{k,out}}{\sqrt{m!V}}\right)^m \left(\frac{b^\dagger_{-k,out}}{\sqrt{n!V}}\right)^n |0,0\rangle_k.$$

(Arbitrary states can be constructed from tensor products of these states.)

Applying this decomposition, the *in* vacuum state leads to

$$|0_{in}\rangle = \bigotimes_k \left\{(1+|\beta_k|^2)^{-1/2}\exp\left[\frac{\beta_k}{V\alpha_k^*}a^\dagger_{k,out}b^\dagger_{out-k}\right]|0,0\rangle_k\right\}.$$

By expanding the exponential we get

$$|0_{in}\rangle = \bigotimes_k \underbrace{\left\{(1+|\beta_k|^2)^{-1/2}\sum_{m_k=0}^{\infty}\left(\frac{\beta_k}{\alpha_k^*}\right)^{m_k}|m_k,m_k\rangle_k\right\}}_{\equiv |0_{in}\rangle_k}.$$

In order to obtain the *inclusive* probability $P_k(m)$ of having m pairs of momentum $(k,-k)$, we need to sum over all the possible numbers of pairs in the other momentum modes. This corresponds to the following definition:

$$P_k(m) = \underbrace{\left|{}_k\langle 0_{\text{in}}|m, m\rangle_k\right|^2}_{\left|\frac{\beta_k}{\alpha_k}\right|^{2m}\frac{1}{1+|\beta_k|^2}} \times \prod_{p\neq k}\underbrace{\sum_{m_p=0}^{\infty}\left|{}_p\langle 0_{\text{in}}|m_p, m_p\rangle_p\right|^2}_{\frac{|\alpha_p|^2}{1+|\beta_p|^2}=1}$$

$$= \frac{1}{1+|\beta_k|^2}\left(\frac{|\beta_k|^2}{1+|\beta_k|^2}\right)^m.$$

This is the so called *discrete Bose–Einstein* (or *geometrical*) distribution, obtained in Problem 79. Since momenta are uncorrelated, the full distribution is just the product of the distributions in each momentum mode.
Source: Fukushima, K., Gelis, F. and Lappi, T. (2009), *Nucl Phys A* 831: 184.

81. Unruh Effect, Hawking Radiation The goal of this problem is to study the relationship between the notions of vacuum and particles for an observer at rest and for an accelerated observer, the final result being a derivation of the *Unruh effect*. A similar approach, sketched at the end of the solution, can be used to derive the *Hawking radiation* of black holes (the analogy is that a fixed observer outside of a black hole feels an acceleration due to the gravitational field).

81.a Show that the coordinates (t, x) in a frame at rest and (τ, u) in a frame with a constant acceleration a in the direction x are related by

$$t = \frac{1+au}{a}\sinh(a\tau), \qquad x = \frac{1+au}{a}\cosh(a\tau).$$

81.b Consider the alternative coordinate $v \equiv a^{-1}\ln(1+au)$. Express (t, x) in terms of (τ, v). Write the expression for the interval ds^2 in terms of (τ, v). Introduce the light-cone coordinates $x^\pm \equiv t \pm x$, $v^\pm \equiv \tau \pm v$. How are $x^\pm$ and $v^\pm$ related?

81.c Consider a massless non-interacting scalar field ϕ in $(1+1)$ dimensions. Write its Fourier decomposition in terms of $x^\pm$ and in terms of $v^\pm$. Note that these two decompositions require two distinct sets of creation and annihilation operators. Find the Bogoliubov coefficients that relate these two sets.

81.d Assume that the system is in the vacuum state of the frame at rest. Show that the accelerated observer sees the following thermal spectrum of particles:

$$\left\langle 0_{\text{rest}}\right|\left(b_k^\dagger b_k\right)_{\text{accelerated}}\left|0_{\text{rest}}\right\rangle = \frac{V}{e^{2\pi k/a}-1}.$$

81.a Let us denote by (t, x) the coordinates in the frame of an observer at rest. Consider now another observer, undergoing an accelerated motion, and denote by τ the proper time of the accelerated observer. With respect to this proper time, we can define

$$u^\mu \equiv \frac{dx^\mu}{d\tau}, \quad \text{with } u^\mu u_\mu = 1,$$

as well as the acceleration

$$a^\mu \equiv \frac{du^\mu}{d\tau}.$$

Note that, since $u^\mu u_\mu = 1$, we have $a^\mu u_\mu = 0$. This implies that in the co-moving frame (i.e., the inertial frame whose velocity is the same as that of the accelerated frame at a given time), where $u^\mu = (1, 0)$, we have $a^\mu = (0, a)$ and therefore $a^\mu a_\mu = -a^2$. Moreover, this norm is the same in all inertial frames.

Now, we consider a motion with a constant acceleration a in the direction x^1 (and we assume that the velocity has null components in the $x^{2,3}$ directions). Therefore, we have

$$(u^0)^2 - (u^1)^2 = 1, \quad \left(\frac{du^0}{d\tau}\right)^2 - \left(\frac{du^1}{d\tau}\right)^2 = -a^2.$$

We can eliminate u^0 by using $u^0 = \sqrt{1 + (u^1)^2}$, and obtain a differential equation that gives $u^1(\tau)$:

$$\frac{du^1}{d\tau} = a\sqrt{1 + (u^1)^2}.$$

(Here, we have chosen the positive square root, i.e., we assume that the acceleration is in the positive x^1 direction.) Solving this equation gives the trajectory of the accelerated object in the frame at rest, parameterized by its proper time:

$$x^1(\tau) = a^{-1}\cosh(a\tau) + \text{const}, \quad x^0(\tau) = a^{-1}\sinh(a\tau) + \text{const'}.$$

(For simplicity, we will take the two integration constants to be zero in the following, which is just a choice of the origin of the system of coordinates.) This trajectory is a hyperbola located in the right quadrant, as shown in Figure 5.6. Before going on, let us mention a few properties of these accelerated motions:

- These trajectories start at $\tau = -\infty$ with a velocity equal to the speed of light in the negative x^1 direction, and end at $\tau = +\infty$ with a velocity equal to the speed of light in the positive x^1 direction.

- Because of causality, the accelerated observer cannot receive any signal from events such as A, nor send any signal that would arrive at B. In a sense, the accelerated observer is behind a "horizon" that restricts its communications to events located in the right quadrant (shown in light gray) in the figure. This region of spacetime is known as *Rindler space*.

The next step is to determine the mapping between the coordinates (t, x) in the frame at rest and coordinates (τ, u) in the accelerated frame. The time in the accelerated frame is of

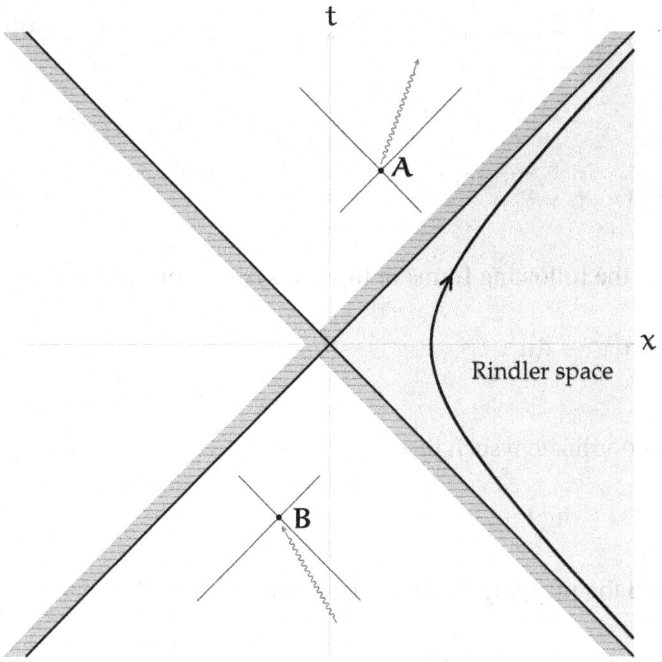

Figure 5.6 Illustration of Rindler space. The hyperbolic line on the right is the trajectory of a uniformly accelerated observer.

course the proper time τ. The spatial coordinate u is naturally defined in such a way that u remains constant (e.g., $u = 0$) for the accelerated observer. In order to find the correspondence between variations of u and variations of x, let us imagine a rigid rod parallel to the x^1 direction, moving along with the accelerated observer. For the accelerated observer, the endpoints of the rigid rod are static, and follow the worldlines $r_0^\mu \equiv (\tau, 0)$ and $r_1^\mu \equiv (\tau, u)$. These are also the coordinates of the endpoints of the rod in the instantaneous co-moving frame (since at the particular instant, the rod is at rest in the co-moving frame). If we define $\Delta^\mu \equiv r_1^\mu - r_0^\mu$, this vector is $\Delta^\mu = (0, u)$ in the co-moving frame. Since the co-moving frame is inertial, a simple Lorentz transformation gives Δ^μ in the frame at rest:

$$\begin{pmatrix} \Delta^0 \\ \Delta^1 \end{pmatrix}_{\substack{\text{frame} \\ \text{at rest}}} = \frac{1}{\sqrt{1-v^2}} \begin{pmatrix} 1 & v \\ v & 1 \end{pmatrix} \begin{pmatrix} 0 \\ u \end{pmatrix} = \begin{pmatrix} u^0 & u^1 \\ u^1 & u^0 \end{pmatrix} \begin{pmatrix} 0 \\ u \end{pmatrix} = \begin{pmatrix} u^1 u \\ u^0 u \end{pmatrix}.$$

In the frame at rest, the endpoints are located at $(x^0(\tau), x^1(\tau))$ (this is the position of the accelerated observer, holding one end of the rod) and at (t, x). Therefore, we have

$$t - x^0(\tau) = u^1 u = \frac{dx^1}{d\tau} u = u \sinh(a\tau),$$

$$x - x^1(\tau) = u^0 u = \frac{du^0}{d\tau} u = u \cosh(a\tau).$$

This gives the mapping between (t, x) and (τ, u):

$$t = \frac{1+au}{a}\,\sinh(a\tau), \quad x = \frac{1+au}{a}\,\cosh(a\tau), \tag{5.36}$$

or conversely

$$\tau = \frac{1}{2a}\ln\left(\frac{x+t}{x-t}\right), \quad u = -a^{-1} + \sqrt{x^2 - t^2}.$$

The squared spacetime interval takes the following forms in the two systems of coordinates:

$$ds^2 = dt^2 - dx^2 = (1+au)^2 d\tau^2 - du^2.$$

81.b It is useful to define another coordinate ν such that

$$du = (1+au)d\nu, \quad \text{i.e., } \nu = a^{-1}\ln(1+au).$$

In terms of (τ, ν) the interval ds^2 and the mapping between (t, x) and (τ, ν) read

$$ds^2 = (1+au)^2\,(d\tau^2 - d\nu^2) = e^{2a\nu}(d\tau^2 - d\nu^2),$$
$$t = a^{-1}\,e^{a\nu}\,\sinh(a\tau), \quad x = a^{-1}\,e^{a\nu}\,\cosh(a\tau).$$

(Thus, the mapping between (t, x) and (τ, ν) is a conformal transformation.) Let us mention also the *light-cone coordinates*, which will prove useful later:

$$x^+ \equiv t + x, \quad x^- \equiv t - x, \qquad \nu^+ \equiv \tau + \nu, \quad \nu^- \equiv \tau - \nu.$$

Their main advantage in the present context is that the mapping between $x^\pm$ and $\nu^\pm$ does not mix the $+$ and $-$ coordinates:

$$x^+ = a^{-1}\,e^{a\nu^+}, \quad x^- = -a^{-1}\,e^{-a\nu^-}.$$

81.c Consider now a massless non-interacting scalar field, whose action in a general system of coordinates is given by

$$\mathcal{S} = \frac{1}{2}\int d^2x\,\sqrt{-g}\,g^{\mu\nu}\,(\partial_\mu\phi)(\partial_\nu\phi),$$

where $-g$ is the determinant of $g_{\mu\nu}$. Since the mapping between (t, x) and (τ, ν) is conformal, the product $g^{\mu\nu}\sqrt{-g}$ is the same in both systems of coordinates, in $(1+1)$ dimensions. Indeed, the metric tensors in the two systems of coordinates are proportional,

$$g_{\mu\nu} = \Omega(x) g'_{\mu\nu},$$

which implies $g^{\mu\nu} = \Omega^{-1} g'^{\mu\nu}$ and $\sqrt{-g} = \Omega^{d/2}\sqrt{-g'}$ (the latter is $\Omega\sqrt{-g'}$ in two dimensions). (In $d \neq 2$, $g^{\mu\nu}\sqrt{-g}$ is not invariant and the calculation would be more involved, but the final result still holds.) Therefore, the action in the system at rest and in the accelerated system reads

$$\mathcal{S} = \frac{1}{2}\int dt dx \left((\partial_t\phi)^2 - (\partial_x\phi)^2\right) = \frac{1}{2}\int d\tau dv \left((\partial_\tau\phi)^2 - (\partial_v\phi)^2\right).$$

The corresponding classical equations of motion are Klein–Gordon equations,

$$(\partial_t^2 - \partial_x^2)\phi = (\partial_\tau^2 - \partial_v^2)\phi = 0,$$

implying that the free solutions are plane waves in both cases. Therefore, the field operators can be decomposed as follows:

$$\begin{aligned}\phi(t,x) &= \int \frac{dp}{2\pi}\frac{1}{\sqrt{2|p|}}\left(e^{i(|p|t-px)}\,a_p^\dagger + e^{-i(|p|t-px)}\,a_p\right),\\ \phi(\tau,v) &= \int \frac{dk}{2\pi}\frac{1}{\sqrt{2|k|}}\left(e^{i(|k|\tau-kv)}\,b_k^\dagger + e^{-i(|k|\tau-kv)}\,b_k\right).\end{aligned} \tag{5.37}$$

(Note that we have adopted the same normalization as in the Problem 80, which is a bit more convenient when discussing Bogoliubov transformations.) The important point here is that we need two distinct sets of creation and annihilation operators. Indeed, as we shall see shortly, the notions of vacuum and of particles are different for the two observers. The vacuum is defined in both cases as the state which is annihilated by all the annihilation operators,

$$a_p\left|0_{\text{rest}}\right\rangle = 0, \quad b_k\left|0_{\text{accelerated}}\right\rangle = 0,$$

but these vacuum states may not coincide.

Now, our task is to find the relationship between the a_p, $a_p^\dagger$ and the b_k, $b_k^\dagger$, in order to show that the vacuum state in the frame at rest does not appear to be empty for the accelerated observer. However, a complication in this identification is that the transformation (5.36) mixes positive and negative frequencies. This is where the light-cone coordinates become useful. First, we can rewrite (5.37) by separating the positive and negative momenta:

$$\phi(t,x) = \left(\int_0^{+\infty} + \int_{-\infty}^0\right)\frac{dp}{2\pi}\frac{1}{\sqrt{2|p|}}\left(e^{i(|p|t-px)}\,a_p^\dagger + e^{-i(|p|t-px)}\,a_p\right),$$

and we change $p \to -p$ in the second term, which leads to

$$\phi(t,x) = \int_0^{+\infty}\frac{dp}{2\pi}\frac{1}{\sqrt{2|p|}}\left(e^{ipx^-}\,a_p^\dagger + e^{-ipx^-}\,a_p + e^{ipx^+}\,a_{-p}^\dagger + e^{-ipx^+}\,a_{-p}\right).$$

Likewise, we have also

$$\phi(\tau,v) = \int_0^{+\infty}\frac{dk}{2\pi}\frac{1}{\sqrt{2|k|}}\left(e^{ikv^-}\,b_k^\dagger + e^{-ikv^-}\,b_k + e^{ikv^+}\,b_{-k}^\dagger + e^{-ikv^+}\,b_{-k}\right).$$

Since the mapping $x^\pm \to v^\pm$ does not mix the $+$ and $-$ coordinates, the relationship between the two sets of creation and annihilation operators will not mix operators carrying momenta of opposite signs. Requesting that the sector of positive momenta agrees gives

$$\int_0^{+\infty}\frac{dp}{2\pi}\frac{1}{\sqrt{2|p|}}\left(e^{ipx^-}\,a_p^\dagger+e^{-ipx^-}\,a_p\right)=\underbrace{\int_0^{+\infty}\frac{dk}{2\pi}\frac{1}{\sqrt{2|k|}}\left(e^{ikv^-}\,b_k^\dagger+e^{-ikv^-}\,b_k\right)}_{\equiv\,\phi(v^-)}.$$

Expressing the b's in terms of the a's can be done by applying a Fourier transform on v^-, since

$$\int_{-\infty}^{+\infty}dv^-\;e^{-ikv^-}\;\phi(v^-)=\frac{1}{\sqrt{2|k|}}\begin{cases}b_k^\dagger & \text{if } k>0\\ b_{-k} & \text{if } k<0\end{cases}.$$

Applying the same transform to the left-hand side gives

$$\begin{aligned}&\int_{-\infty}^{+\infty}dv^-\;e^{-ikv^-}\int_0^{+\infty}\frac{dp}{2\pi}\frac{1}{\sqrt{2|p|}}\left(e^{ipx^-}\,a_p^\dagger+e^{-ipx^-}\,a_p\right)\\&\quad=\int_0^{+\infty}\frac{dp}{\sqrt{2|p|}}\left(\gamma(k,p)\,a_p^\dagger+\gamma(k,-p)\,a_p\right),\end{aligned}$$

with

$$\gamma(k,p)\equiv\int_{-\infty}^{+\infty}\frac{dv^-}{2\pi}\;e^{-ikv^-}e^{ipx^-}\qquad(\text{with }x^-=-a^{-1}e^{-av^-}).$$

(These integrals are expressible in terms of Bessel functions.) Therefore, we have

$$\begin{aligned}b_k^\dagger&=\int_0^{+\infty}dp\,\sqrt{\tfrac{k}{p}}\left(\gamma(k,p)\,a_p^\dagger+\gamma(k,-p)\,a_p\right)=\int_0^{+\infty}dp\,\left(\alpha_{kp}\,a_p^\dagger+\beta_{kp}\,a_p\right),\\ b_k&=\int_0^{+\infty}dp\,\sqrt{\tfrac{k}{p}}\left(\gamma(-k,p)\,a_p^\dagger+\gamma(-k,-p)\,a_p\right)=\int_0^{+\infty}dp\,\left(\beta_{kp}^*\,a_p^\dagger+\alpha_{kp}^*\,a_p\right),\end{aligned}$$

with the following Bogoliubov coefficients:

$$\alpha_{kp}\equiv\sqrt{\tfrac{k}{p}}\,\gamma(k,p),\quad \beta_{kp}\equiv\sqrt{\tfrac{k}{p}}\,\gamma(k,-p).$$

(To see that the decomposition of b_k is consistent with that of $b_k^\dagger$, note that $\gamma(-k,-p)=\gamma^*(k,p)$.) Let us also calculate the commutator

$$\begin{aligned}\left[b_k,b_{k'}^\dagger\right]&=\int_{p,p'\geq 0}\left[\beta_{kp}^*\,a_p^\dagger+\alpha_{kp}^*\,a_p,\alpha_{k'p'}\,a_{p'}^\dagger+\beta_{k'p'}\,a_{p'}\right]\\&=\int_{p,p'\geq 0}\Big(\beta_{kp}^*\beta_{k'p'}\underbrace{\left[a_p^\dagger,a_{p'}\right]}_{-2\pi\delta(p-p')}+\alpha_{kp}^*\alpha_{k'p'}\underbrace{\left[a_p,a_{p'}^\dagger\right]}_{2\pi\delta(p-p')}\Big)\\&=2\pi\int_{p\geq 0}\left(\alpha_{kp}^*\alpha_{k'p}-\beta_{kp}^*\beta_{k'p}\right).\end{aligned}$$

Therefore, for the b's to have a properly normalized commutation relation, we must have

$$\int_{p\geq 0} \left(\alpha^*_{kp}\alpha_{k'p} - \beta^*_{kp}\beta_{k'p}\right) = \delta(k-k'). \tag{5.38}$$

(This is a generalization of the relation $|\alpha_p|^2 - |\beta_p|^2 = 1$ obtained in the simpler setting of Problem 80.)

81.d We now have all the tools to calculate the spectrum of particles the accelerated observer would see in the rest frame vacuum state:

$$\begin{aligned}\left\langle 0_{\text{rest}}\middle|b^\dagger_k b_k\middle|0_{\text{rest}}\right\rangle &= \int_{p,p'\geq 0} \left\langle 0_{\text{rest}}\middle|\left(\alpha_{kp}\, a^\dagger_p + \beta_{kp}\, a_p\right)\left(\beta^*_{kp'}\, a^\dagger_{p'} + \alpha^*_{kp'}\, a_{p'}\right)\middle|0_{\text{rest}}\right\rangle \\ &= \int_{p,p'\geq 0} \beta_{kp}\beta^*_{kp'} \underbrace{\left\langle 0_{\text{rest}}\middle|a_p a^\dagger_{p'}\middle|0_{\text{rest}}\right\rangle}_{2\pi\delta(p-p')} = 2\pi \int_0^{+\infty} dp \left|\beta_{kp}\right|^2. \end{aligned} \tag{5.39}$$

This result is non-zero whenever b_k contains an admixture of $a^\dagger_p$. In order to calculate this integral, note that

$$\gamma(k,p) = \int_{-\infty}^{+\infty} \frac{dv^-}{2\pi}\, e^{-ikv^-} e^{-ipa^{-1}e^{-av^-}},$$

$$\begin{aligned}\gamma(k,-p) &= \int_{-\infty}^{+\infty} \frac{dv^-}{2\pi}\, e^{-ikv^-} e^{ipa^{-1}e^{-av^-}} = \int_{-\infty}^{+\infty} \frac{dv^-}{2\pi}\, e^{-ikv^-} e^{-ipa^{-1}e^{-a(\overbrace{v^- + i\pi a^{-1}}^{w^-})}} \\ &= \int_{-\infty}^{+\infty} \frac{dw^-}{2\pi}\, e^{-ik(w^- - i\pi a^{-1})} e^{-ipa^{-1}e^{-aw^-}} = e^{-\pi k a^{-1}}\gamma(k,p).\end{aligned}$$

(There is a caveat here: instead of writing $-p = e^{-i\pi}p$ we could have chosen to write $-p = e^{-i\pi(2n+1)}$, which would change the prefactor $e^{-\pi k a^{-1}}$ into $e^{-(2n+1)\pi k a^{-1}}$. This ambiguity indicates that the function $\gamma(k,p)$ is multi-valued and has a branch cut on the semi-axis $p < 0$. While staying on the same Riemann sheet, there are two ways to approach $p < 0$. As can be seen below, we have made the choice that guarantees that the integral of $|\beta_{kp}|^2$ is positive.) Let us now use this extra information in (5.38), evaluated at the special point $k = k'$ (note that $2\pi\delta(0) = V$):

$$\begin{aligned}\frac{V}{2\pi} &= \int_{p\geq 0} \frac{k}{p}\left(\gamma^*(k,p)\gamma(k,p) - \gamma^*(k,-p)\gamma(k,-p)\right) \\ &= \left(e^{2\pi k/a} - 1\right) \underbrace{\int_{p\geq 0} \frac{k}{p}\gamma^*(k,-p)\gamma(k,-p)}_{\int_0^{+\infty} dp\, \left|\beta_{kp}\right|^2}.\end{aligned}$$

Comparing with (5.39), we see that

$$\left\langle 0_{\rm rest}\middle| b_k^\dagger b_k \middle| 0_{\rm rest}\right\rangle = \frac{V}{e^{2\pi k/a}-1}.$$

Therefore, from the point of view of the accelerated observer, the vacuum state defined in the frame at rest appears to be filled with a thermal distribution of particles (here, a Bose–Einstein distribution, since we considered scalar fields), with a temperature $T \equiv \frac{a}{2\pi}$. This is known as the *Unruh effect*. Note that an observer with a non-constant acceleration would also see a non-empty distribution of particles, but not in thermal equilibrium.

Hawking Radiation: By analogy with the Unruh effect, one can also obtain the radiation observed by a static observer outside a black hole, known as *Hawking radiation* (in a simplified $(1+1)$-dimensional setup, where we take into account only the time and the radial coordinate). In the context of black holes, the static observer is the accelerated one (this observer feels the gravitational force, which is equivalent to an acceleration in general relativity), while the unaccelerated observer would be a freely falling one.

Let us first describe several possible choices of coordinates for the static observer. The simplest system of coordinates is to take the usual time t and radius r. The Schwarzchild metric around a black hole of mass M gives the following line interval in terms of these coordinates:

$$ds^2 = \left(1-\frac{2M}{r}\right)dt^2 - \left(1-\frac{2M}{r}\right)^{-1} dr^2.$$

These coordinates are not very convenient when writing the action of a scalar field. As in the discussion of the Unruh effect, it is preferable to use a system of coordinates where the metric is conformally flat. For this we replace the radius r by a radial coordinate ρ defined by

$$d\rho \equiv \left(1-\frac{2M}{r}\right)^{-1} dr, \qquad \text{i.e., } \rho = r - 2M + 2M \ln\left|\frac{r}{2M}-1\right|,$$

which leads to the line interval

$$ds^2 = \left(1-\frac{2M}{r}\right)\left(dt^2 - d\rho^2\right).$$

Note that the horizon located at $r = 2M$ is at $\rho = -\infty$ in these coordinates (thus, this system of coordinates maps only the region of space outside of the horizon). In terms of (t,ρ), the action of a scalar field in $(1+1)$ dimensions reads simply

$$\mathcal{S} = \frac{1}{2}\int dt d\rho \left((\partial_t\phi)^2 - (\partial_\rho\phi)^2\right).$$

Here also, it is convenient to introduce light-cone coordinates $\nu^\pm \equiv t \pm \rho$ (we use the notation $\nu^\pm$ to stress the analogy with the Unruh effect – recall that the observer at rest with respect to the black hole is the one that feels an acceleration).

Let us now give without proof the expression for the light-cone coordinates of a freely falling observer (i.e., the unaccelerated one):

$$x^+ \equiv 4M\, e^{\nu^+/(4M)}, \quad x^- \equiv -4M\, e^{-\nu^-/(4M)}. \tag{5.40}$$

(These are known as *Kruskal coordinates*.) The region outside the horizon corresponds to the quadrant $x^+ \geq 0, x^- \leq 0$. Note also that $x^+x^- = 16M^2\, e^{\rho/(2M)}$, which vanishes when

$\rho \to -\infty$. Therefore, the horizon of the black hole is located on the two semi-axes $x^+ = 0$ and $x^- = 0$. It turns out that, in the Kruskal coordinates, there is no singularity at the horizon, and they can also be used below the horizon.

Note now that (5.40) are formally identical to the relationship between the light-cone coordinates of the static and accelerated observers, as discussed for the Unruh effect, provided we interpret $(4M)^{-1}$ as the acceleration. Therefore, we can exactly reproduce the same steps without any modification: (1) decompose the fields in Fourier modes in the two systems of coordinates, (2) find the Bogoliubov transformation that relates the two sets of creation and annihilation operators, (3) calculate the expectation value of the number operator $b_k^\dagger b_k$ of one system in the vacuum state of the other system. This leads to

$$\left\langle 0_{\text{kruskal}}\right|(b_k^\dagger b_k)_{\substack{\text{static}\\ \text{obs.}}}\left|0_{\text{kruskal}}\right\rangle = \frac{V}{e^{8\pi M k}-1},$$

i.e., a thermal equilibrium distribution with a temperature $T = (8\pi M)^{-1}$ (this is known as the *Hawking temperature*). Let us end with a remark about why we use the Kruskal vacuum in the last equation. Loosely speaking, in the presence of a black hole, other vacuum states, defined by non-freely falling observers, would have a very large energy density near the black hole (because excitations that are not freely falling must have an energy that compensates for the gravitational potential). This energy density would act as a source term in Einstein's equations and therefore have a strong back-reaction on the metric. Therefore, the physical setup of quantum fields on top of an unperturbed Schwarzchild metric can only be achieved with the Kruskal vacuum.

82. Anderson Localization In a periodic atomic lattice, the electronic eigenfunctions take the form of Bloch states, whose energies span continuous bands. An electron whose energy lies in such a band can propagate freely over macroscopic distances (whether the material is a conductor or an insulator depends on whether the Fermi energy lies inside a band or between two bands). The goal of this problem is to show that, in a disordered medium, electrons are localized and cannot propagate.

82.a Check that $\gamma^0 = \begin{pmatrix} 1 & 0 \\ 0 & -1 \end{pmatrix}, \gamma^i = \begin{pmatrix} 0 & \sigma^i \\ -\sigma^i & 0 \end{pmatrix}$ provide a valid representation of the Dirac matrices. Using this representation, show that the two upper components are the dominant ones in the non-relativistic limit, and show that the Dirac equation becomes

$$i\partial_t \phi = \left(m - eA^0 - \frac{(\boldsymbol{\nabla} - ie\mathbf{A})^2}{2m} + e\frac{\mathbf{B}\cdot\boldsymbol{\sigma}}{2m}\right)\phi.$$

82.b Show that the two-point function $G(x, y) \equiv \left\langle \phi(x)\phi^\dagger(y)\right\rangle$ obeys a Dyson equation of the form $G = G_0 + ie\, G_0 A^0 G$ in the case of a purely electric external field.

82.c Consider a medium in which the potential A^0 has local Gaussian fluctuations with a variance $\left\langle A^0(x)A^0(y)\right\rangle = \sigma\,\delta(\mathbf{x} - \mathbf{y})$. Write the Dyson equation satisfied by G at order e^2. By calculating the self-energy that appears in this equation, show that the denominator of the dressed propagator has an imaginary part, which leads to an exponential damping of electronic waves.

82.a To check that the announced matrices are a valid representation of the Dirac matrices, we must verify that $\{\gamma^\mu, \gamma^\nu\} = 2\, g^{\mu\nu}$. This is completely straightforward using the fact that $\sigma^i\sigma^j + \sigma^j\sigma^i = 2\,\delta^{ij}$. With this representation in mind, we write the four-component Dirac spinors as a combination of two two-component objects:

$$u_s(p) \equiv \begin{pmatrix} \phi(p) \\ \chi(p) \end{pmatrix}.$$

Consequently, the Dirac equation $(\not{p} - m)u_s(p) = 0$ can be split into two separate equations for ϕ and χ,

$$(E_p - m)\phi - (p \cdot \sigma)\chi = 0, \quad (E_p + m)\chi - (p \cdot \sigma)\phi = 0,$$

from which we get

$$\chi = \frac{1}{E_p + m}\,(p \cdot \sigma)\,\phi \underset{p \ll m}{\approx} \frac{p \cdot \sigma}{2m}\,\phi.$$

Therefore, if we denote $\phi(x) \equiv \phi(p)\, e^{-ip\cdot x}$, we obtain

$$i\partial_t\phi(x) = \left(m + \frac{(p\cdot\sigma)^2}{2m}\right)\phi(p)\, e^{-ip\cdot x} = \left(m - \frac{\nabla^2}{2m}\right)\phi(x),$$

which is the non-relativistic Schrödinger equation (the term in m could be removed if we factor out an explicit factor e^{-imt} from the wavefunction). Thus, we see that, in this representation, the two upper components of a Dirac spinor obey the Schrödinger equation, while the two lower components are suppressed by a power of p/m (and in addition they are not independent).

This equation, valid for a free fermion, can be promoted to an equation for the evolution of a non-relativistic fermion in an external electromagnetic field by simply replacing the ordinary derivatives by covariant ones: $\partial^\mu \to D^\mu \equiv \partial^\mu - ieA^\mu$. But in doing so, one should keep in mind that $\boldsymbol{\nabla}$ and $\mathbf{A}$ do not commute, which imposes that we do this replacement before eliminating the Pauli matrices:

$$\begin{aligned} i\partial_t\phi &= \left(m - eA^0 - \frac{\left((\boldsymbol{\nabla} - ie\mathbf{A})\cdot\sigma\right)^2}{2m}\right)\phi \\ &= \left(m - eA^0 - \frac{(\boldsymbol{\nabla} - ie\mathbf{A})^2 + e(A^i\nabla^j + \nabla^i A^j)\epsilon^{ijk}\sigma^k}{2m}\right)\phi \\ &= \underbrace{\left(m - eA^0 - \frac{(\boldsymbol{\nabla} - ie\mathbf{A})^2}{2m} + e\frac{\mathbf{B}\cdot\sigma}{2m}\right)}_{\equiv\,\mathcal{H}}\phi, \end{aligned}$$

where $\mathbf{B}$ is the external magnetic field. From this Schrödinger equation, we read the Hamiltonian $\mathcal{H}$ that drives the evolution of this fermion in the non-relativistic limit.

82.b When the external field is a pure electric field, we may choose a gauge in which $\mathbf{A} = 0$, so that the above Hamiltonian simplifies to

$$\mathcal{H} = m - \frac{\nabla^2}{2m} - eA^0.$$

Consider the time-ordered two-point correlation function of the non-relativistic field ϕ,

$$G(x, y) \equiv \big\langle T\, \phi(x)\phi^\dagger(y)\big\rangle = \theta(x^0 - y^0)\big\langle\phi(x)\phi^\dagger(y)\big\rangle - \theta(y^0 - x^0)\big\langle\phi^\dagger(y)\phi(x)\big\rangle.$$

Applying a time derivative to this definition, and recalling the equal-time anti-commutator $\{\phi(x), \phi^\dagger(y)\} = \delta(\boldsymbol{x}-\boldsymbol{y})$ (this is the restriction to the two upper components of the canonical anti-commutation relation of Dirac spinors), we obtain

$$\Big(i\frac{\partial}{\partial x^0} - \mathcal{H}\Big)\, G(x, y) = i\,\delta(x - y).$$

We can also define the free correlation function G_0, defined as a Green's function of the free Hamiltonian, $\mathcal{H}_0 \equiv m - \frac{\nabla^2}{2m}$. Since $\mathcal{H} = \mathcal{H}_0 - eA^0$, the equation for G can be rewritten as follows:

$$\big(i\partial_{x^0} - \mathcal{H}_0\big)G = i\delta(x - y) - eA^0G.$$

This can be solved formally in the form of a Dyson equation,

$$G = G_0 + ieG_0A^0G.$$

82.c Diagrammatically, the expansion of the solution of the Dyson equation for the propagator reads

If the external potential is fixed, the next step is to perform the geometrical sum contained in this solution (which may be a rather difficult task if the potential A^0 has a non-trivial spatial dependence).

Another interesting situation is when this potential fluctuates, with a probability distribution $W[A^0]$. In this case, the above solution must be averaged over all the configurations of A^0, weighted by the distribution $W[A^0]$. Diagrammatically, this amounts to connecting all the crosses in the above representation by links and vertices as allowed by $W[A^0]$. For instance, a random time-independent external potential with only local Gaussian fluctuations would have the following distribution:

$$W[A^0] = \exp\Big(-\frac{1}{2}\int d^3x\, \frac{A^0(x)A^0(x)}{2\sigma}\Big),$$

for which the only possible connection is a link

$$\big\langle A^0(x)A^0(y)\big\rangle = \sigma\,\delta(\boldsymbol{x} - \boldsymbol{y}).$$

(This is independent of time, and the delta function is a spatial one.) With such Gaussian fluctuations of the external field, the expectation value of the propagator takes the following form:

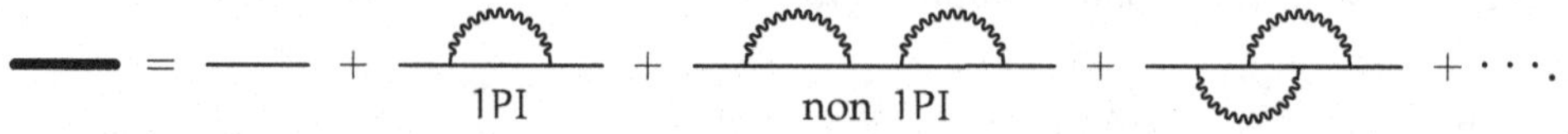

This has the form of a propagator dressed by self-energy corrections. This plain perturbative expansion can be improved by calculating the self-energy at a certain order in e^2 and by re-summing its repeated insertions on the free propagator (this is a geometrical series). When doing this, it is important to include only one-particle-irreducible contributions in the self-energy in order to avoid double counting. If we restrict ourselves to the order-e^2 self-energy, the equation for the dressed propagator is a Dyson equation that reads

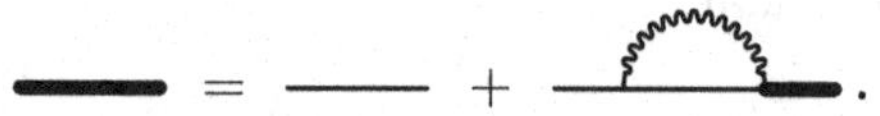

This is simple algebraic equation in momentum space:

$$G(p) = G_0(p) + G_0(p)\big(-i\Sigma(p)\big)G(p), \quad \text{i.e.,} \quad G(p) = \frac{i}{p^0 - m - \frac{\mathbf{p}^2}{2m} - \Sigma(p)},$$

and the self-energy is given by

$$\begin{aligned}
\Sigma(p) &= -ie^2\sigma \int \frac{d^4k}{(2\pi)^4}\, 2\pi\,\delta(k^0)\, \frac{i}{p^0 + k^0 - m - \frac{(\mathbf{p}+\mathbf{k})^2}{2m}} \\
&= 2e^2\sigma m \int \frac{k^2 dk\, d\cos\theta}{(2\pi)^2} \frac{1}{\underbrace{2m(p^0 - m)}_{\approx p^2 \text{ on-shell}} - p^2 - k^2 - 2pk\cos\theta} \\
&= -e^2\frac{\sigma m}{p} \int \frac{k\,dk}{(2\pi)^2} \underbrace{\ln\left(\frac{k+2p}{k-2p}\right)}_{i\pi\theta(2p-k)+\text{real terms}} = -i\frac{e^2\sigma m p}{2\pi} + \text{real terms.}
\end{aligned}$$

Note that the real part of the self-energy, which we have not calculated here, is ultraviolet divergent. This could be remedied on physical grounds by noting that the inverse inter-atomic distance should certainly act as a cutoff in this integral since the external field is not an uncorrelated noise on atomic scales.

However, the main result here is the fact that the self-energy has an imaginary part proportional to the strength of the fluctuations of the external field. The physical implication is that fermions cannot propagate in such a disordered medium, since their propagator in coordinate space is exponentially damped. The damping length is given by the inverse of the imaginary part of Σ and is thus inversely proportional to the strength σ of the fluctuations of the external field. Moreover, this imaginary part exists at all values of the momentum p, which means that there is no energy range where electrons can propagate freely in this material. This phenomenon, known as *Anderson localization*, is a very important result of solid-state physics that implies that disordered media are insulators (unlike periodic media, which may have a conduction band, i.e., an energy range in which the electronic states – known as Bloch states – are completely delocalized over macroscopic sizes).

83. Scaling Theory of Anderson Localization Problem 82 corresponds to the limit of strong disorder, in which the motion of electrons is damped regardless of the dimension of space. The situation regarding Anderson localization is in fact richer than what is suggested by that simple study, and depends crucially on the dimension. As we shall see, it is possible to gain a semi-quantitative understanding of this question without considering any specific model, by using scaling arguments and ideas inspired from the renormalization group. For this, we consider a cubic sample of size L in d dimensions. We denote by g its conductance (the inverse of its resistance), and we define $\beta \equiv d(\ln g)/d(\ln L)$.

83.a Explain why we have the following two limits:

$$\lim_{g\to 0} \beta = \ln(g), \qquad \lim_{g\to +\infty} \beta = d-2.$$

83.b Use these limits to discuss the behavior of the conductance as $L \to \infty$. Show that there is a qualitative difference between $d = 1, 2$ and $d = 3$.

83.a Let us consider a cubic sample of material of length L. In d spatial dimensions, its volume is therefore L^d and its transverse section is L^{d-1}. We denote by g its electrical conductance, i.e., the inverse of its resistance R. This parameter depends on the size L (this is quite obvious from the fact that a thick wire conducts better than a thin one), and it is convenient to introduce a β function that controls its variation with L:

$$\beta(g) \equiv L\frac{d\ln(g)}{dL}.$$

(The implicit assumption here is that the β function depends on the conductance, but not explicitly on the system size. In other words, it may depend on the system size only through the value of the conductance.)

Consider first the limit of strong disorder, studied in Problem 82. In this limit, the conductance goes to zero exponentially with the system size (in any dimension d):

$$g \propto e^{-L/\xi},$$

where the length ξ is given by the inverse of the imaginary part of the self-energy calculated in that problem. This formula follows from the fact that electrons cannot propagate beyond a distance ξ, so the conductance is non-zero only for systems whose length is smaller than ξ. Therefore, in this limit, we have

$$\beta(g) \underset{g\to 0}{\approx} -\frac{L}{\xi} = \ln(g) + \text{const.}$$

The opposite limit, $g \to +\infty$, is the limit of good conductors. In this situation, the resistance of a sample of length L and transverse section A is given by

$$R \propto \frac{L}{A} = g^{-1}L^{2-d},$$

where the unwritten prefactor is the resistivity (i.e., the inverse of the conductivity). Since $R = g^{-1}$, we obtain the following behavior for the β function in this limit:

$$\beta(g) \underset{g\to+\infty}{\approx} d-2.$$

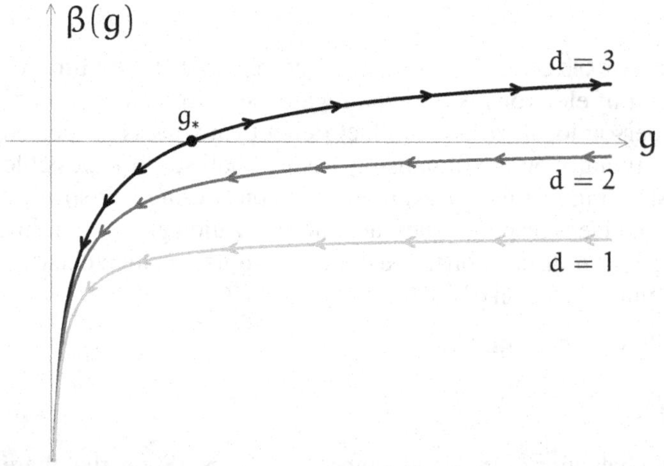

Figure 5.7 Behavior of the β function for the conductance in $d = 1, 2, 3$ dimensions. The arrows indicate the flow towards larger sizes.

83.b Assuming a monotonous behavior of the β function between the two limits we have considered, we obtain the curves sketched in Figure 5.7. In this figure, we show the β functions in $d = 1, 2, 3$ spatial dimensions. We have also indicated by arrows on the curves the direction of the flow when the size L increases (this follows directly from the sign of the β function). Interestingly, there are qualitative differences between $d = 1, 2$ and $d = 3$:

- **$d = 1, 2$:** the β function is always negative, implying that the conductance goes to zero as $L \to \infty$. Therefore, one- and two-dimensional materials become insulators when their size becomes infinite, no matter how weak the disorder is.
- **$d = 3$:** the β function vanishes at some critical value g_*, is negative for $g < g_*$ and positive for $g > g_*$. Moreover, this critical point is repulsive. Therefore, a three-dimensional material with a sufficiently small disorder (such that $g > g_*$ at a certain scale) may be a conductor when $L \to \infty$, while for a stronger disorder (such that $g < g_*$ at a certain scale) it becomes an insulator when $L \to \infty$.

Source: Abrahams, E., Anderson, P. W., Licciardello, D. C. and Ramakrishnan, T. V. (1979), *Phys Rev Lett* 42: 673.

Index

Printed in the United States
by Baker & Taylor Publisher Services